Finite Element Simulations with ANSYS Workbench 15

Huei-Huang Lee

Department of Engineering Science
National Cheng Kung University, Taiwan

SDC
Publications

SDC Publications
P.O. Box 1334
Mission, KS 66222
913-262-2664
www.SDCpublications.com
Publisher: Stephen Schroff

ISBN-13: 978-1-58503-907-4

ISBN-10: 1-58503-907-1

Printed and bound in the United States of America.

Contents

Contents

Chapter 11 Modal Analyses 381

Chapter 12 Transient Structural Simulations 411

Chapter 13 Nonlinear Simulations 460

Chapter 14 Nonlinear Materials 517

Chapter 15 Explicit Dynamics 560

Index 591

Preface

Use of the Book

This book is developed mainly for graduate and senior undergraduate students. It may be used in courses such as Computer-Aided Engineering or Finite Element Simulations.

Why ANSYS?

ANSYS has been a synonym of finite element simulations. I've been using ANSYS as a teaching platform for more than 20 years. The reasons I prefer ANSYS to other CAE softwares are due to its multiphysics capabilities, completeness of on-line documentations, and popularity among both academia and industry. Equipping engineering students with multiphysics capabilities is becoming a necessity. Complete documentations allow students to go steps further after taking an introductory CAE course. Popularity among academia and industry implies that an engineering student, after his graduation from college, can work with a standard software without any further training. Recent years, I have another reason to advocate this software, namely the user-friendliness.

ANSYS Workbench

ANSYS Workbench has evolved for years but matured enough in recent years. Before the Workbench is practical useful, I have been using ANSYS Classic, nowadays dubbed Mechanical APDL. The unfriendly APDL imposes unnecessary constraints, making the software very difficult to harness. As a result, students or engineers often refrain themselves within limited applications; for example, working on component simulations rather than assembly simulations. The Workbench adds friendliness on the top of the power of APDL, releasing many unnecessary constraints.

Why a New Tutorial?

Preparing a tutorial for the Workbench needs much more effort than that for the APDL, due to the graphic nature of the interface. The most complete tutorial, to my knowledge so far, is the training tutorials prepared by ANSYS Inc. However, they may not be suitable for use as a college textbook for the following reasons. First, the cases used are either too complicated or too trivial. Many of the cases are geometrically too complicated for students to create from scratch. The students need to rely on the geometry files accompanied with the tutorials. Students usually obtain a better comprehension by working from scratch. Second, the tutorial covers too little on theories of finite element methods and solid mechanics while too much on software operation details, which may not be necessary for a one-semester university course.

Organization of the Book

We'll explain the organization of the book in Section 1.1. Here is a quick overview of the book.

Using a case study, Section 1.1 walks through a typical Workbench simulation procedure. As more concepts or tools are needed, specific chapters or sections are pointed out, in which an in-depth discussion will be provided. The rest of Chapter 1 provides introductory backgrounds essential for the discussion in later chapters. These backgrounds include a concise knowledge of structural mechanics, equations that govern the behavior of a mechanical or structural system (Section 1.2), the finite element methods that solve these governing equations (Section 1.3), and the failure criteria of materials (Section 1.4). Chapter 1 is the only chapter that doesn't have any hands-on exercises because, in the very beginning of a semester, students may not be ready to access the software facilities yet.

Starting from Chater 2, a learning-by-doing approach is used throughout the book. Chapters 2 and 3 cover 2D geometric modeling and simulations. Chapters 4-7 introduce 3D geometric modeling and simulations. The following two chapters dedicate to two useful topics: Chapter 8 to optimization and Chapter 9 to meshing. Chapter 10 deals with buckling and its related topic: stress stiffening. Chapters 11 and 12 discuss dynamic simulations, Chapter 11 on modal analysis while Chapter 12 on transient structural analysis. Up to this point, the discussions are mostly on linear problems. Although several nonlinear simulations have been carried out, their nonlinear behaviors were not been discussed further. Chapters 13 and 14 discuss nonlinear simulations in a more in-depth way. Chapter 15 introduces an exciting topic: explicit dynamics, which is becoming a necessary discipline for a CAE engineer.

Features of the Book

Comprehensive and friendly are the ultimate goals of this book. To achieve these goals, following features are incorporated in the book.

Real-World Cases. There are 45 step-by-step hands-on exercises in the book, each completed in a single section. These exercises are designed from 27 real-world cases, carefully chosen to ensure that they are neither too trivial nor too complicated. Many of them are industrial or research projects; pictures of prototypes are presented whenever available. The size of the problems are not too large so that they can be performed in an academic teaching version of the ANSYS, which has a limitation on the numbers of nodes (30,000) and elements (30,000). They are not too complicated so that the students can build each project step by step from scratch themselves. Throughout the book, the students may not need any supplement files to work on these exercises.

Theoretical Background. Relevant background knowledge is provided whenever necessary, such as solid mechanics, finite element methods, structural dynamics, nonlinear solution methods (e.g., Newton-Raphson methods), nonlinear materials (e.g., plasticity models), explicit integration methods, etc. To be efficient, the teaching methods are conceptual rather than mathematical; concise, yet rather comprehensive. The last four chapters (chapters 12-15) cover more advanced topics, each chapter having an opening section that gives basics of that topic in an efficient way to facilitate the subsequent learning.

How the Workbench internally solves a model is also illustrated throughout the book. Understanding these procedures, at least conceptually, is useful for a CAE engineer.

Key concepts are inserted whenever necessary. Must-know concepts, such as structural error, finite element convergence, stress singularity, are taught by using designed hands-on exercises, rather than by abstract lecturing. For example, how finite element solutions converge to their analytical solutions, as the meshes become finer, is illustrated by guiding the students to plot convergence curves. That way, the students should have strong knowledge of the finite elements convergence behaviors. And, after hours of laborious working, they will not forget it for the rest of their life. Step-by-step guiding the students to plot curves to illustrate important concepts is one of the featuring teaching methods in this book.

Learning by Hands-on Exercises. A learning approach emphasizing hands-on exercises spreads through the entire book. In my own experience, this is the best way to learn a complicated software such as ANSYS Workbench. A typical chapter, such as Chapter 3, consists of 6 sections, the first two sections providing two step-by-step examples, the third section trying to give a more systematic view of the chapter subject, the following two sections providing more exercises in a not-so-step-by-step way, and the final section providing review problems.

Demo Videos. Each of the 45 step-by-step exercises has been screen-recorded and uploaded to YouTube for free access (see Companion Webpage, next page).

ANSYS On-line References. One of the objectives of this book is to be a guiding book toward the huge repository of ANSYS on-line documentation, a well of knowledge for many students and engineers. The on-line documentation includes a theory reference, an element reference, and many examples. Whenever helpful, we point to a location in the on-line documentation as a further study for the students.

End-of-Chapter Keywords. Keywords are summarized at the ending section of each chapter in a quiz form. One goal of this book is to help the students comprehend terminologies and use them efficiently. That is not so easy for some students. For example, whenever asked "What are shape functions?" most of the students cannot satisfyingly define the terminology. Yes, many textbooks spend pages teaching students what the shape functions are, but the challenge is how to define or describe a term in less than two lines of words. This part of the book demonstrates how to define or describe a term in an efficient way; for example, "Shape functions serve as interpolating functions, to calculate continuous displacement fields from discrete nodal displacements."

To Instructors: How I Use the Book

I use this book in a 3-credit 18-week course Computer-Aided Engineering. The progress is one chapter per week, except Chapter 1, which takes 2 weeks. Each week, after a classroom introduction of a chapter using the lecture slides (see Companion Webpage, next page), I set up a discussion forum in an e-learning system maintained by the university. After completing the exercises of the chapter, the students are required to discuss in the forum.

In the forum, the students may post their questions, help or answer other students' questions, or share their comments. In addition to take part in the discussion, I rate each posted article with 0-5 stars; the sum of the ratings becomes the grade of a student's performance for the week. The weekly discussion closes before next classroom hours.

The course load is not light. Nevertheless, most of the students were willing to spend hours working on these step-by-step exercises, because these exercises are tangible, rather than abstract. Students of this generation are usually better in picking up knowledge through tangible software exercises rather than abstract lecturing. Further, the students not only feel comfortable to post questions in the forum but also enjoy helping other students or sharing their comments with others.

At the end of the semester, each student is required to turn in a project. Students who are currently working as engineers usually choose topics related to their jobs. Students who are working on their theses usually choose topics related to their studies. They are also allowed to repeat a project found in any references, as long as they go through all details by themselves. The purpose of the final project is to ensure that students are capable of carrying out a project independently, which is the goal of the course, not just following the step-by-step procedures in the book.

To Students: How My Students Use the Book

Many students, when following the steps in the book, often made mistakes and ended up with different results from those in the book. In many cases they cannot figure out which steps the mistakes were made. In these case, they have to redo the exercise from the beginning. It is not uncommon that they redid the exercise several times and finally saw the beautiful results.

What I want to say is that you may come across the same situation, but you are not wasting your time when you redo the exercises. You are learning from the mistakes. Each time you fix a mistake, you gain more insight. After you obtain the reasonable results, redo it and try to figure out if there are other ways to accomplish the same results. That's how I learned finite element simulations when I was a young engineer.

Finite element methods and solid mechanics are the foundation of mechanical or structural simulations. If you haven't taken these courses, plan to take them after you complete this course. If you've already taken them and feel not comfortable enough, review them.

Companion Webpage

A dedicate webpage for this book is maintained by the author:

http://myweb.ncku.edu.tw/~hhlee/Myweb_at_NCKU/ANSYS15.html

The webpage contains links to many resources, including the YouTube videos that demonstrate the steps of each exercise in the book and the finished project files of each exercise.

As to the finished project files, if everything works smoothly, you may not need them at all. Every project can be built from scratch according to the steps described in the book. The author provides these project files just in some cases you need them. For examples, when you run into troubles and you don't want to redo from the beginning, you may find these files useful. Or, when you have troubles following the geometry details in the book, you may need to look up the geometry details from project files.

Another reason we provide the finished project files is as follows. It is strongly suggested that, in the beginning of an exercise when previously saved project files are needed, you use the project files provided by the author rather than your own files so that you are able to obtain results that have minimum differences in numerical values from those in the book.

Also note that these finished project files are saved with the Workbench GUI command **File/Archive...**; i.e., they are in **WBPZ** compressed format. To open a compressed project file, please use the Workbench GUI command **File/Restore Archive...**

Companion DVD

To save your time of downloading the project files and the video files, we also provide a DVD for each hard copy of the book. The DVD contains all the resources in the webpage. The finished projects files are uncompressed and the video files are in MOV format (rather than links to YouTube), which can be played in a PC or a Macintosh.

Numbering and Self-Referencing System

To efficiently present the material, the writing of this book is not always done in a traditional format. Chapters and sections are numbered in a traditional way. Each section is further divided into subsections, for example, the 3th subsection of the 2nd section of Chapter 1 is denoted as "1.2-3." Textboxes in a subsection are ordered with numbers. A number is enclosed by a pair of square brackets (e.g., [4]). When needed, we may refer to that textbox such as "1.2-3[4]." When referring to a textbox in the same subsection, we drop the subsection identifier; for the foregoing example, we simply write "[4]." Equations are numbered in a similar way, except that the equation number is enclosed by a pair of round brackets (parentheses) rather than square brackets. For example, "1.2-3(1)" refers to the 1st equation in the Subsection 1.2-3. Numbering notations are summarized as follows (some of these notations are empasized again in the shaded textboxes of pages 7, 9, 54):

1.2-3	The number after a hyphen is a subsection number.
[1], [2], ...	Square brackets are used to number textboxes.
(1), (2), ...	Round brackets are used to number equations
(a), (b), ...	Lower case letters are used to number items in the text.
Reference[Refs 1, 2]	Superscripts indicate reference numbers.
Mechanical	Bold face are used to highlight Workbench keywords.
Round-cornered textboxes	A round-cornered textbox indicates that mouse or keyboard actions are needed.
Sharp-cornered textboxes	A sharp-cornered textbox is used for commentary only; no mouse or keyboard actions are needed in that step.
#	A symbol # is used to indicate the last textbox of a subsection.

Acknowledgement

I feel thankful to students who had ever sat in my classroom, listening to my lectures. It is my students, past and present, that motivated me to give birth to this book.

Many of the cases presented in this book are from my students' final projects. Some are industry cases while others are thesis-related research topics. Without these real-world cases, the book would never be so useful. The following is a list of the names who contributed to the cases in this book.

"Pneumatic Finger" (1.1 and 9.1) is contributed by Che-Min Lin and Chen-Hsien Fan, ME, NCKU.
"Microgripper" (2.6 and 13.3) is contributed by P.W. Shih, ME, NCKU.
"Cover of Pressure Cylinder" (4.2 and 9.2) is contributed by M. H. Tsai, ME, NCKU.
"Lifting Fork" (4.3 and 12.2) is contributed by K.Y. Lee, ES, NCKU.
"LCD Display Support" (4.5 and 5.4) is contributed by Y.W. Lee, ES, NCKU.
"Bellows Tube" (6.1) is contributed by W. Z. Liu, ME, NCKU.
"Flexible Gripper" (7.1 and 8.1) is contributed by Shang-Yun Hsu, ME, NCKU.
"3D Truss" (7.2) is contributed by T. C. Hung, ME, NCKU.
"Snap Lock" (13.4) is contributed by C. N. Chen, ME, NCKU.

Many of the original ideas of these projects came from the academic advisors of the above students. I also owe them a debt of thanks. Specifically, the project "Pneumatic Finger" is an unpublished work led by Prof. Chao-Chieh Lan of the Department of ME, NCKU. The project "Microgripper" originates from a work led by Prof. Ren-Jung Chang of the Department of ME, NCKU.

Thanks to Dr. Shen-Yeh Chen, the CEO of FEA-Opt Technology, for letting me use his article "An Unofficial History of ANSYS," as an appendix at the end of Chapter 1.

S.Y. Kan, the chief structural engineer of the Taipei 101, has reviewed some of texts and corrected some mistakes.

Much of information about the ANSYS Workbench are obtained from the training tutorials prepared by ANSYS Inc. I didn't specifically cite them in the text, but I appreciate these well-compiled training tutorials.

Thanks to Mrs. Lilly Lin, the CEO, and Mr. Nerow Yang, the general manager, of Taiwan Auto Design, Co., the partner of ANSYS, Inc. in Taiwan. The couple, my long-term friends, provided much of substantial support during the writing of this book.

Thanks to Professor Sheng-Jye Hwang, of the ME Department, NCKU, and Professor Durn-Yuan Huang, of Chung Hwa University of Medical Technology. They are my long-term research partners. Together, we have accomplished many projects, and, in carrying out these projects, I've learned much from them.

Lastly, thanks to my family, including my wife, my son, and the pet dogs, for their patience and sharing the excitement with me.

Huei-Huang Lee

Associate Professor
Department of Engineering Science
National Cheng Kung University, Tainan, Taiwan
e-mail: hhlee@mail.ncku.edu.tw
webpage: myweb.ncku.edu.tw/~hhlee

Chapter 1
Introduction

Purpose of This Chapter

ANSYS is a software implementation of finite element simulations on structural, mechanical, fluid, electromagnetic, etc. This book discusses only on structural and mechanical simulations. This chapter (a) shows the procedure of a typical structural simulation using Workbench, (b) explains the organization of this book, (c) reviews solid mechanics, (d) and gives a brief introduction of finite element methods.

About Each Section

The procedure of structural simulations and the organization of this book are illustrated in Section 1.1 by using a case study. Section 1.1 also try to motivate students learning the topics in Sections 2, 3, and 4.

Section 1.2 presents equations that govern the behavior of a structural system. This section reviews many terminologies, such as displacements, stresses, and strains, which are used throughout the book.

ANSYS solves these governing equations by finite element methods. Section 1.3 introduces the basic ideas and the procedure of the finite element methods. The introduction is conceptual rather than theoretical or mathematical. These concepts should be adequate for the purpose of understanding the topics in the later chapters.

One goal of structural simulations is to predict whether or not a system would fail. We usually compare the calculated stresses with certain critical values. If the calculated stresses are larger than the critical values, then the system is said to fail. Section 1.4 discusses the theory behind these failure criteria.

This is the Only Chapter without Hands-On Exercises

All chapters of this book use learning-by-doing approach, except this chapter. There are no hands-on exercises in this chapter. The main reason is that an overall picture is usually helpful before any hands-on exercises. A secondary reason is that, in the first week of a semester, students may not be ready to access the software facility yet.

Section 1.1

Case Study: Pneumatically Actuated PDMS Fingers[Ref 1]

1.1-1 Problem Description

About the Pneumatic Fingers

The pneumatic fingers [1] are designed as part of a surgical parallel robot system which is remotely controlled by a surgeon through the Internet[Ref 2].

 The robot fingers are made of a PDMS-based (polydimethylsiloxane) elastomer material. Geometry of a typical finger is shown in [2], in which 14 air chambers are cut from the material.

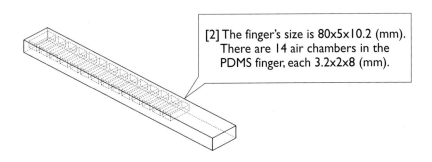

[2] The finger's size is 80x5x10.2 (mm). There are 14 air chambers in the PDMS finger, each 3.2x2x8 (mm).

[1] A robot hand has five fingers, remotely controlled by a surgeon through internet.

The air chambers locate closer to the upper face than the bottom face so that when air pressure is applied, the finger bends downward [3, 4]. Note that, in [4], the chambers are visible because only half of a finger is modeled.

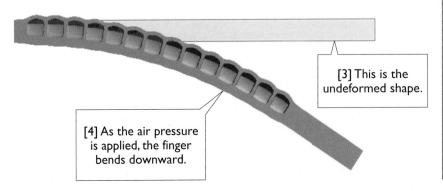

[3] This is the undeformed shape.

[4] As the air pressure is applied, the finger bends downward.

Order of Textboxes

In this book, textboxes within a subsection are ordered with numbers enclosed by pairs of square brackets. When you read figures, please follow the order; the order is significant. The numbers also serve as reference numbers when referred from other part of text. A symbol # is used to indicate the last textbox of a subsection (e.g., [5], next page) so that you don't leave out any textboxes.

About the PDMS Elastomer

The mechanical properties of PDMS elastomers depend on their ingredients. The chart on the right shows the stress-strain curve of the PDMS elastomer used in this case[Ref 3] [5]. It shows that the material exhibits linear stress-strain relationship up to a strain of about 0.6. Within this range, the Young's modulus (the slope of the straight line) is calculated to be 2.0 MPa (1.2 MPa/0.6). In addition, the Poisson's ratio is 0.48, obtained from other experimental data.

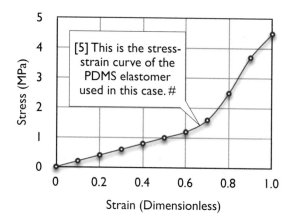

[5] This is the stress-strain curve of the PDMS elastomer used in this case. #

Purpose of the Simulation

The purpose of the simulation is to assess the *efficiency* of the design, defined as the magnitude of the actuation (the vertical deflection) under a working air pressure up to 0.18 MPa. More specifically, we want to plot a pressure-versus-deflection chart (1.1-8[5], page 15; also see 1.1-12[1, 2], page 19).

1.1-2 Workbench GUI

You can launch Workbench by selecting Workbench from the **Start** menu (of a **Windows** system) [1]. The contents of the **Start** menu in your computer may be different from here, depending on your installation. A **Workbench GUI** (graphic user interface) then shows up [2]. The Workbench GUI is a gateway to Workbench applications; i.e., all the Workbench application can be accessed via Workbench GUI. The Workbench applications that will be used in this book are: **Project Schematic**, **Engineering Data**, **DesignModeler**, **Mechanical**, and **Design Exploration**.

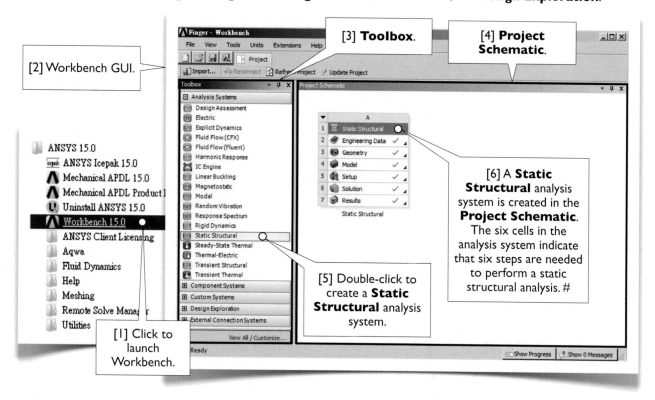

[2] Workbench GUI.

[3] **Toolbox**.

[4] **Project Schematic**.

[1] Click to launch Workbench.

[5] Double-click to create a **Static Structural** analysis system.

[6] A **Static Structural** analysis system is created in the **Project Schematic**. The six cells in the analysis system indicate that six steps are needed to perform a static structural analysis. #

On the left of the Workbench GUI is a **Toolbox** [3], and on the right is a Project Schematic window [4]. **Toolbox** lists many built-in **Analysis Systems**. The contents of **Toolbox** in your computer may be different from here, depending on your installation. In this book, "analysis" and "simulation" are often interchangeable, for example, "static structural analysis" is synonymous to "static structural simulation." In this case, we will perform a **Static Structural** analysis.

Double-clicking **Static Structural** [5] in **Toolbox** creates a **Static Structural** analysis system [6] in the **Project Schematic** window. The six cells in the analysis system indicate that six steps are needed to perform a static structural analysis: (a) prepare engineering data, (b) create a geometric model, (c) divide the geometric model into a finite element mesh, or called a finite element model, (d) setup loads and supports, (e) solve the finite element model, and (f) view the results.

Double-clicking each cell will bring up a relevant application to process that step.

> In this book, to facilitate the readability, an Workbench keyword is usually bold faced. In cases that bold-face does not help the readability, then the bold-face may be omitted.

1.1-3 Prepare Engineering Data

Double-clicking **Engineering Data** cell [1] brings **Engineering Data** up on the Workbench GUI [2]. We now specify material properties for the PDMS elastomer. The material is modeled as a linear **Isotropic Elasticity** material, for which we need to input a Young's modulus (2.0 MPa) and a Poisson's ratio (0.48) [3-5]. Finally, click **Project** to quit **Engineering Data** application and return to **Project Schematic** [6].

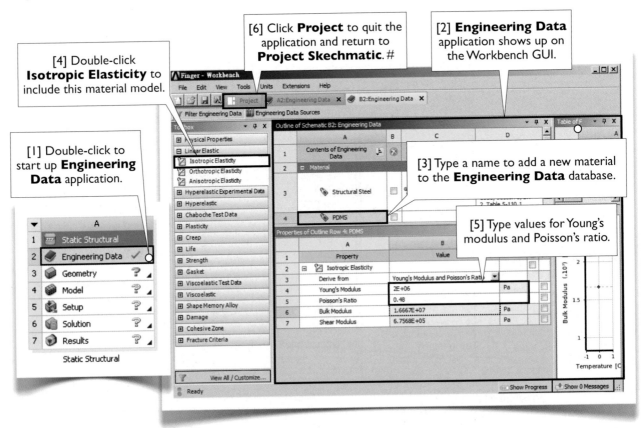

[6] Click **Project** to quit the application and return to **Project Skechmatic**. #

[2] **Engineering Data** application shows up on the Workbench GUI.

[4] Double-click **Isotropic Elasticity** to include this material model.

[1] Double-click to start up **Engineering Data** application.

[3] Type a name to add a new material to the **Engineering Data** database.

[5] Type values for Young's modulus and Poisson's ratio.

1.1-4 Create Geometric Model

Double-clicking **Geometry** cell [1] brings up DesignModeler application [2]. Functions of DesignModeler are similar to any CAD (computer-aided design) software such as **SolidWorks**, **Creo** (formerly **Pro/ENGINEER**), etc., except that DesignModeler is specifically designed to create geometric models for use in ANSYS Workbench simulations. In our case, we want to construct a geometric model for the pneumatic finger [3]. Due to the symmetry, we will model only one half of the pneumatic finger.

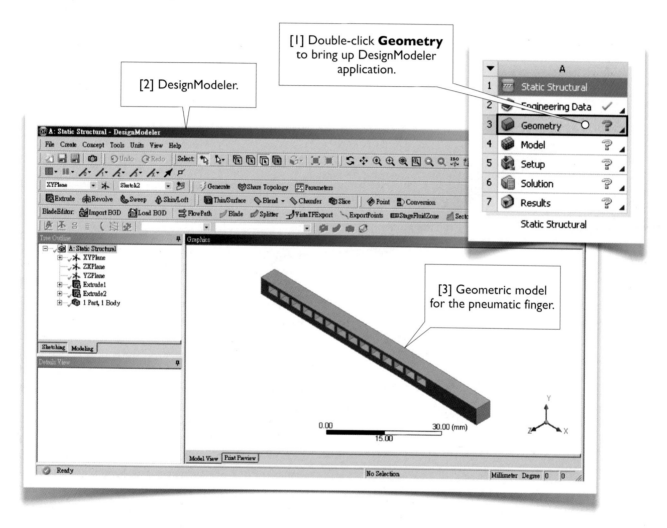

[1] Double-click **Geometry** to bring up DesignModeler application.

[2] DesignModeler.

[3] Geometric model for the pneumatic finger.

Creating the geometric model in our case [3] is relatively simple. It can be viewed as a two-step task, and each step consists of a 2D sketching and a 3D extrusion, summarized as follows.

Step 1. Create an 80 mm x 5 mm x 5.1 mm solid.
    Step 1.1. Sketch an 80 mm x 5 mm rectangle [4].
    Step 1.2. Extrude the sketch 5.1 mm to create a solid [5].
Step 2. Create fourteen 3.2 mm x 2 mm x 4 mm air chambers.
    Step 2.1. Sketch fourteen 3.2 mm x 2 mm rectangles [6].
    Step 2.2. Extrude the sketch 4 mm deep to cut material from the solid [7].

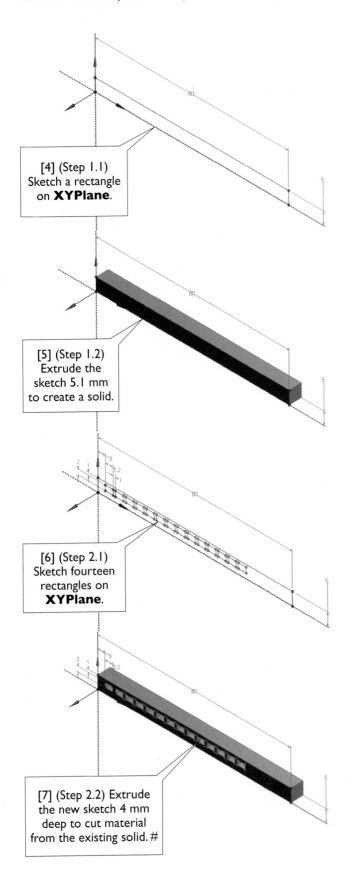

[4] (Step 1.1) Sketch a rectangle on **XYPlane**.

[5] (Step 1.2) Extrude the sketch 5.1 mm to create a solid.

[6] (Step 2.1) Sketch fourteen rectangles on **XYPlane**.

[7] (Step 2.2) Extrude the new sketch 4 mm deep to cut material from the existing solid. #

2D and 3D Simulations

Workbench supports 2D and 3D simulations. For 3D simulations, Workbench supports three types of geometric bodies: *solid bodies* (which have volumes), *surface bodies* (which do not have volumes, but have surface areas), and *line bodies* (which do not have volumes or surface areas, but have length). Thin shell structures are often modeled as surface bodies (Chapter 6); Beam or frame structures are often modeled as line bodies (Chapter 7). Solid, surface, and line bodies can be mixed up in a 3D model. For 2D simulations, Workbench supports solid bodies only (Chapter 3). A 2D model must be created entirely on **XYPlane**.

More on Geometric Modeling

Creating a geometric model is often more involved than that in this case, but still can be viewed as a series of two-step operations demonstrated in this case: drawing a sketch and then using the sketch to create a 3D body by one of the tools, such as extrusion, revolution, sweeping, and skin/lofting (4.4-9, page 193).

Geometric modeling is the first step toward the success of finite element simulations. Chapters 2 and 4 dedicate to geometric modeling techniques.

Chapter 2 introduces sketching methods. Some of the sketches created in Chapter 2 are reused in Chapter 3 to demonstrate the creation of 2D solid models. Simulations of these 2D solid models are then performed in Chapter 3.

Chapters 4-7 devote to 3D geometric modeling and linear static simulations. Chapter 4 discusses the creation of 3D solid models and Chapter 5 demonstrates 3D simulations using the solid models created in Chapter 4. Chapter 6 demonstrates 3D surface modeling and simulations, and Chapter 7 to 3D line modeling and simulations.

In the real-world, there are no such things as surface/line geometries, therefore surface/line models (as well as 2D solid models) are aptly called *conceptual models*.

1.1-5 Divide Geometric Model into Finite Elements

The procedure that Workbench solves a problem can be viewed as two major steps: (a) establishing governing equations and (b) solving the governing equations. The geometric model of a real-world problem is often too complicated to establish governing equations directly. A basic idea of the finite element methods is to divide the entire geometric model into many geometrically simpler shapes called the *finite elements*. The elements are connected by *nodes*. Governing equations for each element then can be easily established, and the system of equations for all elements can be solved simultaneously. We will discuss this idea further in Section 1.3.

The dividing of a geometric model into elements is called *meshing*, and the collection of the elements is called a *finite element mesh*, or loosely called a *finite element model*. In this book, we'll use a more rigorous definition: *a finite element model consists of a mesh and its environment conditions*, which we will introduce in the 1.1-6, next page.

Double-clicking **Model** cell in the analysis template [1] brings up a **Mechanical GUI** [2]. The rest of the simulation will be done in **Mechanical**; i.e., the functions of **Mechanical** include meshing, setup of loads and supports, solution, and viewing results [3].

Workbench can perform meshing task under your control [4]. In **Details of Mesh**, a **Statistics** displays the number of nodes and elements [5].

More of Meshing

Quality of meshing cannot be overemphasized. Although it is possible to let ANSYS Workbench perform the meshing automatically, its quality is not guaranteed. Tasks of achieving a high quality mesh is not trivial; it needs much background knowledge and experience. Chapter 9 demonstrates many meshing techniques.

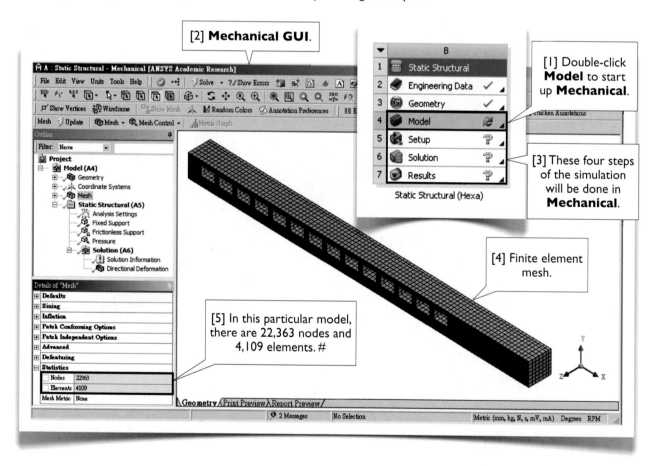

[2] **Mechanical GUI**.

[1] Double-click **Model** to start up **Mechanical**.

[3] These four steps of the simulation will be done in **Mechanical**.

[4] Finite element mesh.

[5] In this particular model, there are 22,363 nodes and 4,109 elements. #

Static Structural (Hexa)

1.1-6 Set up Loads and Supports

In the real-world, every object is part of the world, interacting with other part of the world. When we take an object apart for simulation, we are cutting it away from the rest of the world. The cutting surfaces of the model are called the *boundary* of the model. The choice of the boundary is arbitrary AS LONG AS we can specify the *boundary conditions* on ALL of the boundary surfaces. In Workbench, all conditions applying on the finite element mesh and affecting the response of the model are called the *environment conditions*. Environment conditions include boundary conditions as well as conditions that are not specified on the boundaries. Temperature change INSIDE the body is an example of environment conditions that are not specified on boundaries; another example is gravitational forces.

In our case, things surrounding the half model of the pneumatic finger are: pressurized air in the chambers [1], the material connecting the root of the finger [2], the material on the other side of the model [3], and the atmosphere air around the rest of boundary surfaces.

Modeling the pressurized air in the chambers is straightforward: specify a pressure of 0.18 MPa for all the surfaces of the chambers [1]. The root of the finger is modeled as fixed support [2]. The plane of symmetry is modeled as frictionless support [3]. Note that a plane of symmetry is equivalent to a surface of frictionless support. Finally, assuming the atmospherical air has little effect on the model, we simply neglect it and model all boundary surfaces surrounded by atmospherical air as free boundaries, boundaries with no forces acting on them.

As mentioned (1.1-5, last page), a finite element model is defined as a finite element mesh plus its environment conditions. We will stick on this definition throughout the book. Make sure that you can distinguish these three terms: *geometric model*, *finite element mesh*, and *finite element model*.

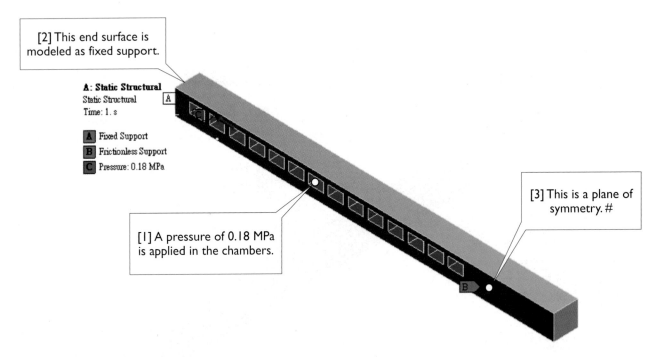

[2] This end surface is modeled as fixed support.

A: Static Structural
Static Structural [A]
Time: 1. s

[A] Fixed Support
[B] Frictionless Support
[C] Pressure: 0.18 MPa

[1] A pressure of 0.18 MPa is applied in the chambers.

[3] This is a plane of symmetry. #

More of Environment Conditions

Modeling environment conditions is sometimes not so easy as in this case. The challenge mostly comes from the need of domain knowledge. It is not possible to perform a structural analysis if an engineer doesn't have enough domain knowledge of structural mechanics.

We will start to introduce environment conditions in Chapters 3. From then on, each chapter will involve some demonstrations of environment conditions.

1.1-7 Solve Finite Element Model

To solve a finite element model, simply click **Solve** in **Mechanical GUI** [1]. The time to complete a simulation depends on its problem size (number of nodes and number of elements), number of time steps, and nonlinearities.

Inside Workbench

As mentioned (1.1-5, page 12), the solution procedure can be viewed as two major steps: establishing governing equations and solving the governing equations. Section 1.2 is an overview of structural mechanics, summarizing the governing equations. How does ANSYS Workbench establish and solve these governing equations? The answer is: *finite element methods*. Section 1.3 is a summary of finite element methods, to quickly equip the students with enough concepts of finite element methods so that they can proceed the learning for the rest of the book.

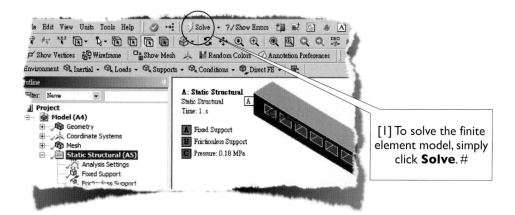

[1] To solve the finite element model, simply click **Solve**. #

1.1-8 View Results

After solving a problem, numerical results are stored in databases, available for your request. In our case, we are concerned about the vertical deflection [1-3]. The deformation can be animated [4]. A useful information is a deflection-versus-pressure chart [5, 6] (next page), in which the deflection is measured at the tip of the finger [3].

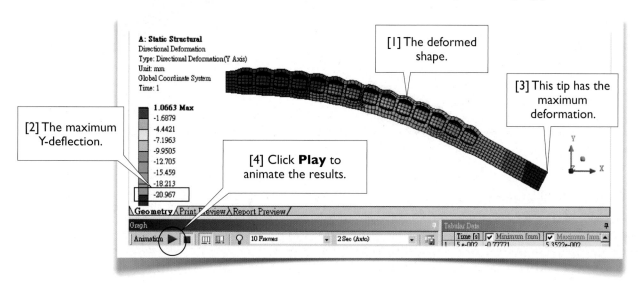

[1] The deformed shape.

[3] This tip has the maximum deformation.

[2] The maximum Y-deflection.

[4] Click **Play** to animate the results.

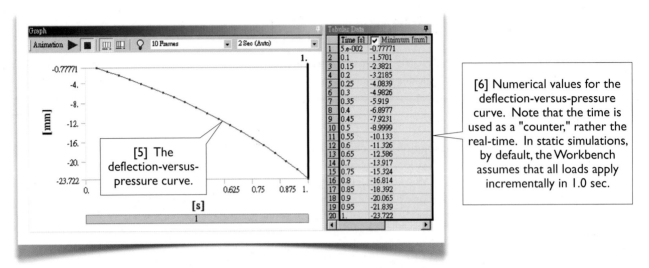

[6] Numerical values for the deflection-versus-pressure curve. Note that the time is used as a "counter," rather the real-time. In static simulations, by default, the Workbench assumes that all loads apply incrementally in 1.0 sec.

In 1.1-3[4] (page 9), we assume the stress-strain relationship of the PDMS material is linear. This implies that the strains are less than 0.6 (1.1-1[5], page 8). This assumption has to be checked at this point. The results show that the maximum strain is about 0.64 [7]. Note that we also assumed the compressive behavior is the same as tensile behavior, but this is usually not true for an elastomer under such a large deformation. More accurate material models (called a hyperelasricity model) for the elastomer will be discussed in Chapter 14.

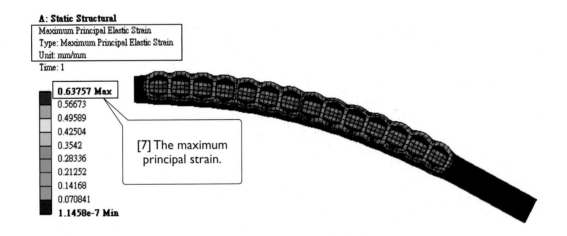

Lastly, the stress is reviewed. It shows that the maximum von Mises stress, defined in Eq. 1.4-5(15) (page 42), is 1.45 MPa [8] (next page). The experimental data (1.1-1[5], page 8) reveal that the material can withstand up to 4.5 MPa without failure.

Failure Criteria

The purpose of checking the stresses is to make sure the material doesn't fail under specified load. What is von Mises stress [8]? The experimental data (1.1-1[5], page 8) are produced by a uniaxial tensile test, but the stress state in the pneumatic finger, as in any real-world situation, is 3D by nature. How can we compare a 3D stress state with a uniaxial one, and make a judgement about the failure of the material? Section 1.4 will discuss failure criteria of materials.

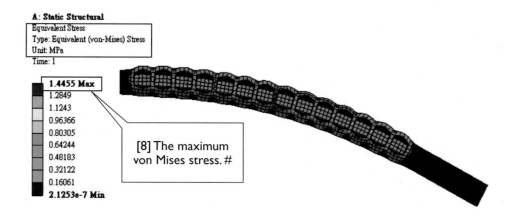

A: Static Structural
Equivalent Stress
Type: Equivalent (von-Mises) Stress
Unit: MPa
Time: 1

1.4455 Max
1.2849
1.1243
0.96366
0.80305
0.64244
0.48183
0.32122
0.16061
2.1253e-7 Min

[8] The maximum von Mises stress. #

1.1-9 Buckling and Stress-Stiffening

The more tension of a guitar string, the more force you need to deflect the string laterally. In technical words, the string's lateral stiffness increases as the longitudinal tensile stress increases; i.e., the longitudinal tensile stress causes the string stiffer in lateral direction. By lateral direction, we mean the direction orthogonal to the longitudinal direction. The increase of lateral stiffness can also be justified by the fact that the string's vibrating frequency (pitch) increases with the increase of its tension. (Stiffer strings have higher frequencies.) This effect is called the *stress-stiffening effect: a structure's lateral stiffness increases with the increase of its longitudinal tensile stress.*

Is the opposite also true? That is, the lateral stiffness of a structure decreases with the increase of its axial compressive stress. For example, does a column's lateral stiffness decrease when subject to an axial compressive force? The answer is YES. An even more dramatic phenomenon is that, as the compression is increasing and the lateral stiffness is decreasing, the lateral stiffness will eventually reach zero and the structure is said to be in an unstable state. That is, a tiny lateral force would deflect the structure infinitely. This phenomenon is called *buckling.*

The buckling must be considered in a compressive structural component where its lateral dimension is much smaller than the longitudinal dimension; for examples, slender columns subject to axial loads, thin-walled pipes subject to a circumferential twist.

Back to the pneumatic finger. Instead of air pressure applied on the chambers' surfaces, we now apply an upward lifting at the finger tip [1]. The upper surface, which is essentially a layer of thin PDMS film, would undergo compressive stress [2]. Our concern then is to know the magnitude of the lifting force that will cause the thin film buckle.

Simulations of buckling and stress-stiffening will be covered in Chapter 10.

[2] The upper surface would undergo compressive stress. #

[1] If we apply an upward lifting at the end surface...

1.1-10 Dynamic Simulations

Consider the pneumatic finger again. If the load (air pressure) is applied very fast, the deformation would also occur very fast. When a body moves or deforms very fast, two effects must be taken into account: *inertia effect* and *damping effect*. Combination of these two effects are called *dynamic effects*. When the dynamic effects are considered in a simulation, it is called a *dynamic simulation*.

Imagine that the pressure in the chambers is increased from zero to 0.18 MPa in just only 0.1 seconds. The pressure is applied so fast that the deformation must also be very fast and that dynamic effects must play an important role in the structure's behavior. The figure and the chart below show the time-varying deflections of the finger tip under such a loading condition [1-4]. The curve [4] shows that the deflections in a dynamic simulation can be much larger than those obtained in a static simulation. Furthermore, the vibration lasts for several seconds. This, as a surgical application, is not desirable.

The foregoing simulation, a structure subject to dynamic loads, is called a *transient structural simulation*. Chapters 12 and 15 will cover transient structural simulations, Chapter 12 discussing *implicit methods* while Chapter 15 introducing *explicit methods*.

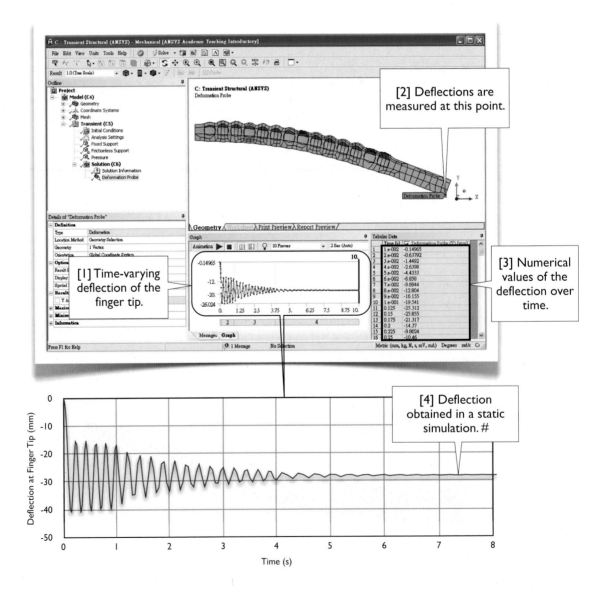

1.1-11 Modal Analysis

A special case of dynamic simulations is the simulation of *free vibrations*, the vibrations of a structure without any external loading (prestress is allowed). Consider that you deflect a structure and then release, causing the structure vibrate under no external forces. We want to know the nature of this free vibration of the structure. The simulation is called a *modal analysis*. The results of modal analysis include the *natural frequencies* and the *vibration modes* of the structure [1-4]. The figure below [1-4] shows the four lowest natural frequencies and the corresponding vibration modes of the pneumatic finger.

A modal analysis is much less expensive (in terms of engineer's work hours and computing time) than a transient analysis and is often performed before a transient analysis, to obtain preliminary dynamic characteristics of a structure. Chapter 11 will discuss modal analysis and its applications.

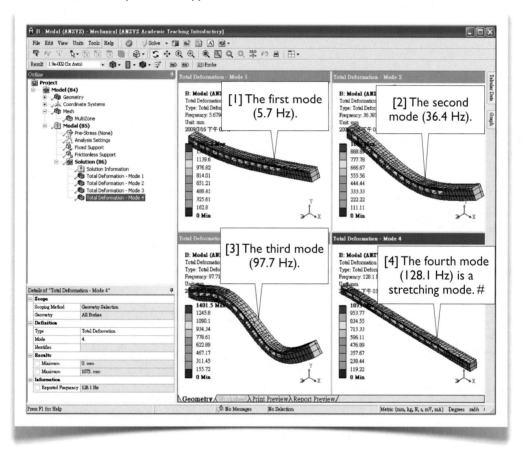

1.1-12 Structural Nonlinearities

When the responses (deflection, stress, strain, etc.) of a structure is linearly proportional to the loads, the structure is called a *linear structure* and the simulation is called a *linear simulation*. Otherwise the structure is called a *nonlinear structure* and the simulation is called a *nonlinear simulation*.

Structural nonlinearities commonly come from three sources. (a) Due to large deformation. This is called *geometry nonlinearity*. (b) Due to topological change of the structure. This is called *topology nonlinearity*. A common case of topology change is the change of contact status, and is called *contact nonlinearity*. (c) Due to nonlinear stress-strain relationship of the material. This is called *material nonlinearity*.

In our case, the stress-strain relationship of the PDMS is reasonably linear (within the range of operational air pressures) and there is no contact between any parts; therefore, there is no material nonlinearity or contact nonlinearity. This case, however, has a deflection so large that it exhibits certain degree of geometry nonlinearity.

The curve [1] in the figure shown below is reproduced from the curve in 1.1-8[5], page 15. It shows a nonlinear relationship between the deflection and the pressure. For comparison, we also include a linear solution [2] in the figure. The linear solution is obtained by turning **Large Deflection** off [3] and solving the model again.

A comparison between the nonlinear solution [1] and the linear solution [2] concludes that, in our case, the error would be significant if geometry nonlinearity were not taken into account.

Solving a nonlinear problem is often challenging (and sometimes frustrating). Real-world problems often involve more-or-less nonlinearities. We will experience nonlinear simulations as early as in Section 3.2, without detail discussions of nonlinear solution controls. Chapters 13 and 14 will dedicate to the discussion of nonlinear simulations, Chapter 13 discussing general nonlinear solution methods, covering geometry nonlinearity and contact nonlinearity, while Chapter 14 discussing material nonlinearity.

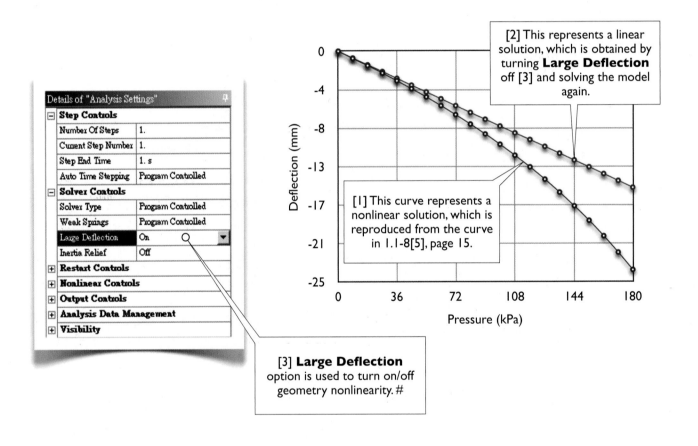

[2] This represents a linear solution, which is obtained by turning **Large Deflection** off [3] and solving the model again.

[1] This curve represents a nonlinear solution, which is reproduced from the curve in 1.1-8[5], page 15.

[3] **Large Deflection** option is used to turn on/off geometry nonlinearity. #

References

1. This case study is adapted from an unpublished work led by Prof. Chao-Chieh Lan of the Department of Mechanical Engineering, National Cheng Kung University, Taiwan.
2. Jeong, O. K and Konishi, S., "All PDMS Pneumatic Microfinger With Bidirectional Motion and Its Application," *Journal of Microelectomechanical Systems*, Vol. 15, No. 4, August 2006, pp. 896-903.
3. Draheim, J, Kamberger, R., and Wallrabe, U, "Process and material properties of polydimethylsiloxane (PDMS) for Optical MEMS," *Sensors and Actuators A: Physical*, Vol. 151, Issue 2, April 2009, pp. 95-99.

Section 1.2

Structural Mechanics: A Quick Review

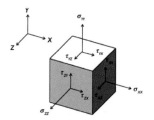

This section (a) defines a structural analysis problem so that the students have a clear picture about the input and the output of a structural analysis system; (b) introduces basic terminologies that will be used throughout the book, such as displacements, stresses, and strains; (c) and summarizes equations that govern the behavior of structures so that the students know what equations Workbench is solving.

To simplify the discussion, this review is limited on homogeneous, isotropic, linear static structural analyses. That is, (a) the material is assumed to be homogeneous, isotropic, and linearly elastic. Assumption of linear elasticity implies that Hooke's law is applicable; (b) the deformation is small enough so that we assume a linear relationship between the displacements and the strains; (c) there are no changes of topology; specifically, there are no changes of contact status during the deformation; (d) the deformation is slow enough so that the dynamic effects are not taken into account.

The concepts introduced in this section can be generalized to include non-homogeneous, anisotropic, nonlinear, dynamic problems.

1.2-1 Structural Analysis Problems

Many engineering analysis problems (e.g., structural, mechanical, flow, electromagnetic) can be defined as the process of finding the *responses* of a *problem domain* subject to some *environmental conditions*.

In structural problems, the problem domain consists of *bodies* (solid, surface, or line bodies); the environmental conditions include *loads* and *supports*; the responses can be described by the *displacements*, *stresses*, or *strains*.

For the pneumatic fingers case (Section 1.1), the problem domain is a solid body made of the PDMS elastomer. There are two support conditions: fixed support at one of the end faces and frictionless support at the face of symmetry. There is only one loading condition: the air pressure applied on the faces of the air chambers.

Note that these environmental conditions are applied on boundary faces, so these conditions are also called *boundary conditions*. Environmental conditions may not be applied on boundary surfaces. Common environmental conditions that do not apply on boundary surfaces include temperature changes and inertia forces; these loads distribute over the volumes (rather than boundary faces) of the problem domain.

1.2-2 Displacement

Deformation of a body can be described by a *displacement field* $\{u\}$ [1-5] (next page). Note that the quantity $\{u\}$ is a function of positions and, since it is a vector, we may express the displacement with three components,

$$\{u\} = \left\{ \begin{array}{ccc} u_x & u_y & u_z \end{array} \right\} \tag{1}$$

All three components are, of course, functions of positions. The SI unit for displacements is meter (m).

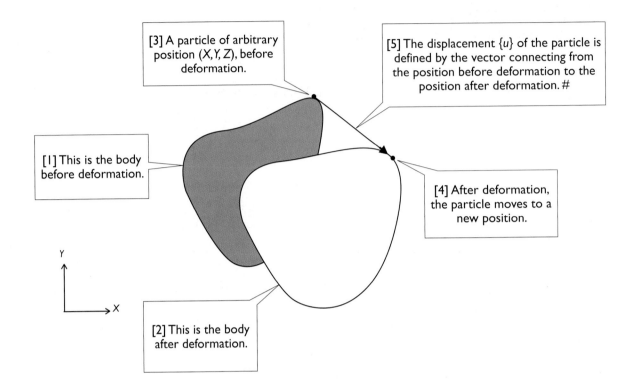

[3] A particle of arbitrary position (X, Y, Z), before deformation.

[5] The displacement {u} of the particle is defined by the vector connecting from the position before deformation to the position after deformation. #

[1] This is the body before deformation.

[4] After deformation, the particle moves to a new position.

[2] This is the body after deformation.

1.2-3 Stress

Concepts of displacements are relatively easy to understand, since the displacement can be defined by a vector, and most of college students are familiar with the mathematics of vectors. In contrast, the concepts of stress are not so obvious.

Stresses are quantities to describe the intensity of force in a body (either solid or fluid). Its unit is force per unit area (i.e., N/m^2 in SI). It is a position-dependent quantity.

Imagine that your arms are pulled by your friends with two forces of the same magnitude but opposite directions. What are the stresses in your arms? Assuming the magnitude of the forces is P and the cross-sectional area of your arms is A, then you may answer, "the stresses are P/A, everywhere in my arms." This case is simple and the answer is good enough. For an one-dimensional case like this, the stress σ may be defined as $\sigma = P/A$, where P is the applied force and A is the cross sectional area.

In 3D cases, things are much more complicated. Now, imagine that you are buried in the soil by your friends, and your head is deep below the ground surface. How do you describe the force intensity (i.e., stress) on your head?

If the soil is replaced by still water, than the answer would be much simpler. The magnitude of the pressure (stress) on the top of your head would be the same as the pressure on your cheeks, and the direction of the pressure would always be perpendicular to the surface where the pressure applies. You've learned these in your high school. And you've learned that the magnitude of the pressure is $\sigma = \rho g h$, where ρ is the mass density of the water, g is the gravitational acceleration, and h is the depth of your head. In general, to describe the force intensity at a certain position in still water, we place an infinitesimally small body at that position, and measure the force per unit surface area on that body.

In the soil (which is a solid material rather than water), the behavior is quite different First, the magnitude of the pressure on the top of your head may not be the same as that on your cheeks. Second, the direction of pressure is not necessarily perpendicular to the surface where the pressure applies. However, the above definition of stresses for water still holds.

Definition of Stress

The stress at a certain point can be defined as *the force per unit area acting on the boundary faces of an infinitesimally small body centered at that point* [1]. The stress values may be different at different faces. And the small body can be any shape. However, for the purpose of describing the stress, we usually use an infinitesimally small cube [2] of which each edge is parallel to a coordinate axis. If we can find the stresses on a small cube, we then can calculate the stresses on any other shapes of small body.

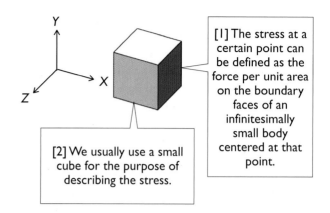

[1] The stress at a certain point can be defined as the force per unit area on the boundary faces of an infinitesimally small body centered at that point.

[2] We usually use a small cube for the purpose of describing the stress.

X-Face, Y-Face, and Z-Face

Each of the six faces of the cube can be assigned an identifier, namely *X-face, Y-face, Z-face, negative-X-face, negative-Y-face,* and *negative-Z-face,* respectively [3-6]. Note that *X-face* has *X*-axis as its outer normal vector, and so on.

[6] Negative-X-face (opposite of the X-face).

[4] Y-face.

[3] X-face.

[5] Z-face.

Stress Components

Let $\vec{p}_X$ be the force per unit area acting on the *X*-face. In general, $\vec{p}_X$ may not be normal or parallel to the *X*-face. We may decompose $\vec{p}_X$ into *X*-, *Y*-, and *Z*-component, and denote $\sigma_{XX}, \tau_{XY},$ and τ_{XZ} respectively [7]. The first subscript (*X*) is used to indicate the **face** on which the stress components act, while the second subscript (*X, Y,* or *Z*) is used to indicate the **direction** of the stress components. Note that σ_{XX} is normal to the face, while $\tau_{XY},$ and τ_{XZ} are parallel to the face. Therefore, σ_{XX} is called a **normal stress,** while $\tau_{XY},$ and τ_{XZ} are called **shear stresses.** We usually use the symbol σ for a normal stress and τ for a shear stress.

Similarly, let $\vec{p}_Y$ be the force per unit area acting on the *Y*-face and we may decompose $\vec{p}_Y$ into a normal component (σ_{YY}) and two shear components (τ_{YX} and τ_{YZ}) [8]. Also, let $\vec{p}_Z$ be the force per unit area acting on the *Z*-face and we may decompose $\vec{p}_Z$ into a normal component (σ_{ZZ}) and two shear components (τ_{ZX} and τ_{ZY}) [9]. Organized in a matrix form, these stress components may be written as

$$\{\sigma\} = \begin{pmatrix} \sigma_{XX} & \tau_{XY} & \tau_{XZ} \\ \tau_{YX} & \sigma_{YY} & \tau_{YZ} \\ \tau_{ZX} & \tau_{ZY} & \sigma_{ZZ} \end{pmatrix} \quad (1)$$

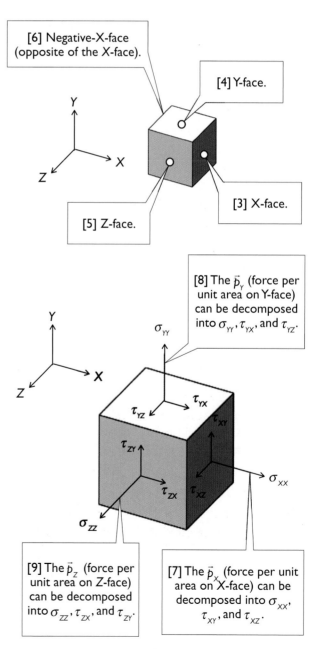

[8] The $\vec{p}_Y$ (force per unit area on Y-face) can be decomposed into $\sigma_{YY}, \tau_{YX},$ and τ_{YZ}.

[9] The $\vec{p}_Z$ (force per unit area on Z-face) can be decomposed into $\sigma_{ZZ}, \tau_{ZX},$ and τ_{ZY}.

[7] The $\vec{p}_X$ (force per unit area on X-face) can be decomposed into $\sigma_{XX}, \tau_{XY},$ and τ_{XZ}.

Stress Components on Other Faces

It can be proven that the stress components on the negative-X-face, negative-Y-face, and negative-Z-face can be derived from the 9 stress components in Eq. (1). For example, on the negative-X-face, the stress components have exactly the same stress values as those on the X-face but with opposite directions [10]. Similarly, the stress components on the negative-Y-face have the same stress values as those on the Y-face but with opposite directions [11], and the stress components on the negative-Z-face have the same stress values as those on the Y-face but with opposite directions [12].

The proof can be done by taking the cube as free body and applying the force equilibria in X, Y, and Z directions respectively.

On an arbitrary face (which may not be parallel or perpendicular to an axis), the stress components also can be calculated from the 9 stress components in Eq. (1). This can be done by using a Mohr's circle (1.4-2[5], page 37).

Symmetry of Shear Stresses

It also can be proven that the shear stresses are symmetric; i.e.,

$$\tau_{XY} = \tau_{YX}, \quad \tau_{YZ} = \tau_{ZY}, \quad \tau_{ZX} = \tau_{XZ} \qquad (2)$$

The proof can be done by taking the cube as free body and applying the moment equilibria in X, Y, and Z directions respectively.

Stress State

We now conclude that 3 normal stress components and 3 shear stress components are needed to describe the **stress state** at a certain point, which may be written in a vector form

$$\{\sigma\} = \left\{ \begin{array}{cccccc} \sigma_X & \sigma_Y & \sigma_Z & \tau_{XY} & \tau_{YZ} & \tau_{ZX} \end{array} \right\} \qquad (3)$$

Note that, for more concise, we use σ_X in place of σ_{XX}, σ_Y in place of σ_{YY}, and σ_Z in place of σ_{ZZ}.

These 6 components are, of course, functions of position.

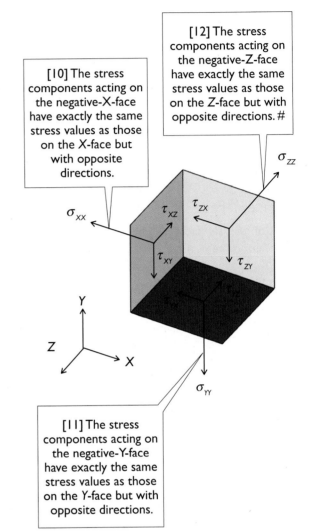

[10] The stress components acting on the negative-X-face have exactly the same stress values as those on the X-face but with opposite directions.

[12] The stress components acting on the negative-Z-face have exactly the same stress values as those on the Z-face but with opposite directions. #

[11] The stress components acting on the negative-Y-face have exactly the same stress values as those on the Y-face but with opposite directions.

1.2-4 Strain

Strains are quantities to describe how the material in a body is stretched and distorted. Or, for operational purpose, strains are defined as displacements of a point relative to its neighboring points. Although the notations of strain components are very similar to those of stress components, the concepts of strains are even more difficult to comprehend.

Let's consider 2D cases first. The concepts can be extended to 3D cases. Consider a point A and its neighboring points B and C, which are respectively along X-axis and Y-axis [1]. Suppose that, after deformation, ABC displaces to a new configuration $A'B'C'$ [2]. Keep in mind that, in this section, we assume the deformation is infinitesimally small. Under the small deformation assumption, the normal strains in X-axis and Y-axis can be defined respectively as

$$\varepsilon_X = \frac{A'B' - AB}{AB} \ \text{(dimensionless)} \tag{1}$$

$$\varepsilon_Y = \frac{A'C' - AC}{AC} \ \text{(dimensionless)} \tag{2}$$

The strains defined in (1) and (2) represent the stretch at the point A in X-direction and Y-direction respectively. Stretch are not the only deformation modes; there are other deformation modes: changes of angles; e.g., from $\angle CAB$ to $\angle C'A'B'$, which is defined as the shear strain in XY-plane,

$$\gamma_{XY} = \angle CAB - \angle C'A'B' \ \text{(rad)} \tag{3}$$

Note that the normal strains (1, 2) and the shear strain (3) are all dimensionless, since the radian is also regarded as dimensionless.

In the above illustration, we consider only 2D cases. In general, the stretching may also occurs in Z-direction and the shearing may also occurs in YZ-plane and ZX-plane. Therefore, we need six strain components to completely describe the stretching and shearing of the material at a point:

$$\{\varepsilon\} = \left\{ \ \varepsilon_X \quad \varepsilon_Y \quad \varepsilon_Z \quad \gamma_{XY} \quad \gamma_{YZ} \quad \gamma_{ZX} \ \right\} \tag{4}$$

These 6 components are, of course, functions of position.

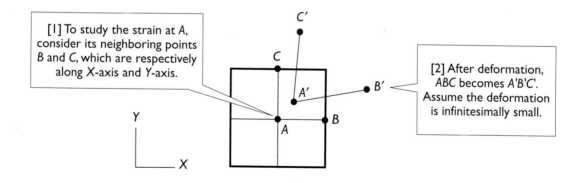

[1] To study the strain at A, consider its neighboring points B and C, which are respectively along X-axis and Y-axis.

[2] After deformation, ABC becomes $A'B'C'$. Assume the deformation is infinitesimally small.

Why They Are Called Normal/Shear Strains?

The definitions in Eqs. (1), (2), and (3) do not explain why they are called "normal" strains and "shear" strain respectively. To clarify this, let's redefine normal and shear strains using a different but equivalent way.

First, we translate and rotate $A'B'C'$ such that A' coincides with A and $A'C'$ aligns with AC [3]. Now the vector BB' is the "absolute" displacement (displacement excluding rigid body motion) of a neighboring point B which is on X-axis [4]. This displacement BB' can be decomposed into two components: BD and DB', the former is called the normal component, while the latter is called the shear component. They are so named because BD is normal to the X-face and DB' is parallel to the X-face. The normal strain and shear strain on X-face are then defined respectively by dividing the components with the original length,

$$\varepsilon_X = \frac{BD}{AB} \text{ (dimensionless)} \tag{5}$$

$$\gamma_{XY} = \frac{DB'}{AB} \text{ (rad)} \tag{6}$$

Note that, under the assumption of small deformation, the definition in Eq. (5) is the same as Eq. (1), while the definition in Eq. (6) is the same as Eq. (3). Also note that there are two subscripts in the shear strain γ_{XY}. The first subscript X is the face where the shearing occurs, while the second subscript Y is the direction of the shearing.

Similarly, we may translate and rotate $A'B'C'$ such that A' coincides with A and $A'B'$ aligns with AB [5]. Now the vector CC' is the "absolute" displacement of a neighboring point C which is on Y-axis [6]. This displacement CC' can be decomposed into two components: CE and EC', the former is the normal component, while the latter is the shear component. The normal strain and shear strain on Y-face is then defined by

$$\varepsilon_Y = \frac{CE}{AC} \text{ (dimensionless)} \tag{7}$$

$$\gamma_{YX} = \frac{EC'}{AC} \text{ (rad)} \tag{8}$$

Note that, under the assumption of small deformation, the definition in Eq. (7) is the same as Eq. (2), while the definition in Eq. (8) is the same as Eq. (3). From Eqs. (3, 6, 8), we may write

$$\gamma_{XY} = \gamma_{YX} = \text{change of a right angle in } XY\text{-plane (rad)} \tag{9}$$

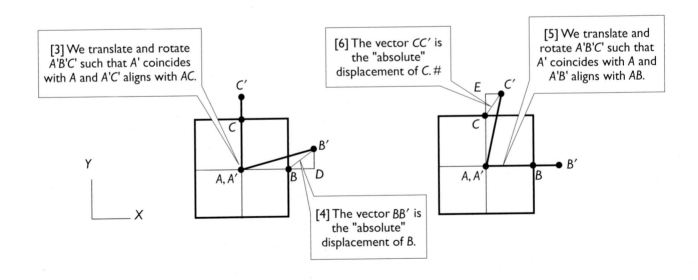

[3] We translate and rotate A'B'C' such that A' coincides with A and A'C' aligns with AC.

[6] The vector CC' is the "absolute" displacement of C. #

[5] We translate and rotate A'B'C' such that A' coincides with A and A'B' aligns with AB.

[4] The vector BB' is the "absolute" displacement of B.

1.2-5 Governing Equations

Let's summarize what we have concluded so far. In structural analysis, we can use following quantities (all or part of them) to describe the response of a structure subject to environmental conditions:

$$\{u\} = \left\{ \begin{array}{ccc} u_X & u_Y & u_Z \end{array} \right\}$$ Copy of 1.2-2(1), page 20

$$\{\sigma\} = \left\{ \begin{array}{cccccc} \sigma_X & \sigma_Y & \sigma_Z & \tau_{XY} & \tau_{YZ} & \tau_{ZX} \end{array} \right\}$$ Copy of 1.2-3(3), page 23

$$\{\varepsilon\} = \left\{ \begin{array}{cccccc} \varepsilon_X & \varepsilon_Y & \varepsilon_Z & \gamma_{XY} & \gamma_{YZ} & \gamma_{ZX} \end{array} \right\}$$ Copy of 1.2-4(4), page 24

These 15 quantities are not independent each other. Relations according to physical principles and mathematics exist among these quantities. To solve for these 15 quantities, we must establish 15 equations. These equations are called governing equations; they govern the structure's behaviors. These equations, will be introduced in the rest of this section, include 3 equilibrium equations, 6 strain-displacement relations, and 6 stress-strain relations.

1.2-6 Equilibrium Equations

The stress components in Eq. 1.2-3(3) (page 23) must satisfy the principle of force equilibrium:

$$\sum F_X = 0, \quad \sum F_Y = 0, \quad \sum F_Z = 0 \qquad (1)$$

If we apply Eqs. (1) on a point INSIDE the structural body, we can obtain three equilibrium equations involving the stress components:

$$\frac{\partial \sigma_X}{\partial X} + \frac{\partial \tau_{XY}}{\partial Y} + \frac{\partial \tau_{XZ}}{\partial Z} + b_X = 0$$

$$\frac{\partial \tau_{YX}}{\partial X} + \frac{\partial \sigma_Y}{\partial Y} + \frac{\partial \tau_{YZ}}{\partial Z} + b_Y = 0 \qquad (2)$$

$$\frac{\partial \tau_{ZX}}{\partial X} + \frac{\partial \tau_{ZY}}{\partial Y} + \frac{\partial \sigma_Z}{\partial Z} + b_Z = 0$$

where b_X, b_Y, b_Z are components of body forces (forces distributed in the body, with SI unit N/m^3). If we apply Eqs. (1) on a point ON the boundary surface of the structural body, the three equilibrium equations will have the form:

$$\sigma_X n_X + \tau_{XY} n_Y + \tau_{XZ} n_Z + S_X = 0$$

$$\tau_{YX} n_X + \sigma_Y n_Y + \tau_{YZ} n_Z + S_Y = 0 \qquad (3)$$

$$\tau_{ZX} n_X + \tau_{ZY} n_Y + \sigma_Z n_Z + S_Z = 0$$

where S_X, S_Y, S_Z are components of surface forces (forces distributed on the boundary, with SI unit N/m^2), and n_X, n_Y, n_Z are components of the unit normal vector on the boundary surface.

We will not discuss the derivations of Eqs. (2, 3) further; they can be found in any textbooks of Solid Mechanics[Refs 1, 2]. Here, we want to emphasize again that the equilibrium equations originate from Eqs. (1). Also note that, in order to derive Eqs. (2, 3), the stress components in 1.2-3[7-12] (pages 22-23) must be expanded to include differential terms.

1.2-7 Strain-Displacement Relations

There exist geometric relations between displacement and strain. Under the assumption of small deformation, the relations are linear:

$$\varepsilon_X = \frac{\partial u_X}{\partial X}, \quad \varepsilon_Y = \frac{\partial u_Y}{\partial Y}, \quad \varepsilon_Z = \frac{\partial u_Z}{\partial Z}$$

$$\gamma_{XY} = \frac{\partial u_X}{\partial Y} + \frac{\partial u_Y}{\partial X}, \quad \gamma_{YZ} = \frac{\partial u_Y}{\partial Z} + \frac{\partial u_Z}{\partial Y}, \quad \gamma_{ZX} = \frac{\partial u_Z}{\partial X} + \frac{\partial u_X}{\partial Z}$$

(1)

Again, we will not discuss the derivations of Eq. (1) further; they can be found in any textbooks of Solid Mechanics[Refs 1, 2]. Here, we want to emphasize that Eq. (1) is derived from mathematics (geometry) without applying any physical principles. You also see that the strain-displacement relation in Eq. (1) is consistent with what we have illustrated in 1.2-4 (pages 24-25). Second, if the assumption of small deformation is removed, a strain-displacement relation still exists but is no longer linear; there will be some high-order differential terms in the equations.

1.2-8 Stress-Strain Relations

To solve the response of a structure, it is practical to assume a relation between stress and strain. Experiments show that a linear relation between stress and strain often can be adopted; it is called the Hooke's law[Refs 1, 2, 3]:

$$\varepsilon_X = \frac{\sigma_X}{E} - v\frac{\sigma_Y}{E} - v\frac{\sigma_Z}{E}$$

$$\varepsilon_Y = \frac{\sigma_Y}{E} - v\frac{\sigma_Z}{E} - v\frac{\sigma_X}{E}$$

$$\varepsilon_Z = \frac{\sigma_Z}{E} - v\frac{\sigma_X}{E} - v\frac{\sigma_Y}{E}$$

$$\gamma_{XY} = \frac{\tau_{XY}}{G}, \quad \gamma_{YZ} = \frac{\tau_{YZ}}{G}, \quad \gamma_{ZX} = \frac{\tau_{ZX}}{G}$$

(1)

Although Eq. (1) is purely an assumption (no physical or mathematical principles applied), it is proved to be very useful and often accurate enough, depending on the material and the application. Eq. (1) is called a *material model*; it characterizes the behavior of a material, independent of the geometry and environmental conditions.

There are three material parameters in Eq. (1): the Young's modulus E, the Poisson's ratio v, and the shear modulus G. In SI, the Young's modulus and the shear modulus have units of pascal (Pa) and the Poisson's ratio is dimensionless. It can be shown that these three quantities are not independent to each other; they satisfy the relation[Ref 3]

$$G = \frac{E}{2(1+v)}$$

(2)

We conclude that, for an isotropic, linearly elastic material, any two of E, v, and G can be used to describe the stress-strain relation. In Workbench, it requires the input of Young's modulus and Poisson's ratio to define an **Isotropic Elasticity** model (1.1-3[4, 5], page 9).

Thermal Effects

Consider that the temperature changes over the structural body. Since a temperature change of ΔT induces a strain of $\alpha \Delta T$, in which α is the *coefficient of thermal expansion*, this *thermal strain* should be added to Eq. (1) (last page); i.e.,

$$\varepsilon_X = \frac{\sigma_X}{E} - v\frac{\sigma_Y}{E} - v\frac{\sigma_Z}{E} + \alpha \Delta T$$

$$\varepsilon_Y = \frac{\sigma_Y}{E} - v\frac{\sigma_Z}{E} - v\frac{\sigma_X}{E} + \alpha \Delta T$$

$$\varepsilon_Z = \frac{\sigma_Z}{E} - v\frac{\sigma_X}{E} - v\frac{\sigma_Y}{E} + \alpha \Delta T$$

$$\gamma_{XY} = \frac{\tau_{XY}}{G}, \quad \gamma_{YZ} = \frac{\tau_{YZ}}{G}, \quad \gamma_{ZX} = \frac{\tau_{ZX}}{G}$$

(3)

Orthotropic Elasticity

For orthotropic materials (14.1-1[2], page 518), in which there exist three mutual orthogonal planes of material symmetry, the Hooke's law can be generalized to[Refs 1, 2, 3]

$$\varepsilon_X = \frac{\sigma_X}{E_X} - v_{YX}\frac{\sigma_Y}{E_Y} - v_{ZX}\frac{\sigma_Z}{E_Z} + \alpha_X \Delta T$$

$$\varepsilon_Y = \frac{\sigma_Y}{E_Y} - v_{ZY}\frac{\sigma_Z}{E_Z} - v_{XY}\frac{\sigma_X}{E_X} + \alpha_Y \Delta T$$

$$\varepsilon_Z = \frac{\sigma_Z}{E_Z} - v_{XZ}\frac{\sigma_X}{E_X} - v_{YZ}\frac{\sigma_Y}{E_Y} + \alpha_Z \Delta T$$

$$\gamma_{XY} = \frac{\tau_{XY}}{G_{XY}}, \quad \gamma_{YZ} = \frac{\tau_{YZ}}{G_{YZ}}, \quad \gamma_{ZX} = \frac{\tau_{ZX}}{G_{ZX}}$$

(4)

where E_X, E_Y, E_Z are Young's moduli in their respective directions, G_{XY}, G_{YZ}, G_{ZX} are the shear moduli in their respective planes, and v_{XY}, v_{YZ}, v_{ZX} are the Poisson's ratios in their respective planes. The first subscript in each of the Poisson's ratios refers to the direction of the load, and the second to the direction of the contraction. For example, v_{XY} represents the amount of contraction in *Y*-direction, when the material is stretched in *X*-direction.

1.2-9 Summary

We now have 15 equations, including three equilibrium equations, either (INSIDE the structural body)

$$\frac{\partial \sigma_X}{\partial X} + \frac{\partial \tau_{XY}}{\partial Y} + \frac{\partial \tau_{XZ}}{\partial Z} + b_X = 0$$

$$\frac{\partial \tau_{YX}}{\partial X} + \frac{\partial \sigma_Y}{\partial Y} + \frac{\partial \tau_{YZ}}{\partial Z} + b_Y = 0$$

Copy of 1.2-6(2), page 26

$$\frac{\partial \tau_{ZX}}{\partial X} + \frac{\partial \tau_{ZY}}{\partial Y} + \frac{\partial \sigma_Z}{\partial Z} + b_Z = 0$$

or (ON the boundary surface)

$$\sigma_X n_X + \tau_{XY} n_Y + \tau_{XZ} n_Z + S_X = 0$$

$$\tau_{YX} n_X + \sigma_Y n_Y + \tau_{YZ} n_Z + S_Y = 0$$

Copy of 1.2-6(3), page 26

$$\tau_{ZX} n_X + \tau_{ZY} n_Y + \sigma_Z n_Z + S_Z = 0$$

and six equations describing the strain-displacement relation

$$\varepsilon_X = \frac{\partial u_X}{\partial X}, \quad \varepsilon_Y = \frac{\partial u_Y}{\partial Y}, \quad \varepsilon_Z = \frac{\partial u_Z}{\partial Z}$$

Copy of 1.2-7(1), page 27

$$\gamma_{XY} = \frac{\partial u_X}{\partial Y} + \frac{\partial u_Y}{\partial X}, \quad \gamma_{YZ} = \frac{\partial u_Y}{\partial Z} + \frac{\partial u_Z}{\partial Y}, \quad \gamma_{ZX} = \frac{\partial u_Z}{\partial X} + \frac{\partial u_X}{\partial Z}$$

and six equations describing the stress-strain relation

$$\varepsilon_X = \frac{\sigma_X}{E} - v\frac{\sigma_Y}{E} - v\frac{\sigma_Z}{E} + \alpha\Delta T$$

$$\varepsilon_Y = \frac{\sigma_Y}{E} - v\frac{\sigma_Z}{E} - v\frac{\sigma_X}{E} + \alpha\Delta T$$

Copy of 1.2-8(3), page 28

$$\varepsilon_Z = \frac{\sigma_Z}{E} - v\frac{\sigma_X}{E} - v\frac{\sigma_Y}{E} + \alpha\Delta T$$

$$\gamma_{XY} = \frac{\tau_{XY}}{G}, \quad \gamma_{YZ} = \frac{\tau_{YZ}}{G}, \quad \gamma_{ZX} = \frac{\tau_{ZX}}{G}$$

In theory, these 15 equations can be solved for 15 unknown quantities:

$$\{u\} = \left\{ \begin{matrix} u_X & u_Y & u_Z \end{matrix} \right\}$$

Copy of 1.2-2(1), page 20

$$\{\sigma\} = \left\{ \begin{matrix} \sigma_X & \sigma_Y & \sigma_Z & \tau_{XY} & \tau_{YZ} & \tau_{ZX} \end{matrix} \right\}$$

Copy of 1.2-3(3), page 23

$$\{\varepsilon\} = \left\{ \begin{matrix} \varepsilon_X & \varepsilon_Y & \varepsilon_Z & \gamma_{XY} & \gamma_{YZ} & \gamma_{ZX} \end{matrix} \right\}$$

Copy of 1.2-4(4), page 24

In practice, only a few extremely simple "textbook problems" can be solved analytically. Most of real-world problems are too complicated to solve analytically. For a linear problem described in this section, the complexity comes from geometry and environmental conditions. Numerical methods are the only feasible methods. The finite element methods, which has been the most successful numerical methods for *boundary-value problems* (such as the problems described in this section), are implemented in ANSYS Workbench to solve the governing equations.

Remark

By "governing equations" of a structural simulation problem, we sometimes mean the equilibrium equations: Eq. 1.2-6(2) (page 26) governs the behavior in the body and Eq. 1.2-6(3) (page 26) governs the behavior on the boundary. The two sets of equations constitute a boundary value problems.

It is possible to replace the stress components in the equilibrium equations by strain components using Eq. 1.2-8(3) (page 28), and in turn replace the strain components by displacement components using Eq. 1.2-7(1) (page 27). The result is a set of three equilibrium equations involving three displacement components, and we can solve the three differential equations for the three displacement components.

That is how ANSYS Workbench solves a structural problem (see Eq. 1.3-1(1), next page).

References

1. Cook, R. D. and Young, W. C., *Advanced Mechanics of Materials*, Macmillan, 1985.
2. Haslach, H. W. Jr. and Armstrong, R. W., *Deformable Bodies and Their Material Behavior*, John Wiley & Sons, Inc., 2004; Chapter 9 Plasticity.
3. Beer, F. P., Johnston, E. R. Jr., and DeWolf, J. T., *Mechanics of Materials*, 3rd Ed., McGraw Hill, 2002.

Section 1.3

Finite Element Methods:
A Concise Introduction

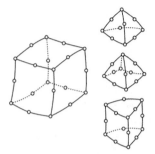

This section (a) introduces terminologies relevant to finite element methods, such as degrees of freedom, shape functions, stiffness matrix, etc., that will be used throughout the book; (b) introduces the basic procedure of finite element methods, so that the students have better understanding about how the Workbench performs a simulation.

1.3-1 Basic Procedure

As mentioned (near the end of 1.2-9, page 29), most of real-world problems are too complicated to be solved analytically, because of the complexity of geometry or environmental conditions. If we add complexity of nonlinearity and dynamic effects into the problems, then their analytical solutions are practically unreachable.

A basic idea of finite element methods is to divide the entire structural body into many small and geometrically simple bodies, called *elements*, so that equilibrium equations of each element can be written down, and all the equilibrium equations are then solved simultaneously. The elements have finite sizes (contrasting to the infinitesimal sizes of elements used in Calculus), thus the name of *Finite Element Methods*.

The elements are assumed to be connected by *nodes* located on the elements' edges and vertices. Another idea is to solve unknown discrete values (e.g., displacements at the nodes) rather than to solve unknown functions (e.g., displacement fields). Since the displacement on each node is a vector and has three components (in 3D cases), the number of total unknown quantities to be solved is three times the number of nodes [1].

The types of elements available in the ANSYS Workbench and their specific configuration of nodes will be given in 1.3-3, pages 33-35.

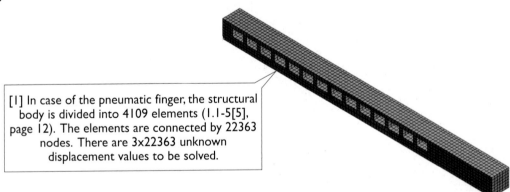

[1] In case of the pneumatic finger, the structural body is divided into 4109 elements (1.1-5[5], page 12). The elements are connected by 22363 nodes. There are 3x22363 unknown displacement values to be solved.

The nodal displacement components, collectively denoted by a vector {D}, are called the *degrees of freedom* (DOFs) of the structure. They are so called because these values fully define the response of a structure. In a static case, the system of equilibrium equations has following form (also see the remark at the end of Section 1.2, last page)

$$[K]\{D\} = \{F\} \tag{1}$$

The size of Eq. (1) is determined by the number of degrees of freedom, which is, in 3D cases, three times the number of nodes. The vector $\{F\}$ is the external forces acting on the nodes, which is calculated from the environmental conditions. Physical meaning of the matrix $[K]$ can be understood by thinking of the structure as a spring, $\{F\}$ as external force, and $\{D\}$ as the deformation of the spring. In this analogy, $[K]$ would be the spring constant, or the stiffness of the spring. In finite element methods, $[K]$ is called the *stiffness matrix* of the structure. More precisely, physical meaning of the values on its ith column is the forces required on all the DOFs to make the ith DOF a unit displacement while restrain the other DOFs from any displacements.

Note that, for a linear structure, $[K]$ is a constant matrix; while for a nonlinear cases (1.1-12, pages 18-19), $[K]$ is a function of $\{D\}$, which will be discussed in Chapters 13 and 14. For dynamic cases (1.1-10 and 1.1-11, pages 17-18), dynamic effects have to be added to Eq. (1), which will be discussed in Chapters 11 and 12.

After the discrete nodal displacements $\{D\}$ in Eq. (1) are solved, the displacement fields $\{u\}$ are calculated by interpolating the nodal displacements, either linearly or quadratically [2] (see Eq. 1.3.2(2)). These interpolating functions are called *shape functions*. Concepts of shape functions are crucial in the finite element methods; they will be discussed further in 1.3-2.

As soon as the displacement fields become known, the strain fields can be calculated by Eq. 1.2-7(1) (page 27), the strain-displacement relation. The stress fields in turn can be calculated by Eq. 1.2-8(3) (page 28), the stress-strain relation.

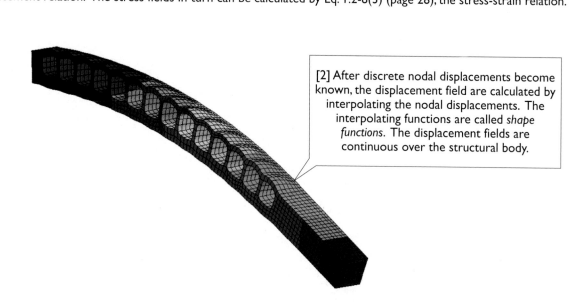

[2] After discrete nodal displacements become known, the displacement field are calculated by interpolating the nodal displacements. The interpolating functions are called *shape functions*. The displacement fields are continuous over the structural body.

1.3-2 Shape Functions and the Order of Element

As mentioned (1.3-1[2]), the displacement fields $\{u\}$ are calculated by interpolating the nodal displacements. The interpolating functions are called shape functions. In other words, the shape functions are used to establish a relation between the displacement fields and the nodal displacements.

As an example, consider a 2D 4-node quadrilateral element [1]. The nodal displacements, collectively denoted by a vector $\{d\}$, has 8 components

$$\{d\} = \left\{ \begin{array}{cccccccc} d_1 & d_2 & d_3 & d_4 & d_5 & d_6 & d_7 & d_8 \end{array} \right\} \tag{1}$$

The displacement fields $\{u\}$ can be calculated by interpolating the nodal displacements $\{d\}$

$$\{u\} = [N]\{d\} \tag{2}$$

where [N] is called the *matrix of shape functions*. The role of the shape functions is the interpolating functions from the nodal displacements {d} to the displacement fields {u}. Let's summarize again: the components of nodal displacements {d} are discrete values and the components of displacement fields {u} are continuous functions of (X, Y, Z). The shape functions in [N] are to bridge the continuous functions {u} and the discrete values {d}.

Now, let's examine the dimension of the [N] matrix using this example. Since {d} has 8 components and {u} has 3 components (Eq. 1.2-2(1), page 20), the matrix [N] must be of dimension 3x8.

Since {u} contains functions of (X, Y, Z) and {d} contains discrete values, therefore [N] must contains functions of (X, Y, Z). Besides, for the element shown [1], since the interpolating points are on the vertices of the element, the shape functions must be a linear form. When the shape functions are linear, the element is called a *linear element*, *first-order element*, or *lower-order element* [2].

Often, using quadratic polynomials as shape functions can be more efficient. In such cases, a node is added on the middle of each edge of the element; the added nodes are called the *midside nodes*; the element is called a *quadratic element*, *second-order element*, or *higher-order element*. In this book, we will not use the term *linear element*, to avoid confusing with the terms such as linear material.

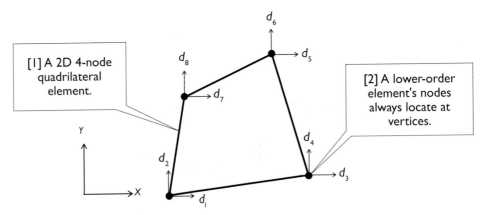

[1] A 2D 4-node quadrilateral element.

[2] A lower-order element's nodes always locate at vertices.

1.3-3 Workbench Element Shapes

ANSYS literally provides hundreds of element types[Ref 1]. Workbench, however, directly supports only a portion of them[Ref 2], other element types can be accessed by using APDL. To identify these element types, each element type is assigned a code name (e.g., SOLID186, PLANE183, etc.).

In 2D cases, the element shapes available are quadrilateral (4-sided) or triangular (3-sided). In 3D cases, the available shapes include hexahedral (6-faced), triangle-based prism (5-faced), quadrilateral-based pyramid (5-faced), and tetrahedral (4-faced). By default, the Workbench automatically chooses appropriate element types from an element library[Ref 2] according to the types of the structural bodies. Current version of the Workbench supports only 4 body types: 3D solid body, 2D solid body, 3D surface body, and 3D line body. 3D surface bodies are geometrically 2D but spatially 3D, while 3D line bodies are geometrically 1D but spatially 3D. How the Workbench chooses an element type according to the body type is illustrated as follows.

3D Solid Bodies

Workbench meshes a 3D solid body with SOLID186[Ref 3], a 3D 20-node second-order structural solid elements [1]. The element is hexahedral in its most general shape. By combining some of nodes, the element can degenerate to a triangle-based prism [2], quadrilateral-based pyramid [3], or tetrahedron [4]. Capability of degeneration is useful since it allows different shapes of elements mixed up in a body. If a body is to be meshed with tetrahedral elements exclusively, Workbench meshes the body with SOLID187[Ref 4], which is a 3D 10-node tetrahedral second-order structural solid element and has a shape exactly the same as the one degenerated from SOLID186 [4]. Workbench allows an option to drop off elements' midside nodes; in that case the edges become straight and the element becomes first-order.

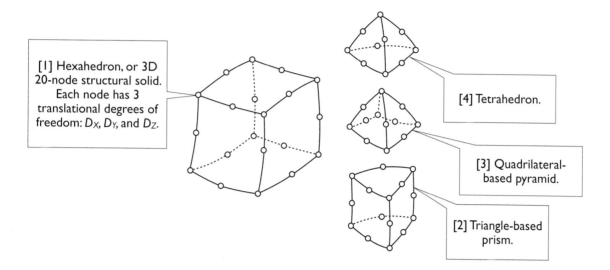

[1] Hexahedron, or 3D 20-node structural solid. Each node has 3 translational degrees of freedom: D_X, D_Y, and D_Z.

[4] Tetrahedron.

[3] Quadrilateral-based pyramid.

[2] Triangle-based prism.

2D Solid Bodies

Workbench meshes a 2D solid body with PLANE183[Ref 5], a 2D 8-node second-order structural solid element [5]. The element is quadrilateral in its general shape. By combining some of nodes, the element can degenerate to a triangle [6]. If you choose to drop midside nodes, the edges become straight, and the Workbench meshes the body with PLANE182[Ref 6], a 2D 4-node first-order structural solid element. It is important to remember that all 2D solid elements must be arranged on **XYPlane**.

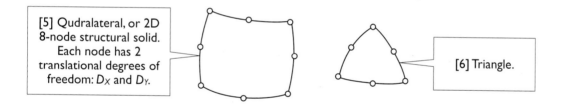

[5] Qudralateral, or 2D 8-node structural solid. Each node has 2 translational degrees of freedom: D_X and D_Y.

[6] Triangle.

3D Surface Bodies

Workbench meshes a 3D surface body with SHELL181[Ref 7], a 3D 4-node first-order structural shell element [7], or SHELL281[Ref 8], a 3D 8-node second-order structural shell. The element is quadrilateral in its general shape, but can degenerate to a triangle [8]. Note that shell elements, although planar by themselves, can be arranged in a 3D space.

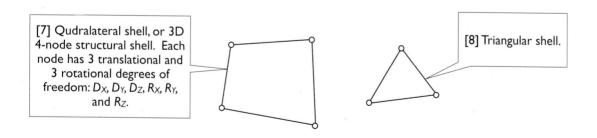

[7] Qudralateral shell, or 3D 4-node structural shell. Each node has 3 translational and 3 rotational degrees of freedom: D_X, D_Y, D_Z, R_X, R_Y, and R_Z.

[8] Triangular shell.

3D Line Bodies

For a 3D line body, the Workbench meshes it with BEAM188[Ref 9], a 3D 2-node first-order beam element [9]. Note that beam elements can be arranged in 3D space. Note also that beam elements are not necessarily arranged in a straight line or on a plane.

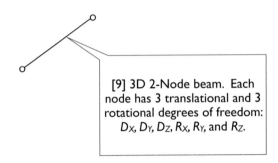

[9] 3D 2-Node beam. Each node has 3 translational and 3 rotational degrees of freedom: D_X, D_Y, D_Z, R_X, R_Y, and R_Z.

References

1. ANSYS Documentation//Mechanical APDL//Element Reference//I. Element Library
2. ANSYS Help System//Mechanical APDL//Theory Reference//1.4.1. Elements Used by the ANSYS Workbench Product (This page is available in version 13 and before but removed since version 14.)
3. ANSYS Documentation//Mechanical APDL//Element Reference//I. Element Library//SOLID186
4. ANSYS Documentation//Mechanical APDL//Element Reference//I. Element Library//SOLID187
5. ANSYS Documentation//Mechanical APDL//Element Reference//I. Element Library//PLANE183
6. ANSYS Documentation//Mechanical APDL//Element Reference//I. Element Library//PLANE182
7. ANSYS Documentation//Mechanical APDL//Element Reference//I. Element Library//SHELL181
8. ANSYS Documentation//Mechanical APDL//Element Reference//I. Element Library//SHELL281
9. ANSYS Documentation//Mechanical APDL//Element Reference//I. Element Library//BEAM188

Section 1.4
Failure Criteria of Materials

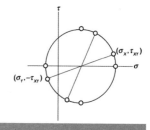

Achieving functionality, safety, and reliability are often the main purposes of structural simulations. Deformation usually relates to the functionality, while stress to the safety and reliability. Which stress should we look into to ensure that a structure doesn't fail? Normal stress? Shear stress? Or something else? How can we say a stress value is too large? What critical values should the stresses be compared with? In short, what are the failure criteria of materials? This section intends to answer these questions.

1.4-1 Ductile versus Brittle Materials

From structural mechanics point of view, stress-strain relation is the most important characteristics of a material. We usually obtain stress-strain relations from standardized uniaxial tensile test data. Two examples of stress-strain relation are shown in the figure below [1, 2]. On the left, the material exhibits a large amount of strain before it fractures [3]; it is called a *ductile material*. On the right, the material's fracture strain is relatively small [4]; it is called a *brittle material*. Fracture strain is a measure of ductility. There are some essential differences between these two types of materials.

Failure Points: Yield Point or Fracture Point?

Mild steel is a typical ductile material. For ductile materials, there often exists an obvious yield point [5], beyond which the deformation would be too large so that the material is no longer reliable or functional; the failure is accompanied by excess deformation. Therefore, for ductile materials, we are concerned about whether the material reaches the yield point. The yield point is characterized by a yield stress σ_y [5]. This is the critical stress we want to compare with. But, with which stress to compare? σ_x? σ_y? σ_z? τ_{xy}? τ_{yz}? τ_{zx}? Or something else? 1.4-4 and 1.4-5 (pages 38-42) will answer this question.

Cast iron and ceramics are two examples of brittle materials. For brittle materials, there usually doesn't exist obvious yield point, and we are concerned about their fracture point. The fracture point is characterized by a fracture stress σ_f [4]. This is the critical stress we want to compare with. But, again, with which stress to compare? σ_x? σ_y? σ_z? τ_{xy}? τ_{yz}? τ_{zx}? Or something else? 1.4-3 (page 38) will answer this question.

Failure Modes: Tensile Failure or Shear Failure?

The fracture of brittle materials is mostly due to *tensile failure*; the yielding of ductile materials is mostly due to *shear failure*[Refs 1, 2]. The tensile failure of brittle materials is easy to understand: the failure always occurs after cracking, induced by tensile stresses. The shear failure of ductile materials can be justified in a standard uniaxial tensile test, in which the failure is accompanied by a necking phenomenon and a cone-shape breaking surface. It is important to remember that a material often fail due to a mix-up of both mechanisms.

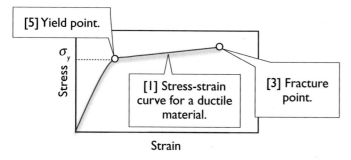

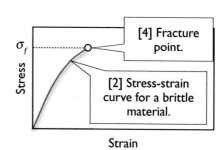

1.4-2 Principal Stresses

We mentioned (1.2-3, page 22) that, at a certain point, the stresses are different in faces of different directions. It then naturally raises a question: In which direction of face does the normal stress reach its maximum? And in which direction of face does the shear stress reach its maximum?

To present the concepts efficiently, consider a 2D case [1], and let's call the X-direction the base direction. In the base direction, the stress can be expressed with a stress pair (σ_X, τ_{XY}). Likewise, in an arbitrary direction, the stress can be expressed with a stress pair (σ, τ). Let's now try to find a relationship between normal stress σ and shear stress τ: how does σ vary with τ?

First, we mark the stress pair (σ_X, τ_{XY}) in the σ-τ space [2]. Second, noting that the stress $(\sigma_Y, -\tau_{XY})$ is also a stress pair, whose direction forms 90° (counter-clockwise) with the base direction, we mark the stress pair $(\sigma_Y, -\tau_{XY})$ in the σ-τ space [3]. Similarly, we could draw other stress pairs in the σ-τ space [4].

First, we want to point out that the collection of these points forms a circle in the σ-τ space [5], called the *Mohr's circle*. The details (including the sign conventions of stresses) can be found in any textbook of Solid Mechanics[Refs 2, 3]. Here, we only emphasize a concept: *a Mohr's circle represents a stress state*. With this useful concept, finding the maximum normal stress and maximum shear stress becomes straightforward.

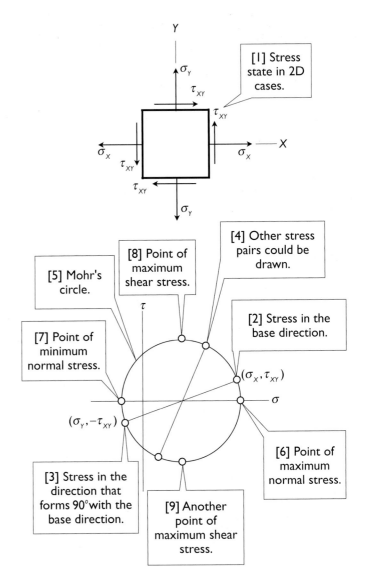

[1] Stress state in 2D cases.

[4] Other stress pairs could be drawn.

[8] Point of maximum shear stress.

[5] Mohr's circle.

[2] Stress in the base direction.

[7] Point of minimum normal stress.

[6] Point of maximum normal stress.

[3] Stress in the direction that forms 90° with the base direction.

[9] Another point of maximum shear stress.

A stress state defines a Mohr's circle, and vice versa. Further, the stress values $(\sigma_X, \sigma_Y, \tau_{XY})$ fully defines a Mohr's circle. Once we have a Mohr's circle, the points of maximum normal stress, minimum normal stress, and maximum shear stress can be located. The maximum normal stress is located at the right-most of the Mohr's circle [6]. The minimum normal stress is located at the left-most of the Mohr's circle [7]. The *maximum shear stress* are located at the upper-most and the lower-most of the Mohr's circle [8, 9]. Note that at points of maximum and minimum normal stresses, the shear stress vanishes.

The maximum normal stress [6] is called the *maximum principal stress* and denoted by σ_1; The minimum normal stress [7] is called the *minimum principal stress* and denoted by σ_3. Their corresponding directions are called *principal direction*. At any point of a 3D solid, there are three principal directions and three principal stresses. The *medium principal stress* is denoted by σ_2. The maximum principal stress is often a positive value, a tension; the minimum principal stress is often a negative value, a compression.

Given $(\sigma_X, \sigma_Y, \tau_{XY})$ to define a Mohr's circle, we can easily calculate the values and their corresponding directions of the principal stresses and the maximum shear stress. We will not derive these formulas in this book; Workbench can report for you on your request. Finally, make sure you understand the concepts and can generalize them to the 3D cases on your own.

1.4-3 Failure Criterion for Brittle Materials

As mentioned (1.4-1, page 36), the failure of brittle materials is a tensile failure. In other words, a brittle material fractures because its tensile stress reaches the fracture strength σ_f of the material (1.4-1[4], page 36). Thus, we may state a failure criterion for brittle materials as follows: At a certain point of a body, if the maximum principal stress reaches the fracture strength of the material, it will fail. In short, a point of material fails if

$$\sigma_1 \geq \sigma_f \tag{1}$$

1.4-4 Tresca Criterion for Ductile Materials

Also mentioned (1.4-1, page 36), the failure of ductile materials is a shear failure. In other words, a ductile material yields because its shear stress reaches the shear strength τ_y of the material. Thus, we may state a failure criterion for ductile materials as follows: At a certain point of a body, if the maximum shear stress reaches the shear strength of the material, it will fail. In short, a point of material fails if

$$\tau_{max} \geq \tau_y \tag{1}$$

where the maximum shear stress τ_{max}, from the geometry of the Mohr's circle (1.4-2[5], page 37), is simply the radius of the circle. Noting that the diameter of the circle is $(\sigma_1 - \sigma_3)$, we may write down

$$\tau_{max} = \frac{\sigma_1 - \sigma_3}{2} \tag{2}$$

In a uniaxial tensile test, the material yields when undergoing its yield stress σ_y in the axial direction. Let the axial direction be X-direction, then the stress state $(\sigma_X, \sigma_Y, \tau_{XY})$ is $(\sigma_y, 0, 0)$, Its maximum principal stress is σ_y while the minimum principal stress is zero [1]. Thus, when the material yields, its shear stress (shear strength) is

$$\tau_y = \frac{\sigma_y}{2} \tag{3}$$

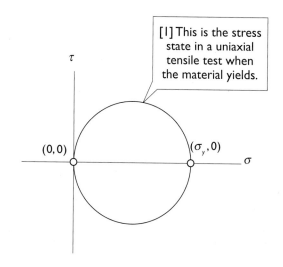

[1] This is the stress state in a uniaxial tensile test when the material yields.

Substituting Eqs. (2, 3) into (1), we have the criterion

$$\frac{\sigma_1 - \sigma_3}{2} \geq \frac{\sigma_y}{2} \tag{4}$$

Simplifying the above equation, we reach a conclusion that the material yields if

$$\sigma_1 - \sigma_3 \geq \sigma_y \tag{5}$$

The quantity on the left-hand side, $(\sigma_1 - \sigma_3)$, is called the *stress intensity*, which the Workbench can report for you on your request. This failure criterion is called the *maximum shear stress criterion*, or *Tresca Criterion*, first proposed by Henri Tresca (1814-1885), a French mechanical engineer, in 1864.

1.4-5 Von Mises Criterion for Ductile Materials

The Tresca criterion (Eq. 1.4-4(5), last page) is easy to understand but often not accurate enough for predicting the yielding of many ductile materials, particularly, metals. The theory discussed in 1.4-4 (last page) is too simplified and deviates from real-world situations. For example, the medium principal stress σ_2 plays no role in Eq. 1.4-4(5) (last page), but that is not always true. A more sophisticated theory, called von Mises criterion, often predicts yielding more accurately than Tresca criterion.

Hydrostatic Stress and Deviatoric Stress

Without loss of generality, a stress state

$$\{\sigma\} = \begin{Bmatrix} \sigma_X & \tau_{XY} & \tau_{XZ} \\ \tau_{YX} & \sigma_Y & \tau_{YZ} \\ \tau_{ZX} & \tau_{ZY} & \sigma_Z \end{Bmatrix}$$

can be expressed in a form of principal stresses if we choose the principal axes as coordinate axes,

$$\{\sigma\} = \begin{Bmatrix} \sigma_1 & 0 & 0 \\ 0 & \sigma_2 & 0 \\ 0 & 0 & \sigma_3 \end{Bmatrix} \tag{1}$$

Note that, as mentioned (1.4-2, page 37), in the principal directions, all the shear stresses vanish.

Define the hydrostatic stress as the average of the normal stresses,

$$p = \frac{\sigma_1 + \sigma_2 + \sigma_3}{3} \tag{2}$$

The stress state (1) can be decomposed into two parts

$$\{\sigma\} = \begin{Bmatrix} \sigma_1 & 0 & 0 \\ 0 & \sigma_2 & 0 \\ 0 & 0 & \sigma_3 \end{Bmatrix} = \begin{Bmatrix} p & 0 & 0 \\ 0 & p & 0 \\ 0 & 0 & p \end{Bmatrix} + \begin{Bmatrix} \sigma_1 - p & 0 & 0 \\ 0 & \sigma_2 - p & 0 \\ 0 & 0 & \sigma_3 - p \end{Bmatrix} \tag{3}$$

or, written in a more compact form,

$$\{\sigma\} = \{\sigma^p\} + \{\sigma^d\} \tag{4}$$

The first part of the right-hand-side is the *hydrostatic stress*, and the second part is called the *deviatoric stress*, the stress deviating from the hydrostatic stress.

A material deformation can be thought of a superposition of a dilation (volumetric change) and a distortion. The hydrostatic stress contributes exclusively to the dilation, while the deviatoric stress contributes exclusively to the distortion. The dilation plays no role in shear failure; it is the distortion that causes shear failure. In short, the deviatoric stress

$$\{\sigma^d\} = \begin{Bmatrix} \sigma_1 - p & 0 & 0 \\ 0 & \sigma_2 - p & 0 \\ 0 & 0 & \sigma_3 - p \end{Bmatrix} \tag{5}$$

can be used to establish a shear failure criterion. However, it is not a scalar value. How can we use it to compare with a uniaxial yield stress σ_y? We need more elaborate theory to derive a useful criterion.

Von Mises Yield Criterion

In 1913, Richard von Mises, an Austria-Hungary born scientist, proposed a theory for predicting the yielding of ductile materials. The theory states that the yielding occurs when the deviatoric strain energy density (or deviatoric energy for short) reaches a critical value.

For linearly elastic materials, the total strain energy *density* is half of the inner product of the stress and the strain,

$$w = \frac{1}{2}\{\sigma\}\cdot\{\varepsilon\} \tag{6}$$

This total strain energy can be decomposed into two parts: energy w^p caused by the hydrostatic stress and energy w^d caused by the deviatoric stress,

or

$$w = w^p + w^d \tag{7}$$

$$w^d = w - w^p \tag{8}$$

The von Mises criterion can be stated as follows: the yielding occurs when

$$w^d \geq w^{yd} \tag{9}$$

where w^{yd} is the deviatoric energy when the material yields in its uniaxial tension test. In the following discussion, we will express w^p and w^d in terms of stresses.

Deviatoric Energy in Uniaxial Tensile Test

In a uniaxial tension test, when yielding occurs, the stress state (1.4-4[1], page 38) is

$$\begin{Bmatrix} \sigma_y & 0 & 0 \\ 0 & 0 & 0 \\ 0 & 0 & 0 \end{Bmatrix} = \begin{Bmatrix} \sigma_y/3 & 0 & 0 \\ 0 & \sigma_y/3 & 0 \\ 0 & 0 & \sigma_y/3 \end{Bmatrix} + \begin{Bmatrix} 2\sigma_y/3 & 0 & 0 \\ 0 & -\sigma_y/3 & 0 \\ 0 & 0 & -\sigma_y/3 \end{Bmatrix}$$

or, written in a more compact form,

$$\{\sigma^y\} = \{\sigma^{yp}\} + \{\sigma^{yd}\}$$

The first part of the right-hand-side is the hydrostatic stress and the second part is the deviatoric stress. The Hooke's law, Eq. 1.2-8(1) (page 27), can be used to obtain the strains. The strains corresponding to the total stress and the hydrostatic stress are respectively

$$\{\varepsilon^y\} = \frac{\sigma_y}{E}\begin{Bmatrix} 1 & 0 & 0 \\ 0 & -v & 0 \\ 0 & 0 & -v \end{Bmatrix} \text{ and } \{\varepsilon^{yp}\} = \frac{(1-2v)\sigma_y}{3E}\begin{Bmatrix} 1 & 0 & 0 \\ 0 & 1 & 0 \\ 0 & 0 & 1 \end{Bmatrix}$$

Calculated using Eq. (6), the energies corresponding to the total stress and the hydrostatic stress are respectively

$$w^y = \frac{1}{2}\{\sigma^y\}\cdot\{\varepsilon^y\} = \frac{\sigma_y^2}{2E} \text{ and } w^{yp} = \frac{1}{2}\{\sigma^{yp}\}\cdot\{\varepsilon^{yp}\} = \frac{(1-2v)\sigma_y^2}{6E}$$

By using Eq. (8), the deviatoric energy is

$$w^{yd} = w^y - w^{yp} = \frac{(1+v)\sigma_y^2}{3E} \tag{10}$$

Deviatoric Energy in General 3D Cases

Now, we consider the general 3D stress state, Eqs. (3, 4). The strains corresponding the total stress $\{\sigma\}$ and the hydrostatic stress $\{\sigma^p\}$ are, using Eq. 1.2-8(1) (page 27), respectively

$$\{\varepsilon\} = \frac{1}{E}\left\{ \begin{matrix} \sigma_1 - v(\sigma_2 + \sigma_3) & 0 & 0 \\ 0 & \sigma_2 - v(\sigma_3 + \sigma_1) & 0 \\ 0 & 0 & \sigma_3 - v(\sigma_1 + \sigma_2) \end{matrix} \right\} \quad \text{and} \quad \{\varepsilon^p\} = \frac{(1-2v)p}{E}\left\{ \begin{matrix} 1 & 0 & 0 \\ 0 & 1 & 0 \\ 0 & 0 & 1 \end{matrix} \right\}$$

Calculated using Eq. (6), the energies corresponding to the total stress and the hydrostatic stress are respectively

$$w = \frac{1}{2}\{\sigma\}\cdot\{\varepsilon\} = \frac{1}{2E}\left[\sigma_1\left(\sigma_1 - v\sigma_2 - v\sigma_3\right) + \sigma_2\left(\sigma_2 - v\sigma_3 - v\sigma_1\right) + \sigma_3\left(\sigma_3 - v\sigma_1 - v\sigma_2\right) \right]$$

and

$$w^p = \frac{1}{2}\{\sigma^p\}\cdot\{\varepsilon^p\} = \frac{3(1-2v)p^2}{2E}$$

From Eqs. (8, 2), the deviatoric energy is

$$w^d = w - w^p$$

$$= \frac{1}{2E}\left[\sigma_1\left(\sigma_1 - v\sigma_2 - v\sigma_3\right) + \sigma_2\left(\sigma_2 - v\sigma_3 - v\sigma_1\right) + \sigma_3\left(\sigma_3 - v\sigma_1 - v\sigma_2\right) \right] - \frac{3(1-2v)}{2E}\left(\frac{\sigma_1 + \sigma_2 + \sigma_3}{3}\right)^2 \quad (11)$$

After some manipulations, it can be simplified as follows

$$w^d = \frac{1+v}{6E}\left[\left(\sigma_1 - \sigma_2\right)^2 + \left(\sigma_2 - \sigma_3\right)^2 + \left(\sigma_3 - \sigma_1\right)^2 \right] \quad (12)$$

Derivation of Eq. (12)

$$w^d = \frac{1}{2E}\left[\sigma_1\left(\sigma_1 - v\sigma_2 - v\sigma_3\right) + \sigma_2\left(\sigma_2 - v\sigma_3 - v\sigma_1\right) + \sigma_3\left(\sigma_3 - v\sigma_1 - v\sigma_2\right) \right] - \frac{3(1-2v)}{2E}\left(\frac{\sigma_1 + \sigma_2 + \sigma_3}{3}\right)^2$$

$$= \frac{1}{2E}\left[\sigma_1^2 + \sigma_2^2 + \sigma_3^2 - 2v\left(\sigma_1\sigma_2 + \sigma_2\sigma_3 + \sigma_3\sigma_1\right) \right] - \frac{1-2v}{6E}\left[\sigma_1^2 + \sigma_2^2 + \sigma_3^2 + 2\left(\sigma_1\sigma_2 + \sigma_2\sigma_3 + \sigma_3\sigma_1\right) \right]$$

$$= \frac{1}{6E}\left[3\left(\sigma_1^2 + \sigma_2^2 + \sigma_3^2\right) - 6v\left(\sigma_1\sigma_2 + \sigma_2\sigma_3 + \sigma_3\sigma_1\right) - \left(1-2v\right)\left(\sigma_1^2 + \sigma_2^2 + \sigma_3^2\right) - \left(2-4v\right)\left(\sigma_1\sigma_2 + \sigma_2\sigma_3 + \sigma_3\sigma_1\right) \right]$$

$$= \frac{1}{6E}\left[\left(2+2v\right)\left(\sigma_1^2 + \sigma_2^2 + \sigma_3^2\right) - \left(2+2v\right)\left(\sigma_1\sigma_2 + \sigma_2\sigma_3 + \sigma_3\sigma_1\right) \right]$$

$$= \frac{1+v}{3E}\left[\sigma_1^2 + \sigma_2^2 + \sigma_3^2 - \sigma_1\sigma_2 - \sigma_2\sigma_3 - \sigma_3\sigma_1 \right]$$

$$= \frac{1+v}{6E}\left[\left(\sigma_1 - \sigma_2\right)^2 + \left(\sigma_2 - \sigma_3\right)^2 + \left(\sigma_3 - \sigma_1\right)^2 \right]$$

Von Mises Stress (Equivalent Stress)

Substituting Eqs. (10, 12) into the von Mises Yield criterion, Eq. (9), we conclude that the material yields when

$$\frac{1+v}{6E}\left[\left(\sigma_1-\sigma_2\right)^2+\left(\sigma_2-\sigma_3\right)^2+\left(\sigma_3-\sigma_1\right)^2\right]\geq\frac{1+v}{3E}\sigma_y^2 \tag{13}$$

or, in a more concise form

$$\sqrt{\frac{1}{2}\left[\left(\sigma_1-\sigma_2\right)^2+\left(\sigma_2-\sigma_3\right)^2+\left(\sigma_3-\sigma_1\right)^2\right]}\geq\sigma_y \tag{14}$$

The quantity on the left-hand-side is termed *von Mises stress* or *effective stress*, and denoted by σ_e; in ANSYS, it is also referred to as *equivalent stress*,

$$\sigma_e=\sqrt{\frac{1}{2}\left[\left(\sigma_1-\sigma_2\right)^2+\left(\sigma_2-\sigma_3\right)^2+\left(\sigma_3-\sigma_1\right)^2\right]} \tag{15}$$

To have more insight of Eq. (14), let's plot Eq. (14) in σ_1-σ_2-σ_3 space and consider only equal sign. It will be a cylindrical surface aligned with the axis $\sigma_1=\sigma_2=\sigma_3$ and with a radius of $\sqrt{2}\sigma_y$ (see 14.1-4[1], page 521). It is called the *von Mises yield surface*. Condition of Eq. (14) is equivalent to say that the material fails when the stress state is on or outside the von Mises yield surface. When $\sigma_1=\sigma_2=\sigma_3$, the material is under hydrostatic pressure. It is the portion of stress that deviates from the axis $\sigma_1=\sigma_2=\sigma_3$ that contributes to the failure of the material.

Equivalent Strain

The *equivalent strain*, or von Mises strain, ε_e is defined by[Ref 4]

$$\varepsilon_e=\frac{1}{1+v'}\sqrt{\frac{1}{2}\left[\left(\varepsilon_1-\varepsilon_2\right)^2+\left(\varepsilon_2-\varepsilon_3\right)^2+\left(\varepsilon_3-\varepsilon_1\right)^2\right]} \tag{16}$$

Where v', the effective Poisson's ratio, defaults to the Poisson's ratio of the material, 0.5, or 0, depending on various applications (for details, see Ref 4).

References

1. Haslach, H. W. Jr. and Armstrong, R. W., *Deformable Bodies and Their Material Behavior*, John Wiley & Sons, Inc., 2004; Chapter 9 Plasticity.
2. Cook, R. D. and Young, W. C., *Advanced Mechanics of Materials*, Macmillan, 1985.
3. Beer, F. P., Johnston, E. R. Jr., and DeWolf, J. T., *Mechanics of Materials*, 3rd Ed., McGraw Hill, 2002.
4. ANSYS Documentation//Mechanical APDL//Theory Reference//2.4.1. Combined Strains

Section 1.5

Review

1.5-1 Keywords (Part I)

Choose a letter for each keyword from the list of descriptions

1. (   ) APDL
2. (   ) Boundary Conditions
3. (   ) Brittle Materials
4. (   ) Buckling
5. (   ) Degenerated Element
6. (   ) Degree of Freedom
7. (   ) DesignModeler
8. (   ) Displacement
9. (   ) Ductile Materials
10. (   ) Dynamic Simulations

11. (   ) Engineering Data
12. (   ) Environment Conditions
13. (   ) Failure Criteria of Materials
14. (   ) Finite Element
15. (   ) Finite Element Mesh
16. (   ) Finite Element Model
17. (   ) First-Order Element
18. (   ) Free Boundaries
19. (   ) Governing Equations
20. (   ) Isotropic Elasticity

Answers:

1. (E)   2. (K)   3. (R)   4. (N)   5. (P)   6. (I)   7. (C)   8. (T)   9. (Q)   10. (S)
11. (A)   12. (L)   13. (O)   14. (B)   15. (G)   16. (H)   17. (J)   18. (M)   19. (D)   20. (F)

List of Descriptions

(A) An application of the Workbench GUI. Material properties are the most common type of engineering data. Loads and boundary conditions also can be stored as engineering data.

(B) A small portion of a problem domain. Its geometry is so simple that the governing equations can be pre-formulated in terms of discrete nodal degrees of freedom.

(C) An application of the Workbench GUI. Its is similar to any other feature-based CAD software, except that it is specifically used to create geometric models for use in the ANSYS Workbench simulations.

(D) A set of equations governing the behavior of an engineering system (e.g., a structural system), usually in a form of differential equations.

(E) ANSYS Parametric Design Language. A set of text commands to drive ANSYS software.

(F) In **Engineering Data**, this term is used for materials whose stress-strain relation can be described by Hooke's law and characterized by two material parameters: Young's modulus and Poisson's ratio.

(G) A collection of elements and nodes.

(H) A finite element mesh plus its environment conditions.

(I) In the finite element methods, a term used for the discrete nodal values. In Workbench structural simulations, they are nodal displacements. In 3D, the number of degrees of freedom is 3 times the number of nodes.

(J) If linear polynomials are used as shape functions, the element is called a first-order element, or a lower-order element, in which nodes are on the vertices of the element.

(K) Conditions applied on the boundaries of a finite element mesh.

(L) Conditions applied on the boundaries or interior of a finite element mesh.

(M) Boundaries that are free of boundary conditions. ANSYS assumes a zero pressure (stress normal to the surface) on a free boundary.

(N) When the compression in a structure is large enough such that its lateral stiffness vanishes, the structure becomes unstable, or buckled.

(O) For a brittle material, it fails if the maximum principal stress reaches the fracture stress. For a ductile material, Tresca criterion or von Mises criterion may be used. The Tresca criterion states that the material fails if the stress intensity reaches the yield stress. The von Mises criterion states that the material fails if the von Mises stress reaches the yield stress.

(P) A 3D solid element, hexahedral in its natural shape, may combine some of nodes to form a triangle-based prism, quadrilateral-based pyramid, or tetrahedron; they are called degenerated elements. A 2D solid element, quadrilateral in its natural shape, may degenerate to a triangle.

(Q) If the strain is large before it is stretched up to fracture, the material is said to be ductile. The fracture is mostly due to a shear failure. There is usually an obvious yield point in its stress-strain curve.

(R) If the strain is small before it is stretched up to fracture, the material is said to be brittle. The fracture is mostly due to a tensile failure. There is usually no obvious yield point in its stress-strain curve.

(S) Structural simulations in which dynamic effects are included.

(T) The displacement of a particle in a deformed body is the vector connecting from its initial position to its final position.

1.5-2 Keywords (Part II)

Choose a letter for each keyword from the list of definitions

21. (   ) Linear Simulations   31. (   ) Stiffness
22. (   ) Mechanical   32. (   ) Stiffness Matrix
23. (   ) Modal Analysis   33. (   ) Strain
24. (   ) Node   34. (   ) Strain State
25. (   ) Nonlinear Simulations   35. (   ) Stress
26. (   ) Procedure of Finite Element Method   36. (   ) Stress Intensity
27. (   ) Project Schematic   37. (   ) Stress State
28. (   ) **Principal Stress**   38. (   ) Stress Stiffening
29. (   ) Second-Order Element   39. (   ) Von Mises Stress
30. (   ) Shape Functions   40. (   ) Workbench GUI

Answers:

21.(S)   22.(E)   23.(J)   24.(G)   25.(T)   26.(N)   27.(R)   28.(L)   29.(K)   30.(I)
31.(P)   32.(H)   33.(C)   34.(D)   35.(F)   36.(M)   37.(B)   38.(Q)   39.(O)   40.(A)

List of Definitions

(A) A gateway to ANSYS applications, including **Project Schematic, Engineering Data, DesignModeler, Mechanical, Design Exploration**, etc.

(B) Stress state, of a point in a body, is a group of values describing the force intensity on the point in all directions. In 3D, three independent directions are needed to complete the description.

(C) Strain, of a point in a body, in a certain direction, or face, is a vector describing the stretch and twist in that direction. The vector can be decomposed into two components: one that is normal to the face and one that is parallel to the face.

(D) Strain state, of a point in a body, is a group of values describing the stretch and twist in all directions. In 3D, three independent directions are needed to complete the description.

(E) An application of the Workbench GUI. It performs structural and mechanical simulations, including meshing, setting up environment conditions, solving finite element models, and viewing results.

(F) The stress at a certain point is the force per unit area acting on the boundary surfaces of an infinitesimally small body centered at that point.

(G) Nodes are portion of an element that overlaps with other elements. The overlapping implies that the elements share the same degrees of freedom values on the nodes.

(H) The matrix that describes the linear relation between displacement vector and the force vector. Physical meaning of the ith column is the forces required on each of DOF's to maintain a unit displacement on the ith DOF and zero displacements on the other DOFs.

(I) In finite element methods, shape functions are used as interpolating functions to calculate continuous displacement fields from discrete nodal displacements. Linear and quadratic polynomials are commonly used as shape functions.

(J) Also called free vibration analysis. It is a special case of dynamic analysis, in which the structure is free of external forces. Results of modal analysis include natural frequencies and their corresponding vibration modes.

(K) If quadratic polynomials are used as shape functions, the element is called a second-order element, or a higher-order element, in which nodes are on the vertices as well on the middle of the edges.

(L) At a point in a body, different directions have different stress values. There exist directions in which the normal components are in their extremities and the shear components vanish; the directions and the corresponding normal stresses are called the principal directions and the principal stresses respectively. In 3D, there are 3 principal stresses, they are denoted as, in order starting from the largest, σ_1, σ_2, and σ_3 respectively. It is often used in a criterion for a brittle material: when the maximum principal stress is larger than the fracture strength, the material fails.

(M) Stress intensity is the difference between the maximum principal stress and the minimum principal stress. It equals to twice the maximum shear stress. It is used in Tresca failure criterion for a ductile material: when the stress intensity is larger than the yield strength, it is equivalently to say that the shear stress is larger than the yielding shear strength and, thus, the material fails.

(N) (a) Calculate stiffness matrix for each element according to the element's geometry and its material properties. (b) Add up the element matrices to form a global stiffness matrix. The force vector is also calculated, according to the loading conditions. (c) Eq. 1.3-1(1) (page 31) is solved for the nodal displacements. (d) For each element, the displacement fields are calculated according to Eq. 1.3-2(2) (page 32), the strain fields are calculated according to Eq. 1.2-7(1) (page 27), and the stress fields are calculated according to Eq. 1.2-8(3) (page 28).

(O) Von Mises stress is defined in Eq. 1.4-5(15) (page 42). It is used in von Mises failure criterion for a ductile material: when the von Mises stress is larger than the yield strength, it is equivalently to say that the distortion strain energy density is larger than the yielding distortion strain energy density and, thus, the material fails.

(P) The stiffness of a structure is the forces required to deform the structure.

(Q) The phenomenon that, when a structure member subject to a tensile stress, its lateral stiffness increases with the increase of tensile stress.

(R) An application of the Workbench GUI. Its function is to lay out simulation systems and their data flows.

(S) When the responses of a system are linearly proportional to the loads, it is called a linear system and the simulation is called a linear simulation.

(T) When the responses of a system are not linearly proportional to the loads, it is called a nonlinear system and the simulation is called a nonlinear simulation.

Section 1.6

Appendix: An Unofficial History of ANSYS

Shen-Yeh Chen, Ph. D.

About the Author

Dr. Shen-Yeh Chen earned his BS degree of civil engineering from National Chung Hsing University, Taiwan in 1990, and his MS and Ph.D. degrees of structural mechanics from Arizona State University, USA in December of 1997.

He began to work for Honeywell Engines and Systems in Phoenix, Arizona in February of 1998. Shen-Yeh was a heavy user of ANSYS and LS-DYNA during his service in Honeywell. Soon he became a cross-department and cross-campus expert of ANSYS and LS-DYNA in Honeyewell, and was very active on the internet (the XANSYS group) of the ANSYS internal users community. Dr. Chen published several popular ANSYS macros and articles during that time, and developed many in-house codes for Honeywell, including an optimizer to couple with ANSYS and LS-DYNA.

In August of 2002, Dr. Chen took a VP position offered from CADMEN, an ANSYS distributor at Taiwan. In 2005, Dr. Chen established his own company, FEA-Opt Technology (www.FEA-Optimization.com). About 80% of the revenue for FEA-Opt Technology is related to ANSYS and LS-DYNA. In 2006, he released a general purpose design optimization software, SmartDO. Today FEA-Opt Technology is well-known as a very successful CAE consulting firm in Taiwan, and its SmartDO is gaining more and more market share since 2006.

Dr. John Swanson holds B.S. and M.S. degrees in mechanical engineering from Cornell University. He holds a Ph.D. in applied mechanics from the University of Pittsburgh, obtained in night school with Westinghouse support.

In 1963, Dr. John Swanson worked at Westinghouse Astronuclear Labs in Pittsburgh, responsible for stress analysis of the components in NERVA nuclear reactor rockets. He used computer codes to model and predict transient stresses and displacements of the reactor system due to thermal and pressure loads. Swanson continued to develop 3D analysis, plate bending, nonlinear analysis for plasticity and creep, and transient dynamic analysis, in the next several years, using a finite element heat conduction program that was developed by Wilson at Aerojet. The old Westinghouse codes included a 2D/axisymmetric one also, possibly called FEATS (according to Kohnke). John wanted to combine these codes to remove the duplication, like equation solvers and some postprocessing.

Swanson believed an integrated, general-purpose FEA code could be used to do complex calculations that engineers typically did manually, such as heat transfer analysis. It would save money and time for Westinghouse and other companies.

Westinghouse didn't support the idea, and Swanson left the company in 1969. Before he left, he made sure that all code work had been sent to COSMIC, so that he could pick it up again from the outside.

Swanson Analysis Systems, Inc was incorporated in the middle of 1970 at Swanson's home. The offices were part of Swanson's home (There was no garage) in Pittsburgh. At the same time, Westinghouse realized that they needed John, so they hired him as a consultant. John said sure, but with the proviso that whatever he put into STASYS, the Westinghouse code, he could also put into ANSYS. Westinghouse had no trouble with this, as they just wanted to solve their problems. So this consulting kept bread on the table for the Swansons, and at the same time brought forth further improvements to ANSYS.

He developed his program using a keypunch and a time-shared mainframe at U.S. Steel. The first version of ANSYS was coded by the end of 1970, and the ANSYS program was first released soon after that. Westinghouse was the first customer, running as a data center. The data center was at the Telecommunications Center on the Parkway East, on the east side of Pittsburgh. According to Dr. Swanson, the name ANSYS was because the copyright lawyers assured Swanson that ANSYS was just a name, and did not stand for anything. Understandable, during that period all programs were "written" on punch card. When installing the program on the customer's computer, it meant carried a relatively big case of punch cards to the customer's place, and fed them into the machine.

Dr. Peter Kohnke met John Swanson first about early 1971. Swanson offered Peter a job in the fall of 71, but Peter did not accept. As that time Peter was a brand new father at the time, and Westinghouse looked a lot more secure than SASI. Peter told John he was interested and would accept in the spring of 72, but then he had hired Gabe DeSalvo, and did not have the resources to hire Peter also. But that fall John did, and then hired Peter. Dr. Peter Kohnke's start date was 1/1/73.

When Peter started work, he asked what John wanted him to do. John told Peter that he developed code, did technical support, wrote manuals, gave seminars and did systems work so that the program would run on a variety of systems. John said he needed relief, so Peter should pick one or two of them. Ultimately, Peter did all of them, except systems work.

In around 1970, users can ran ANSYS 2.x on a CDC 6600 machine over the Cybernet timesharing network. That time only fixed format input was available. The users would work up the input listing off-line, key it onto a tape cassette, log on, submit the run about quitting time for the best computer rates and stop by the CDC data center next morning to find out what went wrong. In 1973, ANSYS ran on three kinds of hardware: CDC, Univac, and IBM. And around 1973 the USS mainframe that they used to develop code was the US Steel CDC 6500.

The first minicomputer that ANSYS ran on was a MODCOMP 4 (or IV?). VAX came later. Being a small company, everyone did everything. When a "mini" computer was delivered, everyone helped wrestle it off of the truck. When printout paper was delivered, everyone helped unload the boxes.

In 1975, MITS began to build and sell the first PC ever in human history, the Altair. That, of course, did not have anything to do with ANSYS yet. The so-called PC was just a few switches and lights on the front board, and input had to be done in a binary fashion (no keyboard and monitor, of course). What was worse, was that you have to assemble it by yourself. And, it usually didn't work. Although Altair was rather popular, nobody really knew what to do with this machine. One former customer said that, the most popular activity on Altair, was to figure out what to do with this machine. At the same time, Microsoft built the BASIC language for Altair.

1977, Apple I was born.

In around 1979, Revision 3.0, ANSYS run on a VAX 11-780 minicomputer. ANSYS evolved from fixed format input to purely command line driven and monocolor (green) on a Tektronix 4010 or 4014 vector graphics monitor. For a descent size model, the hidden lines plots could take 20-30 minutes. All of the nodes and elements were created separately without the benefit of importing CAD geometry. NGEN, EGEN, RPnnn, were used extensively. There was a geometry preprocessor, PREP7.

1980, we had Apple II.

In around 1980, John Swanson bought a Radio Shack TRS-80 machine, and planned to build a commercial version on it. However, later John returned the machine because Radio Shack left out (a socket for) a floating point processor. John decided that Finite Element Analysis probably should utilize a floating point processor, so he got his money back for that one.

Also around 1980, Rev 4 on an VAX 11-780 system was great, according to some old users. The chasm between batch and interactive running pretty much disappeared and file management was a very easy thing. No more element hard coding, the post processing got hugely better and you could mix batch and interactive running as you saw fit. Big dynamic transient runs or substructuring over night, post-processing and plotting next morning. Emag capabilities were first introduced at Rev 4.1.

Also in 1980, Microsoft signed contract with IBM to provide the OS, PC DOS, for its up coming PC. This OS, however, was not created by Microsoft. Microsoft bought it from an engineer for 50K USD, which was named the QDOS - the Quick and Dirty Operation System.

1981, IBM PC was born. This computer was created using the off the shelf technology, and an open architecture. The original reasons were to push the product to the market ASAP, so that IBM could catch up with the PC market. However, the BIOS was proprietary. Later Compaq reverse-engineer the BIOS and created a fully IBM PC compatible BOIS. This ignited the PC cloning market and war. The booming of PC market directly changed the meaning of computing. PC price dropped 30% at one month. And, it was the booming of cloned IBM PC that really brought money into Microsoft.

1984, the revolutionary Macintosh was born. Macintosh was far advanced then the IBM PC family at that time. The concept of GUI in the OS level and WYSIWYG was not possible on IBM PC until almost one decade later. However, the market of Macintosh did not pick up very soon, which caused the Steve Job's leave from Apple computer.

However, later the sales of Macintosh began to take off, which proved that Steve Job's vision had all been right. Macintosh saved Apple, and was directly responsible for the phenomena of Apple craze and fans.

A PC version of ANSYS was also available at around version 4.0 too in about 1984. It was running on a Intel 286, with interactive command line input and limited graphics on the screens, like elements and nodes. No Motif GUI yet. In the first release on ANSYS on PC's, preprocessing, solution and post processing were performed in separate programs.

"Design Optimization" was introduced at Rev 4.2 (1985). This is also the release at which "Macro length is no longer limited to 400 characters."

FLOTRAN started as a graduate (PhD) project by Rita J. Schnipke in the University of Virginia circa 1986. After grad school Rita started (or helped start) Compuflo which was later sold to ANSYS in 1992. Rita later started her own shop which is in Charlottesville VA called Blue Ridge Numerics. They make CFDesign, a finite element based CFD code (www.cfdesign.com).

1988 at an ANSYS conference in California, IBM was there pushing their first unix machine, the "RT". It was slow. They asked Dr. Swanson if he would make a comment on it. He said "RT" must stand for Real Turkey.

SASI first started working with Compuflo (FLOTRAN) in 1989. At ANSYS Rev 5.0 and FLOTRAN V2.1A, SASI had what they called a "seamless interface" between the two programs (1993). FLOTRAN was "fully integrated" into ANSYS at Rev 5.1 (1994).

In 1993, Version 5.0 was released. And the version 5.1 later has a Motif GUI, which would remained the similar layout up to 6.0.

Swanson Analysis Systems, Inc., was sold to TA Associates in 1994. The new company name, ANSYS, Inc., was announced at AUTOFACT '94 in Detroit.

According to many different people in the old SASI, John Swanson treated the people there pretty well. In contrast to the old "sandwich" jokes, John never passed up a chance to go to a restaurant. Indeed, for many years, John invited the entire staff to a restaurant regularly for the staff meetings.

At one time, Kohnke told Swanson that he ran the company like a benevolent dictatorship, and later Swanson told Kohnke that he liked that characterization.

Many people had said that John Swanson had an amazing overall understanding as well as detailed knowledge of the ANSYS code. Kohnke told a small story in an email to the author: " Sometime in the late 70's, a bug came my way. I wrestled with it for maybe half a day without making real progress. Then I went to John's office to ask him if he had any ideas. After I explained the bug, John thought about it for about 3 seconds (literally!) and said: 'Didn't you make a change in XXX about 6 months ago that would have a bearing on this?' In a nutshell, he was correct and I was then able to resolve the bug! John's knowledge and understanding of the code was always amazing to me."

1995, Windows 95 was published. Windows 95 was an important milestone for Microsoft. It bridged between the old DOS OS into the new NT technology. The birth of Windows 95 finally made it more and more acceptable for engineering community to use PC as a heavy duty calculation machine like workstations.

In 1996, ANSYS 5.3 was published, with support for LS-DYNA. The feature of ANSYS/LS-DYNA in ANSYS 5.3 was still in the beginning stage.

On June 20, 1996, ANSYS Inc. common stock began trading on Nasdaq under ANSS after being 26 years a privately held company. The IPO generated more than $41 million.

1998, ANSYS began to ship ANSYS/ed to university labs and paper reviewers. One of the copy arrived at the Structures Lab of Civil Engineering Department in Arizona State University, and that was the first time the author knew about ANSYS.

In the same year, on ANSYS's Annual report, it said "John is retiring from his direct role at ANSYS Inc., but will continue his association as a key consultant, mentoring all of us for many years to come."

On August 31, 2000, ANSYS acquired ICEM CFD.

January 2001, ANSYS announced the release of CADfix (International TechneGroup Incorporated) for ANSYS version 5.6.2 and 5.7. CADfix was to address the issue of importing CAD model into ANSYS with automatic geometric data repair.

November 2001, ANSYS acquired CADOE S. A., an independent software vendor that specializes in the CAD/CAE market. In the same month, ANSYS announced a strategic OEM partnership with SAS LLC, a provider of NASTRAN simulation software and services. The alliance was focused on the joint development of a new NASTRAN computer-aided engineering solution that will be distributed exclusively by ANSYS Inc.

November 2001, ANSYS announced AI*Environment. AI*Environment combines ICEM CFD Engineering's pre- and post-processor technologies.

December 2001, ANSYS 6.0 was released. In this version, the Sparse solver was greatly improved. Efficient and reliable large scale model analysis (say, 1M DOF) finally became practical. The graphics screen of ANSYS was also painted blue in 6.0, which came out to be a great disappointment to a lot of users.

In April of 2002, ANSYS 6.1 was released. The familiar Motif GUI was replaced by a Tcl/tk developed interface. It runs on 64-bit Intel Itanium architecture with Windows XP.

In February 26, 2003, ANSYS acquired CFX. ANSYS also announce that the functionality of Flotran will be "capped" at 8.1. That is, there will be more development of Flotran after 8.1. Expect for the Multiphysics platform, Flotran will be replaced by CFX.

March 2004, ANSYS Announces ParaMesh 2.3.

May 2004, ANSYS 8.1 and CFX 5.7 was released. According to the statement from ANSYS, the functionality of Flotran is "capped" since this version. That is, there will be no more revision of Flotran after 8.1.

June 2004, ICEM CFD 5.0 was released.

Jan 5, 2005, ANSYS announced that it has acquired Century Dynamics. Century Dynamics' main product, AUTODYN, includes computational structural dynamics finite element solvers (FE), finite volume solvers for fluid dynamics (CFD), mesh-free particle solvers for high velocity, large deformation and fragmentation problems (SPH), and multi-solver coupling for multiphysics solutions including coupling between FE, CFD and SPH methods.

February 16, 2006, ANSYS Signs Definitive Agreement to Acquire Fluent.

As of March of 2012, Dr. Peter Kohnke is still working for ANSYS, but has cut back to 4 days a week. He turned 71 in December of 2012.

Dr. John Swanson is officially retired and living in The Villages, in Florida. But he still programs for ANSYS under contract on a varying schedule, on projects such as High performance mesh interpolation, Symmetric Multi Processing, 64 bit conversions, and APDL enhancements.

John is currently on the Board of Trustees of the University of Pittsburgh and The ASME Foundation and served two six-year terms as a Trustee of Washington and Jefferson College. He is a member of the Engineering College Council at Cornell University. His support of colleges and universities includes the donation of research laboratories to the Engineering Schools at Cornell, the University of Pittsburgh and (with Janet, his wife) the Veterinary School at Cornell. He gave the naming gift for the John A. Swanson Science Center at Washington and Jefferson College. The John A. Swanson School of Engineering at the University of Pittsburgh is named in his honor. Swanson recently invested in Applied Quantum Technology (AQT), a California startup company with the objective of reducing the cost of PV Solar Power by another factor of two. He serves on the AQT Board of Directors.

What is a data center?

I actually heard the word "data center" first time from Dr. Kohnke. For my age, I am more familiar with workstation and PC and xbox and psp and so on. I used VMS when I was in college. And I remember later soon the whole room was occupied by PC and Macs. So I asked Dr. Kohnke what is a data center anyway? And this is what he told me.

Long ago (in the computer age), computers were relatively rare (and expensive). So, if you invested in one, you wanted to get the maximum use of it. You would run it day and night. This is unlike our pc's now (like our cars), which frankly just sit there most of the time, waiting for our command. Before it was very much a shared and continually used resource.

So, ANSYS got its start in that environment. Users at Westinghouse wanted to run ANSYS, so Westinghouse made a contract with SASI, where SASI would supply the code and support, and Westinghouse would pay to SASI so much per unit of time royalty for the time that the computer was actually being used to run ANSYS. As a result, many users were using the same machine (one after the other--parallel did not exist then). This was called a "data center."

The next step was external users. Knowing that even as a large company, you might not need your computer full time, so you would have your sales people out there selling time on it, often only at night, to external users.

Other companies sprang up that had no user base of their own, but only a computer(s), sales people, and external customers. The better ones also offered their own very good technical support. These were also called data centers.

And of course the whole concept of a data center disappeared as computers got cheaper and faster.

The "lease" jumped to the concept that the clock was not important; this "new" contract had SASI being paid so much per month, regardless of actual usage. This is closer to how things are now.

Acknowledgement

The author wants to thank the help from many engineers and scientists in the xansys internet group. Some of the former employees of ANSYS also contribute greatly to this article, and many of them prefer not to be named. I also received emails from different people, and I usually tried to verify before I used them. Although I am trying to keep all the statement as accurate as possible, I really can not guarantee the correctness of any information in this article.

Many of us, including the author in the xansys group, especially want to thank Dr. John Swanson, who invented ANSYS, and changes the life of many engineers forever in certain ways.

I have lived in the States for totally 10 years. I lived there, educated there, married there, had my son there, bought my first new car there, and had my first house there. It totally changes my life, my thinking (and my head) and everything. And I have to say most of my financial basis was built on the existence of ANSYS and another program LS-DYNA.

This article is also available on the web site www.FEA-Optimization.com. Anyone is welcome to distribute this article anyway he or she wants, as long as the original article remains unchanged. Comments and suggestion should be forwarded to the authors directly. I will be glad to update this file continuously.

Chapter 2
Sketching

A complex 3D geometry can be viewed as a series of adding/removing material of simple solid bodies. Each solid body is often created by first drawing a 2D sketch, and then using the sketch to generate a 3D solid body with tools such as **Extrude**, **Revolve**, **Sweep**, **Loft**, etc.

Purpose of This Chapter

This chapter provides exercises for the students so that they can be acquainted with sketching using DesignModeler. Profiles of several mechanical parts are sketched in this chapter. Each sketch is then used to generate a 3D model using either **Extrude** or **Revolve**. The use of these 3D tools is so trivial that we may focus on 2D sketching techniques.

About Each Section

Each mechanical part will be completed in a section. Section 2.1 sketches a cross section of W16x50; the cross section is then extruded to generate a 3D beam. Section 2.2 sketches a triangular plate; the sketch is then extruded to generate a 3D solid model. Section 2.3 does not provide a hands-on case. Rather, it overviews the sketching tools in a systematic way, attempting to complement what were missed in the first two sections. Sections 2.4, 2.5, and 2.6 provide three additional exercises, in which we'll purposely leave out some steps for the students to figure out the details.

Section 2.1

W16x50 Beam

2.1-1 About the W16x50 Beam

In this section, we will create a W16x50 [1-4] steel beam. The beam has a a length of 10 ft.

[1] Wide-flange I-shape section.

[2] Nominal depth 16 in.

[3] Weight 50 lb/ft.

[4] W16x50 cross section. #

W16x50

7.07"

.380"

.628"

16.25"

R.375"

2.1-2 Start Up DesignModeler

[5] **Project Schematic** window.

[2] **Workbench GUI** shows up.

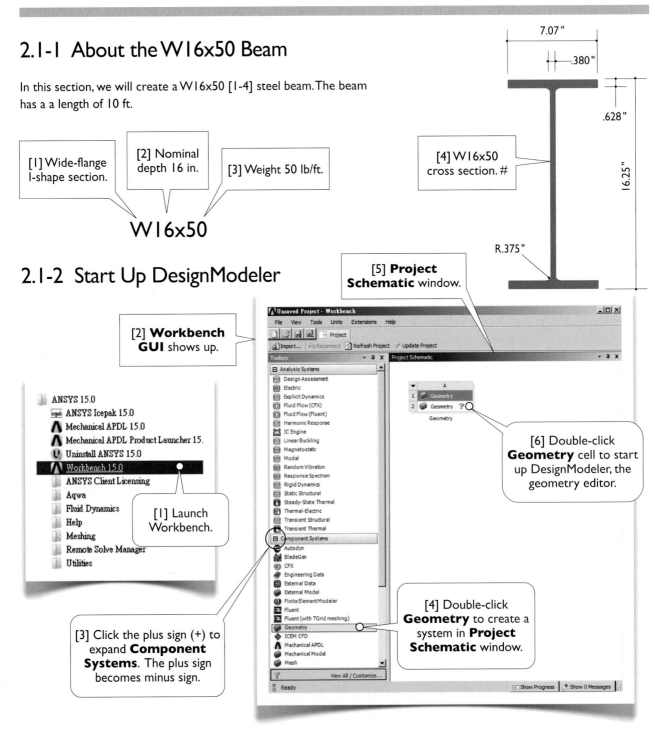

ANSYS 15.0
ANSYS Icepak 15.0
Mechanical APDL 15.0
Mechanical APDL Product Launcher 15.
Uninstall ANSYS 15.0
Workbench 15.0
ANSYS Client Licensing
Aqwa
Fluid Dynamics
Help
Meshing
Remote Solve Manager
Utilities

[1] Launch Workbench.

[6] Double-click **Geometry** cell to start up DesignModeler, the geometry editor.

[4] Double-click **Geometry** to create a system in **Project Schematic** window.

[3] Click the plus sign (+) to expand **Component Systems**. The plus sign becomes minus sign.

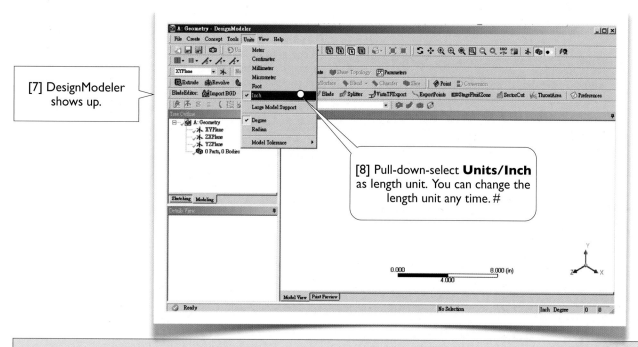

[7] DesignModeler shows up.

[8] Pull-down-select **Units/Inch** as length unit. You can change the length unit any time. #

In this book, a round-cornered textbox (e.g., [1, 3, 4, 6, 8]) is used to indicate that mouse or keyboard ACTIONS are needed in that step. A sharp-cornered textbox (e.g., [2, 5, 7]) is used for commentary only, no mouse or keyboard actions are needed in that step.

2.1-3 Draw a Rectangle on **XYPlane**

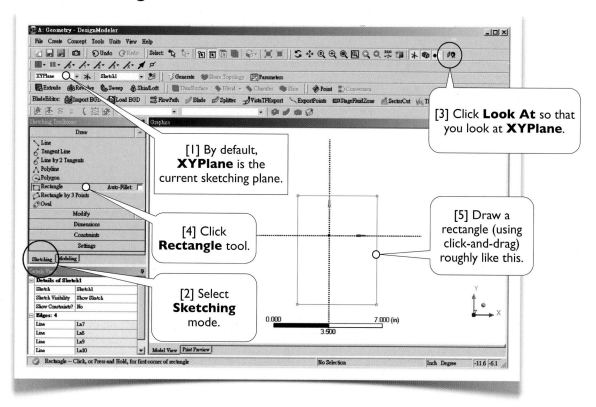

[3] Click **Look At** so that you look at **XYPlane**.

[1] By default, **XYPlane** is the current sketching plane.

[4] Click **Rectangle** tool.

[5] Draw a rectangle (using click-and-drag) roughly like this.

[2] Select **Sketching** mode.

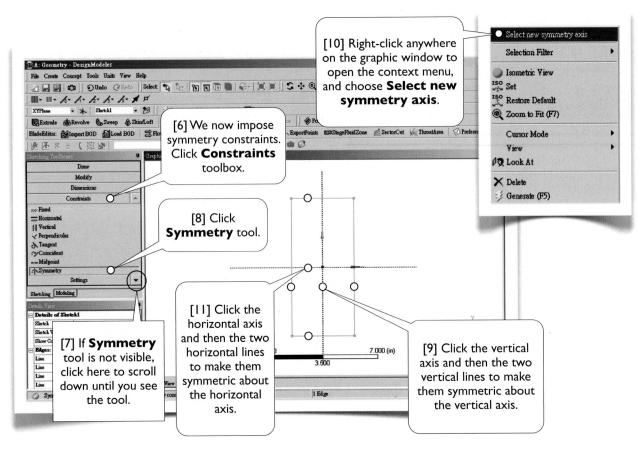

[10] Right-click anywhere on the graphic window to open the context menu, and choose **Select new symmetry axis**.

[6] We now impose symmetry constraints. Click **Constraints** toolbox.

[8] Click **Symmetry** tool.

[7] If **Symmetry** tool is not visible, click here to scroll down until you see the tool.

[11] Click the horizontal axis and then the two horizontal lines to make them symmetric about the horizontal axis.

[9] Click the vertical axis and then the two vertical lines to make them symmetric about the vertical axis.

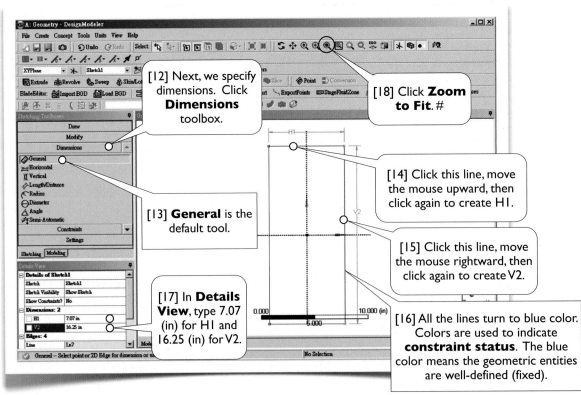

[12] Next, we specify dimensions. Click **Dimensions** toolbox.

[18] Click **Zoom to Fit**. #

[13] **General** is the default tool.

[14] Click this line, move the mouse upward, then click again to create H1.

[15] Click this line, move the mouse rightward, then click again to create V2.

[17] In **Details View**, type 7.07 (in) for H1 and 16.25 (in) for V2.

[16] All the lines turn to blue color. Colors are used to indicate **constraint status**. The blue color means the geometric entities are well-defined (fixed).

2.1-4 Set up Sketching Options

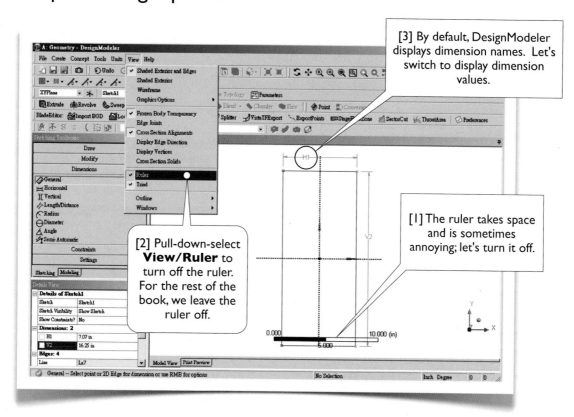

[3] By default, DesignModeler displays dimension names. Let's switch to display dimension values.

[1] The ruler takes space and is sometimes annoying; let's turn it off.

[2] Pull-down-select **View/Ruler** to turn off the ruler. For the rest of the book, we leave the ruler off.

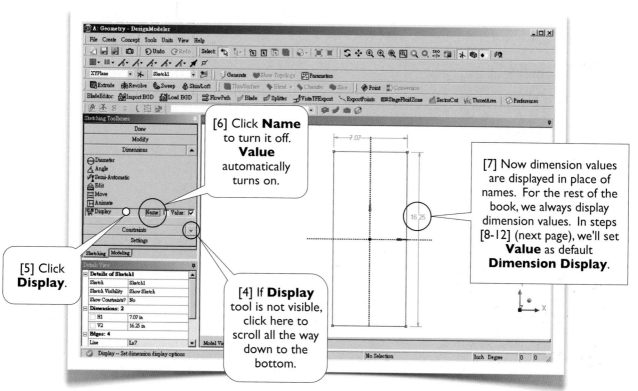

[6] Click **Name** to turn it off. **Value** automatically turns on.

[7] Now dimension values are displayed in place of names. For the rest of the book, we always display dimension values. In steps [8-12] (next page), we'll set **Value** as default **Dimension Display**.

[5] Click **Display**.

[4] If **Display** tool is not visible, click here to scroll all the way down to the bottom.

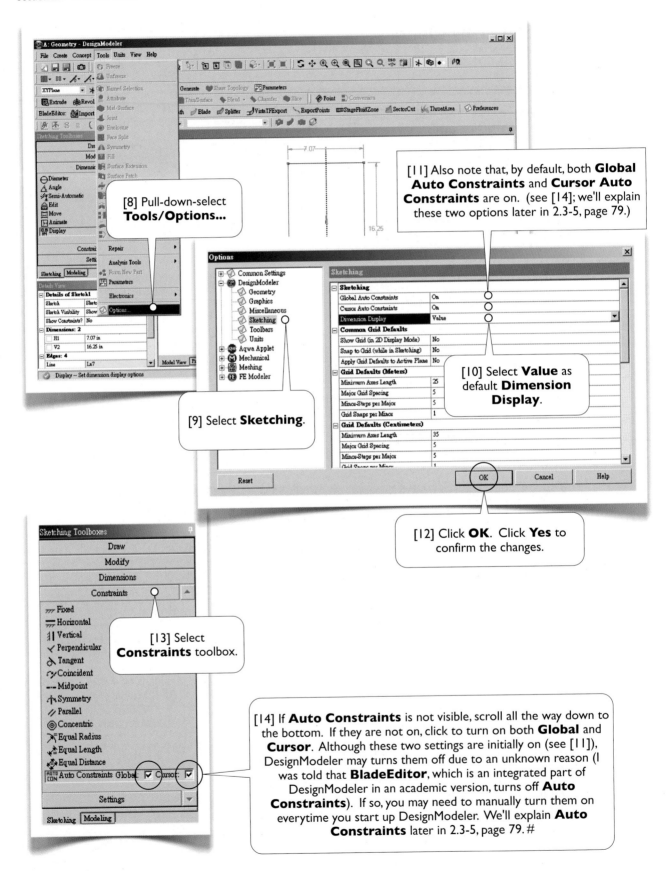

[8] Pull-down-select **Tools/Options...**

[11] Also note that, by default, both **Global Auto Constraints** and **Cursor Auto Constraints** are on. (see [14]; we'll explain these two options later in 2.3-5, page 79.)

[9] Select **Sketching**.

[10] Select **Value** as default **Dimension Display**.

[12] Click **OK**. Click **Yes** to confirm the changes.

[13] Select **Constraints** toolbox.

[14] If **Auto Constraints** is not visible, scroll all the way down to the bottom. If they are not on, click to turn on both **Global** and **Cursor**. Although these two settings are initially on (see [11]), DesignModeler may turns them off due to an unknown reason (I was told that **BladeEditor**, which is an integrated part of DesignModeler in an academic version, turns off **Auto Constraints**). If so, you may need to manually turn them on everytime you start up DesignModeler. We'll explain **Auto Constraints** later in 2.3-5, page 79. #

2.1-5 Draw a Polyline

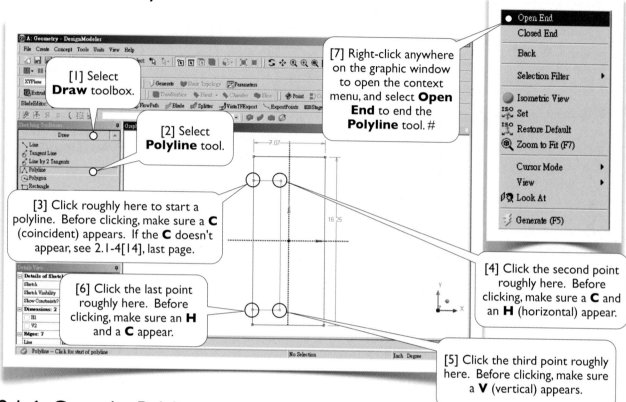

[1] Select **Draw** toolbox.

[2] Select **Polyline** tool.

[7] Right-click anywhere on the graphic window to open the context menu, and select **Open End** to end the **Polyline** tool. #

[3] Click roughly here to start a polyline. Before clicking, make sure a **C** (coincident) appears. If the **C** doesn't appear, see 2.1-4[14], last page.

[6] Click the last point roughly here. Before clicking, make sure an **H** and a **C** appear.

[4] Click the second point roughly here. Before clicking, make sure a **C** and an **H** (horizontal) appear.

[5] Click the third point roughly here. Before clicking, make sure a **V** (vertical) appears.

2.1-6 Copy the Polyline

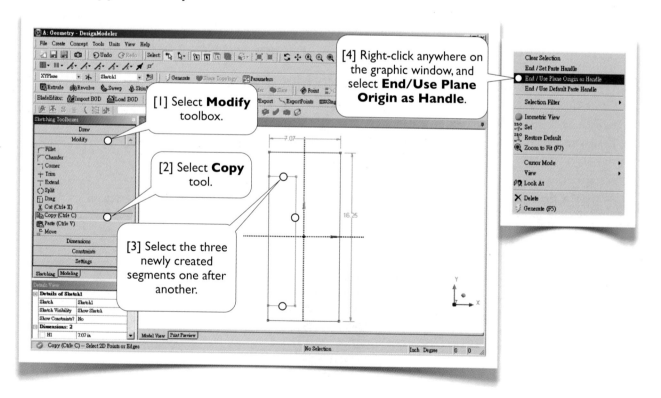

[1] Select **Modify** toolbox.

[2] Select **Copy** tool.

[3] Select the three newly created segments one after another.

[4] Right-click anywhere on the graphic window, and select **End/Use Plane Origin as Handle**.

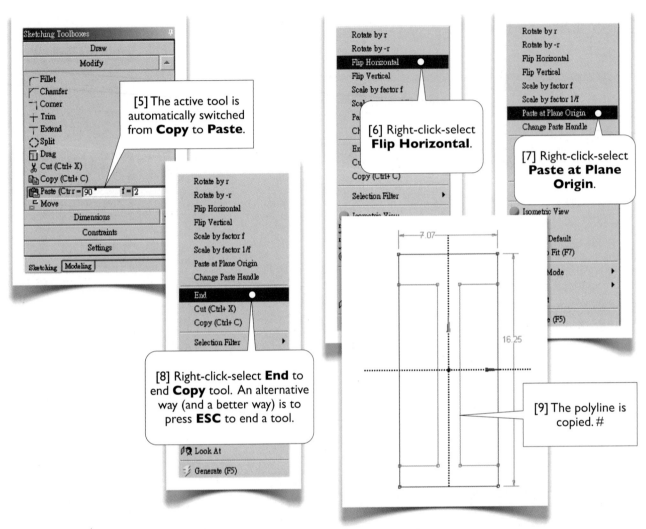

[5] The active tool is automatically switched from **Copy** to **Paste**.

[6] Right-click-select **Flip Horizontal**.

[7] Right-click-select **Paste at Plane Origin**.

[8] Right-click-select **End** to end **Copy** tool. An alternative way (and a better way) is to press **ESC** to end a tool.

[9] The polyline is copied. #

2.1-7 Basic Mouse Operations in Sketching Mode

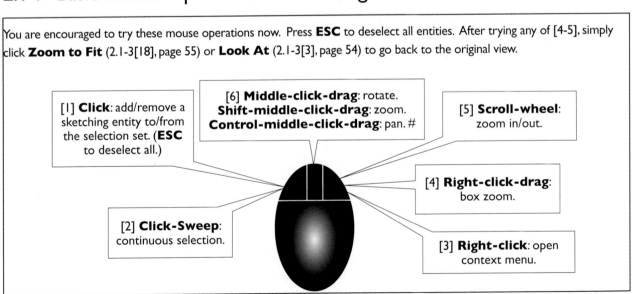

You are encouraged to try these mouse operations now. Press **ESC** to deselect all entities. After trying any of [4-5], simply click **Zoom to Fit** (2.1-3[18], page 55) or **Look At** (2.1-3[3], page 54) to go back to the original view.

[1] **Click**: add/remove a sketching entity to/from the selection set. (**ESC** to deselect all.)

[6] **Middle-click-drag**: rotate. **Shift-middle-click-drag**: zoom. **Control-middle-click-drag**: pan. #

[5] **Scroll-wheel**: zoom in/out.

[4] **Right-click-drag**: box zoom.

[2] **Click-Sweep**: continuous selection.

[3] **Right-click**: open context menu.

2.1-8 Trim Away Unwanted Segments

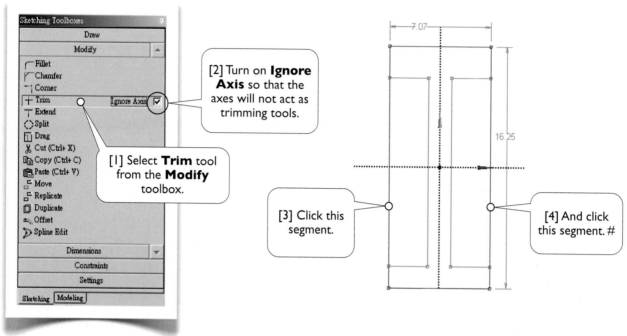

[2] Turn on **Ignore Axis** so that the axes will not act as trimming tools.

[1] Select **Trim** tool from the **Modify** toolbox.

[3] Click this segment.

[4] And click this segment. #

2.1-9 Impose Symmetry Constraints

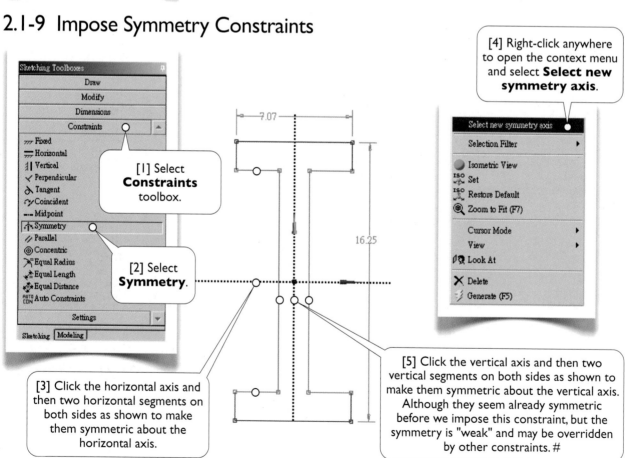

[4] Right-click anywhere to open the context menu and select **Select new symmetry axis**.

[1] Select **Constraints** toolbox.

[2] Select **Symmetry**.

[3] Click the horizontal axis and then two horizontal segments on both sides as shown to make them symmetric about the horizontal axis.

[5] Click the vertical axis and then two vertical segments on both sides as shown to make them symmetric about the vertical axis. Although they seem already symmetric before we impose this constraint, but the symmetry is "weak" and may be overridden by other constraints. #

2.1-10 Specify Dimensions

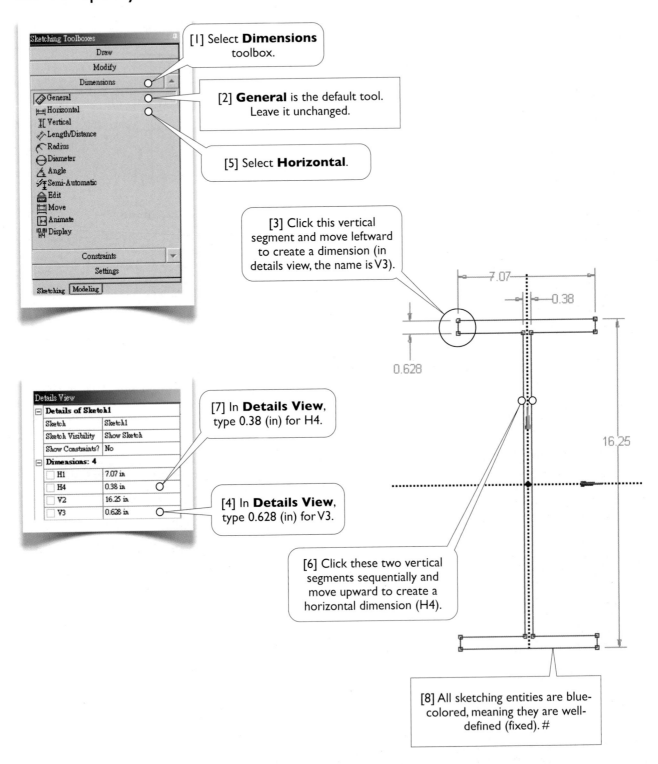

[1] Select **Dimensions** toolbox.

[2] **General** is the default tool. Leave it unchanged.

[5] Select **Horizontal**.

[3] Click this vertical segment and move leftward to create a dimension (in details view, the name is V3).

[7] In **Details View**, type 0.38 (in) for H4.

[4] In **Details View**, type 0.628 (in) for V3.

[6] Click these two vertical segments sequentially and move upward to create a horizontal dimension (H4).

[8] All sketching entities are blue-colored, meaning they are well-defined (fixed). #

2.1-11 Add Fillets

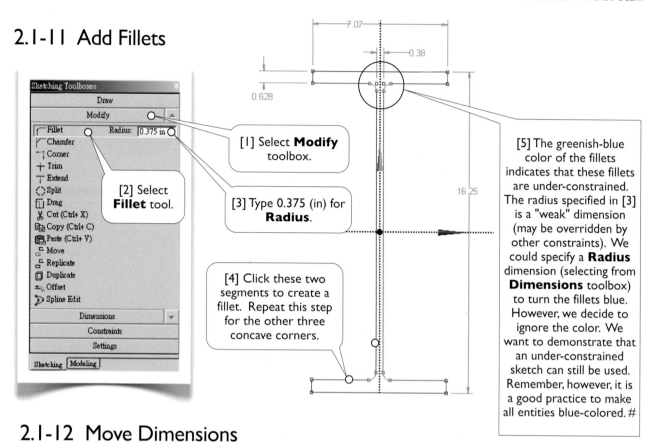

[1] Select **Modify** toolbox.

[2] Select **Fillet** tool.

[3] Type 0.375 (in) for **Radius**.

[4] Click these two segments to create a fillet. Repeat this step for the other three concave corners.

[5] The greenish-blue color of the fillets indicates that these fillets are under-constrained. The radius specified in [3] is a "weak" dimension (may be overridden by other constraints). We could specify a **Radius** dimension (selecting from **Dimensions** toolbox) to turn the fillets blue. However, we decide to ignore the color. We want to demonstrate that an under-constrained sketch can still be used. Remember, however, it is a good practice to make all entities blue-colored. #

2.1-12 Move Dimensions

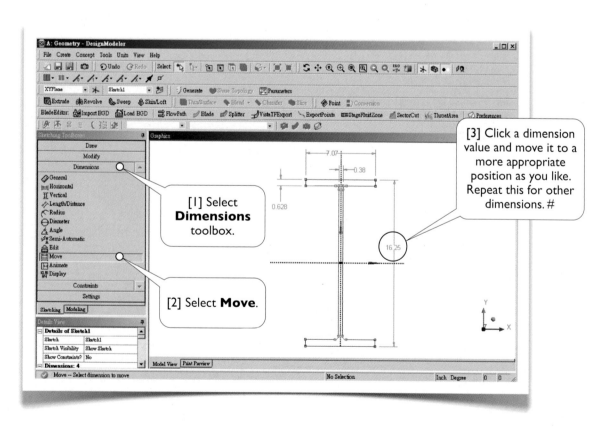

[1] Select **Dimensions** toolbox.

[2] Select **Move**.

[3] Click a dimension value and move it to a more appropriate position as you like. Repeat this for other dimensions. #

2.1-13 Generate 3D Solid Body

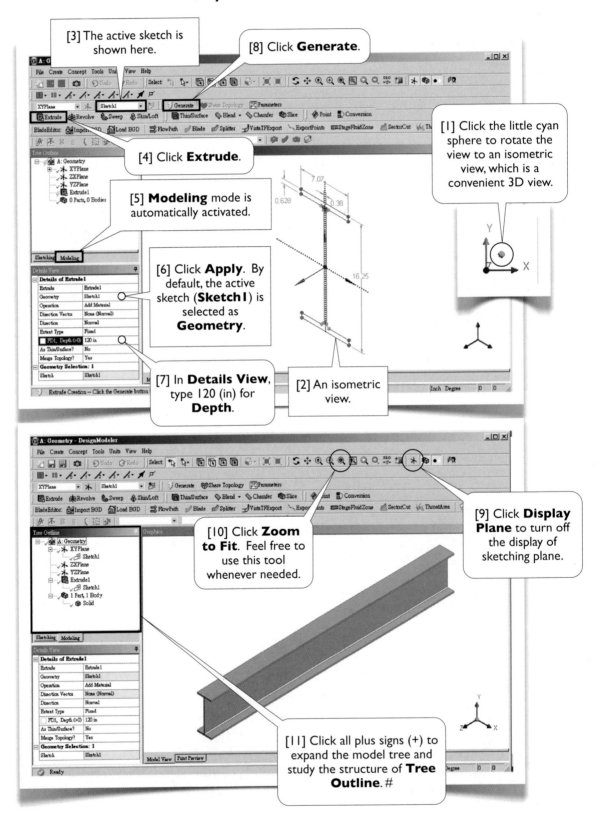

[3] The active sketch is shown here.

[8] Click **Generate**.

[1] Click the little cyan sphere to rotate the view to an isometric view, which is a convenient 3D view.

[4] Click **Extrude**.

[5] **Modeling** mode is automatically activated.

[6] Click **Apply**. By default, the active sketch (**Sketch1**) is selected as **Geometry**.

[7] In **Details View**, type 120 (in) for **Depth**.

[2] An isometric view.

[9] Click **Display Plane** to turn off the display of sketching plane.

[10] Click **Zoom to Fit**. Feel free to use this tool whenever needed.

[11] Click all plus signs (+) to expand the model tree and study the structure of **Tree Outline**. #

63

2.1-14 Save Project and Exit Workbench

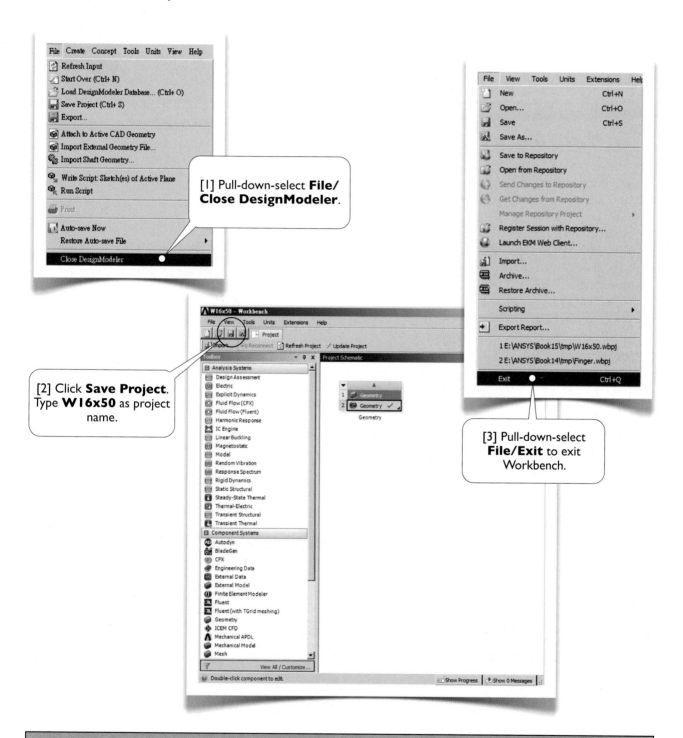

[1] Pull-down-select **File/ Close DesignModeler**.

[2] Click **Save Project**. Type **W16x50** as project name.

[3] Pull-down-select **File/Exit** to exit Workbench.

Supporting Files

To download the finished project files or view the demo videos, please see **Companion Webpage** and **Companion DVD** in the Preface, page 4.

Section 2.2

Triangular Plate

2.2-1 About the Triangular Plate

The triangular plate [1, 2], its thickness 10 mm, is made to withstand tensile forces on three side faces [3].

In this section, we'll sketch a profile of the plate on **XYPlane** and then extrude a thickness of 10 mm along Z-axis to generate a 3D solid body.

In Section 3.1, we will use this sketch again to generate a 2D solid model, which is then used for a static structural simulation to assess the stress under design loads.

The 2D solid model will be used again in Section 8.2 to demonstrate a design optimization procedure.

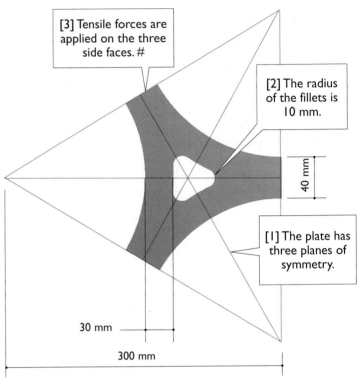

[3] Tensile forces are applied on the three side faces. #

[2] The radius of the fillets is 10 mm.

40 mm

[1] The plate has three planes of symmetry.

30 mm

300 mm

2.2-2 Start up DesignModeler

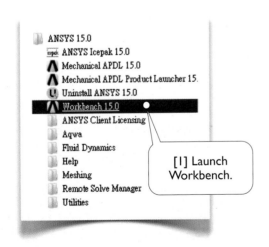

[1] Launch Workbench.

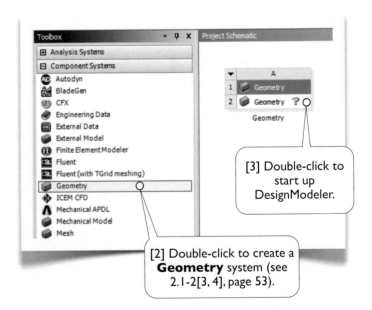

[3] Double-click to start up DesignModeler.

[2] Double-click to create a **Geometry** system (see 2.1-2[3, 4], page 53).

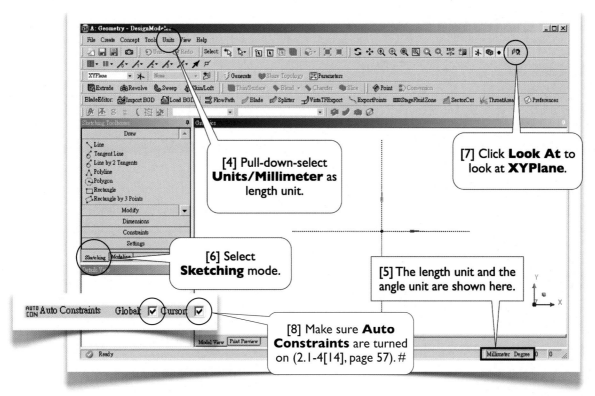

[4] Pull-down-select **Units/Millimeter** as length unit.

[7] Click **Look At** to look at **XYPlane**.

[6] Select **Sketching** mode.

[5] The length unit and the angle unit are shown here.

[8] Make sure **Auto Constraints** are turned on (2.1-4[14], page 57). #

2.2-3 Draw a Triangle on XYPlane

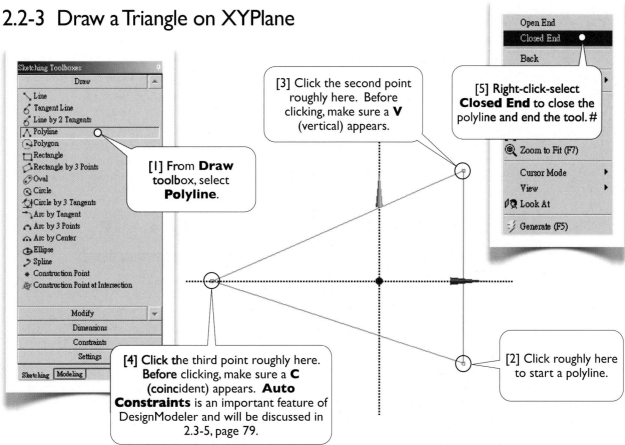

[1] From **Draw** toolbox, select **Polyline**.

[3] Click the second point roughly here. Before clicking, make sure a **V** (vertical) appears.

[5] Right-click-select **Closed End** to close the polyline and end the tool. #

[4] Click the third point roughly here. **Before** clicking, make sure a **C** (coincident) appears. **Auto Constraints** is an important feature of DesignModeler and will be discussed in 2.3-5, page 79.

[2] Click roughly here to start a polyline.

2.2-4 Make the Triangle Regular

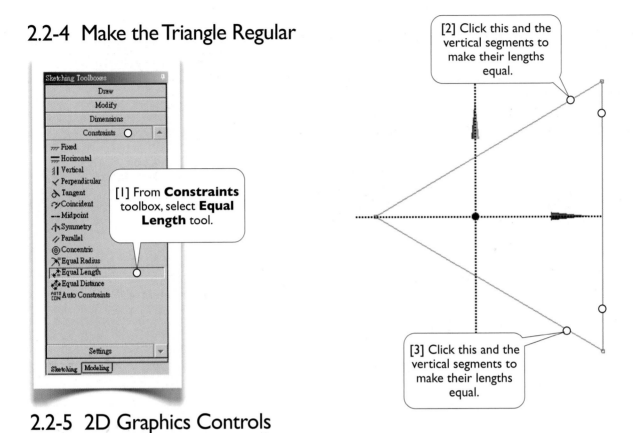

[1] From **Constraints** toolbox, select **Equal Length** tool.

[2] Click this and the vertical segments to make their lengths equal.

[3] Click this and the vertical segments to make their lengths equal.

2.2-5 2D Graphics Controls

Tools for 2D graphics controls are available in a toolbar [1-9]; click tools in [3-5] to switch them on/off. Feel free to use these tools whenever needed. Try to click each tool now; they don't modify the model. Note that, a better way for **Pan**, **Zoom**, and **Box Zoom** is using mouse shortcuts, given in 2.1-7 (page 59) and 2.3-4 (page 78).

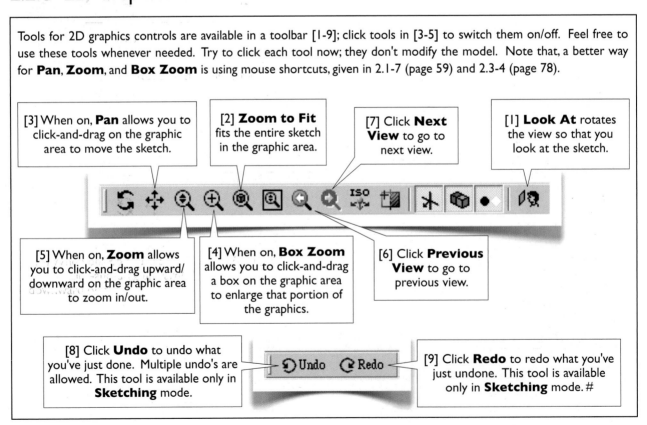

[3] When on, **Pan** allows you to click-and-drag on the graphic area to move the sketch.

[2] **Zoom to Fit** fits the entire sketch in the graphic area.

[7] Click **Next View** to go to next view.

[1] **Look At** rotates the view so that you look at the sketch.

[5] When on, **Zoom** allows you to click-and-drag upward/downward on the graphic area to zoom in/out.

[4] When on, **Box Zoom** allows you to click-and-drag a box on the graphic area to enlarge that portion of the graphics.

[6] Click **Previous View** to go to previous view.

[8] Click **Undo** to undo what you've just done. Multiple undo's are allowed. This tool is available only in **Sketching** mode.

[9] Click **Redo** to redo what you've just undone. This tool is available only in **Sketching** mode. #

2.2-6 Specify Dimensions

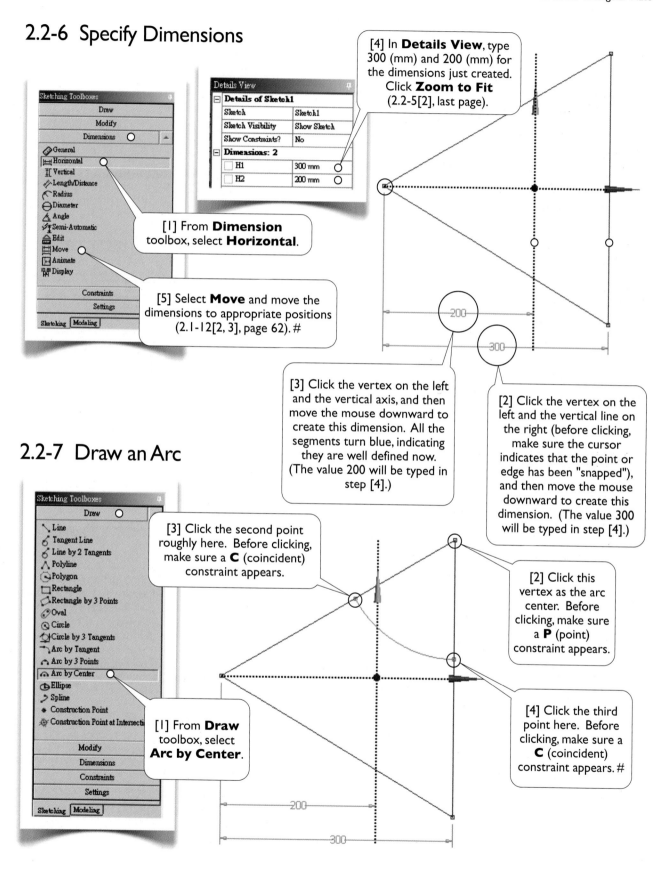

[4] In **Details View**, type 300 (mm) and 200 (mm) for the dimensions just created. **Click Zoom to Fit** (2.2-5[2], last page).

[1] From **Dimension** toolbox, select **Horizontal**.

[5] Select **Move** and move the dimensions to appropriate positions (2.1-12[2, 3], page 62). #

[3] Click the vertex on the left and the vertical axis, and then move the mouse downward to create this dimension. All the segments turn blue, indicating they are well defined now. (The value 200 will be typed in step [4].)

[2] Click the vertex on the left and the vertical line on the right (before clicking, make sure the cursor indicates that the point or edge has been "snapped"), and then move the mouse downward to create this dimension. (The value 300 will be typed in step [4].)

2.2-7 Draw an Arc

[3] Click the second point roughly here. Before clicking, make sure a **C** (coincident) constraint appears.

[2] Click this vertex as the arc center. Before clicking, make sure a **P** (point) constraint appears.

[1] From **Draw** toolbox, select **Arc by Center**.

[4] Click the third point here. Before clicking, make sure a **C** (coincident) constraint appears. #

68

2.2-8 Replicate the Arc

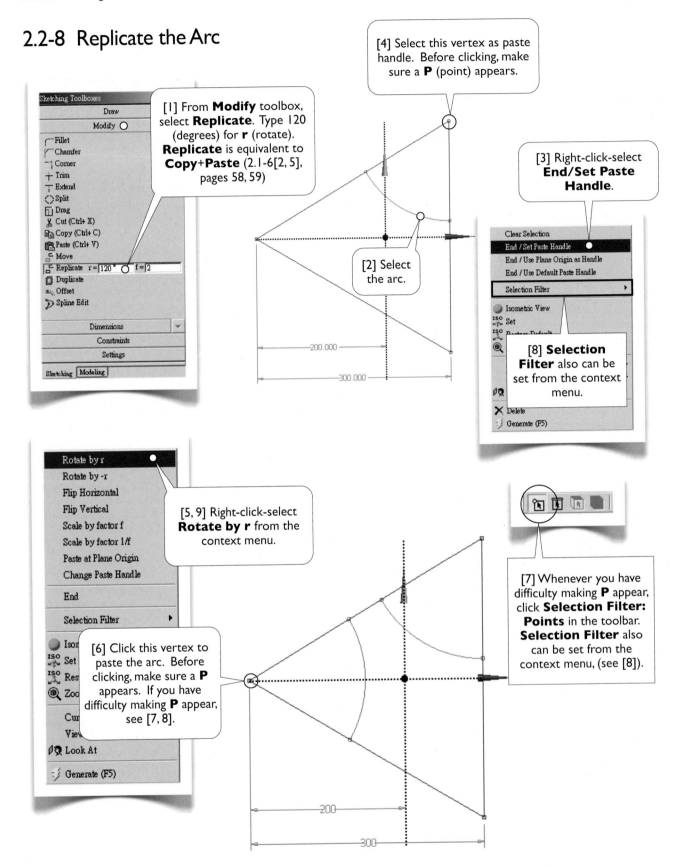

[4] Select this vertex as paste handle. Before clicking, make sure a **P** (point) appears.

[1] From **Modify** toolbox, select **Replicate**. Type 120 (degrees) for **r** (rotate). **Replicate** is equivalent to **Copy+Paste** (2.1-6[2, 5], pages 58, 59)

[3] Right-click-select **End/Set Paste Handle**.

[2] Select the arc.

[8] **Selection Filter** also can be set from the context menu.

[5, 9] Right-click-select **Rotate by r** from the context menu.

[7] Whenever you have difficulty making **P** appear, click **Selection Filter: Points** in the toolbar. **Selection Filter** also can be set from the context menu, (see [8]).

[6] Click this vertex to paste the arc. Before clicking, make sure a **P** appears. If you have difficulty making **P** appear, see [7, 8].

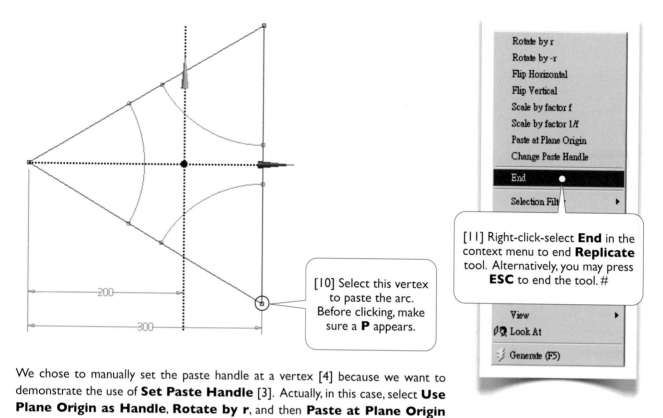

[10] Select this vertex to paste the arc. Before clicking, make sure a **P** appears.

Rotate by r
Rotate by -r
Flip Horizontal
Flip Vertical
Scale by factor f
Scale by factor 1/f
Paste at Plane Origin
Change Paste Handle
End
Selection Filt

[11] Right-click-select **End** in the context menu to end **Replicate** tool. Alternatively, you may press **ESC** to end the tool. #

View
Look At
Generate (F5)

We chose to manually set the paste handle at a vertex [4] because we want to demonstrate the use of **Set Paste Handle** [3]. Actually, in this case, select **Use Plane Origin as Handle**, **Rotate by r**, and then **Paste at Plane Origin** may be more convenient (see 2.1-6, pages 58, 59).

2.2-9 Trim Away Unwanted Segments

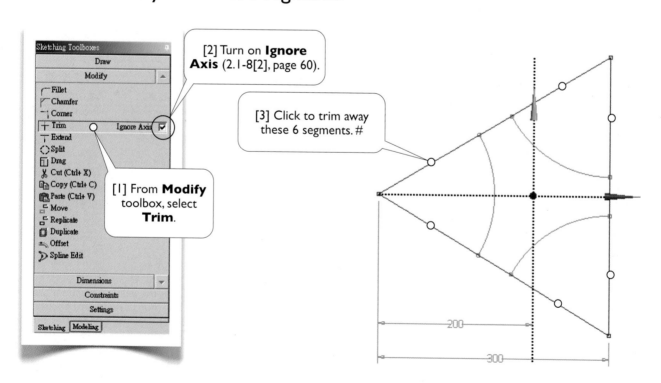

[1] From **Modify** toolbox, select **Trim**.

[2] Turn on **Ignore Axis** (2.1-8[2], page 60).

[3] Click to trim away these 6 segments. #

70

2.2-10 Impose Constraints

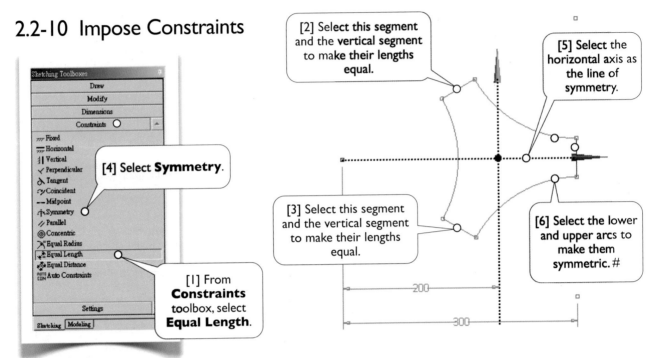

[2] Select **this segment** and the **vertical segment** to make their lengths equal.

[5] Select the horizontal axis as the line of symmetry.

[4] Select **Symmetry**.

[1] From **Constraints** toolbox, select **Equal Length**.

[3] Select this segment and the vertical segment to make their lengths equal.

[6] Select the lower and upper arcs to make them symmetric. #

Constraint Status

The three straight lines turn blue, indicating they are well-defined, while the three arcs remain greenish-blue, indicating they are not well-defined yet (under-constrained). Other color codes are: black for fixed; red for over-constrained; gray for inconsistency.

2.2-11 Specify Dimension for Side Edges

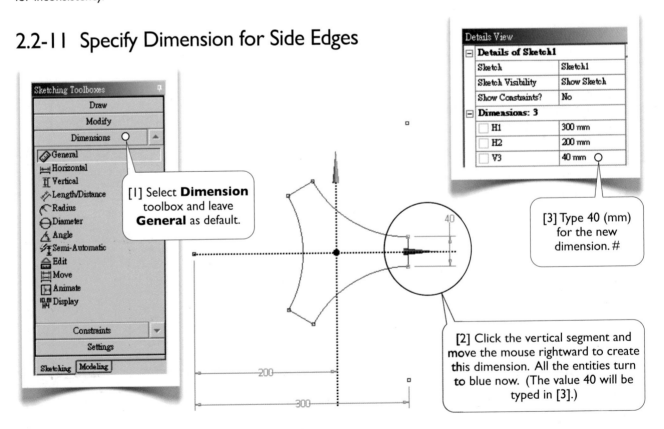

[1] Select **Dimension** toolbox and leave **General** as default.

Details View	
☐ **Details of Sketch1**	
Sketch	Sketch1
Sketch Visibility	Show Sketch
Show Constraints?	No
☐ **Dimensions: 3**	
☐ H1	300 mm
☐ H2	200 mm
☐ V3	40 mm

[3] Type 40 (mm) for the new dimension. #

[2] Click the vertical segment and move the mouse rightward to create this dimension. All the entities turn to blue now. (The value 40 will be typed in [3].)

2.2-12 Create Offset

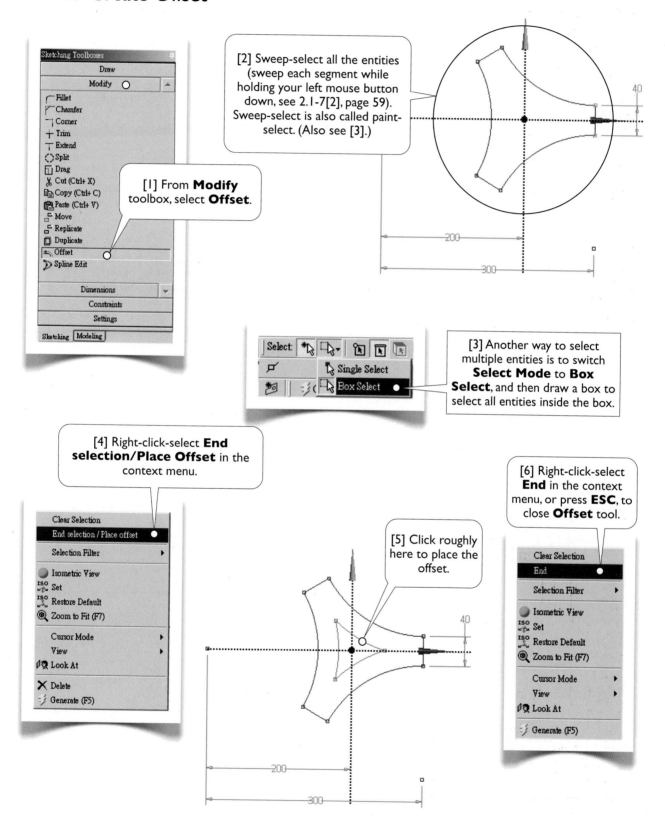

[1] From **Modify** toolbox, select **Offset**.

[2] Sweep-select all the entities (sweep each segment while holding your left mouse button down, see 2.1-7[2], page 59). Sweep-select is also called paint-select. (Also see [3].)

[3] Another way to select multiple entities is to switch **Select Mode** to **Box Select**, and then draw a box to select all entities inside the box.

[4] Right-click-select **End selection/Place Offset** in the context menu.

[5] Click roughly here to place the offset.

[6] Right-click-select **End** in the context menu, or press **ESC**, to close **Offset** tool.

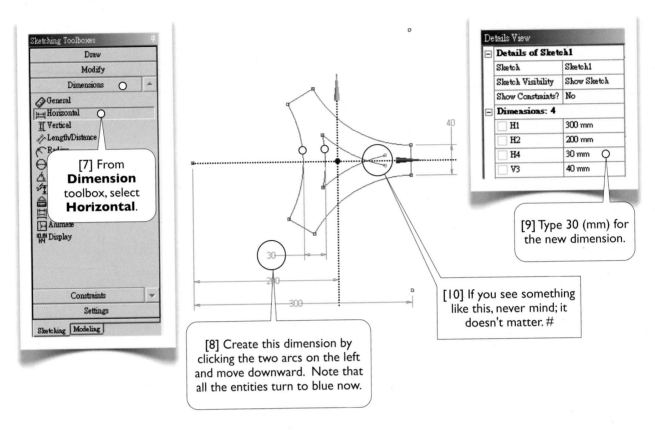

[7] From **Dimension** toolbox, select **Horizontal**.

[9] Type 30 (mm) for the new dimension.

[10] If you see something like this, never mind; it doesn't matter. #

[8] Create this dimension by clicking the two arcs on the left and move downward. Note that all the entities turn to blue now.

2.2-13 Create Fillets

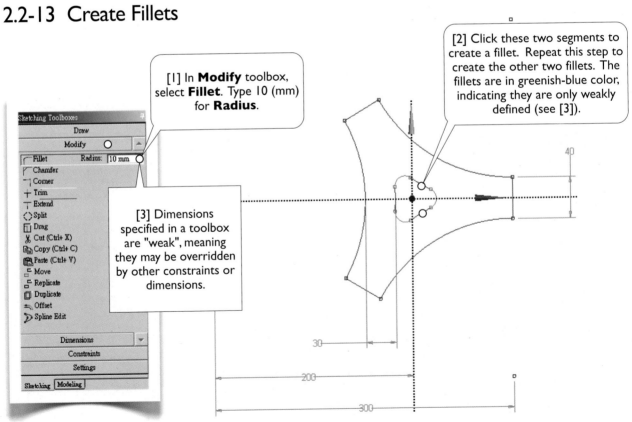

[1] In **Modify** toolbox, select **Fillet**. Type 10 (mm) for **Radius**.

[2] Click these two segments to create a fillet. Repeat this step to create the other two fillets. The fillets are in greenish-blue color, indicating they are only weakly defined (see [3]).

[3] Dimensions specified in a toolbox are "weak", meaning they may be overridden by other constraints or dimensions.

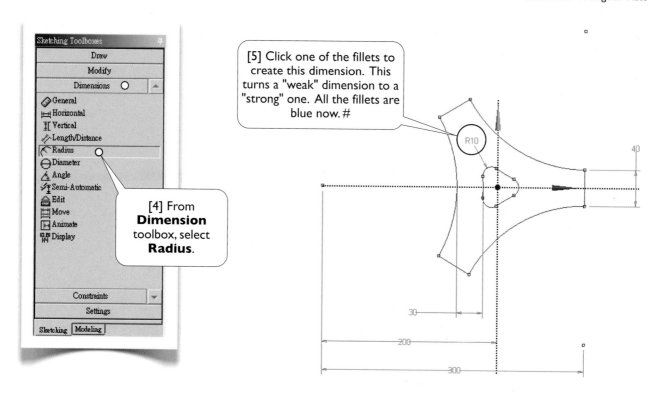

[5] Click one of the fillets to create this dimension. This turns a "weak" dimension to a "strong" one. All the fillets are blue now. #

[4] From **Dimension** toolbox, select **Radius**.

2.2-14 Extrude to Create 3D Solid

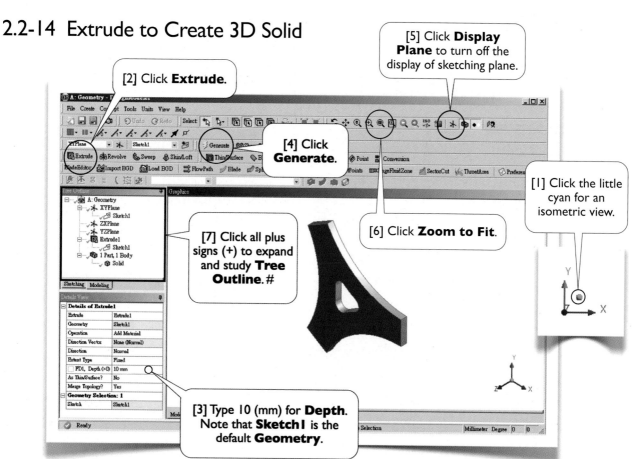

[5] Click **Display Plane** to turn off the display of sketching plane.

[2] Click **Extrude**.

[4] Click **Generate**.

[1] Click the little cyan for an isometric view.

[7] Click all plus signs (+) to expand and study **Tree Outline**. #

[6] Click **Zoom to Fit**.

[3] Type 10 (mm) for **Depth**. Note that **Sketch1** is the default **Geometry**.

2.2-15 Save the Project and Exit Workbench

[2] Pull-down-select **File/Close DesignModeler**.

[1] Click **Save Project**. Type **Triplate** as project name.

[3] In Workbench GUI, pull-down-select **File/Exit** to exit Workbench. #

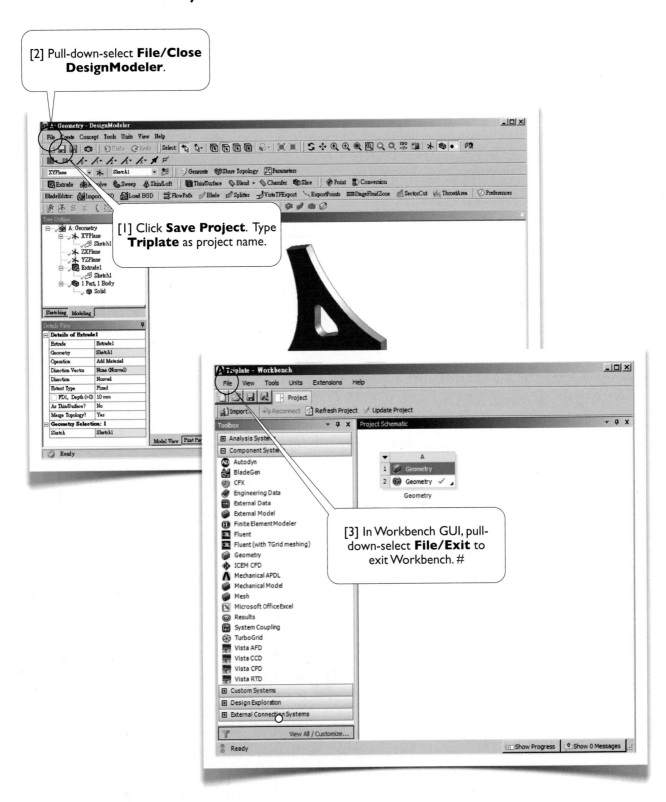

Section 2.3

More Details

2.3-1 DesignModeler GUI

DesignModeler GUI is divided into several areas [1-7]. On the top are pull-down menus and toolbars [1]; on the bottom is a status bar [7]. In-between are several "window panes." Separators [8] between window panes can be dragged to resize the window panes. You even can make a window pane "float" by dragging or double-clicking its title bar. To return to its original position, simply double-click its title bar again.

    Tree Outline [3] shares the same area with **Sketching Toolboxes** [4]. To switch between **Modeling** mode and **Sketching** mode, simply click a "mode tab" [2]. **Details View** [6] shows the detail information of the objects highlighted in **Tree Outline** [3] or **Graphics Window** [5], the former displaying a **Model Tree** (see explanation next page) while the latter displaying the geometric model. Note that, we focus on 2D functions of DesignModeler in this chapter and will discuss 3D functions in Chapter 4.

[1] Pull-down menus and toolbars.

[4] **Sketching Toolboxes**, in **Sketching** mode.

[3] **Tree Outline**, in **Modeling** mode.

[5] Graphics window.

[2] Mode tabs.

[8] Separators allow you to resize window panes. #

[6] **Details View**.

[7] Status bar.

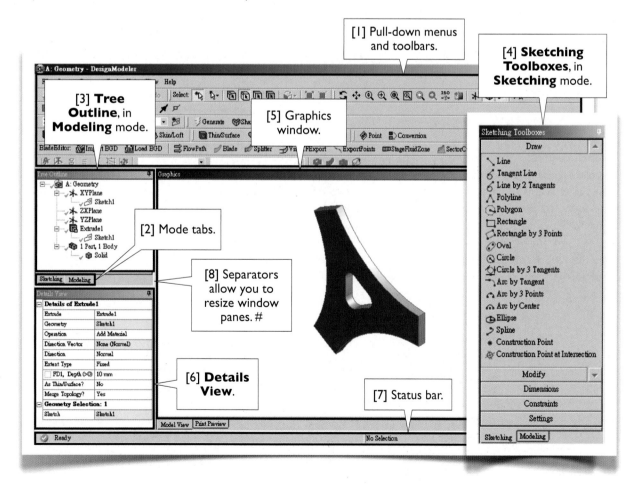

Model Tree

Tree Outline [3] contains an outline of a *model tree*, the data structure of the geometric model. Each *branch* of the tree is called an *object*, which may contain one or more objects. At the bottom of the model tree is a **part** branch, which is the only object that will be exported to **Mechanical** for simulations. By right-clicking an object and selecting a tool from the *context menu*, you can operate on the object, such as delete, rename, duplicate, etc.

The order of the objects is relevant. DesignModeler renders the geometry according to the order of objects in the model tree. New objects are normally added one after another. If you want to insert a new object BEFORE an existing object, right-click the existing object and select **Insert/...** from the context menu. After insertion, DesignModeler will re-render the geometry.

2.3-2 Sketching Planes

A sketch must be created on a *sketching plane*, or simply called a *plane*; each plane may contain multiple sketches. In the beginning of a DesignModeler session, three planes are automatically created: **XYPlane**, **YZPlane**, and **ZXPlane**. Currently active plane is shown on the toolbar [1]. You can create new planes as many as needed [2]. There are several ways of creating a new plane [3]. In this chapter, since we always create sketches on **XYPlane**, we will not discuss how to create sketching planes now and will discuss it in Chapter 4.

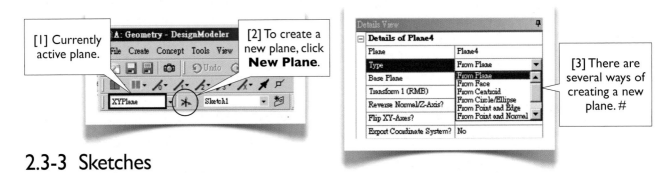

[1] Currently active plane.

[2] To create a new plane, click **New Plane**.

[3] There are several ways of creating a new plane. #

2.3-3 Sketches

A sketch consists of *points* and *edges*; edges may be straight lines or curves. Dimensions and constraints may be imposed on points and edges. As mentioned (2.3-2), multiple sketches may be created on a plane. To create a new sketch on a plane on which there are yet no sketches, you simply switch to **Sketching** mode and draw any geometric entities on it. Later, if you want to add a new sketch on that plane, you have to click **New Sketch** [1]. Exactly one plane and one sketch is active at a time [2-5]; newly created points and edges are added to the active sketch, and newly created sketches are added to the active plane. In this chapter, we almost exclusively work with a single sketch; the only exception is Section 2.6, in which a second sketch is used (2.6-4[1, 2], page 101). When a new sketch is created, it becomes the active sketch. More on creating sketches will be discussed in Chapter 4.

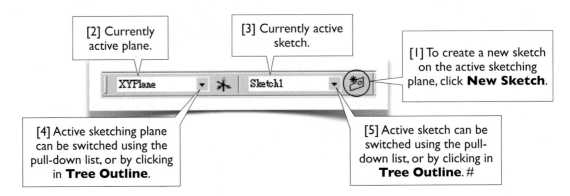

[2] Currently active plane.

[3] Currently active sketch.

[1] To create a new sketch on the active sketching plane, click **New Sketch**.

[4] Active sketching plane can be switched using the pull-down list, or by clicking in **Tree Outline**.

[5] Active sketch can be switched using the pull-down list, or by clicking in **Tree Outline**. #

2.3-4 Sketching Toolboxes

When you switch to **Sketching** mode by clicking the mode tab (2.3-1[2], page 76), you see **Sketching Toolboxes** (2.3-1[4], page 76). There are five **Sketching Toolboxes**: **Draw**, **Modify**, **Dimensions**, **Constraints**, and **Settings** [1-5]. Most of the tools in the toolboxes are self-explained. The best way to learn these tools is to try them out individually. During the tryout, whenever you want to clean up the graphics window, pull-down-select **File/Start Over**. These sketching tools will be explained, starting from 2.3-6 (pages 79) up to the end of this section.

Before we discuss these sketching tools, let's emphasize some tips relevant to sketching.

Pan, Zoom, and Box Zoom

Besides **Pan** tool (2.2-5[3], page 67), a sketch can be panned by dragging your mouse while holding down both the control key and the middle mouse button (2.1-7[6], page 59). Besides **Zoom** tool (2.2-5[5], page 67) a sketch can be zoomed in/out by simply rolling forward/backward your mouse wheel (2.1-7[5], page 59); the cursor position is the "zoom center." Besides **Box Zoom** tool (2.2-5[4], page 67), box zoom can also be done by dragging a rectangle in the graphics window using the right mouse button (2.1-7[4], page 59). When you get used to these mouse shortcut, you usually don't need **Pan**, **Zoom**, and **Box Zoom** tools any more.

Context Menu

While most of operations can be done by commands in pull-down menus or toolbars, many operations either require or are more efficient using a context menu. The context menu can be popped-up by right-clicking an entity in the graphics window or an object in the model tree. Try to explore whatever available in the context menu.

Status Bar

The status bar (2.3-1[7], page 76) contains instructions on each operations. Look at the instruction whenever helpful. When a draw tool is in use, the coordinates of your mouse pointer are shown in the status bar.

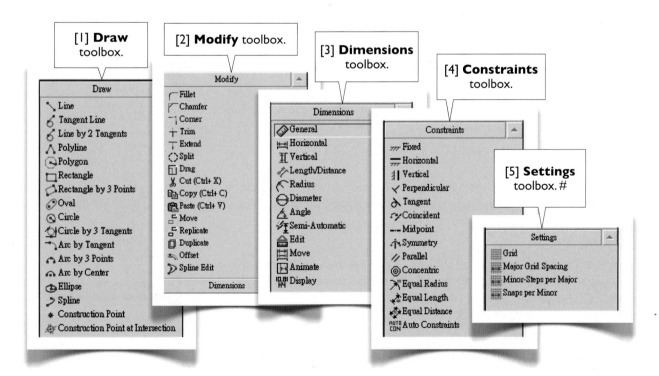

[1] **Draw** toolbox.

[2] **Modify** toolbox.

[3] **Dimensions** toolbox.

[4] **Constraints** toolbox.

[5] **Settings** toolbox. #

2.3-5 Auto Constraints[Refs 1, 2]

By default, DesignModeler is in **Auto Constraints** mode, both globally and locally (see 2.1-4[14], page 57). DesignModeler attempts to detect the user's intentions and try to automatically impose constraints on sketching entities. The following cursor symbols indicate the kind of constraints that are applied:

C - The cursor is coincident with a line.
P - The cursor is coincident with a point.
T - The cursor is a tangent point.
⊥ - The cursor is a perpendicular foot.
H - The line is horizontal.
V - The line is vertical.
// - The line is parallel to another line.
R - The radius is equal to another radius.

Both **Global** and **Cursor** modes are based on all entities of the active plane (not just the active sketch). The difference is that **Cursor** mode only examines the entities nearby the cursor, while **Global** mode examines all the entities in the active plane.

While **Auto Constraints** can be useful, they sometimes can lead to problems and add noticeable time on complicated sketches. Turn off them if desired [1].

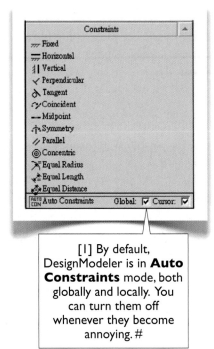

[1] By default, DesignModeler is in **Auto Constraints** mode, both globally and locally. You can turn them off whenever they become annoying. #

2.3-6 Draw Tools[Ref 3] [1]

Line

Draws a line by two clicks.

Tangent Line

Click a point on a curve (e.g., circle, arc, ellipse, or spline) to create a line tangent to the curve at that point.

Line by 2 Tangents

Click two curves to create a line tangent to these two curves. Click a curve and a point to create a line tangent to the curve and connecting to the point.

Polyline

A polyline consists of multiple straight lines. A polyline must be completed by choosing either **Open End** or **Closed End** from the context menu [2].

Polygon

Draws a regular polygon. The first click defines the center and the second click defines the radius of the circumscribing circle.

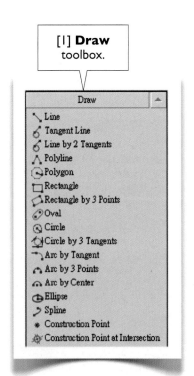

[1] **Draw** toolbox.

Rectangle by 3 Points

The first two points define one side and the third point defines the other side.

Oval

The first two clicks define two centers, and the third click defines the radius.

Circle

The first click defines the center, and the second click defines the radius.

Circle by 3 Tangents

Select three edges (lines or curves) to create a circle tangent to these three edges.

Arc by Tangent

Click a point (usually an end point) on an edge to create an arc starting from that point and tangent to that edge; click a second point to define the other end and the radius of the arc.

Arc by 3 Points

The first two clicks define the two ends of the an, and the third click defines a point in-between the ends.

Arc by Center

The first click defines the center, and two additional clicks define two ends.

Ellipse

The first click defines the center, the second click defines the major radius, and the third click defines the minor radius.

Spline

A spline is either rigid or flexible. A flexible spline can be edited or changed by imposing constraints, while a rigid spline cannot. After defining the last point, you must specify an ending condition [3]: either open end or closed end; either with fit points or without fit points.

Construction Point at Intersection

Select two edges, a construction point will be created at the intersection.

[2] A polyline must be completed by choosing either **Open End** or **Closed End** from the context menu.

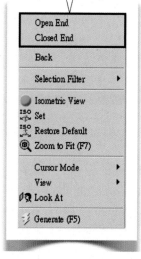

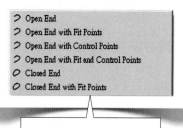

[3] A spline must be complete by specifying an ending condition from the context menu. #

How to delete edges?

To delete edges, select them and choose **Delete** or **Cut** from the context menu. Multiple selection methods (e.g., control-selection or sweep-selection) can be used to select edges. To clean up the graphics window entirely, pull-down-select **File/ Start Over**. A general way of deleting any sketching entities (edges, dimensions, or constraints) is to right-click the entity in **Details View** and issue **Delete**. See 2.3-8[6] (page 84) and 2.3-9[3, 4] (page 85).

How to abort a tool?

Simply press **ESC**.

2.3-7 Modify Tools[Ref 4] [1]

Fillet

Select two edges or a vertex to create a fillet. The radius of the fillet can be specified in the toolbox [2]. Note that this radius value is a weak dimension; i.e., it can be changed by other dimensions or constraints.

Chamfer

Select two edges or a vertex to create an equal-length chamfer. The sizes of the chamfer can be specified in the toolbox.

Corner

Select two edges, and the edges will be trimmed or extended up to the intersection point and form a sharp corner. The clicking points decide which sides to be trimmed.

Trim

Select an edge, and the portion of the edge will be removed up to its intersection with other edge, axis, or point.

Extend

Select an edge, and the edge will be extended up to an edge or axis.

Split

This tool splits an edge into several segments depending on the options from the context menu [3]. **Split Edge at Selection**: Click an edge, and the edge will be split at the clicking point. **Split Edges at Point**: Click a point, and all the edges passing through that point will be split at that point. **Split Edge at All Points**: Click an edge, the edge will be split at all points on the edge. **Split Edge into n Equal Segments**: Click an edge and specify a value n, and the edge will be split equally into n segments.

Drag

Drags a point or an edge to a new position. All the constraints and dimensions are preserved.

Copy

Copies the selected entities to a "clipboard." A **Paste Handle** must be specified using one of the methods in the context menu [4].   After completing this tool, **Paste** tool is automatically activated.

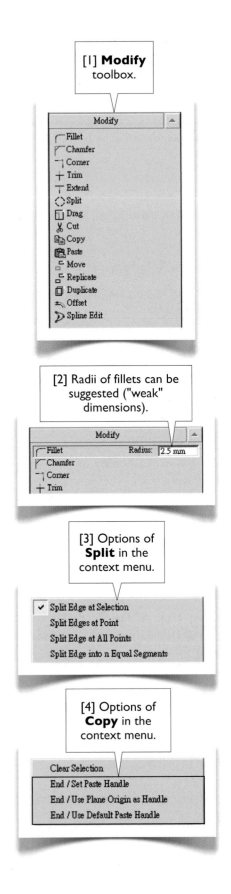

[1] **Modify** toolbox.

[2] Radii of fillets can be suggested ("weak" dimensions).

[3] Options of **Split** in the context menu.

[4] Options of **Copy** in the context menu.

Cut

Similar to **Copy**, except that the copied entities are removed.

Paste

Pastes the entities in the "clipboard" to the graphics window. The click defines the point at which the **Paste Handle** positions. Many options can be chosen from the context menu [5], where the rotating angle r and the scaling factor f can be specified in the toolbox.

Move

Equivalent to a **Cut** followed by a **Paste**. (The original is removed.)

Replicate

Equivalent to a **Copy** followed by a **Paste**. (The original is preserved.)

Duplicate

Similar to **Replicate**. However, **Duplicate** copies entities to the same position in the active plane. **Duplicate** can be used to copy features of a solid body or plane boundaries.

Offset

Creates a set of edges that are offset by a distance from an existing set of edges.

Spline Edit

Used to modify flexible splines. You can insert, delete, drag the fit points, etc [6]. For details, see the reference[Ref 4].

2.3-8 Dimensions Tools[Ref 5] [1]

General

Allows creation of any of the dimension types, depending on what edge and context-menu options are selected. If the selected edge is a straight line, the default dimension is its length [2]. If the selected edge is a circle or arc, the default dimension is its radius [3].

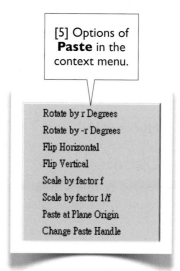

[5] Options of **Paste** in the context menu.

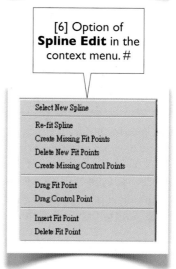

[6] Option of **Spline Edit** in the context menu. #

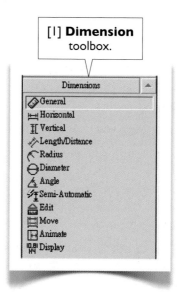

[1] **Dimension** toolbox.

Horizontal

Select two points to specify a horizontal dimension. If you select an edge (instead of a point), the end point near the click will be picked.

Vertical

Similar to **Horizontal**.

Length/Distance

Select two points to specify a distance dimension. You also can select a point and a line to specify the distance between the point and the line.

Radius

Select a circle or arc to specify a radius dimension. If you select an ellipse, the major (or minor) radius will be specified.

Diameter

Select a circle or arc to specify a diameter dimension.

Angle

Select two lines to specify an angle. By varying the selection order and location, you can control which angle you are dimensioning. The end of the lines that you select will be the direction of the hands, and the angle is measured counterclockwise from the first selected hand to the second. Before you click to locate the dimension, if the angle is not what you want, repeatedly choose **Alternate Angle** from the context menu until a correct angle is selected [4].

Semi-Automatic

This tool displays a series of dimensions automatically to help you fully dimension the sketch.

Edit

Click a dimension and this tool allows you to change its name or values.

Move

Click a dimension and move it to a new position.

Animate

Click a dimension to show the animated effects.

Display

Allows you to decide whether to display dimension names, values, or both. In this book, we always choose to display dimension values [5] rather than dimension names.

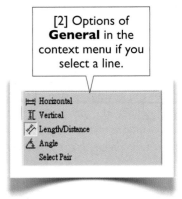

[2] Options of **General** in the context menu if you select a line.

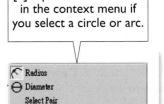

[3] Options of **General** in the context menu if you select a circle or arc.

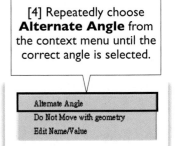

[4] Repeatedly choose **Alternate Angle** from the context menu until the correct angle is selected.

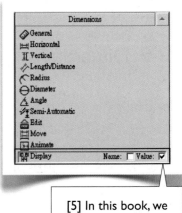

[5] In this book, we always display dimension values.

How to delete dimensions?

To delete a dimension, select the dimension in **Details View**, and choose **Delete** from the context menu [6]. You can delete ALL dimensions by right-click **Dimensions** in **Details View** [7].

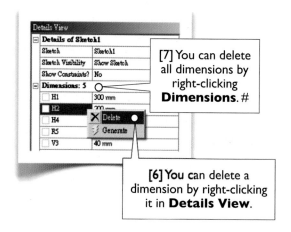

[7] You can delete all dimensions by right-clicking **Dimensions**. #

[6] You can delete a dimension by right-clicking it in **Details View**.

2.3-9 Constraints Tools[Ref 6] [1]

Fixed

Applies on an edge to make it fully constrained if **Fix Endpoints** is selected [2]. If **Fix Endpoints** is not selected, then the edge's endpoints can be changed, but not the edge's position and slope.

Horizontal

Applies on a line to make it horizontal.

Vertical

Applies on a line to make it vertical.

Perpendicular

Applies on two edges to make them perpendicular to each other.

Tangent

Applies on two edges, one of which must be a curve, to make them tangent to each other.

Coincident

Select two points to make them coincident. Or, select a point and an edge to make the edge or its extension pass through the point. There are other possibilities, depending on how you select the entities.

Midpoint

Select a line and a point to make the midpoint of the line coincide with the point.

Symmetry

Select a line or an axis, as the line of symmetry, and then select two entities to make them symmetric about the line of symmetry.

[1] **Constraints** toolbox.

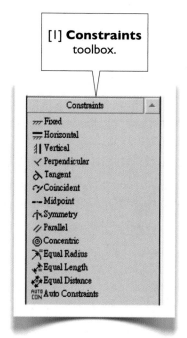

[2] If **Fix Endpoints** is selected, the edge will be fully constrained.

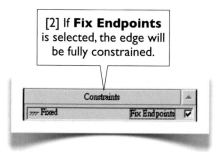

Parallel

Applies on two lines to make them parallel to each other.

Concentric

Applies on two curves, which may be circle, arc, or ellipse, to make their centers coincident.

Equal Radius

Applies on two curves, which must be circle or arc, to make their radii equal.

Equal Length

Applies on two lines to make their lengths equal.

Equal Distance

Applies on two distances to make them equal. A distance can be defined by selecting two points, two parallel lines, or one point and one line.

Auto Constraints

Allows you to turn on/off **Auto Constraints** (2.3-5).

How to delete constraints?

By default, constraints are not displayed in **Details View**. To display constraints, select **Yes** for **Show Constraints?** in **Details View** [3]. To delete a constraint, right-click the constraint and issue **Delete** [4].

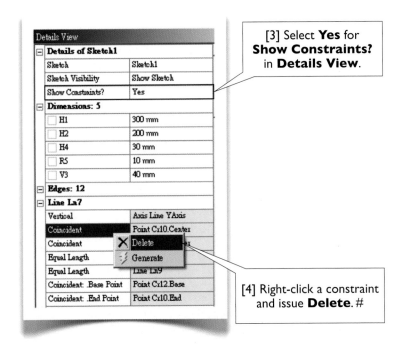

[3] Select **Yes** for **Show Constraints?** in **Details View**.

[4] Right-click a constraint and issue **Delete**. #

2.3-10 Settings Tools[Ref 7] [1]

Grid

Allows you to turn on/off grid visibility and snap capability [2, 3]. The grid is not required to enable snapping.

Major Grid Spacing

Allows you to specify **Major Grip Spacing** [4, 5] if **Show in 2D** is turned on.

Minor-Steps per Major

Allows you to specify **Minor-Steps per Major** [6, 7] if **Show in 2D** is turned on.

Snaps per Minor

Allows you to specify **Snaps per Minor** [8] if **Snap** is turned on.

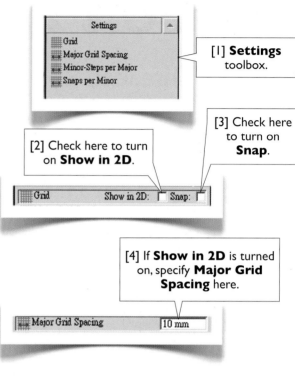

[1] **Settings** toolbox.

[2] Check here to turn on **Show in 2D**.

[3] Check here to turn on **Snap**.

[4] If **Show in 2D** is turned on, specify **Major Grid Spacing** here.

[6] If **Show in 2D** is turned on, specify **Minor-Steps per Major** here.

[8] If **Snap** is turned on, specify **Snaps per Minor** here. #

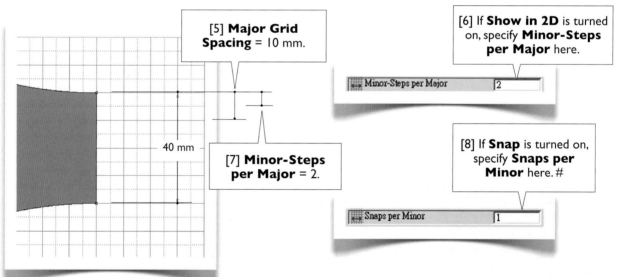

[5] **Major Grid Spacing** = 10 mm.

[7] **Minor-Steps per Major** = 2.

40 mm

References

1. ANSYS Documentation//DesignModeler User's Guide//2D Sketching//Auto Constraints
2. ANSYS Documentation//DesignModeler User's Guide//2D Sketching//Constraints Toolbox//Auto Constraints
3. ANSYS Documentation//DesignModeler User's Guide//2D Sketching//Draw Toolbox
4. ANSYS Documentation//DesignModeler User's Guide//2D Sketching//Modify Toolbox
5. ANSYS Documentation//DesignModeler User's Guide//2D Sketching//Dimensions Toolbox
6. ANSYS Documentation//DesignModeler User's Guide//2D Sketching//Constraints Toolbox
7. ANSYS Documentation//DesignModeler User's Guide//2D Sketching//Settings Toolbox

Section 2.4

M20x2.5 Threaded Bolt

2.4-1 About the M20x2.5 Threaded Bolt[Refs 1, 2]

In this section, we'll create a sketch, revolving the sketch 360° to generate a 3D solid body, a body representing a portion of an M20x2.5 threaded bolt [1-6]. We will use this sketch in Section 3.2 again to generate a 2D solid body, which is then used for a static structural simulation.

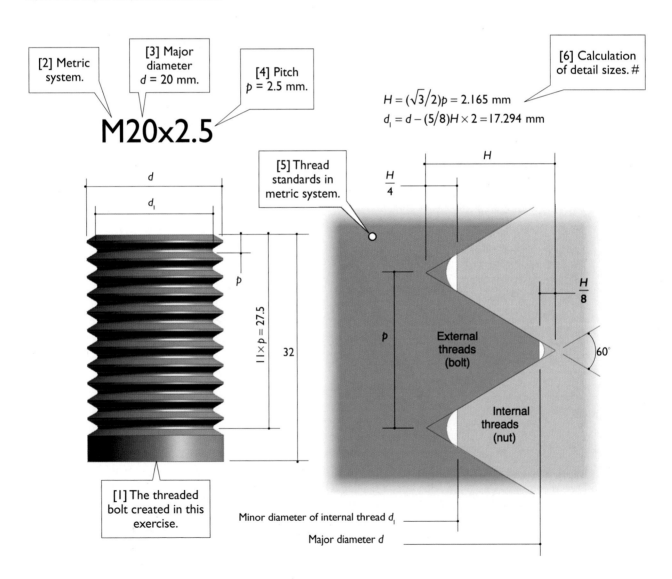

[2] Metric system.

[3] Major diameter d = 20 mm.

[4] Pitch p = 2.5 mm.

[6] Calculation of detail sizes. #

$$H = (\sqrt{3}/2)p = 2.165 \text{ mm}$$
$$d_1 = d - (5/8)H \times 2 = 17.294 \text{ mm}$$

M20x2.5

[5] Thread standards in metric system.

d

d_1

p

$11 \times p = 27.5$

32

$\frac{H}{4}$

H

$\frac{H}{8}$

p

External threads (bolt)

Internal threads (nut)

60°

Minor diameter of internal thread d_1

Major diameter d

[1] The threaded bolt created in this exercise.

87

2.4-2 Draw a Horizontal Line

Launch Workbench and create a **Geometry** System. Save the project as **Threads**. Start up DesignModeler. Select **Millimeter** as length unit. Make sure **Auto Constraints** are turned on (2.1-4[14], page 57).

Draw a horizontal line on **XYPlane**. Specify the dimensions (8.647 mm, 32 mm) as shown [1].

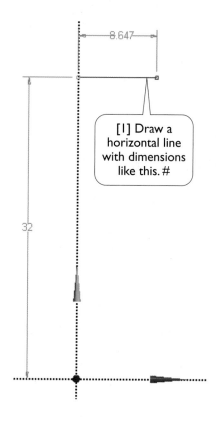

[1] Draw a horizontal line with dimensions like this. #

2.4-3 Draw a Polyline

Draw a polyline (totally 3 segments) and specify dimensions (0.541, 2.165, 2.165, 0.9375, 1.25, and 1.25 mm) as shown [1-2].

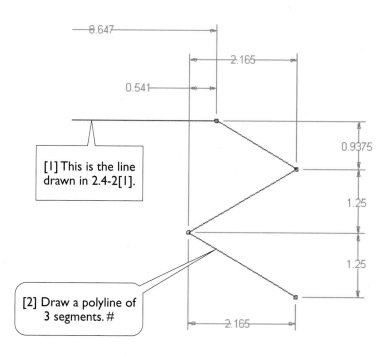

[1] This is the line drawn in 2.4-2[1].

[2] Draw a polyline of 3 segments. #

2.4-4 Draw Fillets

Draw a vertical line and specify dimension [1]. Create a fillet and specify dimension [2, 3].

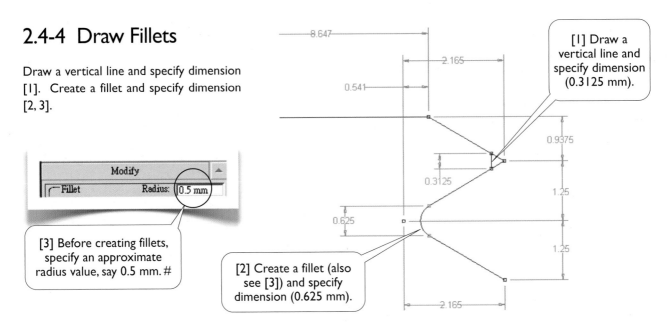

[1] Draw a vertical line and specify dimension (0.3125 mm).

[3] Before creating fillets, specify an approximate radius value, say 0.5 mm. #

[2] Create a fillet (also see [3]) and specify dimension (0.625 mm).

2.4-5 Trim Away Unwanted Segments

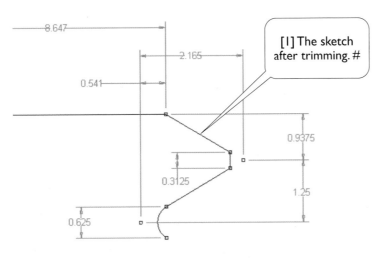

[1] The sketch after trimming. #

2.4-6 Replicate 10 Times

Select all segments except the horizontal line (totally 4 segments), and replicate 10 times. Set the **Paste Handle** as shown [1]. You may need to use **Selection Filter: Points** [2] (2.2-8[7, 8], page 69).

[2] **Selection Filter: Points**. #

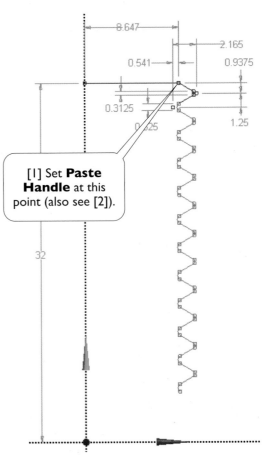

[1] Set **Paste Handle** at this point (also see [2]).

89

2.4-7 Complete the Sketch

Follow steps [1-5] to complete the sketch.

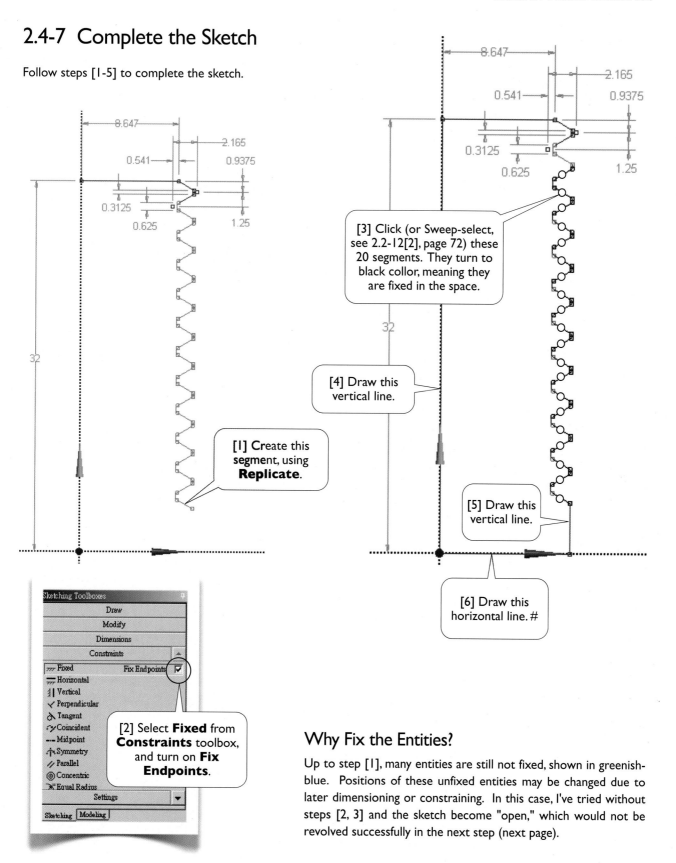

[1] Create this segment, using **Replicate**.

[2] Select **Fixed** from **Constraints** toolbox, and turn on **Fix Endpoints**.

[3] Click (or Sweep-select, see 2.2-12[2], page 72) these 20 segments. They turn to black collor, meaning they are fixed in the space.

[4] Draw this vertical line.

[5] Draw this vertical line.

[6] Draw this horizontal line. #

Why Fix the Entities?

Up to step [1], many entities are still not fixed, shown in greenish-blue. Positions of these unfixed entities may be changed due to later dimensioning or constraining. In this case, I've tried without steps [2, 3] and the sketch become "open," which would not be revolved successfully in the next step (next page).

2.4-8 Revolve to Create 3D Solid

Click **Revolve** to generate a solid of revolution. Select the Y-axis as the axis of revolution. Remember to click **Generate**.

Save the project and exit from Workbench. We will resume this project in Section 3.2.

References

1.   Zahavi, E., *The Finite Element Method in Machine Design*, Prentice-Hall, 1992; Chapter 7. Threaded Fasteners.
2.   Deutschman, A. D., Michels, W. J., and Wilson, C. E., *Machine Design: Theory and Practice*, Macmillan Publishing Co., Inc., 1975; Section 16-6. Standard Screw Threads.

Section 2.5

Spur Gears

Subsections 2.5-1 and 2.5-2 detail the geometry of the spur gears used in this section. If you are not interested in these geometric details for now, you may skip them and jump directly to 2.5-3 (page 94).

2.5-1 About the Spur Gears[Refs 1, 2]

The figure below shows a pair of identical spur gears in mesh [1-4]. Spur gears have their teeth cut parallel to the axis of the shaft on which the gears are mounted, transmiting power between the parallel shafts. To maintain a constant angular velocity ratio, two meshing gears must satisfy a fundamental law of gearing: the shape of the teeth must be such that the common normal [8] at the point of contact between two teeth must always pass through a fixed point on the line of centers[Ref 1] [5]. This fixed point is called the *pitch point* [6].

The angle between the line of action [8] and the common tangent of the pitch circles [7] is known as the *pressure angle*. The parameters defining a spur gear are its pitch radius (r_p = 2.5 in) [3], pressure angle (α = 20°) [8], and number of teeth (N = 20). The teeth are cut with a radius of addendum r_a = 2.75 in [9] and a radius of dedendum r_d = 2.2 in [10]. The shaft has a radius of 1.25 in [11]. All fillets have a radius of 0.1 in [12]. The thickness of the gear is 1.0 in.

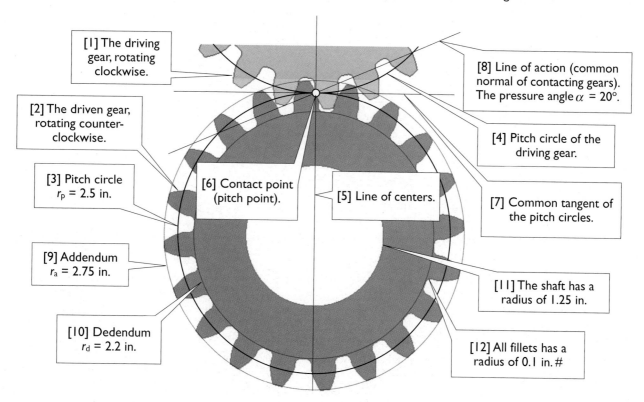

[1] The driving gear, rotating clockwise.

[2] The driven gear, rotating counter-clockwise.

[3] Pitch circle r_p = 2.5 in.

[9] Addendum r_a = 2.75 in.

[10] Dedendum r_d = 2.2 in.

[6] Contact point (pitch point).

[5] Line of centers.

[8] Line of action (common normal of contacting gears). The pressure angle α = 20°.

[4] Pitch circle of the driving gear.

[7] Common tangent of the pitch circles.

[11] The shaft has a radius of 1.25 in.

[12] All fillets has a radius of 0.1 in. #

2.5-2 About Involute Curves[Refs 1, 2]

To satisfy the fundamental law of gearing, gear profiles are usually cut to an *involute curve* [1], which may be constructed by wrapping a string (*BA*) around a *base circle* [2], and then tracing the path (*A-P-F*) of a point (*A*) on the string.

Given the gear's pitch radius r_p and pressure angle α, we can calculated the coordinates of each point on the involute curve. For example, let's calculate the polar coordinates (r, θ) of an arbitrary point A [3] on the involute curve. Note that *BA* and *CP* are tangent lines of the base circle, and *F* is a foot of perpendicular.

Since *APF* is an involute curve and $\overset{\frown}{BCDEF}$ is the base circle, by the definition of involute curve,

$$\overline{BA} = \overset{\frown}{BCDEF} \qquad (1)$$

$$\overline{CP} = \overset{\frown}{CDEF} \qquad (2)$$

In $\triangle OCP$,

$$r_b = r_p \cos \alpha \qquad (3)$$

In $\triangle OBA$,

$$r = \frac{r_b}{\cos \phi} \qquad (4)$$

Or,

$$\phi = \cos^{-1} \frac{r_b}{r} \qquad (5)$$

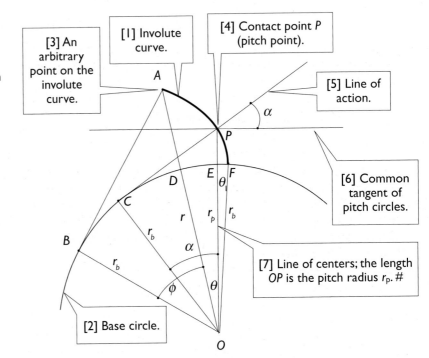

[3] An arbitrary point on the involute curve.

[1] Involute curve.

[4] Contact point *P* (pitch point).

[5] Line of action.

[6] Common tangent of pitch circles.

[7] Line of centers; the length *OP* is the pitch radius r_p. #

[2] Base circle.

To calculate θ, we notice that

$$\overset{\frown}{DE} = \overset{\frown}{BCDEF} - \overset{\frown}{BCD} - \overset{\frown}{EF}$$

Dividing the equation with r_b and using Eq. (1),

$$\frac{\overset{\frown}{DE}}{r_b} = \frac{\overline{BA}}{r_b} - \frac{\overset{\frown}{BCD}}{r_b} - \frac{\overset{\frown}{EF}}{r_b}$$

If radian is used, then the above equation can be written as

$$\theta = (\tan \phi) - \phi - \theta_1 \qquad (6)$$

The last term θ_1 is the angle $\angle EOF$, which can be calculated by dividing Eq. (2) with r_b,

$$\frac{\overline{CP}}{r_b} = \frac{\overset{\frown}{CDEF}}{r_b}, \text{ or } \tan \alpha = \alpha + \theta_1, \text{ or}$$

$$\theta_1 = (\tan \alpha) - \alpha \qquad (7)$$

We'll show how to calculate polar coordinates (r, θ) using Eqs. (3-7). The polar coordinates then can be easily transformed to rectangular coordinates, using *O* as origin and *OP* as y-axis,

$$x = -r \sin \theta, \quad y = r \cos \theta \qquad (8)$$

Numerical Calculations of Coordinates

In our case, the pitch radius r_p = 2.5 in, and pressure angle α = 20°; from Eqs. (3) and (7) respectively,

$$r_b = 2.5\cos 20° = 2.349232 \text{ in}$$

$$\theta_1 = \tan 20° - \frac{20°}{180°}\pi = 0.01490438 \text{ (rad)}$$

The table below lists the calculated coordinates. The values in the first column (r) are chosen such that, except the pitch point (r = 2.5 in), the intermediate points are at the quarter points between r_b (r = 2.349232 in) and r_a (r = 2.75 in). Also note that, when using Eqs. (6) and (7), radian is used as the unit of angles; in the table below, however, we translated the unit to degrees.

r in.	ϕ Eq. (5), degrees	θ Eq. (6), degrees	$x = -r\sin\theta$ in.	$y = r\cos\theta$ in.
2.349232	0.000000	-0.853958	-0.03501	2.3490
2.449424	16.444249	-0.387049	-0.01655	2.4494
2.500000	20.000000	0.000000	0.00000	2.5000
2.549616	22.867481	0.442933	0.01971	2.5495
2.649808	27.555054	1.487291	0.06878	2.6489
2.750000	31.321258	2.690287	0.12908	2.7470

2.5-3 Draw an Involute Curve

Launch Workbench. Create a **Geometry** system. Save the project as **Gear**. Start up DesignModeler. Select **Inch** as length unit. Make sure **Auto Constraints** are turned on (2.1-4[14], page 57). Start to draw sketch on the **XYPlane**.

Using **Construction Point** tool, draw **6 points** and specify dimensions as shown (the vertical **dimensions are** down to the X-axis). If the pitch point [1] is not **blue-colored**, impose a **Coincident** constraint between the **pitch point** and the Y-axis.

Connect these six points using **Spline** tool, leave **Flexible** option on, and finish the spline with **Open End.**

It is equally good that you draw the spline by using **Spline** tool directly without creating construction points first. Select **Open End with Fit Points** from the context menu at the end of **Spline** tool. After dimensioning each points, use **Spline Edit** tool to edit the spline and select **Re-fit Spline** [2] from the context menu to smooth out the spline.

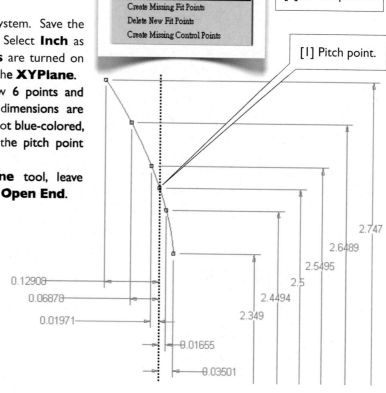

[2] Re-fit spline. #

[1] Pitch point.

2.5-4 Draw Circles

Draw three circles [1-3]. Let the addendum circle "snap" to the outermost construction point [3]. Specify radii for the shaft circle (1.25 in) and the dedendum circle (2.2 in).

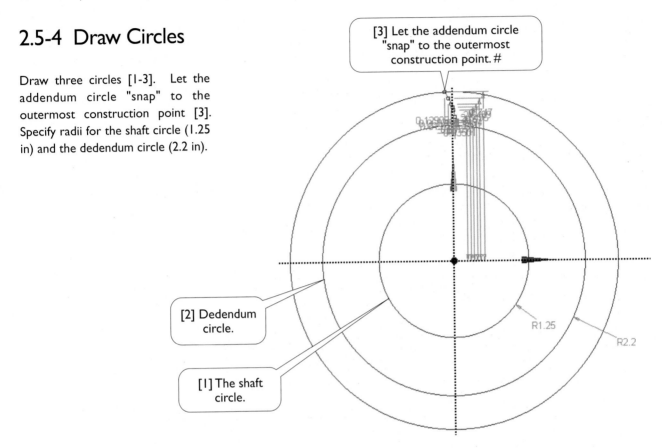

[3] Let the addendum circle "snap" to the outermost construction point. #

[2] Dedendum circle.

[1] The shaft circle.

R1.25

R2.2

2.5-5 Complete the Tooth Profile

Draw a line from the lowest construction point to the dedendum circle, and make it perpendicular to the dedendum circle [1-3]. When drawing the line, avoid a **V** auto-constraint, (since this line is NOT vertical). Draw a fillet [4] of radius 0.1 in to complete the profile of a tooth.

Sometimes, turning off **Display Plane** may be helpful to clear up the graphics window [5]. In this case, all the dimensions referring the plane axes disappear.

[5] Turn off **Display Plane** to clear up the graphics window. #

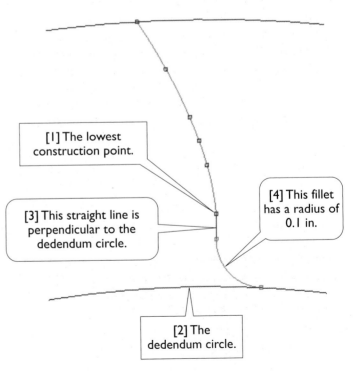

[1] The lowest construction point.

[3] This straight line is perpendicular to the dedendum circle.

[4] This fillet has a radius of 0.1 in.

[2] The dedendum circle.

2.5-6 Replicate the Profile

Activate **Replicate** tool, type 9 (degrees) for **r**. Select the profile (totally 3 segments), **End/Use Plane Origin as Handle**, **Flip Horizontal**, **Rotate by r degrees**, and **Paste at Plane Origin** [1]. End **Replicate** tool by pressing **ESC**.

Note that the gear has 20 teeth, each spaning 18 degrees. The angle between the two pitch points [2] on the left and the right profiles is 9 degrees.

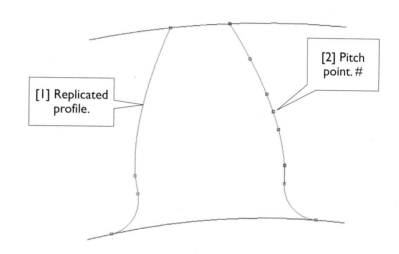

[1] Replicated profile.

[2] Pitch point. #

2.5-7 Replicate the Tooth 19 Times

Activate **Replicate** tool again, type 18 (degrees) for **r**. Select both left and right profiles (totally 6 segments), **End/Use Plane Origin as Handle**, **Rotate by r degrees**, and **Paste at Plane Origin**. Repeat the last two steps (rotate and paste) until completing a full circle (totally 20 teeth).

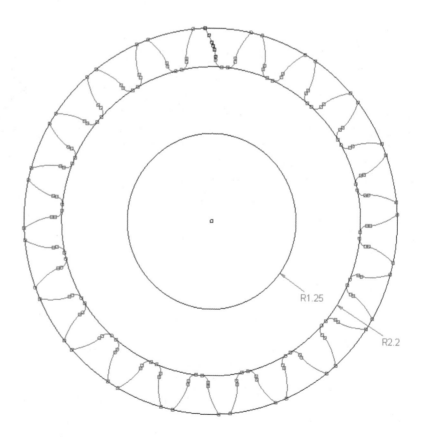

R1.25

R2.2

2.5-8 Trim Away Unwanted Segments

Trim away unwanted portion in the addendum circle and the dedendum circle.

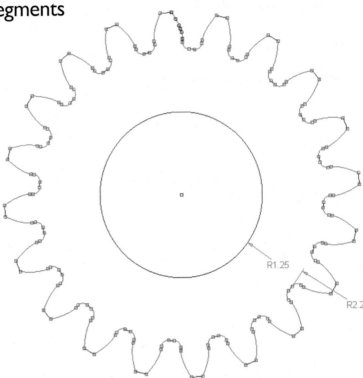

R1.25

R2.2

2.5-9 Extrude to Create 3D Solid

Extrude the sketch 1.0 inch to create a 3D solid. Save the project and exit from Workbench. We will resume this project again in Section 3.4.

It is equally good that you create a single tooth (a 3D solid body) and then duplicate it by using **Create/Pattern** in **Modeling** mode. In this exercise, however, we use **Replicate** in **Sketching** mode because our purpose in this chapter is to practice sketching techniques.

References

1. Deutschman, A. D., Michels, W. J., and Wilson, C. E., *Machine Design: Theory and Practice*, Macmillan Publishing Co., Inc., 1975; Chapter 10. Spur Gears.
2. Zahavi, E., *The Finite Element Method in Machine Design*, Prentice-Hall, 1992; Chapter 9. Spur Gears.

Section 2.6

Microgripper

2.6-1 About the Microgripper[Refs 1, 2]

The microgripper is made of PDMS (polydimethylsiloxane, see 1.1-1[5], page 8), actuated by a shape memory alloy (SMA) actuator [1-3], its motion caused by temperature change, the temperature in turn controlled by electric current. In the lab, the microgripper is tested by gripping a glass bead of a diameter of 30 micrometer [4].

   In this section, we will create a solid model for the microgripper. The model will be used for simulation in Section 13.3 to assess the gripping forces on the glass bead under the actuation of the SMA actuator.

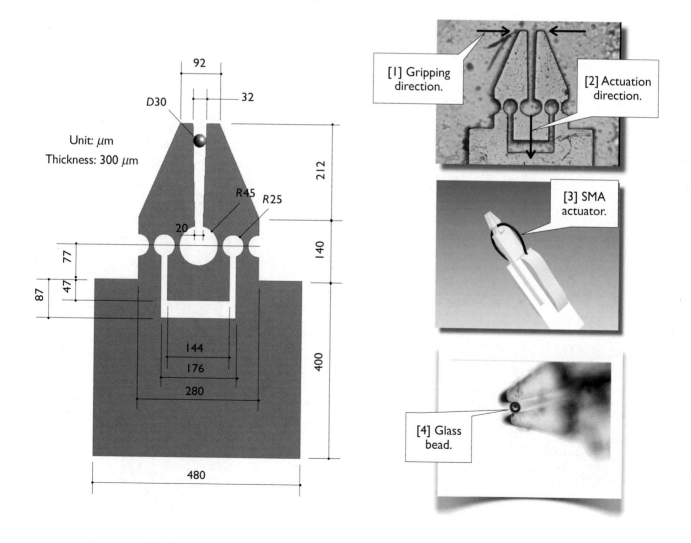

[1] Gripping direction.

[2] Actuation direction.

[3] SMA actuator.

[4] Glass bead.

Unit: μm
Thickness: 300 μm

2.6-2 Create Half of the Model

Launch Workbench. Create a **Geometry** system. Save the project as **Microgripper**. Start up DesignModeler. Select **Micrometer** as length unit. Make sure **Auto Constraints** are turned on.

Draw a sketch on **XYPlane** [1]. Trim away unwanted segments [2]. Note that we drew half of the model, due to the symmetry. Extrude the sketch 150 μm both sides symmetrically (total depth is 300 μm) [3]. We now have a half of the gripper [4].

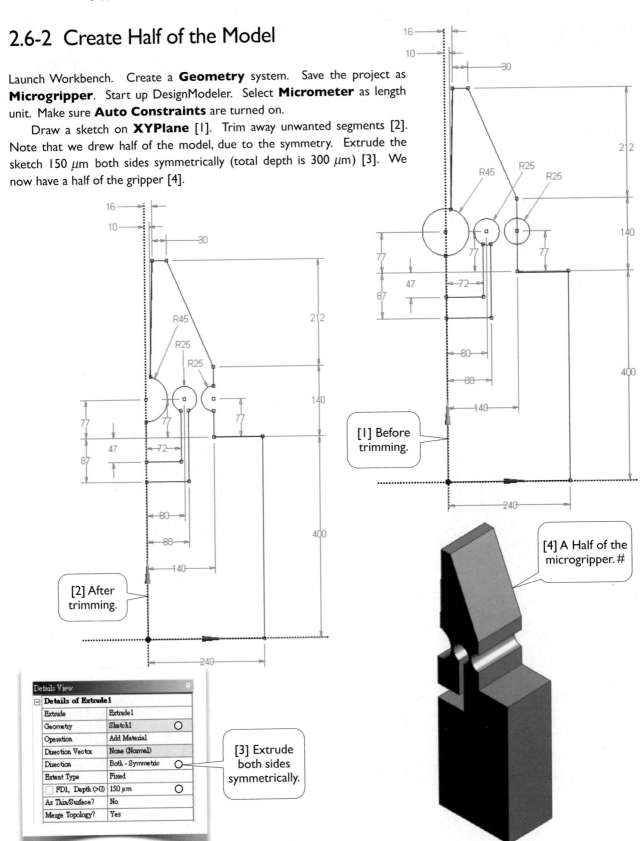

[1] Before trimming.

[2] After trimming.

[4] A Half of the microgripper. #

[3] Extrude both sides symmetrically.

Details of Extrude1	
Extrude	Extrude1
Geometry	Sketch1
Operation	Add Material
Direction Vector	None (Normal)
Direction	Both - Symmetric
Extent Type	Fixed
FD1, Depth (>0)	150 μm
As Thin/Surface?	No
Merge Topology?	Yes

2.6-3 Mirror Copy the Solid Body

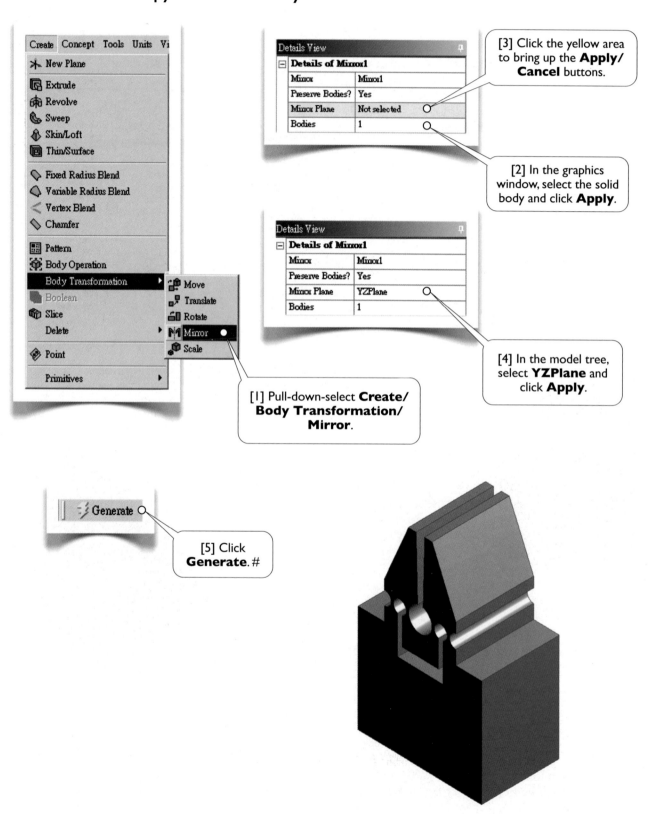

Create Concept Tools Units Vi

- New Plane
- Extrude
- Revolve
- Sweep
- Skin/Loft
- Thin/Surface

- Fixed Radius Blend
- Variable Radius Blend
- Vertex Blend
- Chamfer

- Pattern
- Body Operation
- Body Transformation ▶
- Boolean
- Slice
- Delete ▶
- Point
- Primitives ▶

  - Move
  - Translate
  - Rotate
  - Mirror ●
  - Scale

[3] Click the yellow area to bring up the **Apply/Cancel** buttons.

Details View

Details of Mirror1
Mirror	Mirror1
Preserve Bodies?	Yes
Mirror Plane	Not selected
Bodies	1

[2] In the graphics window, select the solid body and click **Apply**.

Details View

Details of Mirror1
Mirror	Mirror1
Preserve Bodies?	Yes
Mirror Plane	YZPlane
Bodies	1

[4] In the model tree, select **YZPlane** and click **Apply**.

[1] Pull-down-select **Create/Body Transformation/Mirror**.

Generate

[5] Click **Generate**. #

100

2.6-4 Create the Bead

Create a new sketch on **XYPlane** [1, 2] and draw a semicircle as shown [3-6]. Revolve the sketch 360° about the Y-axis to create the glass bead. Note that the two bodies are treated as two parts [7]. Rename the two bodies as **Gripper** and **Bead** respectively [8].

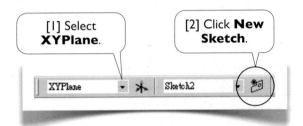

[1] Select **XYPlane**.

[2] Click **New Sketch**.

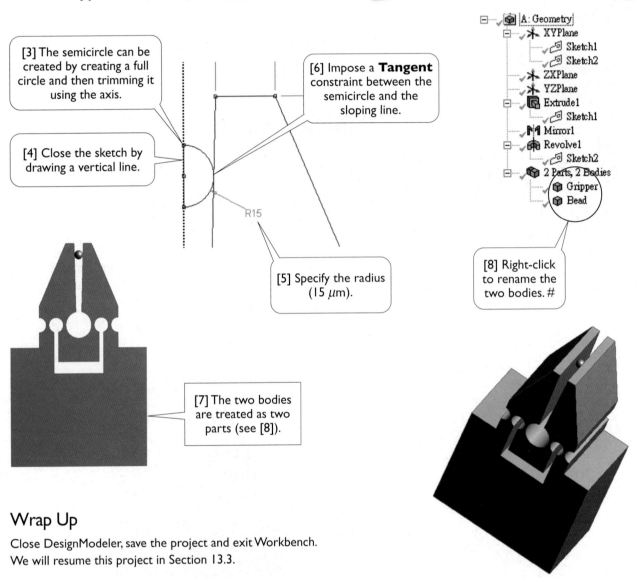

[3] The semicircle can be created by creating a full circle and then trimming it using the axis.

[6] Impose a **Tangent** constraint between the semicircle and the sloping line.

[4] Close the sketch by drawing a vertical line.

R15

[5] Specify the radius (15 μm).

[7] The two bodies are treated as two parts (see [8]).

[8] Right-click to rename the two bodies. #

Wrap Up

Close DesignModeler, save the project and exit Workbench. We will resume this project in Section 13.3.

References

1. Chang, R. J., Lin , Y. C., Shiu, C. C., and Hsieh, Y. T., "Development of SMA-Actuated Microgripper in Micro Assembly Applications," IECON, IEEE, Taiwan, 2007.
2. Shih, P. W., *Applications of SMA on Driving Micro-gripper*, MS Thesis, NCKU, ME, Taiwan, 2005.

Section 2.7

Review

2.7-1 Keywords

Choose a letter for each keyword from the list of descriptions

1. (   ) Auto Constraints
2. (   ) Branch
3. (   ) Constraint Status
4. (   ) Context Menu
5. (   ) Edge
6. (   ) Modeling Mode
7. (   ) Model Tree

8. (   ) Object
9. (   ) Paste Handle
10. (   ) Sketching Mode
11. (   ) Sketching Plane
12. (   ) Sketch
13. (   ) Selection Filter

Answers:

1. (J)   2. (G)   3. (M)   4. (I)   5. (D)   6. (B)   7. (F)   8. (H)
9. (L)   10.(A)   11.(C)   12.(E)   13.(K)

List of Descriptions

(A)   An environment under DesignModeler, its function to draw sketches on a plane.

(B)   An environment under DesignModeler, its function to create 3D or 2D bodies.

(C)   The plane on which a sketch is created. Each sketch must be associated with a plane; each plane may have multiple sketches on it. Usage of planes is not limited for storing sketches.

(D)   In **Sketching** mode, an edge may be a (straight) line or a curve. A curve may be a circle, ellipse, arc, or spline.

(E)   A sketch consists of points and edges. Dimensions and constraints may be imposed on these entities.

(F)   A model tree is the structured representation of a geometry and displayed on **Tree Outline** in DesignModeler. A model tree consists of features and a part branch, in which their order is important. The parts are the only objects exported to **Mechanical**.

(G)   A branch is an object of a model tree and consists one or more objects under itself.

(H)   A leaf or branch of a model tree is called an object.

(I)   The menu that pops up when you right-click your mouse. The contents of the menu depend on what you click.

(J)   While drawing in **Sketching** mode, by default, DesignModeler attempts to detect the user's intentions and try to automatically impose constraints on points or edges. Detection is performed over entities on the active plane, not just active sketch. **Auto Constraints** can be switched on/off in **Constraints** toolbox.

(K)   A selection filter filters one type of geometric entities. When a selection filter is turned on/off, the corresponding type of entities becomes selectable/unselectable. In **Sketching** mode, there are two selection filters, namely points and edges filters. Along with these two filters, face and body selection filters are available in **Modeling** mode.

(L)   A reference point used in a copy/paste operation. The point is defined during copying and will coincide with a specified location when pasting.

(M)   In **Sketching** mode, entities are color coded to indicate their constraint status: greenish-blue for under-constrained; blue and black for well constrained (i.e., fixed in the space); red for over-constrained; gray for inconsistent.

2.7-2 Additional Workbench Exercises

Create Models with Your Own Way

After so many exercises, you should be able to figure out many alternative ways of creating the geometric models in this chapter. Try to re-create the models in this chapter using your own way.

Chapter 3
2D Simulations

In the real-world, everything is 3D; there are no such things as 2D bodies. Some problems, however, can be simplified and simulated in a 2D space. As an example, consider an axisymmetric body subject to axisymmetric loads, in which all the particles with the same radial and axial coordinates (R and Y) share the same behaviors regardless of their tangential coordinate (θ). Thus, we can eliminate the tangential coordinate and reduce the problem to a 2D (in R-Y space) problem. Other 2D cases include *plane stress* problems and *plane strain* problems, which will be defined in Section 3.3.

Reducing a problem to 2D has many advantages over 3D approach, and you should always do it whenever possible. These advantages include (a) simpler to build the geometry, (b) better mesh quality, (c) less computing time, (d) easier display and analysis of the results. In short, the simulation model becomes smaller and easier to handle. Besides, if the problem's nature is indeed 2D, it would not introduce inaccuracy for the solution.

Purpose of This Chapter

Since 2D simulations are usually easier to handle than 3D simulations, we start the learning of simulations by conducting 2D static structural simulations in this chapter, using some of the mechanical parts that we created in Chapter 2. This chapter also serves as a preliminary to 3D simulations, since most of techniques and concepts in this chapter can be extended to 3D simulations.

About Each Section

Sections 3.1 and 3.2 guide the students to perform a 2D simulation in a step-by-step fashion. Section 3.3 looks into more details and tries to provide what we are not able to cover in the first two sections. Section 3.4 provides an additional exercise. Problems in Sections 3.2 and 3.4 involve contact nonlinearities. In-depth discussion of contact nonlinearities will be postponed until Chapter 13. We introduce nonlinearities so early in this chapter is to build some curiosity for learning nonlinear simulations in Chapters 13 and 14. Using a filleted bar subject to tension, Section 3.5 introduces some must-know concepts in finite element simulations, namely stress discontinuity, structural error, finite elements convergence, stress concentration, and stress singularity.

Section 3.1
Triangular Plate

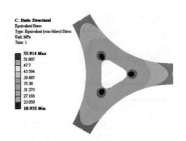

3.1-1 About the Triangular Plate

In this section, we will perform a 2D static structural simulation using the triangular plate created in Section 2.2. The plate is made of steel and designed to withstand a tensile force of 20,000 N on each of its three side faces. The size of the side faces is 40x10 mm; therefore the applied tensile stress on the side faces is 50 MPa. The objective is to investigate the stresses in the plate.

We will model the problem as a 2D plane stress problem. Definition of plane stress will be given in 3.3-1, page 132. For now, what you need to know is that a thin plate subject to in-plane forces can be modeled as a plane stress problem.

There are two planes of symmetry in the model. In the first part of this section, we will analyze the full model without using the symmetries and then, in the second part of this section, reanalyze the model by using the symmetries.

3.1-2 Resume the Project Triplate

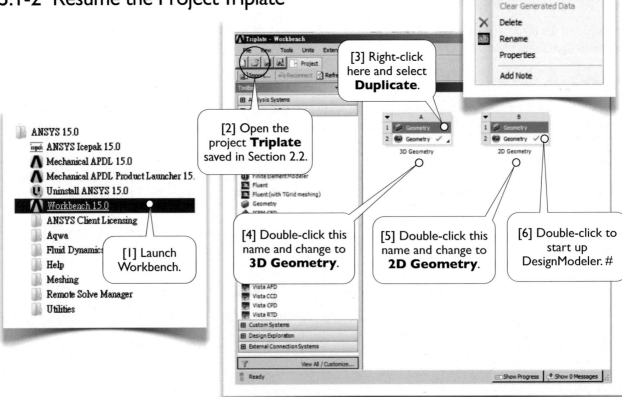

3.1-3 Delete the 3D Body and Create a 2D Body

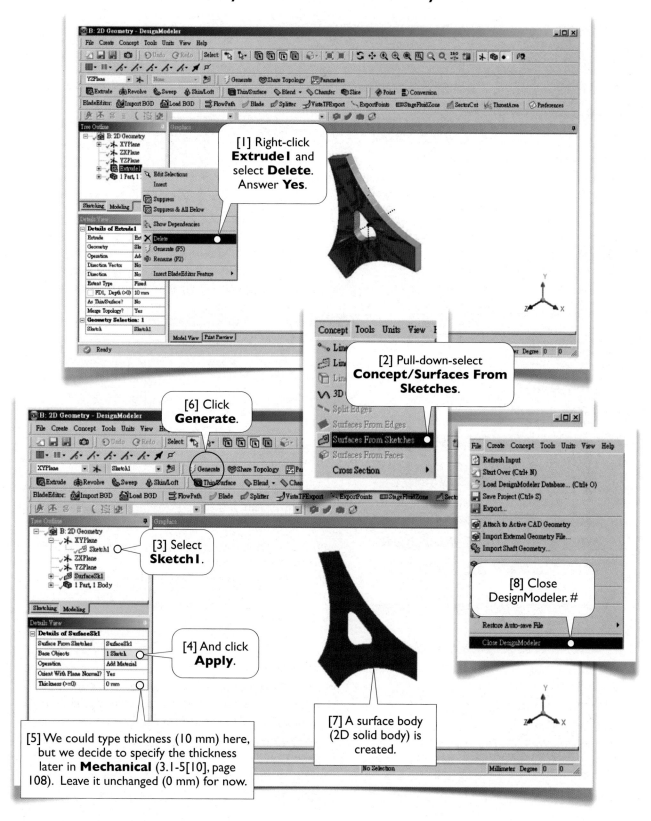

[1] Right-click **Extrude1** and select **Delete**. Answer **Yes**.

[2] Pull-down-select **Concept/Surfaces From Sketches**.

[6] Click **Generate**.

[3] Select **Sketch1**.

[4] And click **Apply**.

[5] We could type thickness (10 mm) here, but we decide to specify the thickness later in **Mechanical** (3.1-5[10], page 108). Leave it unchanged (0 mm) for now.

[7] A surface body (2D solid body) is created.

[8] Close DesignModeler. #

3.1-4 Create Analysis System and Specify Analysis Type

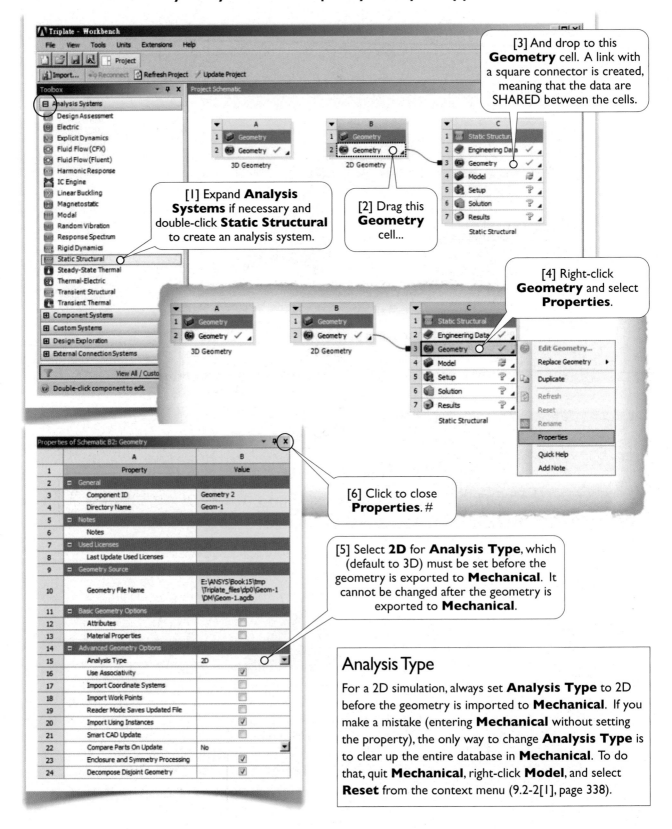

[3] And drop to this **Geometry** cell. A link with a square connector is created, meaning that the data are SHARED between the cells.

[1] Expand **Analysis Systems** if necessary and double-click **Static Structural** to create an analysis system.

[2] Drag this **Geometry** cell...

[4] Right-click **Geometry** and select **Properties**.

[6] Click to close **Properties**. #

[5] Select **2D** for **Analysis Type**, which (default to 3D) must be set before the geometry is exported to **Mechanical**. It cannot be changed after the geometry is exported to **Mechanical**.

Analysis Type

For a 2D simulation, always set **Analysis Type** to 2D before the geometry is imported to **Mechanical**. If you make a mistake (entering **Mechanical** without setting the property), the only way to change **Analysis Type** is to clear up the entire database in **Mechanical**. To do that, quit **Mechanical**, right-click **Model**, and select **Reset** from the context menu (9.2-2[1], page 338).

3.1-5 Start Up Mechanical and Set Up Geometry

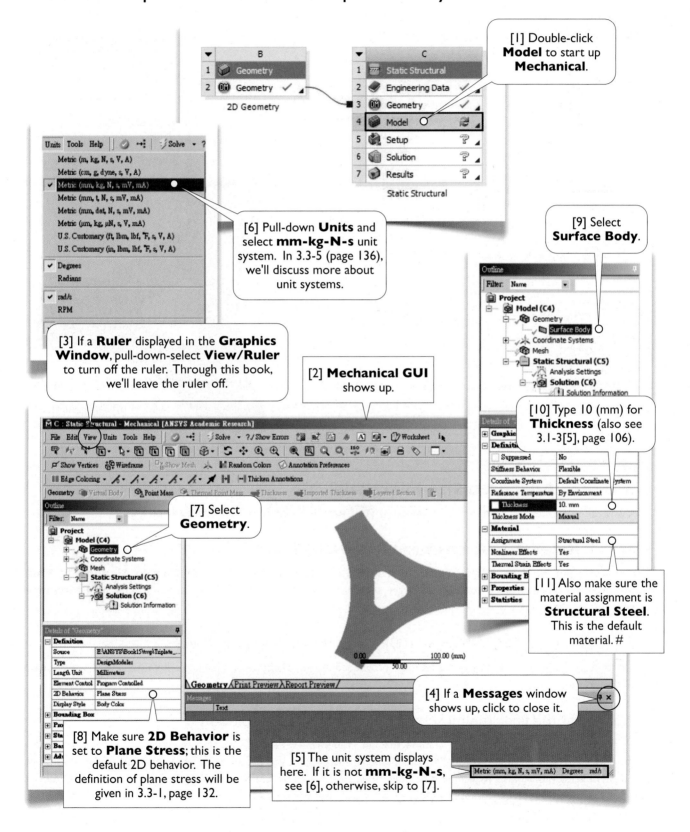

[1] Double-click **Model** to start up **Mechanical**.

2D Geometry

Static Structural

[6] Pull-down **Units** and select **mm-kg-N-s** unit system. In 3.3-5 (page 136), we'll discuss more about unit systems.

[9] Select **Surface Body**.

[3] If a **Ruler** displayed in the **Graphics Window**, pull-down-select **View/Ruler** to turn off the ruler. Through this book, we'll leave the ruler off.

[2] **Mechanical GUI** shows up.

[10] Type 10 (mm) for **Thickness** (also see 3.1-3[5], page 106).

[7] Select **Geometry**.

[11] Also make sure the material assignment is **Structural Steel**. This is the default material. #

[4] If a **Messages** window shows up, click to close it.

[8] Make sure **2D Behavior** is set to **Plane Stress**; this is the default 2D behavior. The definition of plane stress will be given in 3.3-1, page 132.

[5] The unit system displays here. If it is not **mm-kg-N-s**, see [6], otherwise, skip to [7].

3.1-6 Generate Mesh

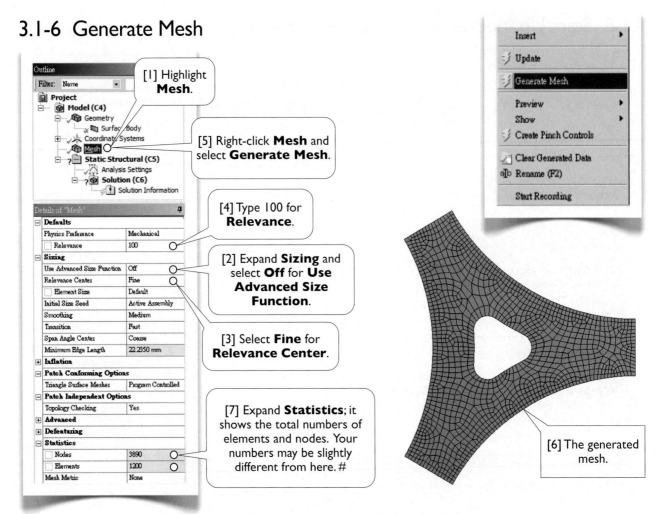

[1] Highlight **Mesh**.

[5] Right-click **Mesh** and select **Generate Mesh**.

[4] Type 100 for **Relevance**.

[2] Expand **Sizing** and select **Off** for **Use Advanced Size Function**.

[3] Select **Fine** for **Relevance Center**.

[7] Expand **Statistics**; it shows the total numbers of elements and nodes. Your numbers may be slightly different from here. #

[6] The generated mesh.

Why turn off **Advanced Size Functions**?

That way, the mesh fineness is entirely determined by **Relevance Center** [3] and **Relevance** [4].

What are **Relevance Center** and **Relevance**?

They together provide a way of global mesh control. **Relevance Center** [3] can be **Coarse**, **Medium**, or **Fine**. **Relevance** [4] ranges from 0 to 100, the larger the finer. In 5.3-1 (page 219), we'll discuss more on **Relevance Center** and **Relevance**.

Is the mesh generated with (Fine, 100) the finest mesh we can have?

No. With many additional mesh control methods, introduced in later chapters and particularly in Chapter 9, we can have a much finer mesh.

Is it true that the finer the mesh, the more accurate the solution?

It is generally true. In this book, we assume that an academic TEACHING version of ANSYS is used, which has a limitation on the number of nodes (30,000) and elements (30,000). If you are using an academic RESEARCH version or an industrial version of ANSYS (which have no limitations on the number of nodes and elements), the only considerations are your computer resources (computing time, memory, disk capacity, etc.).

Is generating mesh always necessary?

Generating a mesh before clicking **Solve** (3.1-8[1], next page) is not really necessary. When you issue **Solve**, Workbench automatically generates a mesh if it does not exist one. However, it is a good practice to set up and preview the mesh before clicking **Solve**. We often mesh the model with default settings first, and then adjust mesh controls to improve the mesh, as discussed in Chapter 9.

3.1-7 Specify Loads

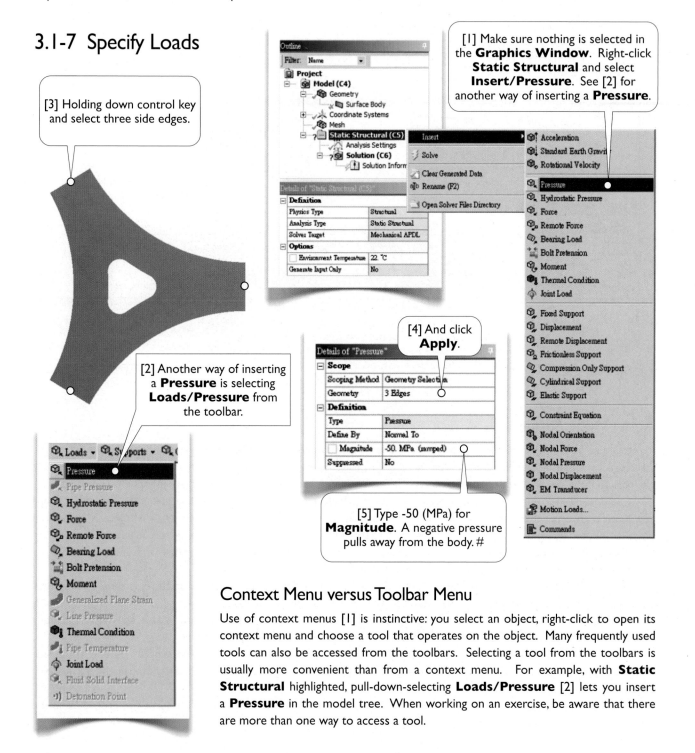

[3] Holding down control key and select three side edges.

[1] Make sure nothing is selected in the **Graphics Window**. Right-click **Static Structural** and select **Insert/Pressure**. See [2] for another way of inserting a **Pressure**.

[2] Another way of inserting a **Pressure** is selecting **Loads/Pressure** from the toolbar.

[4] And click **Apply**.

[5] Type -50 (MPa) for **Magnitude**. A negative pressure pulls away from the body. #

Context Menu versus Toolbar Menu

Use of context menus [1] is instinctive: you select an object, right-click to open its context menu and choose a tool that operates on the object. Many frequently used tools can also be accessed from the toolbars. Selecting a tool from the toolbars is usually more convenient than from a context menu. For example, with **Static Structural** highlighted, pull-down-selecting **Loads/Pressure** [2] lets you insert a **Pressure** in the model tree. When working on an exercise, be aware that there are more than one way to access a tool.

3.1-8 Solve the Model

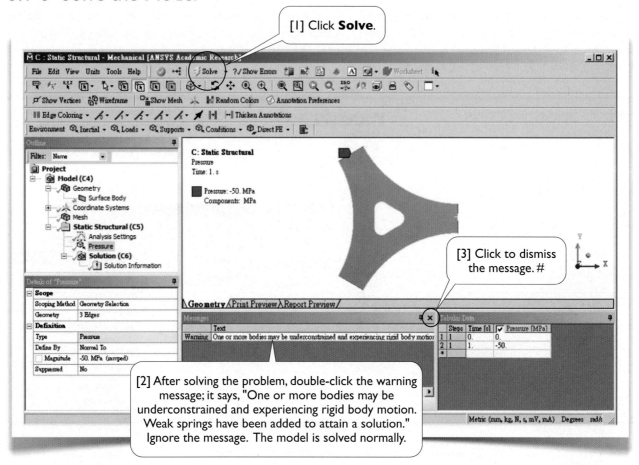

[1] Click **Solve**.

[3] Click to dismiss the message. #

[2] After solving the problem, double-click the warning message; it says, "One or more bodies may be underconstrained and experiencing rigid body motion. Weak springs have been added to attain a solution." Ignore the message. The model is solved normally.

About Weak Springs

If a structure does not have enough supports, the structure is unstable; i.e., any non-zero external forces would cause the structure to move indefinitely; the motion is called a *rigid body motion*. An unstable structure still can achieve static equilibrium if the resultant external force is zero; it is called an unstable equilibrium. This is what happens in our case.

Traditionally, in a static structural analysis, whenever a finite element program detects a structure unstable, it stops and reports an error message. This has been practiced for decades, even if the resultant external force is zero. In a digital computer, the external forces rarely sum up to zero. There is usually a small residual forces left when summing up; this is one of the nature of numerical computations. In other words, an unstable structure will undergo rigid body motion even if the resultant external force is theoretically zero.

In an Workbench's static structural analysis, if the Workbench detects an unstable structure, it simply add **weak springs** to the structure to make the structure capable of withstanding small residual external forces. It is called **weak spring** because the spring constants are very small and negligible. Note that, even with **weak springs**, rigid body motions are small but still inevitable in an unstable structure even the external force is zero.

It is a good practice to provide enough supports. In our case, you might set up a fixed support on one of the side faces and apply pressure of 50 MPa on the other two faces. That way, all deformations reported are relative to the fixed edge.

A better way to model this case is to use the symmetries of the structure. We will present the procedure in the second part of this section, starting from 3.1-11, page 113.

3.1-9 Insert Result Objects

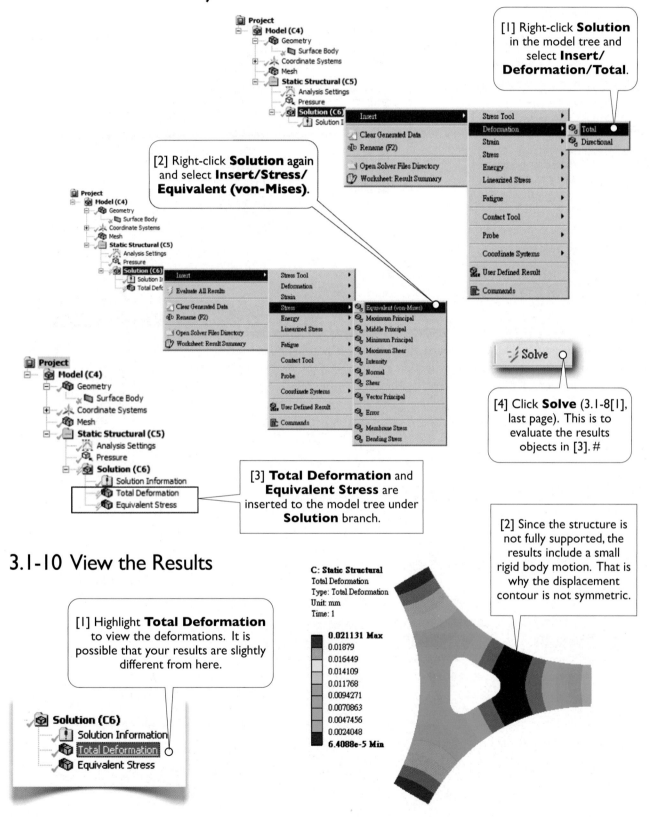

[1] Right-click **Solution** in the model tree and select **Insert/ Deformation/Total**.

[2] Right-click **Solution** again and select **Insert/Stress/ Equivalent (von-Mises)**.

[3] **Total Deformation** and **Equivalent Stress** are inserted to the model tree under **Solution** branch.

[4] Click **Solve** (3.1-8[1], last page). This is to evaluate the results objects in [3]. #

3.1-10 View the Results

[1] Highlight **Total Deformation** to view the deformations. It is possible that your results are slightly different from here.

[2] Since the structure is not fully supported, the results include a small rigid body motion. That is why the displacement contour is not symmetric.

C: Static Structural
Total Deformation
Type: Total Deformation
Unit: mm
Time: 1

0.021131 Max
0.01879
0.016449
0.014109
0.011768
0.0094271
0.0070863
0.0047456
0.0024048
6.4088e-5 Min

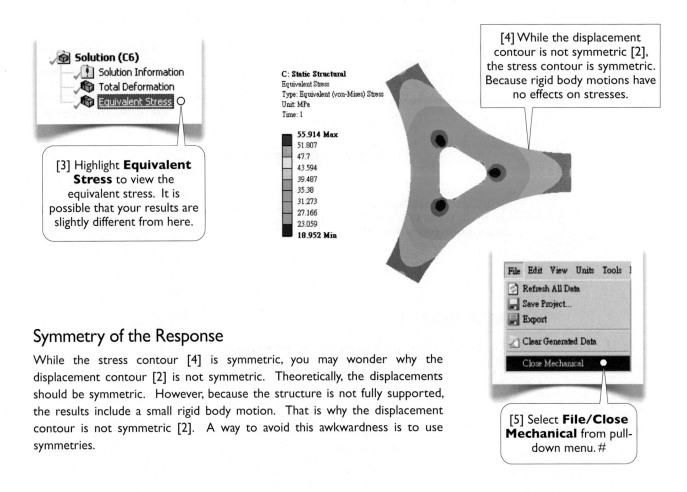

[3] Highlight **Equivalent Stress** to view the equivalent stress. It is possible that your results are slightly different from here.

[4] While the displacement contour is not symmetric [2], the stress contour is symmetric. Because rigid body motions have no effects on stresses.

C: Static Structural
Equivalent Stress
Type: Equivalent (von-Mises) Stress
Unit: MPa
Time: 1

55.914 Max
51.807
47.7
43.594
39.487
35.38
31.273
27.166
23.059
18.952 Min

[5] Select **File/Close Mechanical** from pull-down menu. #

Symmetry of the Response

While the stress contour [4] is symmetric, you may wonder why the displacement contour [2] is not symmetric. Theoretically, the displacements should be symmetric. However, because the structure is not fully supported, the results include a small rigid body motion. That is why the displacement contour is not symmetric [2]. A way to avoid this awkwardness is to use symmetries.

SIMULATION OF SYMMETRIC MODEL

3.1-11 Modify the Geometry

[1] Right-click here and select **Duplicate**.

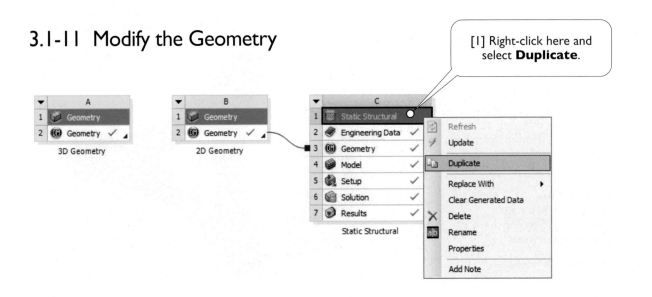

3D Geometry

2D Geometry

Static Structural

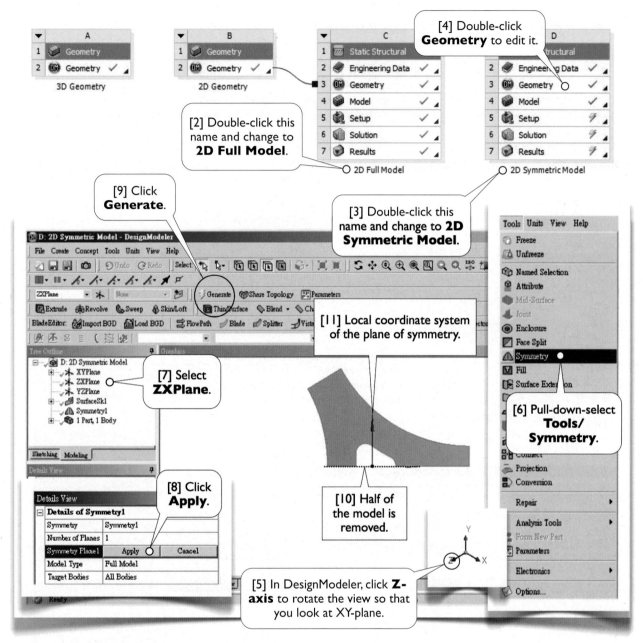

About Coordinate Systems

There is a unique global coordinate system; its directions are shown in the bottom-right corner of the graphics window. Workbench uses R, G, B colors to represent X-, Y-, and Z-axis respectively: red arrow for X-axis, green arrow for Y-axis, and blue arrow for Z-axis. In this book, we use upper-case (X, Y, Z) for both global and local coordinate systems, to be consistent with the notations used in Workbench.

Each plane has its own local coordinate system, using the same color codes. Take **ZXPlane** as an example [11]. Its local XY-plane coincides with global ZX-plane, and the local Z-axis points upward. When we specify a plane as the plane of symmetry, the plane's local XY-plane is used to cut away the portion of the model on the local negative-Z side. The portion of the model on the local positive-Z side remains [11].

The triangular plate has another plane of symmetry. None of the default planes can be used as the plane of symmetry; we need to create one. This plane can be derived from rotating **ZXPlane** by 60 degrees.

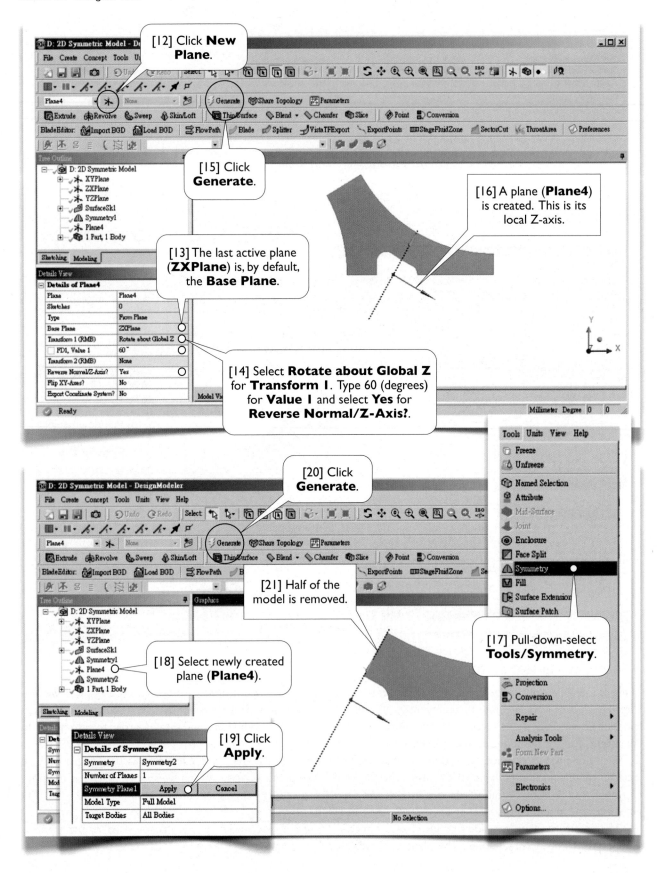

[12] Click **New Plane**.

[15] Click **Generate**.

[16] A plane (**Plane4**) is created. This is its local Z-axis.

[13] The last active plane (**ZXPlane**) is, by default, the **Base Plane**.

[14] Select **Rotate about Global Z** for **Transform 1**. Type 60 (degrees) for **Value 1** and select **Yes** for **Reverse Normal/Z-Axis?**.

[20] Click **Generate**.

[21] Half of the model is removed.

[17] Pull-down-select **Tools/Symmetry**.

[18] Select newly created plane (**Plane4**).

[19] Click **Apply**.

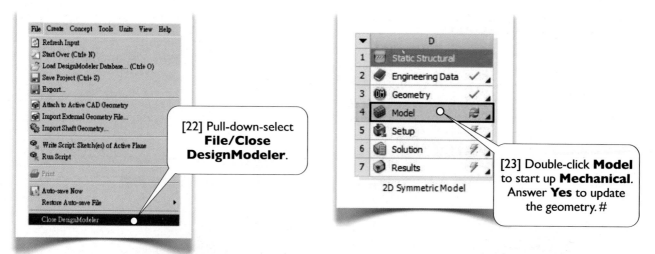

[22] Pull-down-select **File/Close DesignModeler**.

2D Symmetric Model

[23] Double-click **Model** to start up **Mechanical**. Answer **Yes** to update the geometry. #

Do I need to close DesignModeler when working on **Mechanical**, or vice versa?

When starting up, **Mechanical** automatically updates the geometry, including adding symmetry conditions on the model. That's the reason I suggest the newcomers close one application while working on another. That way, the update will be automatic; you won't mess up the work flow.

However, you don't have to close any application when switching to another application, but you have to "refresh" it by yourself. In **Project Schematic** you can right-click **Model** (of an analysis system) and select **Refresh**. Inside **Mechanical** you can right-click **Geometry** (of a model tree) and select **Refresh Geometry**.

Until you become an experienced user, I suggest that you always keep only one application open at a time to make the life simpler.

3.1-12 Solve the New Model

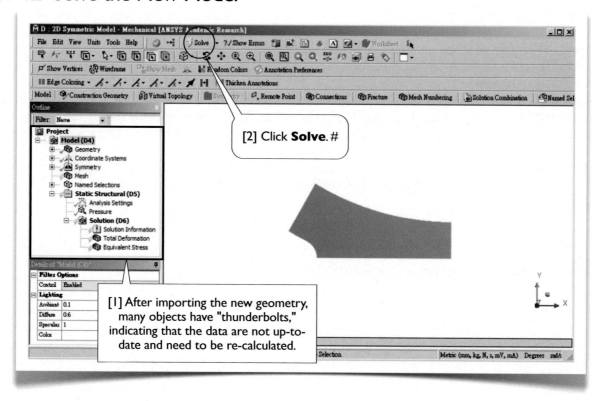

[2] Click **Solve**. #

[1] After importing the new geometry, many objects have "thunderbolts," indicating that the data are not up-to-date and need to be re-calculated.

3.1-13 View the Results

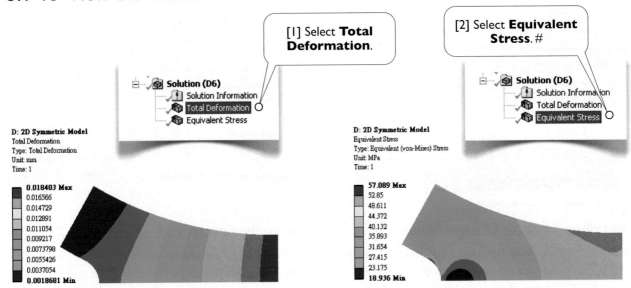

[1] Select **Total Deformation**.

[2] Select **Equivalent Stress**. #

Check Environment Conditions After Modifying Geometry

In this case, you don't need to do anything before solving the new model. In most of other cases, however, whenever you modify your geometry, you may need to redefine the environment conditions. As a good practice, always check your environment conditions each time you modify the geometry before solving it.

Why Different Numerical Results?

Your numerical results may be slightly different from here. This is one of the nature of the finite element methods: different meshes end up with different results. In general, the finer the mesh, the more accurate the results. The question is: how fine should a mesh be, to achieve enough accuracy? We will discuss this must-know concept in Section 3.5. The discussion will be extended to 3D cases in Section 9.3.

Some students may be puzzled about why they obtained a mesh different from the one in the book even they followed EXACT steps in the book. The answer is that the students have no way to follow EXACT steps in the book. For example, a line in the book may be drawn from right to left while you drew it from left to right. It is possible that the direction of the line affects the meshing algorithm in the Workbench.

Limited differences in numerical values are normal, particularly when the mesh are coarse. As the mesh becomes finer, the solution will converge to a theoretical value, independent of mesh variations. This kind of puzzle would disappear.

Wrap Up

Close **Mechanical**. Save the project and exit from Workbench.

Section 3.2

Threaded Bolt-and-Nut

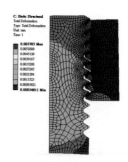

3.2-1 About the Threaded Bolt-and-Nut

The threaded bolt created in Section 2.4 is a part of a bolt-nut-plate assembly [1-4]. The bolt is preloaded with a tension of 10 kN, which is applied by tightening the nut with torque. We want to know the stress at the threads under such a pretension condition.

Pretension is a built-in environment condition in Workbench 3D simulations, in which a pretension can apply on a body or cylindrical surface. It is, however, not applicable for 2D simulations.

In this section, we will make some simplification for a 2D simulation. Assuming a mirror-symmetry between the upper half and lower lower half (which is not exactly true), we model only upper part of the assembly [5, 6, 7]. The plate is not included, to reduce the problem size further and alleviate contact nonlinearity, its contacting surface with the nut replaced by a frictionless support [8].

The pretension is modeled using a uniform force applied on the lower face of the bolt. The results will somewhat deviate from the reality, to be discussed at the end of this section, but the deviation has little effects on the stresses of the threads.

The coefficient of friction between the bolt and the nut is 0.3.

[1] Bolt.

[2] Nut.

[3] Plates.

[4] Section view.

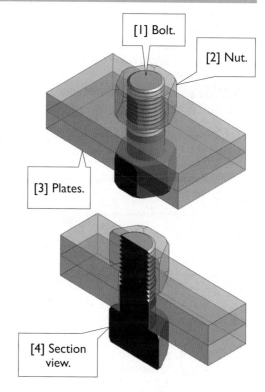

[5] The simulation model.

[6] Axis of symmetry.

[8] Frictionless support. #

[7] Plane of symmetry.

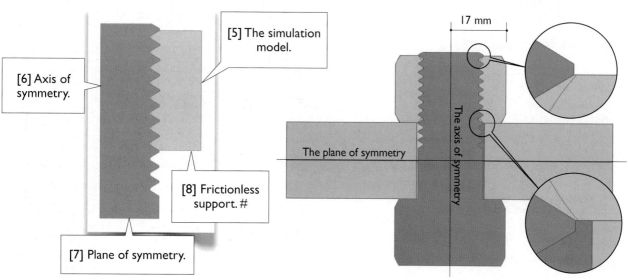

17 mm

The plane of symmetry

The axis of symmetry

3.2-2 Open the Project **Threads**

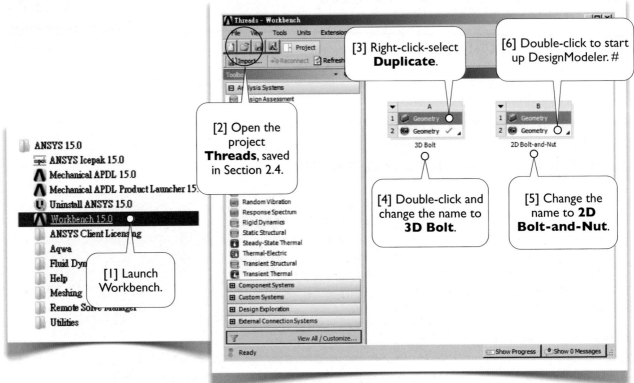

[3] Right-click-select **Duplicate**.

[6] Double-click to start up DesignModeler. #

[2] Open the project **Threads**, saved in Section 2.4.

[4] Double-click and change the name to **3D Bolt**.

[5] Change the name to **2D Bolt-and-Nut**.

[1] Launch Workbench.

3.2-3 Delete the 3D Body and Create a 2D Body

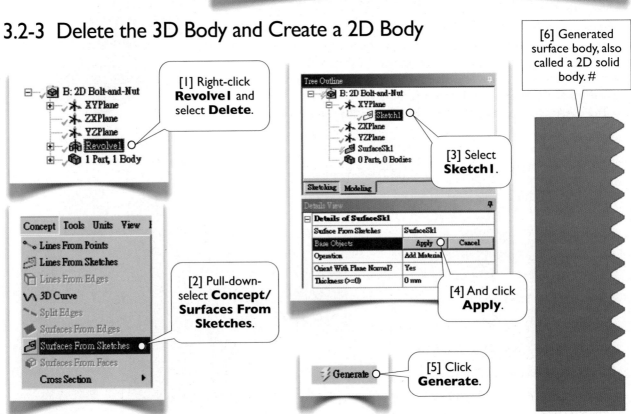

[1] Right-click **Revolve1** and select **Delete**.

[2] Pull-down-select **Concept/ Surfaces From Sketches**.

[3] Select **Sketch1**.

[4] And click **Apply**.

[5] Click **Generate**.

[6] Generated surface body, also called a 2D solid body. #

3.2-4 Create a 2D Body for the Nut

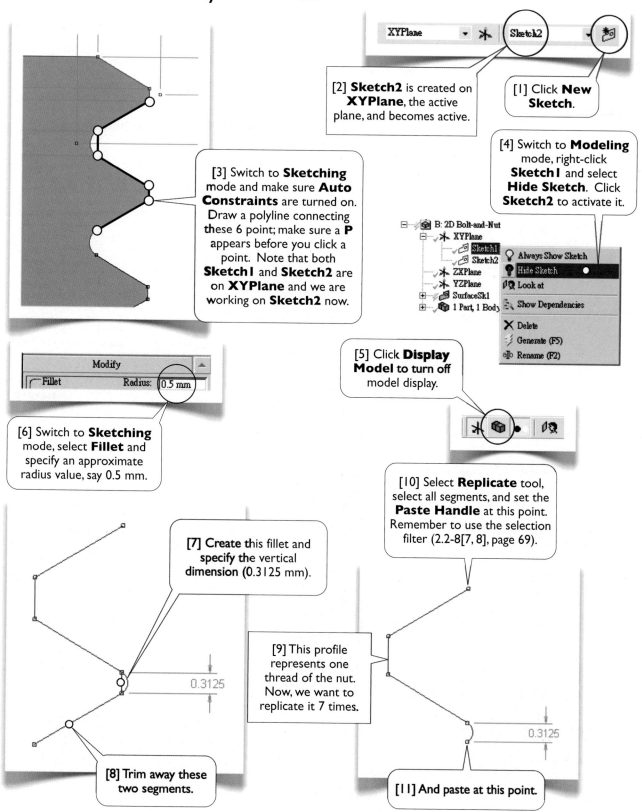

[2] **Sketch2** is created on **XYPlane**, the active plane, and becomes active.

[1] Click **New Sketch**.

[4] Switch to **Modeling** mode, right-click **Sketch1** and select **Hide Sketch**. Click **Sketch2** to activate it.

[3] Switch to **Sketching** mode and make sure **Auto Constraints** are turned on. Draw a polyline connecting these 6 point; make sure a **P** appears before you click a point. Note that both **Sketch1** and **Sketch2** are on **XYPlane** and we are working on **Sketch2** now.

XYPlane

Sketch2

Modify

Fillet   Radius: 0.5 mm

[6] Switch to **Sketching** mode, select **Fillet** and specify an approximate radius value, say 0.5 mm.

B: 2D Bolt-and-Nut
XYPlane
Sketch1
Sketch2
ZXPlane
YZPlane
SurfaceSk1
1 Part, 1 Body

Always Show Sketch
Hide Sketch
Look at
Show Dependencies
Delete
Generate (F5)
Rename (F2)

[5] Click **Display Model** to turn off model display.

[10] Select **Replicate** tool, select all segments, and set the **Paste Handle** at this point. Remember to use the selection filter (2.2-8[7, 8], page 69).

[7] **Create** this fillet and **specify** the vertical dimension (0.3125 mm).

[9] This profile represents one thread of the nut. Now, we want to replicate it 7 times.

0.3125

0.3125

[8] Trim away these two segments.

[11] And paste at this point.

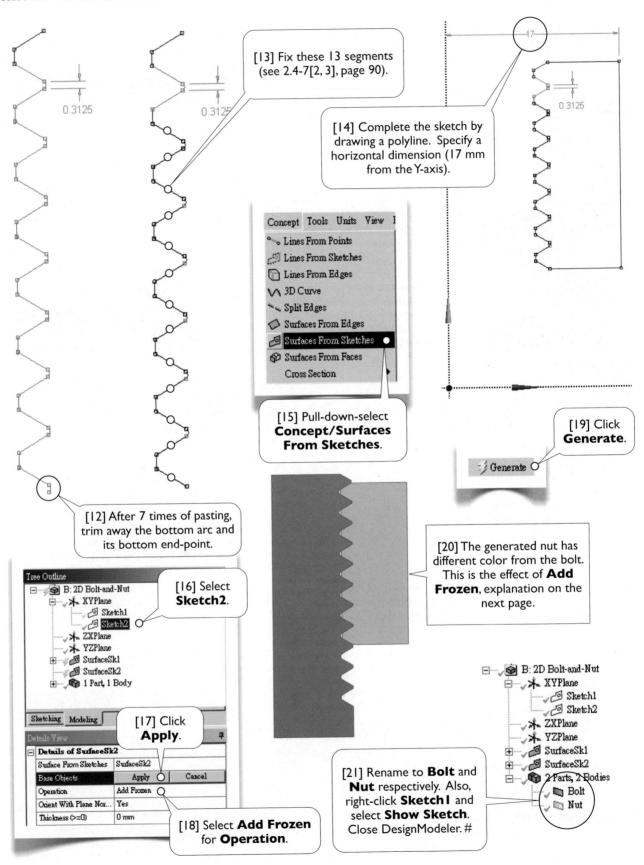

[13] Fix these 13 segments (see 2.4-7[2, 3], page 90).

[14] Complete the sketch by drawing a polyline. Specify a horizontal dimension (17 mm from the Y-axis).

0.3125

0.3125

0.3125

17

Concept   Tools   Units   View
Lines From Points
Lines From Sketches
Lines From Edges
3D Curve
Split Edges
Surfaces From Edges
Surfaces From Sketches
Surfaces From Faces
Cross Section

[15] Pull-down-select **Concept/Surfaces From Sketches**.

[19] Click **Generate**.

Generate

[12] After 7 times of pasting, trim away the bottom arc and its bottom end-point.

[20] The generated nut has different color from the bolt. This is the effect of **Add Frozen**, explanation on the next page.

Tree Outline

B: 2D Bolt-and-Nut
  XYPlane
    Sketch1
    Sketch2
  ZXPlane
  YZPlane
  SurfaceSk1
  SurfaceSk2
  1 Part, 1 Body

[16] Select **Sketch2**.

Sketching   Modeling

Details View

Details of SurfaceSk2

Surface From Sketches	SurfaceSk2	
Base Objects	Apply	Cancel
Operation	Add Frozen	
Orient With Plane Nor...	Yes	
Thickness (>=0)	0 mm	

[17] Click **Apply**.

[18] Select **Add Frozen** for **Operation**.

B: 2D Bolt-and-Nut
  XYPlane
    Sketch1
    Sketch2
  ZXPlane
  YZPlane
  SurfaceSk1
  SurfaceSk2
  2 Parts, 2 Bodies
    Bolt
    Nut

[21] Rename to **Bolt** and **Nut** respectively. Also, right-click **Sketch1** and select **Show Sketch**. Close DesignModeler. #

Add Material versus Add Frozen

With **Add Material** operation mode, the created material adds to the existing body and become an integral part. If you choose **Add Frozen** [18], the created material does not add to the existing one; it becomes another part. This is what we intend: the bolt and nut are separate parts; they are not bonded each other. In Workbench, the most important concept about a part is that each part will be meshed independently.

3.2-5 Create an Analysis System

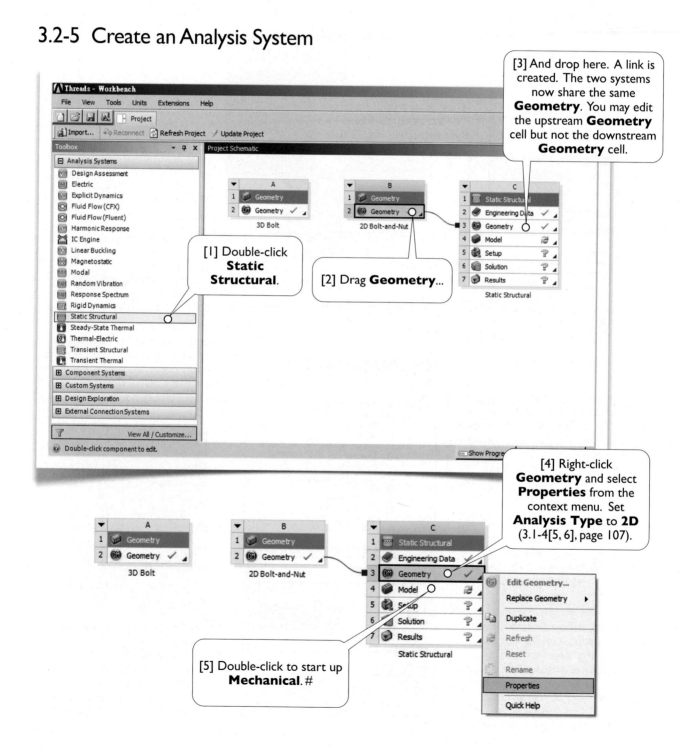

[3] And drop here. A link is created. The two systems now share the same **Geometry**. You may edit the upstream **Geometry** cell but not the downstream **Geometry** cell.

[1] Double-click **Static Structural**.

[2] Drag **Geometry**...

[4] Right-click **Geometry** and select **Properties** from the context menu. Set **Analysis Type** to **2D** (3.1-4[5, 6], page 107).

[5] Double-click to start up **Mechanical**. #

3.2-6 Set Up Geometry in Mechanical

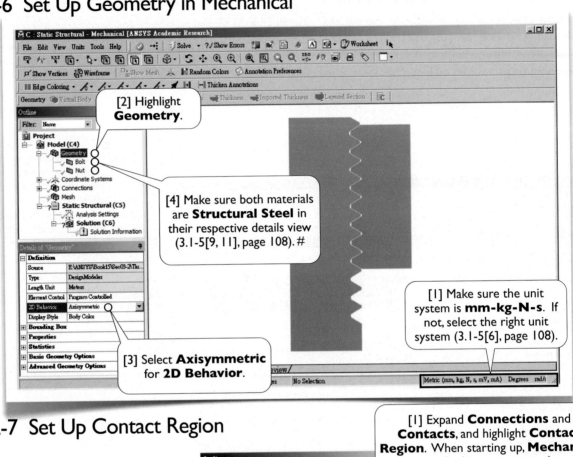

[2] Highlight **Geometry**.

[4] Make sure both materials are **Structural Steel** in their respective details view (3.1-5[9, 11], page 108). #

[1] Make sure the unit system is **mm-kg-N-s**. If not, select the right unit system (3.1-5[6], page 108).

[3] Select **Axisymmetric** for **2D Behavior**.

3.2-7 Set Up Contact Region

[1] Expand **Connections** and **Contacts**, and highlight **Contact Region**. When starting up, **Mechanical** automatically detects contacts between parts and creates contact regions. In this case, a contact region is created.

[2] **Details of Contact Region** shows that 23 edges of the bolt are designated as **Contact** and 24 edges of the nut are **Target**. We'll remove some edges that won't contact, facilitating computation [5].

[4] We also want to change **Type** from **Bonded** to **Frictional** (13.1-8, page 468).

[3] We will change **Behavior** to **Symmetric**, meaning that the contact and target are treated symmetrically, so that we don't need to decide which one is contact and which one is target. We will discuss this in 13.1-9, page 469.

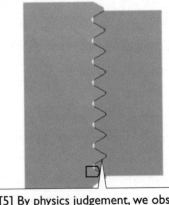

[5] By physics judgement, we observe that only lower faces of the bolt threads will contact with the upper faces of the nut threads. To reduce computing time, we usually keep contact areas minimal; now, let's modify the contact region.

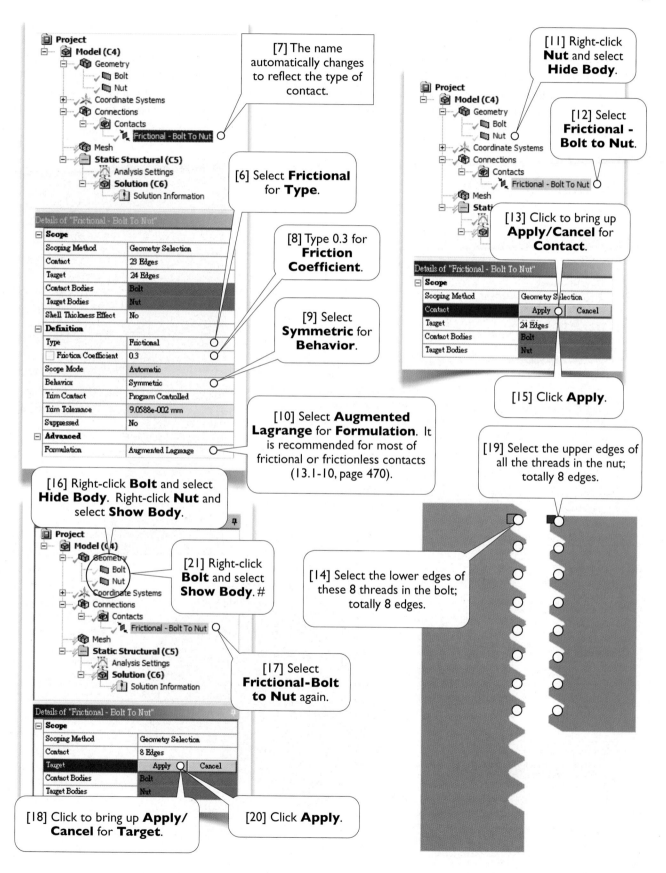

[7] The name automatically changes to reflect the type of contact.

[6] Select **Frictional** for **Type**.

[8] Type 0.3 for **Friction Coefficient**.

[9] Select **Symmetric** for **Behavior**.

[10] Select **Augmented Lagrange** for **Formulation**. It is recommended for most of frictional or frictionless contacts (13.1-10, page 470).

[16] Right-click **Bolt** and select **Hide Body**. Right-click **Nut** and select **Show Body**.

[21] Right-click **Bolt** and select **Show Body**. #

[17] Select **Frictional-Bolt to Nut** again.

[18] Click to bring up **Apply/ Cancel** for **Target**.

[20] Click **Apply**.

[11] Right-click **Nut** and select **Hide Body**.

[12] Select **Frictional - Bolt to Nut**.

[13] Click to bring up **Apply/Cancel** for **Contact**.

[15] Click **Apply**.

[19] Select the upper edges of all the threads in the nut; totally 8 edges.

[14] Select the lower edges of these 8 threads in the bolt; totally 8 edges.

Contacts

During mesh generation (3.2-8), Workbench generates contact elements between the contact edges and the target edges. Contact elements are used to prevent a body (called *contact body*) from penetrating into another body (called *target body*). Therefore, you should set up contact regions wherever contacts may occur. As long as the behavior is **Symmetric** [9], you may choose any one as contact body and the other as target body, Chapter 13 discusses about contacts in details (pages 468-471).

3.2-8 Generate Mesh

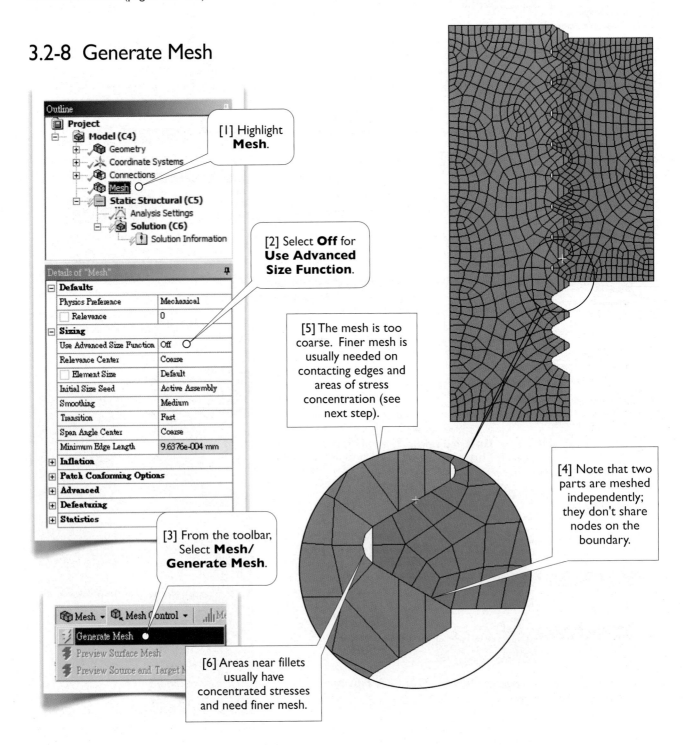

[1] Highlight **Mesh**.

[2] Select **Off** for **Use Advanced Size Function**.

[5] The mesh is too coarse. Finer mesh is usually needed on contacting edges and areas of stress concentration (see next step).

[4] Note that two parts are meshed independently; they don't share nodes on the boundary.

[3] From the toolbar, Select **Mesh/ Generate Mesh**.

[6] Areas near fillets usually have concentrated stresses and need finer mesh.

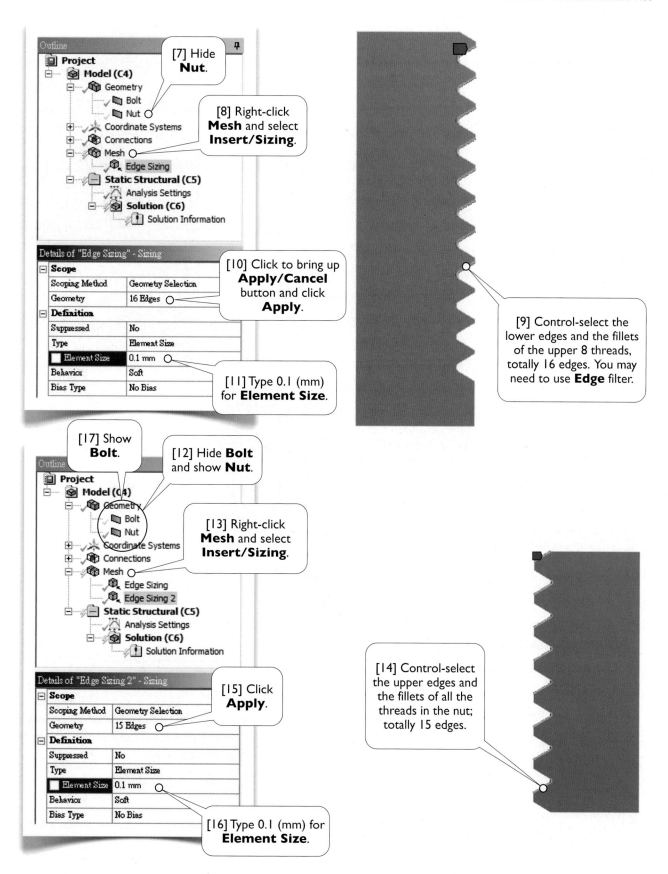

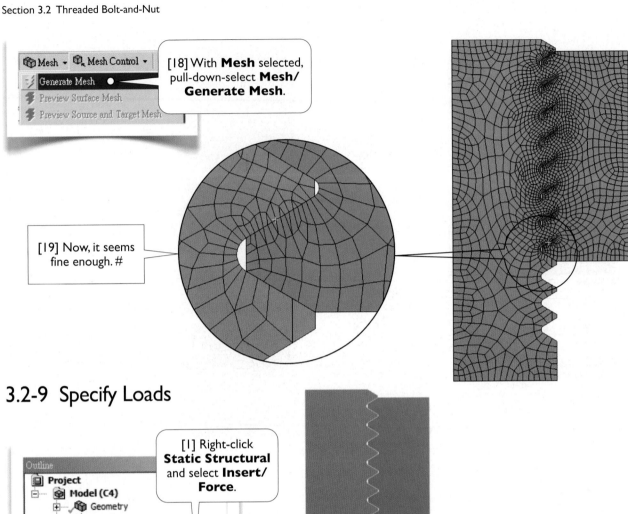

[18] With **Mesh** selected, pull-down-select **Mesh/ Generate Mesh**.

[19] Now, it seems fine enough. #

3.2-9 Specify Loads

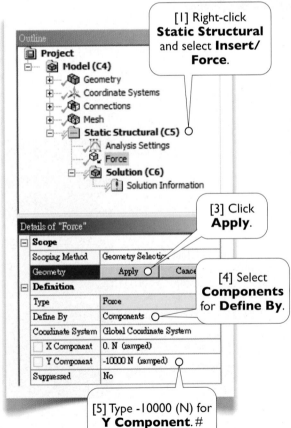

[1] Right-click **Static Structural** and select **Insert/ Force**.

[2] Select this edge.

[3] Click **Apply**.

[4] Select **Components** for **Define By**.

[5] Type -10000 (N) for **Y Component**. #

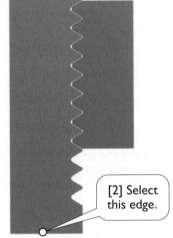

How is the force distributed?

When we specify a force of 10,000 N on the bottom edge of the axisymmetric 2D model, the force will be distributed evenly on an imaginary face which is created by revolving the edge by 360°. In this case, it has exactly the same effect if you specify a (negative) pressure of

$$\frac{10000}{\left(\pi \times 20^2/4\right)} = 31.83 \text{ MPa}$$

If you're not sure how the force distributes, always specify a pressure instead of a force.

3.2-10 Specify Supports

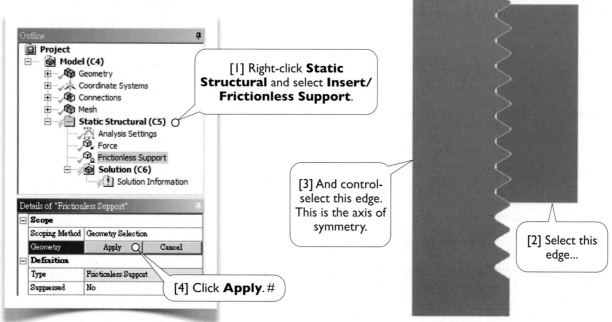

[1] Right-click **Static Structural** and select **Insert/Frictionless Support**.

[3] And control-select this edge. This is the axis of symmetry.

[2] Select this edge...

[4] Click **Apply**. #

Boundary Conditions for the Axis of Symmetry

Since any point on the axis of symmetry (Y-axis) does not move in the radial-direction (X-direction), you must specify a zero X-displacement condition or, equivalently, a frictionless support on the edge [3]. Some software can automatically take care of this boundary condition; however, as a good practice, always explicitly specify this boundary condition. If you leave it as a free boundary, the axis may become a small cylindrical "hole" after deformation.

3.2-11 Solve the Model

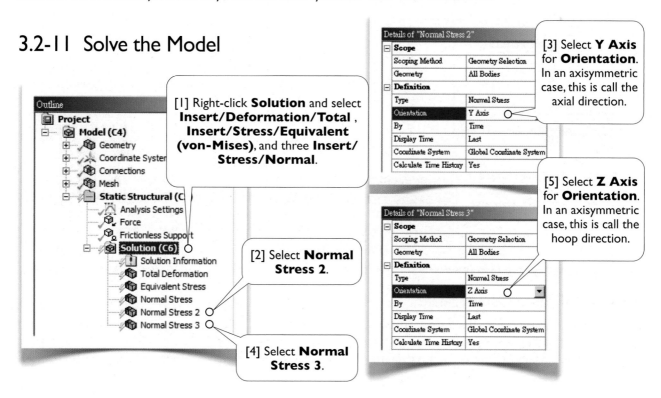

[1] Right-click **Solution** and select **Insert/Deformation/Total** , **Insert/Stress/Equivalent (von-Mises)**, and three **Insert/Stress/Normal**.

[2] Select **Normal Stress 2**.

[4] Select **Normal Stress 3**.

[3] Select **Y Axis** for **Orientation**. In an axisymmetric case, this is call the axial direction.

[5] Select **Z Axis** for **Orientation**. In an axisymmetric case, this is call the hoop direction.

Nonlinear Simulations

Due to the contact status, as the loads increase, the structure's stiffness also changes. The simulation is no longer linear. Workbench will solve the model using nonlinear solution methods. We will discuss nonlinear solution methods in Chapters 13-14. For now, to gain a feeling of nonlinear solution, let's watch how the solution proceeds [7-14].

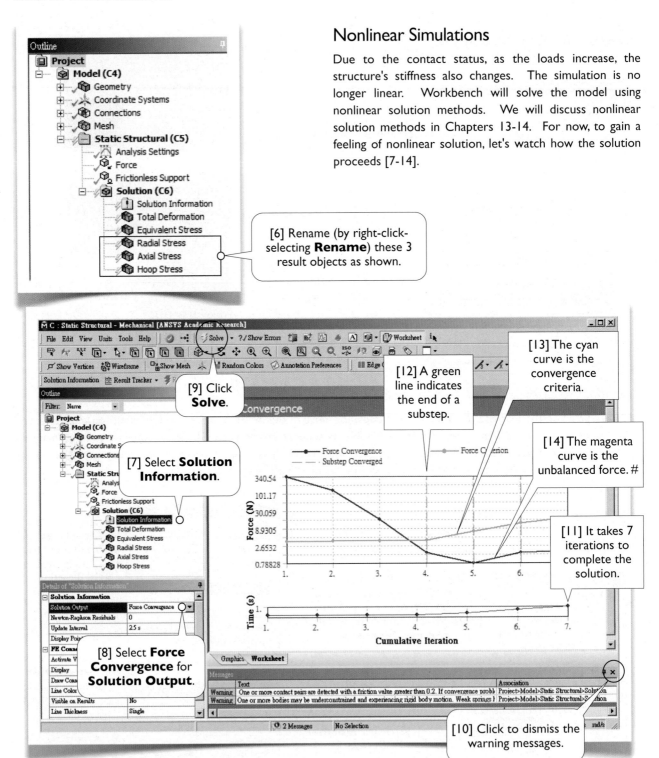

[6] Rename (by right-click-selecting **Rename**) these 3 result objects as shown.

[13] The cyan curve is the convergence criteria.

[12] A green line indicates the end of a substep.

[14] The magenta curve is the unbalanced force. #

[11] It takes 7 iterations to complete the solution.

[9] Click **Solve**.

[7] Select **Solution Information**.

[8] Select **Force Convergence** for **Solution Output**.

[10] Click to dismiss the warning messages.

A Glance of Force Convergence

In this case, Workbench divides the loading (10,000 N) into 4 substeps [12] (i.e., increasing 2,500 N for each substep). The cyan curve is the convergence criteria [13] and the magenta curve is the "unbalanced" force [14]. A substep converges when the unbalanced force is less than the criterion. Details will be given in Chapter 13 (pages 461-467).

3.2-12 View the Results

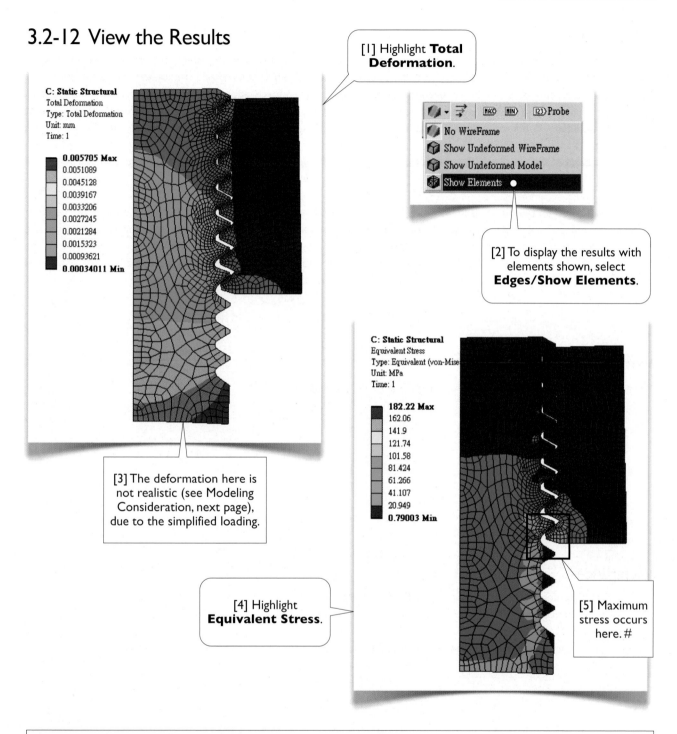

[1] Highlight **Total Deformation**.

C: Static Structural
Total Deformation
Type: Total Deformation
Unit: mm
Time: 1

0.005705 Max
0.0051089
0.0045128
0.0039167
0.0033206
0.0027245
0.0021284
0.0015323
0.00093621
0.00034011 Min

[2] To display the results with elements shown, select **Edges/Show Elements**.

C: Static Structural
Equivalent Stress
Type: Equivalent (von-Mise
Unit: MPa
Time: 1

182.22 Max
162.06
141.9
121.74
101.58
81.424
61.266
41.107
20.949
0.79003 Min

[3] The deformation here is not realistic (see Modeling Consideration, next page), due to the simplified loading.

[4] Highlight **Equivalent Stress**.

[5] Maximum stress occurs here. #

2D Models Must Be in XY Plane[Ref 1]

ANSYS assumes that, for a 2D problem (i.e., plane-stress, plane-strain, or axisymmetric problems), the 2D geometric model MUST lie in the global XY plane. Besides, for an axisymmetric problem, the global Y-axis is always the axis of symmetry and the model MUST be placed on the +X side.

A surface body created in DesignModeler is called a "2D solid body" if it is used in **Mechanical** for a 2D simulation, in which 2D solid elements (1.3-3[5, 6], page 34) are used.

Modeling Considerations

The bottom edge [3] is actually a plane of symmetry; it must remain horizontal and prohibit any vertically displacements. This plane of symmetry might have been modeled as a frictionless support. However, a frictionless support cannot have an out-of-plane loading on it. With this dilemma, we choose to apply a force, and expect an unrealistic deformation near the plane of symmetry [3]. Since we are only concerned about the stress on the threads, and the region of influence of this faulty boundary condition seems not so large to reach the areas that concern us [5], we decide to accept this arrangement.

Remark

The quantities of bolt-and-nut used in daily industrial applications are huge. Their behavior should be carefully investigated. In our preliminary study, it shows that the stresses are distributed so unevenly that most of stresses are taken by a few lower contacting threads. To improve the efficiency of the bolt-and-nut, one way is to allot some of the stresses to the upper contacting threads. In his books[Refs 2, 3], Zahavi has provided several alternatives to reduce the maximum stress. This case is adapted based on an example in his books.

Wrap Up

View other results. Close **Mechanical**, save the project, and exit Workbench.

References

1. ANSYS Documentation//Mechanical APDL//Element Reference//I. Element Library//PLANE182.
2. Zahavi, E. and Barlam, E., *Nonlinear Problems in Machine Design*, CRC Press LLC, 2000; Chapter 10. Threaded Fasteners.
3. Zahavi, E., *The Finite Element Method in Machine Design*, Prentice-Hall, 1992; Chapter 7. Threaded Fasteners.

Section 3.3

More Details

3.3-1 Plane-Stress Problems

Plane-Stress Condition

Consider a plate of ZERO thickness on *XY* plane subject to in-plane forces. The stress state at any point can be depicted in [1]. Note that there are no stresses in *Z*-face; i.e.,

$$\sigma_Z = 0, \quad \tau_{ZY} = 0, \quad \tau_{ZX} = 0 \tag{1}$$

Eq. (1) is called a plane-stress condition. If the plane-stress condition holds everwhere, then it is called a plane-stress problem.

In real world, there is no such thing as ZERO thickness. The triangular plate simulated in Section 3.1 is close to but not exactly a plane-stress problem; the triangular plate has finite thickness of 10 mm. However, since its stresses in *Z*-direction are negligible, we usually assume that the plane-stress condition holds for such a finite thickness plate.

In practice, a problem may assume the plane-stress condition if its thickness direction (*Z*-direction) is not restrained and thus free to expand or contract. As an example, a simply supported beam as shown in [2] is often solved by assuming the plane stress condition, even though its out-of-plane thickness is not zero.

Governing Equations for Plane-Stress Problems

Substituting the plane-stress condition, Eq. (1), into Eq. 1.2-8(1) (page 27), the Hooke's law becomes

$$\varepsilon_X = \frac{\sigma_X}{E} - v\frac{\sigma_Y}{E}$$

$$\varepsilon_Y = \frac{\sigma_Y}{E} - v\frac{\sigma_X}{E}$$

$$\varepsilon_Z = -v\frac{\sigma_X}{E} - v\frac{\sigma_Y}{E} \tag{2}$$

$$\gamma_{XY} = \frac{\tau_{XY}}{G}, \quad \gamma_{YZ} = 0, \quad \gamma_{ZX} = 0$$

Substitution of Eq. (1) into other governing equations (e.g., Eq. 1.2-6(2), page 26) will conclude that all quantities are independent of *Z*. That is, the particles with the same *X* and *Y* coordinates share the same behaviors regardless of their *Z* coordinate. Thus, we can eliminate *Z* coordinate and reduce the problem to a two-dimensional, on *XY* space.

Note that, in Eq. (2), ε_Z is not zero, however, it can be calculated independently from σ_X and σ_Y. The nonzero ε_Z is easy to understand, since *Z*-direction is free to expand or contract.

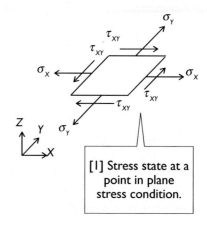

[1] Stress state at a point in plane stress condition.

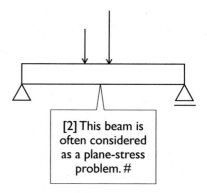

[2] This beam is often considered as a plane-stress problem. #

2D models must be in *XY* plane[Ref 12]

ANSYS assumes that, for a 2D problem (i.e., plane-stress, plane-strain, or axisymmetric problems), the 2D geometric model MUST lie in a global *XY* plane. For an axisymmetric problem, the global *Y*-axis is always the axis of symmetry and the model MUST be placed on the +*X* side.

3.3-2 Plane-Strain Problems

Plane-Strain Condition

Consider a structure of INFINITE LENGTH in Z-direction. The Z-direction is restrained such that no particles can move in Z-direction. Further, all cross-sections perpendicular to the Z-direction have the same geometry, supports, and loads [1]. In such a case, the strain state at any point can be depicted in [2]. Note that there are no strains in Z-face; i.e., $\varepsilon_z = 0$ (otherwise the particle on the Z-face would move in Z-direction) and $\gamma_{ZX} = \gamma_{ZY} = 0$ (otherwise the cube would twist in ZX and ZY planes respectively, and that implies the particles would move in Z-direction),

$$\varepsilon_z = 0, \quad \gamma_{ZX} = 0, \quad \gamma_{ZY} = 0 \qquad (1)$$

Eq. (1) is called a *plane-strain condition*. If the plane-strain condition holds everwhere, then it is called a plane-strain problem.

In real world, there is no such thing as infinite length. In practice, a problem may assume the plane-strain condition if its Z-direction is restrained from expansion or contraction and all cross-sections perpendicular to the Z-direction have the same geometry, supports, and loads. As an example, a pressurized pipe buried under the earth is often considered as a plane-strain problem.

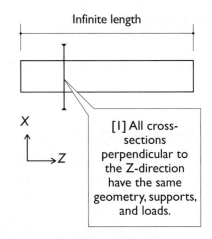

Infinite length

[1] All cross-sections perpendicular to the Z-direction have the same geometry, supports, and loads.

Governing Equation for Plane-Strain Problems

Eq. 1.2-8(1) (page 27), the Hooke's law, can be inverted and rewritten as

$$\sigma_X = \frac{E}{(1+v)(1-2v)}\left[(1-v)\varepsilon_X + v\varepsilon_Y + v\varepsilon_Z\right]$$

$$\sigma_Y = \frac{E}{(1+v)(1-2v)}\left[(1-v)\varepsilon_Y + v\varepsilon_Z + v\varepsilon_X\right]$$

$$\sigma_Z = \frac{E}{(1+v)(1-2v)}\left[(1-v)\varepsilon_Z + v\varepsilon_X + v\varepsilon_Y\right] \qquad (2)$$

$$\tau_{XY} = G\gamma_{XY}, \quad \tau_{YZ} = G\gamma_{YZ}, \quad \tau_{ZX} = G\gamma_{ZX}$$

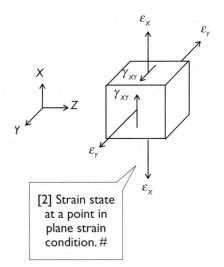

[2] Strain state at a point in plane strain condition. #

The proof of Eq. (2) is listed in 3.3-14 (page 142).

Substitute the plane-strain condition, Eq. (1), into Eq. (2), and the Hooke's law becomes

$$\sigma_X = \frac{E}{(1+v)(1-2v)}\left[(1-v)\varepsilon_X + v\varepsilon_Y\right]$$

$$\sigma_Y = \frac{E}{(1+v)(1-2v)}\left[(1-v)\varepsilon_Y + v\varepsilon_X\right]$$

$$\sigma_Z = \frac{E}{(1+v)(1-2v)}\left[v\varepsilon_X + v\varepsilon_Y\right] \qquad (3)$$

$$\tau_{XY} = G\gamma_{XY}, \quad \tau_{YZ} = 0, \quad \tau_{ZX} = 0$$

Substitution of Eq. (1) into other governing equations (e.g., Eq. 1.2-6(2), page 26) will conclude that all quantities are independent of Z; i.e., the particles with the same X and Y coordinates share the same behaviors regardless of their Z coordinate. Thus, we can eliminate Z coordinate and reduce the problem to a two-dimensional, on XY space.

Note, in Eq. (3), σ_Z is not zero; however, it can be calculated independently from ε_X and ε_Y. The nonzero σ_Z is easy to understand, since Z-direction is restrained from expansion or contraction, it will develop stress to counteract the restriction.

3.3-3 Axisymmetric Problems

Consider a structure of which the geometry, supports, and loads are axisymmetric about the Y-axis. In such a case, all quantities are independent of θ coordinate; i.e., the particles with the same R and Y coordinates share the same behaviors regardless of their θ coordinate. Thus, we may eliminate θ coordinate and reduce the problem to a two-dimensional, on R-Y space.

The strain state at any point can be depicted in [1]. Note that there are no shear strains in θ-face (otherwise the θR-face and the θY-face would twist and the problem is no longer axisymmetric),

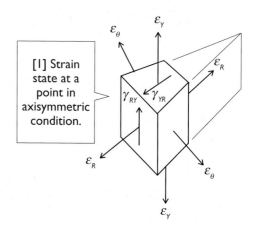

[1] Strain state at a point in axisymmetric condition.

$$\gamma_{\theta R} = 0, \quad \gamma_{\theta Y} = 0 \qquad (1)$$

Eq. (1) implies

$$\tau_{\theta R} = 0, \quad \tau_{\theta Y} = 0 \qquad (2)$$

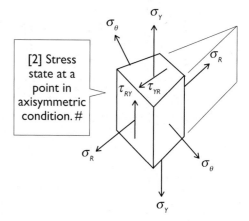

[2] Stress state at a point in axisymmetric condition. #

Eqs. (1) and (2) can be regarded as the *axisymmetric condition*.

Axisymmetric problems are ubiquitous in engineering applications. Many problems are not strictly axisymmetric but can reasonably assume the axisymmetric condition, such as the bolt-and-nut problem simulated in Section 3.2, in which the threads are spiral and the nut is hexagonal.

In an axisymmetric problem, σ_r is called a *radial stress*, σ_θ is called a *hoop stress*, and σ_Y is called an *axial stress* [2].

3.3-4 Mechanical GUI

Mechanical GUI is composed of several areas [1-7] (next page); many of them are similar to those in **DesignModeler GUI** (2.3-1, page 76). On the top are pull-down menus and toolbars [1]; on the bottom is a status bar [7]. In-between are several "window panes" [2-6]. Separators [8] between window panes may be dragged to resize window panes. You even can move or dock a pane by dragging its title bar. Whenever you mess up the workspace, pull-down-select **View/Windows/Reset Layout** to reset the default layout.

Outline [2] displays an outline of a *project tree*, which is a structured representation of the project (to be discussed). **Details** [3] shows the detail information of the object highlighted in the project tree or graphics window. Graphics window [4] displays the geometric model. **Graph** [5] typically shows a result-versus-time plot. **Tabular Data** [6] shows the numerical counterpart of the result-versus-time plot. A set of animation tools are available in the **Graph** window pane; these tools allow you to play, stop, or save the animation.

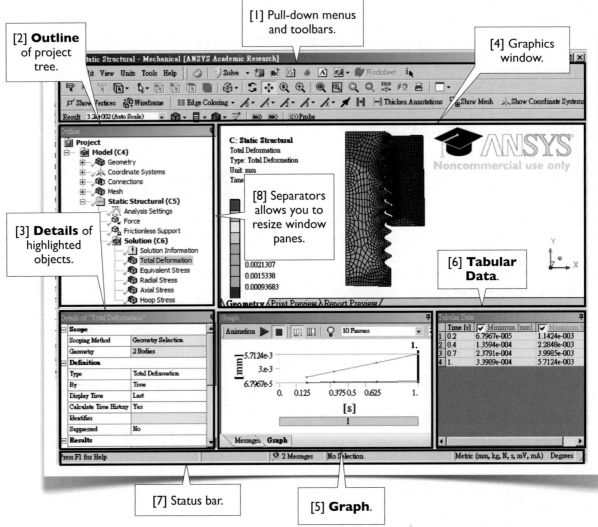

[1] Pull-down menus and toolbars.

[2] **Outline** of project tree.

[4] Graphics window.

[3] **Details** of highlighted objects.

[8] Separators allows you to resize window panes.

[6] **Tabular Data**.

[7] Status bar.

[5] **Graph**.

Project Tree[Refs 1, 11]

A project tree [9] is a structured representation of a project. A project tree may contain one or more simulation models. Often, there is only one simulation model in a project tree. A simulation model may contain one or more **Environment** branches, along with other objects. Each can be renamed. Default name for the **Environment** branch is the name of the analysis system, for example: **Static Structural**. An **Environment** branch contains **Analysis Settings**, several objects that define the environment conditions, and a **Solution** branch, which contains a **Solution Information** and several results objects.

Right-clicking an object (or multiple objects) and selecting a tool from the context menu, you can operate on the object (or objects), such as delete, rename, duplicate, etc.

Unlike the objects of a model tree in DesignModeler, in which their order is important, the order of the objects in a project tree is not relevant.

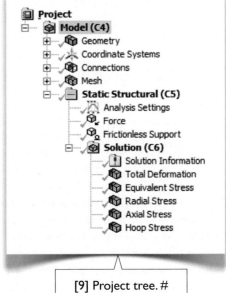

[9] Project tree. #

3.3-5 **Unit Systems**

In DesignModeler, the only unit used is length. In **Mechanical**, units are much more complicated; nevertheless, Workbench takes care of the consistency of unit system, and your responsibility is to select a unit system suitable for your model. Selecting a suitable unit system for your model, in some cases, is crucial. In these cases the solution accuracy may deteriorate due to an accumulation of machine errors.

Choosing Unit Systems: Guideline

As a guideline, select a unit system such that the values stored in the computer have about the same order. For example, if you choose SI unit system for a micro-scale simulation model, you would have the lengths of order 10^{-6} and a Young's modulus of order 10^{11}. That may raise precision issues. On the other hand, if you choose a μMKS unit system, you will have the lengths of order 10^{0} and a Young's modulus of order 10^{5}. That is much better.

Consistent vs. Inconsistent Unit Systems

In **Workbench GUI** (not **Mechanical GUI**) pull-down-select **Units/Unit Systems...**, you will see a list of built-in unit systems [1-5]. Workbench always uses a consistent unit system for internal computations. There are 6 *consistent unit systems* in the list: SI, CGS, NMM, μMKS, BIN, and BFT [5]. Highlight a unit system in the list, you will see the details of that unit system.

Other unit systems are *inconsistent* ones. They are, however, often more convenient to use than consistent ones. When you select an inconsistent unit system for use in **Mechanical**, it internally uses a consistent unit system that is closest to the inconsistent one you've chosen.

Like DesignModeler, **Mechanical** allows you to change the unit system any time, using the pull-down menu **Units** [6]. The internal consistent unit system also changes accordingly.

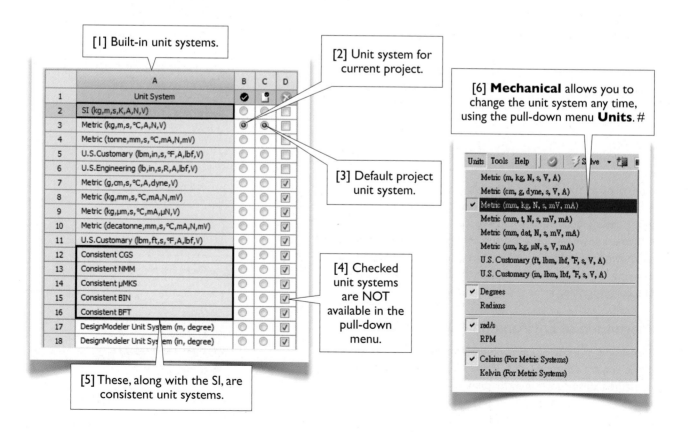

[1] Built-in unit systems.

[2] Unit system for current project.

[3] Default project unit system.

[4] Checked unit systems are NOT available in the pull-down menu.

[5] These, along with the SI, are consistent unit systems.

[6] **Mechanical** allows you to change the unit system any time, using the pull-down menu **Units**. #

3.3-6 Environment Conditions

Highlighting an **Environment** branch (see page 135), you will see a row of environment conditions on the toolbar [1]. These environment conditions may also be accessed through the context menu.

Three groups of environment conditions are frequently used, namely, loads [2], supports [3], and inertial forces [4]. Environment conditions available in each group depend on the dimensionality (2D or 3D) as well as the type of analysis system (e.g., static or dynamic, structural or thermal).

Here, we will introduce the environment conditions available in 2D static structural simulations. Additional environment conditions will be introduced later, starting from Chapter 5. Many of environment conditions are self-explained while others have many useful features. When going through each environment condition, we will point out its location in the ANSYS documentation system, in which many details can be found. You should consult these official documentation whenever needed.

Inside Workbench

Before jumping to individual environment condition, let's describe how Workbench processes the environment conditions.

When an environment condition is applied, it will be eventually transferred to the NODES of the finite element model. For example, if you apply a pressure on a surface, the equivalent nodal forces are calculated and applied on nodes. The support conditions are processed in a similar way.

Consider again Eq. 1.3-1(1) (page 31),

$$[K]\{D\} = \{F\} \qquad \text{Copy of Eq. 1.3-1(1)}$$

The loads and inertia forces are used to set up the vector $\{F\}$, while the supports conditions are used to set up the vector $\{D\}$. After setting up, some of nodal forces and nodal displacements become known values while the others remain unknown. The unknown nodal forces to be solved are called *reaction forces*. For any degree of freedom, if the displacement is known then the corresponding force is unknown, and vice versa.

Magnitude of Environment Conditions[Ref 2]

Magnitude of most of environment conditions can be specify in three ways: a constant value, a time-dependent tabular form, or a mathematical function with time, X, Y, or Z as independent variable.

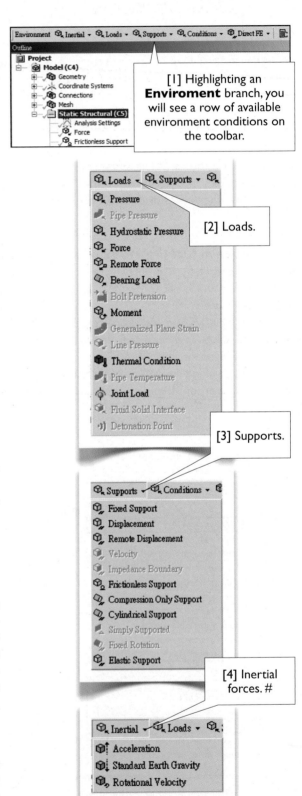

[1] Highlighting an **Enviroment** branch, you will see a row of available environment conditions on the toolbar.

[2] Loads.

[3] Supports.

[4] Inertial forces. #

3.3-7 Loads[Ref 3]

Pressure

Applies on 2D edges or 3D faces. It is possible to define a spatial varying pressure[Ref 4].

Force

Applies on vertices, edges, or faces. If it applies on edges/faces, the force is evenly distributed on the edges/faces.

Thermal Condition

Applies on bodies. The temperature change ΔT (see Eq. 1.2-8(3), page 28) is the difference between specified temperature and the reference temperature, which is part of information of the material properties, default to 22°C.

Bearing Load

Applies on 2D circular edges or 3D cylindrical faces. The total force is distributed on the compressive side of the circular edges or cylindrical faces.

Hydrostatic Pressure

Applies on 2D edges or 3D faces. It simulates pressure that occurs due to fluid weight. A free surface location may be specified, default to the surface at $X = 0$.

Moment

Applies on 2D edges or 3D faces. A statically equivalent pressure distributed on the edges/faces is calculated and applied on the edges/faces.

Remote Force

Applies at a location anywhere in the space. Workbench calculates the equivalent moment and force and applies them on the body. It may be used as an alternative way of building a rigid part and applying a force on it.

Joint Load

Applies on a **Joint**[Ref 5]. You use a joint load to apply a kinematic driving condition in a multi-body dynamic simulation.

3.3-8 Supports[Ref 6]

Fixed Support

Applies on vertices, edges, or faces. Prevents nodes from moving in X- Y- and Z-directions. It also prevents nodes from rotations for beam/shell elements.

Displacement

Applies on vertices, edges, or faces. Displacements in X- Y- and Z-directions can be specified. A zero value prevents nodes from moving in that direction. An unspecified value sets that direction free.

Frictionless Support

Applies on 2D edges or 3D faces. Prevents nodes from moving in the normal direction; allows nodes to freely move in the tangential direction.

Compression Only Support

Applies on 2D edges or 3D faces. The associated body is free to depart from the edges or faces, but cannot to penetrate them. It in effect sets up a frictionless contact region between the body and a rigid support; it introduces contact nonlinearity into the problem.

Cylindrical Support

Applies on 2D circular edges or 3D cylindrical faces. Each of radial, tangential, and axial (3D only) directions can be set free or fixed.

Elastic Support

Applies on 2D edges or 3D faces. A foundation stiffness must be specified to establish the relation between the reaction pressure and the support displacement.

Remote Displacement

Applies at a location anywhere in the space. Workbench calculates the equivalent displacement and rotation and applies them on the body. It may be used as an alternative way of building a rigid part and applying a displacement to it.

3.3-9 Inertial[Ref 3]

Standard Earth Gravity

Applies on bodies. You must select a direction along which the gravitational force applies; it defaults to -Z direction.

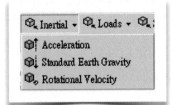

Acceleration

Applies on bodies. You must specify the magnitude and direction of acceleration. The direction is where the bodies accelerate. An "Inertia force" will apply in the opposite direction.

Rotational Velocity

Applies on bodies. You must specify the magnitude and direction of the angular velocity of the bodies. A distributed "inertia force" will apply in the opposite direction of rotation.

3.3-10 Results Objects[Ref 7]

Highlighting **Solution** branch, you will see a row of results tools available on the toolbar [1]. These results tools may also be accessed through the context menu. Most of them are self-explained, but some of them need to be explained.

[2] Special results tools. #

Linearized Stress

Using this tool, you can view stresses along a straight line path. You need to first define a straight line path using **Construction Geometry** under **Model**.

Probe[Ref 8]

It contains tools to explore the results of a point, or maximum/minimum values of the results of a scoped region, along the loading history. In other words, we are concerned about the results across the time domain instead of space domain.

Tools[Ref 7]

It contains special results tools: **Stress Tool**, **Fatigue Tool**, **Contact Tool**, and **Beam Tool** [2].

User Defined Result[Ref 9]

You may define a results expression using keywords. By highlighting **Solution** in the project tree and clicking **Worksheet** on the right side of the GUI, you will see a list of available keywords.

[1] Results toolbar.

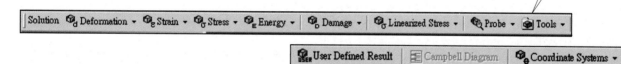

3.3-11 View Results[Ref 7]

To view results, simply highlight a results object. Tools to control the visual effects of the results are shown on the toolbar [1-6].

[2] Click to turn on/off the probe. Results values will display along with your mouse pointer; click to label the value. To remove the label, activate **Label** [3] and press **Delete**.

[3] **Label**.

[5] You can control how the contour displays.

[1] Click to turn on/off the label of maximum/minimum.

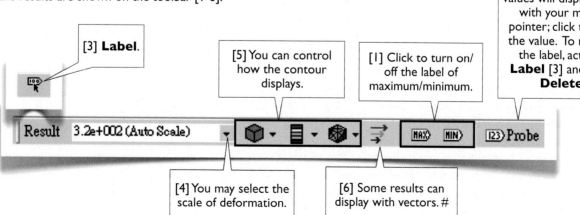

[4] You may select the scale of deformation.

[6] Some results can display with vectors. #

3.3-12 Insert APDL Commands[Ref 10]

Insert Commands [1], if available, allows you to insert APDL Commands. For those who are familiar with APDL, this may be useful, since the current version of Workbench doesn't include all the functionalities provided by APDL. For the newcomers, my suggestion is that you do not worry about APDL for now.

After clicking **Insert Commands** tool, a text editor is opened with several lines of comments telling you WHEN the APDL commands will be executed. For example, an APDL commands object inserted under **Environment** will be executed just before **Solve** command.

[1] Click **Insert Commands** to insert ANSYS APDL commands.

What is APDL?

In the old days, the users operate ANSYS using a set of text command language, called APDL (ANSYS Parametric Design Language). Comparing with modern computer languages, the APDL is not a user-friendly language at all. For many users, use of APDL has been a painful experience.

Later, ANSYS started to provide a graphical user interface (GUI). The users operate ANSYS through pull-down menus, dialogs, etc. Basically, each APDL command has a corresponding operating path in the GUI. Using either APDL or the GUI, the users can use all the functionalities of ANSYS. Again, comparing with **Workbench GUI**, the old ANSYS GUI is not efficient at all. Many experts and school teachers prefer APDL to the old GUI.

It is true that some capabilities of APDL are not directly supported in Workbench. It, however, provides two ways that you may access APDL commands: (a) You can insert APDL commands by clicking **Insert Commands** [1]. (b) You can create a **Mechanical APDL** system [2], which allows APDL files to be read into Workbench.

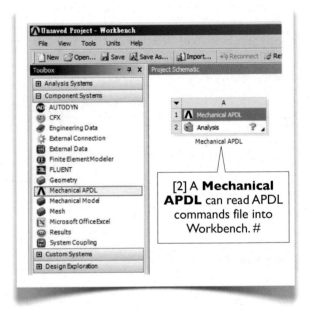

[2] A **Mechanical APDL** can read APDL commands file into Workbench. #

3.3-13 Status Symbols in Tree Outline[Ref 11]

Each object of the project tree has a status symbol, explained below:

- Checkmark indicates branch is fully defined / OK.
- Question mark indicates item has incomplete data (need input).
- Lightning bolt indicates solving is required.
- Exclamation mark means a problem exists.
- "X" means item is suppressed (will not be solved).
- Transparent checkmark means body or part is hidden.
- Green lightning bolt indicates item is currently being evaluated.
- Minus sign means that mapped face meshing failed.
- Check mark with a slash indicates a meshed part/body.
- Red lightning bolt indicates a failed solution.

3.3-14 Appendix: Proof of Eq. 3.3-2(2), page 133

The first 3 and last 3 equations are decoupled, they can be proved independently. Proof of the last 3 equations from the last 3 equations in Eq. 1.2-8(1) (page 27) is trivial. Now, we prove the first 3 equations. The first 3 equations in Eq. 1.2-8(1) can be written in matrix form

$$\left\{\begin{array}{c}\varepsilon_X \\ \varepsilon_Y \\ \varepsilon_Z\end{array}\right\} = \frac{1}{E}\begin{bmatrix} 1 & -v & -v \\ -v & 1 & -v \\ -v & -v & 1\end{bmatrix}\left\{\begin{array}{c}\sigma_X \\ \sigma_Y \\ \sigma_Z\end{array}\right\} \text{ or } \{\varepsilon\} = [D]\{\sigma\}$$

The first 3 equation in Eq. 3.3-2(2) can also be written in matrix form

$$\left\{\begin{array}{c}\sigma_X \\ \sigma_Y \\ \sigma_Z\end{array}\right\} = \frac{E}{(1+v)(1-2v)}\begin{bmatrix} 1-v & v & v \\ v & 1-v & v \\ v & v & 1-v\end{bmatrix}\left\{\begin{array}{c}\varepsilon_X \\ \varepsilon_Y \\ \varepsilon_Z\end{array}\right\} \text{ or } \{\sigma\} = [F]\{\varepsilon\}$$

Then

$$[D][F] = \frac{1}{E}\cdot\frac{E}{(1+v)(1-2v)}\begin{bmatrix} 1 & -v & -v \\ -v & 1 & -v \\ -v & -v & 1\end{bmatrix}\begin{bmatrix} 1-v & v & v \\ v & 1-v & v \\ v & v & 1-v\end{bmatrix} = \begin{bmatrix} 1 & 0 & 0 \\ 0 & 1 & 0 \\ 0 & 0 & 1\end{bmatrix}$$

This completes the proof.

References

1. ANSYS Documentation//Mechanical Applications//Mechanical User's Guide//Objects Reference
2. ANSYS Documentation//Mechanical Applications//Mechanical User's Guide//Setting Up Boundary Conditions// Defining Boundary Condition Magnitude
3. ANSYS Documentation//Mechanical Applications//Mechanical User's Guide//Setting Up Boundary Conditions//Load Type Boundary Conditions
4. ANSYS Documentation//Mechanical Applications//Mechanical User's Guide//Setting Up Boundary Conditions// Spatial Varying Loads and Displacements
5. ANSYS Documentation//Mechanical Applications//Mechanical User's Guide// Setting Connections//Joints
6. ANSYS Documentation//Mechanical Applications//Mechanical User's Guide//Setting Up Boundary Conditions// Support Type Boundary Conditions
7. ANSYS Documentation//Mechanical Applications//Mechanical User's Guide//Using Results//Structural Results
8. ANSYS Documentation//Mechanical Applications//Mechanical User's Guide//Using Results//Result Output//Probe
9. ANSYS Documentation//Mechanical Applications//Mechanical User's Guide//Using Results/User Defined Results
10. ANSYS Documentation//Mechanical Applications//Mechanical User's Guide//Commands Objects
11. ANSYS Documentation//Mechanical Applications//Mechanical User's Guide//Application Interface//Main Windows// Tree Outline
12. ANSYS Documentation//Mechanical APDL//Element Reference//I. Element Library//PLANE182

Section 3.4

Spur Gears

3.4-1 About the Spur Gears

In this section, we'll conduct a simulation for a pair of meshing spur gears introduced in Section 2.5. The goal is to assess the maximum stress during the transmission of a torque of 15,000 lb-in. An engineering judgement reveals that the maximum stress occurs at either contact point [1], or at the root of a tooth due to bending of the tooth [2].

Since no restriction in the depth direction, thus free to expand or contract in the depth direction, the gears is modeled as a plane stress problem (3.3-1, page 132).

The U.S. customary unit system (**in-lbm-lbf-s**) is used in this exercise.

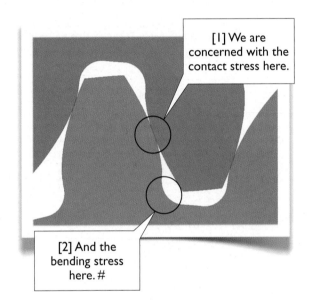

[1] We are concerned with the contact stress here.

[2] And the bending stress here. #

3.4-2 Set Up Project Schematic

Launch Workbench. Open the project **Gear**, saved in Section 2.5 [1]. Duplicate the **Geometry** system [2, 3]. Create a **Static Structural** system by double-clicking it in the **Toolbox** [4]. Create a link between the last two systems to share **Geometry** [5]. Double-click **Geometry** cell in **2D Gear Set** system to edit it [6].

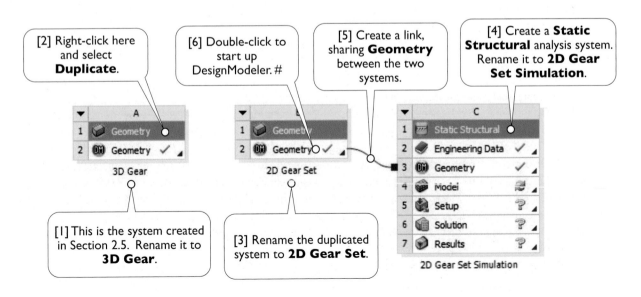

[2] Right-click here and select **Duplicate**.

[6] Double-click to start up DesignModeler. #

[5] Create a link, sharing **Geometry** between the two systems.

[4] Create a **Static Structural** analysis system. Rename it to **2D Gear Set Simulation**.

[1] This is the system created in Section 2.5. Rename it to **3D Gear**.

[3] Rename the duplicated system to **2D Gear Set**.

143

3.4-3 Delete the 3D Body and Create a Surface Body

Delete **Extrude1**. Create a surface body from **Sketch1** by pull-down-selecting **Concept/Surfaces From Sketches**. Remember to click **Generate** [1].

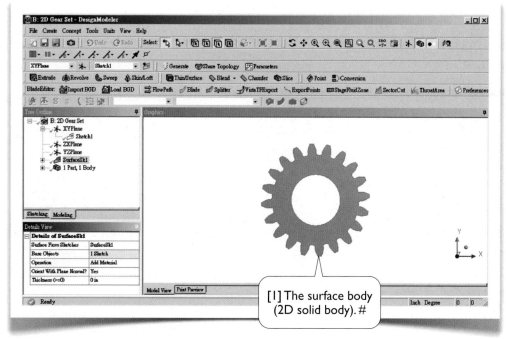

[1] The surface body (2D solid body). #

3.4-4 Duplicate the Gear

Duplicate the gear [1-7]. Rename the two bodies as **Lower Gear** and **Upper Gear** respectively. Close DesignModeler.

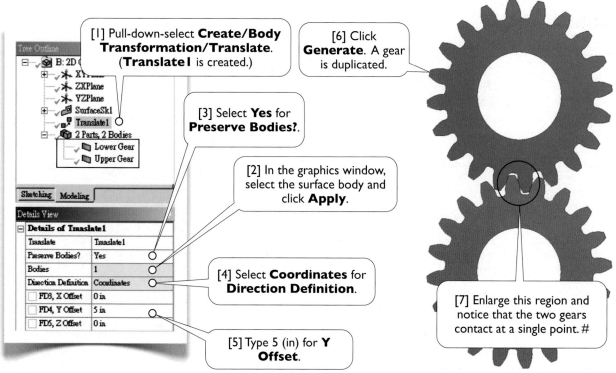

[1] Pull-down-select **Create/Body Transformation/Translate**. (**Translate1** is created.)

[6] Click **Generate**. A gear is duplicated.

[3] Select **Yes** for **Preserve Bodies?**.

[2] In the graphics window, select the surface body and click **Apply**.

[4] Select **Coordinates** for **Direction Definition**.

[7] Enlarge this region and notice that the two gears contact at a single point. #

[5] Type 5 (in) for **Y Offset**.

144

3.4-5 Set Up Geometry in Mechanical

Before entering **Mechanical**, remember to specify **2D** for **Analysis Type** (3.1-4[4-6], page 107). This step is important since, after the geometry attaches to **Mechanical**, you cannot change it any more. Start up **Mechanical** by double-clicking **Model** in **2D Gear Set Simulation** system.

In **Mechanical**, change the units to **in-lbm-lbf-s** [1]. Make sure **2D Behavior** is set to **Plane Stress** [2, 3]. Note that material defaults to **Structural Steel** and thickness defaults to 1.0 (in) [4].

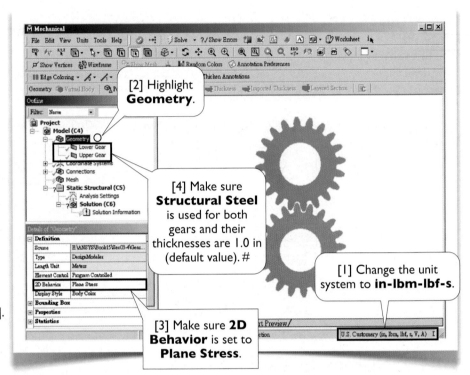

[2] Highlight **Geometry**.

[4] Make sure **Structural Steel** is used for both gears and their thicknesses are 1.0 in (default value). #

[1] Change the unit system to **in-lbm-lbf-s**.

[3] Make sure **2D Behavior** is set to **Plane Stress**.

3.4-6 Set Up Contact Region

Redefine the contact region as follows. In the project tree, highlight **Connections/Contacts/Contact Region** and define **Contact** and **Target** [1-4]. Change the contact type to **Frictionless** [5]. Select **Augmented Lagrange** for **Formulation** [6]; this is generally recommended for frictionless contact (13.1-10, page 470). Finally select **Adjust to Touch** for **Interface Treatment** [7] (13.1-11, page 471).

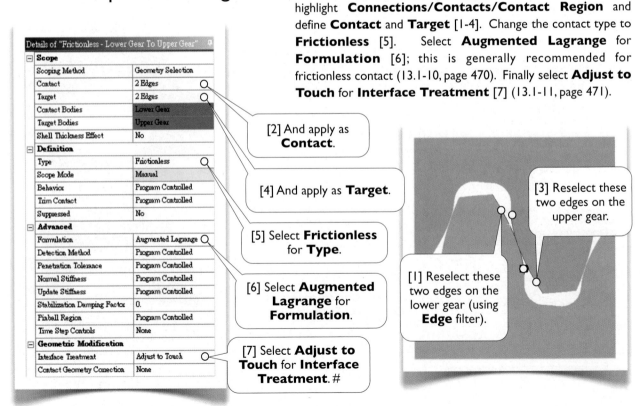

[2] And apply as **Contact**.

[4] And apply as **Target**.

[5] Select **Frictionless** for **Type**.

[6] Select **Augmented Lagrange** for **Formulation**.

[7] Select **Adjust to Touch** for **Interface Treatment**. #

[3] Reselect these two edges on the upper gear.

[1] Reselect these two edges on the lower gear (using **Edge** filter).

3.4-7 Generate Mesh

Generate mesh [1-2]. The mesh with default settings seems too coarse. We need finer mesh on the contact areas and the fillets.   Insert a **Sizing** for the mesh [3-5]. Turn off **Use Advanced Size Function** [6]. The new mesh should be adequate now [7].

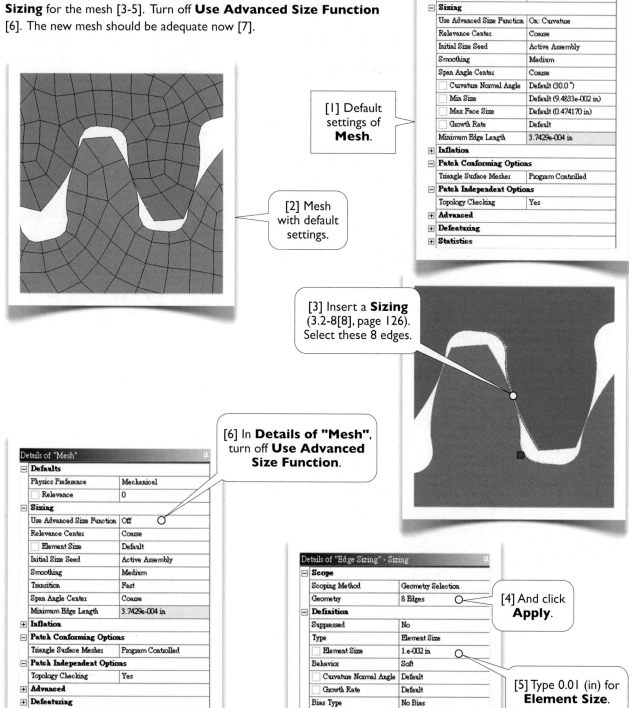

[1] Default settings of **Mesh**.

[2] Mesh with default settings.

[3] Insert a **Sizing** (3.2-8[8], page 126). Select these 8 edges.

[6] In **Details of "Mesh"**, turn off **Use Advanced Size Function**.

[4] And click **Apply**.

[5] Type 0.01 (in) for **Element Size**.

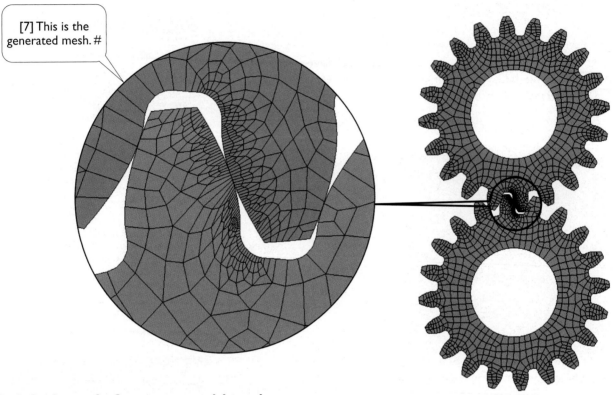

3.4-8 Specify Support and Load

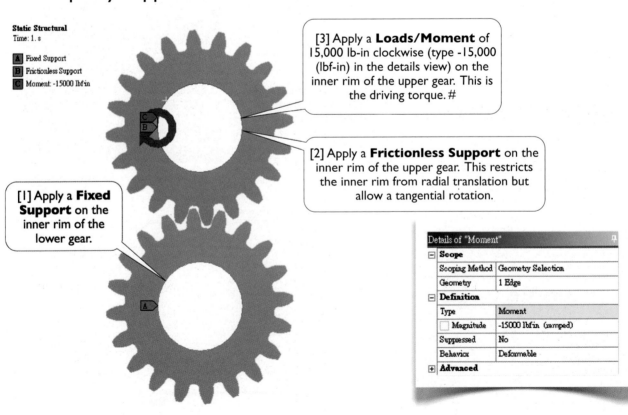

3.4-9 Solve the Model and View the Results

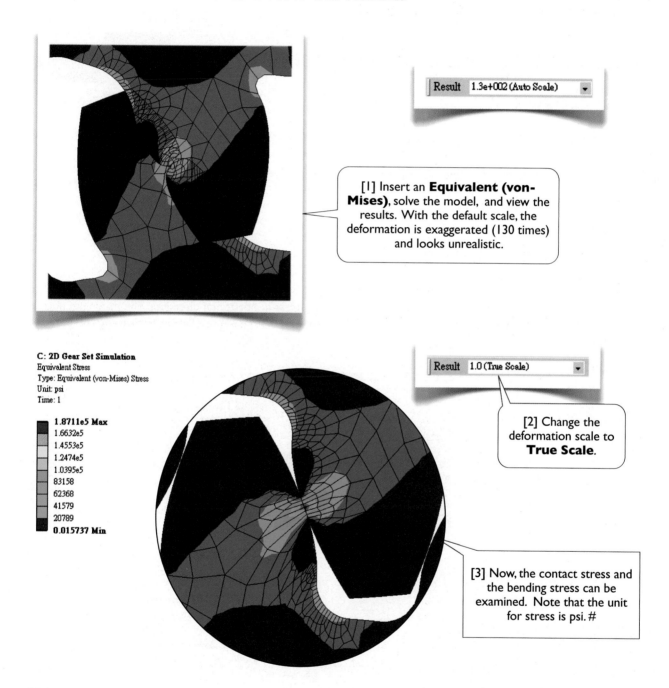

Result   1.3e+002 (Auto Scale)

[1] Insert an **Equivalent (von-Mises)**, solve the model, and view the results. With the default scale, the deformation is exaggerated (130 times) and looks unrealistic.

C: 2D Gear Set Simulation
Equivalent Stress
Type: Equivalent (von-Mises) Stress
Unit: psi
Time: 1

1.8711e5 Max
1.6632e5
1.4553e5
1.2474e5
1.0395e5
83158
62368
41579
20789
0.015737 Min

Result   1.0 (True Scale)

[2] Change the deformation scale to **True Scale**.

[3] Now, the contact stress and the bending stress can be examined. Note that the unit for stress is psi. #

Wrap Up

Close **Mechanical**, save the project, and exit Workbench.

Section 3.5

Structural Error, FE Convergence, and Stress Singularity

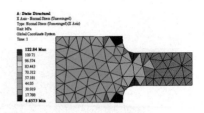

This exercise illustrates some must-know concepts in finite element simulations: (a) stress discontinuity, (b) structural error, (c) finite element convergence, (d) stress concentration, and (e) stress singularity. We use a filleted bar subject to tension to demonstrate these concepts.

3.5-1 About the Filleted Bar

The filleted bar is made of steel with dimensions as shown [1]. The bar is subject to a tension of 50 kN. We are interested in the maximum displacement and the maximum normal stress in horizontal direction. The maximum normal stress occurs near the fillets due to a stress concentration. The unit system used in this exercise is **mm-kg-N-s**.

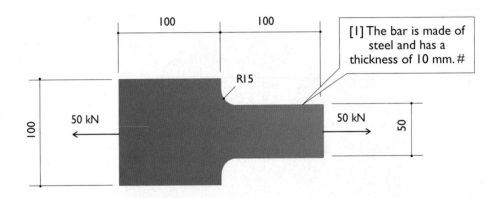

[1] The bar is made of steel and has a thickness of 10 mm. #

PART A. STRESS DISCONTINUITY

3.5-2 Start a New Project

Launch Workbench. Create a **Static Structural** system by double-clicking **Static Structural** in **Toolbox**. Save the project as **Bar**. Start up DesignModeler [1].

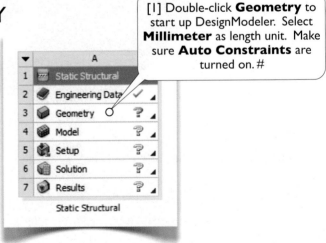

[1] Double-click **Geometry** to start up DesignModeler. Select **Millimeter** as length unit. Make sure **Auto Constraints** are turned on. #

3.5-3 Create a 2D Model in DesignModeler

Create a sketch on **XYPlane** [1], remember to impose **Symmetry** constraints. Create a surface body by pull-down-selecting **Concept/Surfaces from Sketches** [2, 3]. Close DesignModeler.

In the **Project Schematic**, specify **2D** for the **Geometry** (3.1-4[4-6], page 107). Start up **Mechanical** by double-clicking **Model**.

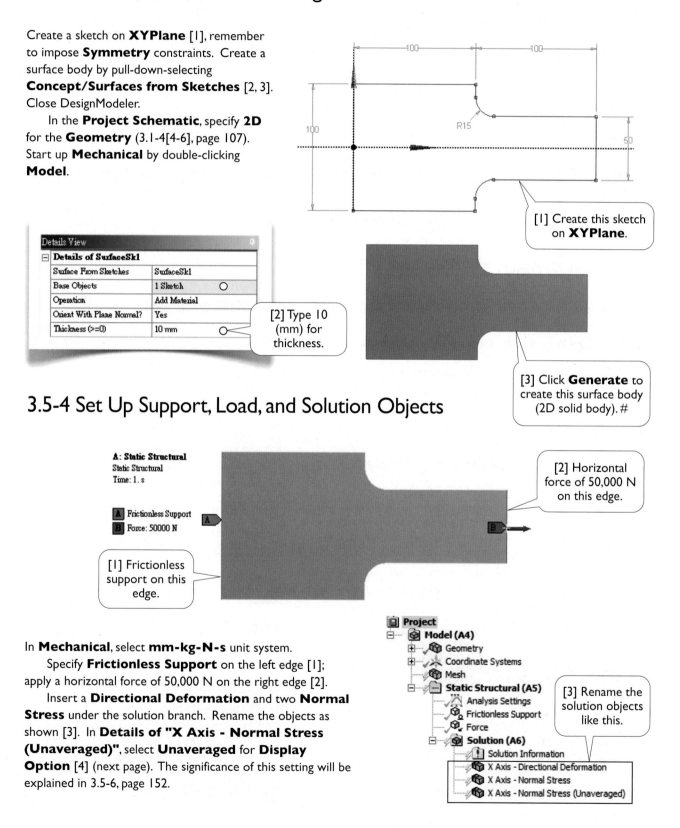

[1] Create this sketch on **XYPlane**.

Details View

Details of SurfaceSk1	
Surface From Sketches	SurfaceSk1
Base Objects	1 Sketch
Operation	Add Material
Orient With Plane Normal?	Yes
Thickness (>=0)	10 mm

[2] Type 10 (mm) for thickness.

[3] Click **Generate** to create this surface body (2D solid body). #

3.5-4 Set Up Support, Load, and Solution Objects

A: Static Structural
Static Structural
Time: 1. s

A Frictionless Support
B Force: 50000 N

[2] Horizontal force of 50,000 N on this edge.

[1] Frictionless support on this edge.

In **Mechanical**, select **mm-kg-N-s** unit system.

Specify **Frictionless Support** on the left edge [1]; apply a horizontal force of 50,000 N on the right edge [2].

Insert a **Directional Deformation** and two **Normal Stress** under the solution branch. Rename the objects as shown [3]. In **Details of "X Axis - Normal Stress (Unaveraged)"**, select **Unaveraged** for **Display Option** [4] (next page). The significance of this setting will be explained in 3.5-6, page 152.

Project
- **Model (A4)**
  - Geometry
  - Coordinate Systems
  - Mesh
  - **Static Structural (A5)**
    - Analysis Settings
    - Frictionless Support
    - Force
    - **Solution (A6)**
      - Solution Information
      - X Axis - Directional Deformation
      - X Axis - Normal Stress
      - X Axis - Normal Stress (Unaveraged)

[3] Rename the solution objects like this.

Details of "X Axis - Directional Deformation"

Scope	
Scoping Method	Geometry Selection
Geometry	All Bodies
Definition	
Type	Directional Deformation
Orientation	X Axis
By	Time
☐ Display Time	Last
Coordinate System	Global Coordinate System
Calculate Time History	Yes
Identifier	
Suppressed	No
⊞ **Results**	

Details of "X Axis - Normal Stress"

Scope	
Scoping Method	Geometry Selection
Geometry	All Bodies
Definition	
Type	Normal Stress
Orientation	X Axis
By	Time
☐ Display Time	Last
Coordinate System	Global Coordinate System
Calculate Time History	Yes
Identifier	
Suppressed	No
Integration Point Results	
Display Option	Averaged
Average Across Bodies	No
⊞ **Results**	

Details of "X Axis - Normal Stress (Unaveraged)"

Scope	
Scoping Method	Geometry Selection
Geometry	All Bodies
Definition	
Type	Normal Stress
Orientation	X Axis
By	Time
☐ Display Time	Last
Coordinate System	Global Coordinate System
Calculate Time History	Yes
Identifier	
Suppressed	No
Integration Point Results	
Display Option	Unaveraged
⊞ **Results**	

[4] Select **Unaveraged** for **Display Option**. #

3.5-5 Set Up Mesh Controls

Highlight **Mesh** in the project tree. In the details view, turn off **Use Advanced Size Function** and type 10 (mm) for **Element Size** [1] and select **Dropped** for **Element Midside Nodes** [2].

Insert a **Mesh Control/Method** (select from toolbar) and select the body as **Geometry** [3]. Select **Triangles** for **Method** [4]. This sets up an all-triangle method.

Generate the mesh [5]. Click **Solve** to solve the model.

Why Coarse Lower-Order Triangular Elements?

The purpose of Part A is to demonstrate *stress discontinuity*, a must-know in finite element solutions. We've set up a mesh that is coarse (about 10 mm of element size), and the elements used are linear (midside nodes dropped [2]) triangular. The purpose of this setup is to exaggerate the stress discontinuity behavior. Stress discontinuity is intrinsic in all the finite element software using displacements as degrees of freedom, and not limited to triangular or lower-order elements.

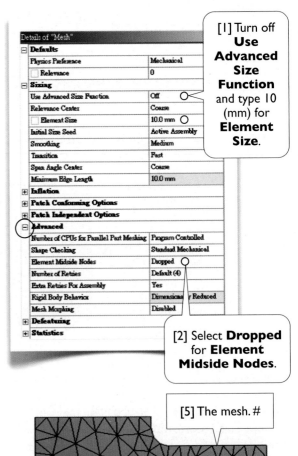

Details of "Mesh"

Defaults	
Physics Preference	Mechanical
☐ Relevance	0
Sizing	
Use Advanced Size Function	Off
Relevance Center	Coarse
☐ Element Size	10.0 mm
Initial Size Seed	Active Assembly
Smoothing	Medium
Transition	Fast
Span Angle Center	Coarse
Minimum Edge Length	10.0 mm
⊞ **Inflation**	
⊞ **Patch Conforming Options**	
⊞ **Patch Independent Options**	
Advanced	
Number of CPUs for Parallel Part Meshing	Program Controlled
Shape Checking	Standard Mechanical
Element Midside Nodes	Dropped
Number of Retries	Default (4)
Extra Retries For Assembly	Yes
Rigid Body Behavior	Dimensionally Reduced
Mesh Morphing	Disabled
⊞ **Defeaturing**	
⊞ **Statistics**	

[1] Turn off **Use Advanced Size Function** and type 10 (mm) for **Element Size**.

[2] Select **Dropped** for **Element Midside Nodes**.

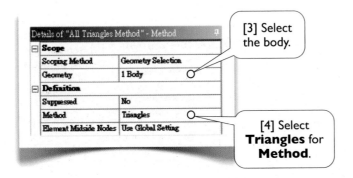

Details of "All Triangles Method" - Method

Scope	
Scoping Method	Geometry Selection
Geometry	1 Body
Definition	
Suppressed	No
Method	Triangles
Element Midside Nodes	Use Global Setting

[3] Select the body.

[4] Select **Triangles** for **Method**.

[5] The mesh. #

3.5-6 View the Results

Displacement Fields

Nodal displacements, calculated by solving Eq. 1.3-1(1) (page 31), are single valued (each node has a unique value). Therefore the displacement fields, calculated using Eq. 1.3-2(2) (page 32), must be continuous over the entire body [1].

The displacement fields are continuous but not necessarily smooth. The use of continuous shape functions within the element guarantees the displacements field piecewise smooth, but not necessarily smooth across the element boundaries.

Stress Fields

The strain fields are then calculated using Eq. 1.2-7(1) (page 27), and stress fields are calculated using Eq. 1.2-8(1) (page 27).

The calculations are element-by-element. The figure in the right [2, 3] is a typical results of stress calculation: a node usually has multiple stress values, since the node may connect to multiple elements, and each element has its own value of stress.

This behavior can be easily understood. Since differentiations of piecewise smooth displacement fields are involved in Eq. 1.2-7(1) (page 27), this ensures the strain fields and the stress fields continuous inside the element but not necessarily continuous across the element boundaries.

By default, stresses are averaged on the nodes, and the stress field is recalculated. After that, the stress field is continuous over the body [4].

Usage of Unaveraged Stresses

The averaged stress fields [4] are visually efficient for human eyes to interpret the results, while unaveraged stress fields [2, 3] provide a way of assessing the solution accuracy.

In general, as the mesh is getting finer, the solution is more accurate, and the stress discontinuity is less obvious. Thus, stress discontinuity can be used to indicate the solution accuracy: the less discontinuous the stress field, the more accurate the solution.

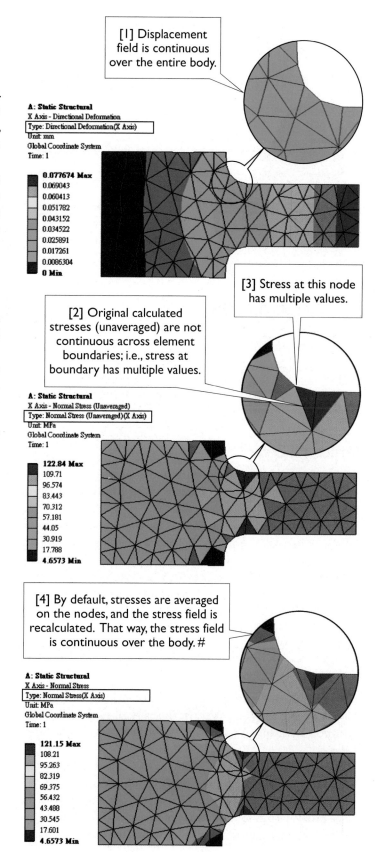

[1] Displacement field is continuous over the entire body.

A: Static Structural
X Axis - Directional Deformation
Type: Directional Deformation(X Axis)
Unit: mm
Global Coordinate System
Time: 1

0.077674 Max
0.069043
0.060413
0.051782
0.043152
0.034522
0.025891
0.017261
0.0086304
0 Min

[2] Original calculated stresses (unaveraged) are not continuous across element boundaries; i.e., stress at boundary has multiple values.

[3] Stress at this node has multiple values.

A: Static Structural
X Axis - Normal Stress (Unaveraged)
Type: Normal Stress (Unaveraged)(X Axis)
Unit: MPa
Global Coordinate System
Time: 1

122.84 Max
109.71
96.574
83.443
70.312
57.181
44.05
30.919
17.788
4.6573 Min

[4] By default, stresses are averaged on the nodes, and the stress field is recalculated. That way, the stress field is continuous over the body. #

A: Static Structural
X Axis - Normal Stress
Type: Normal Stress(X Axis)
Unit: MPa
Global Coordinate System
Time: 1

121.15 Max
108.21
95.263
82.319
69.375
56.432
43.488
30.545
17.601
4.6573 Min

PART B. STRUCTURAL ERROR

3.5-7 Evaluate Structural Error

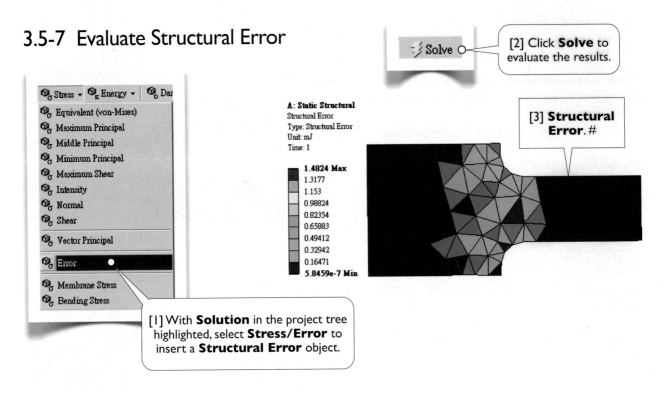

[1] With **Solution** in the project tree highlighted, select **Stress/Error** to insert a **Structural Error** object.

[2] Click **Solve** to evaluate the results.

[3] **Structural Error**.#

Structural Error[Ref 1]

For an element, strain energy calculated using averaged stresses are different from that using unaveraged stresses. The difference between the two values is called **Structural Error** of the element. The finer the mesh, the smaller the structural error.

The structural error can be used for two purposes: (a) As an indicator of global mesh adequacy. In general, we want the values as small as possible. Refining the mesh globally is a way of reducing structural error. (b) As an indicator of the local mesh adequacy. In general, we want the structural error distribution as uniform as possible to optimize the efficiency of computation effort. That is, in the region of large values of structural error we should reduce the element size while in the region of small values we should enlarge the element size.

PART C. FINITE ELEMENT CONVERGENCE

One of the core concepts of the finite element methods is that *the finer the mesh, the more accurate the solution*. Ultimately, the solution will reach an analytical solution as the mesh is fine enough. But, how fast does it approach the analytical solution? This is what we want to answer in this part of the section. It is called the *element convergence behavior*.

The answer depends on what kind of element we are using. We'll draw a conclusion that, for 2D cases, *quadrilateral elements generally converge faster than triangular elements*.

We will compare lower-order triangular with lower-order quadrilateral. To be fair, the comparison is made under the same problem sizes for both types of elements. The problem size is reasonably defined by the number of nodes.

As a representative structural response, the maximum horizontal displacement is recorded for comparison.

3.5-8 Triangular Elements

Repeat the simulation. Change the element size for each run [1], record the number of nodes [2], and the maximum displacement. Tabulate your results like this:

Element Size (mm)	Number of Nodes	Max Displacement (mm)
10	104	0.077674
6	220	0.078106
4	416	0.078279
3	675	0.078442
2	1402	0.078517
1.5	2356	0.078576
1	5039	0.078618
0.7	10223	0.078637
0.6	13714	0.078640

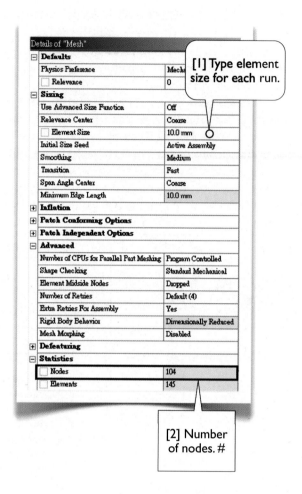

3.5-9 Quadrilateral Elements

Highlight **All Triangles Method** under the **Mesh** branch. In the details view, change the method to **Quadrilateral Dominant** [1], and select **All Quad** for **Free Face Mesh Type** [2]. Repeat the simulation, and tabulate your results like this:

Element Size (mm)	Number of Nodes	Max Displacement (mm)
10	186	0.078294
6	497	0.078448
4	1047	0.078566
3	1838	0.078600
2	4014	0.078628
1.5	7117	0.078639
1	16066	0.078647
0.8	25200	0.078649

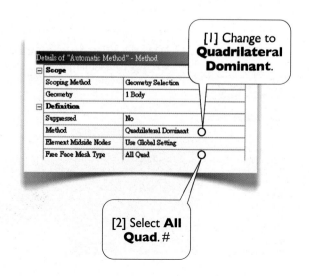

3.5-10 Comparison and Conclusions

Plotting the results in 3.5-8 and 3.5-9 (last page), you should come up with a chart as shown below [1-3]. The two curves share the same horizontal asymptote (Displacement ≈ 0.07864 mm), which is the analytical solution. Additional behaviors can be observed as follows.

First, the quadrilateral element converges faster than the triangular element. The difference seems not dramatic in this particular case. In many other cases, however, the difference can be significant. In Section 9.3 we will investigate the convergence behavior of 3D elements, and we'll conclude that the convergence speed of hexahedra (1.3-3[1], page 34) is faster than prisms or tetrahedra (1.3-3[2, 4], page 34).

Second, all the convergence curves approach the asymptote from below; i.e., displacements calculated by the finite element methods never exceed the analytical solution. In other words, the finite element solutions always underestimate the deformation, or in terms of stiffness (Eq. 1.3-1(1), page 31), the stiffness is always overestimated.

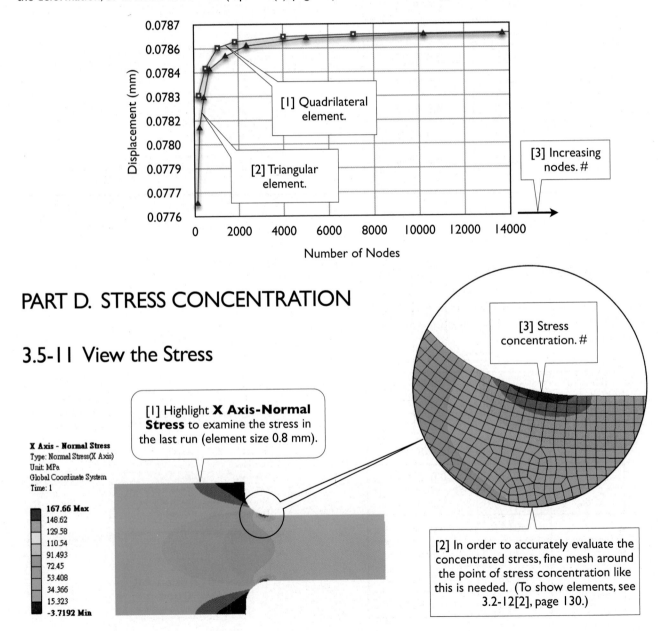

PART D. STRESS CONCENTRATION

3.5-11 View the Stress

[1] Highlight **X Axis-Normal Stress** to examine the stress in the last run (element size 0.8 mm).

X Axis - Normal Stress
Type: Normal Stress(X Axis)
Unit: MPa
Global Coordinate System
Time: 1

167.66 Max
148.62
129.58
110.54
91.493
72.45
53.408
34.366
15.323
-3.7192 Min

[3] Stress concentration. #

[2] In order to accurately evaluate the concentrated stress, fine mesh around the point of stress concentration like this is needed. (To show elements, see 3.2-12[2], page 130.)

3.5-12 Define a Path

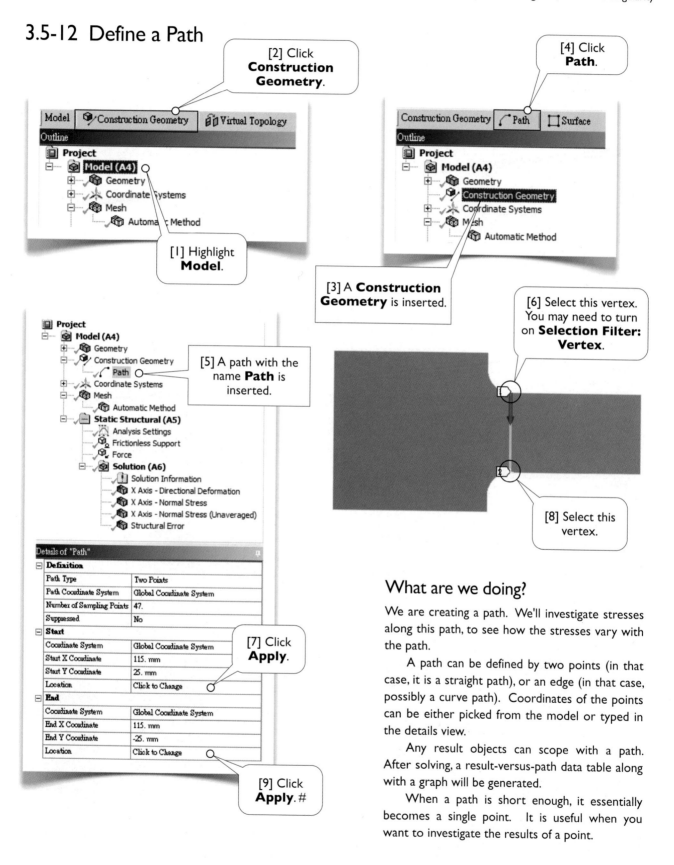

[2] Click **Construction Geometry**.

[4] Click **Path**.

[1] Highlight **Model**.

[3] A **Construction Geometry** is inserted.

[5] A path with the name **Path** is inserted.

[6] Select this vertex. You may need to turn on **Selection Filter: Vertex**.

[8] Select this vertex.

Details of "Path"

Definition	
Path Type	Two Points
Path Coordinate System	Global Coordinate System
Number of Sampling Points	47.
Suppressed	No
Start	
Coordinate System	Global Coordinate System
Start X Coordinate	115. mm
Start Y Coordinate	25. mm
Location	Click to Change
End	
Coordinate System	Global Coordinate System
End X Coordinate	115. mm
End Y Coordinate	-25. mm
Location	Click to Change

[7] Click **Apply**.

[9] Click **Apply**. #

What are we doing?

We are creating a path. We'll investigate stresses along this path, to see how the stresses vary with the path.

A path can be defined by two points (in that case, it is a straight path), or an edge (in that case, possibly a curve path). Coordinates of the points can be either picked from the model or typed in the details view.

Any result objects can scope with a path. After solving, a result-versus-path data table along with a graph will be generated.

When a path is short enough, it essentially becomes a single point. It is useful when you want to investigate the results of a point.

3.5-13 View Stresses Along the Path

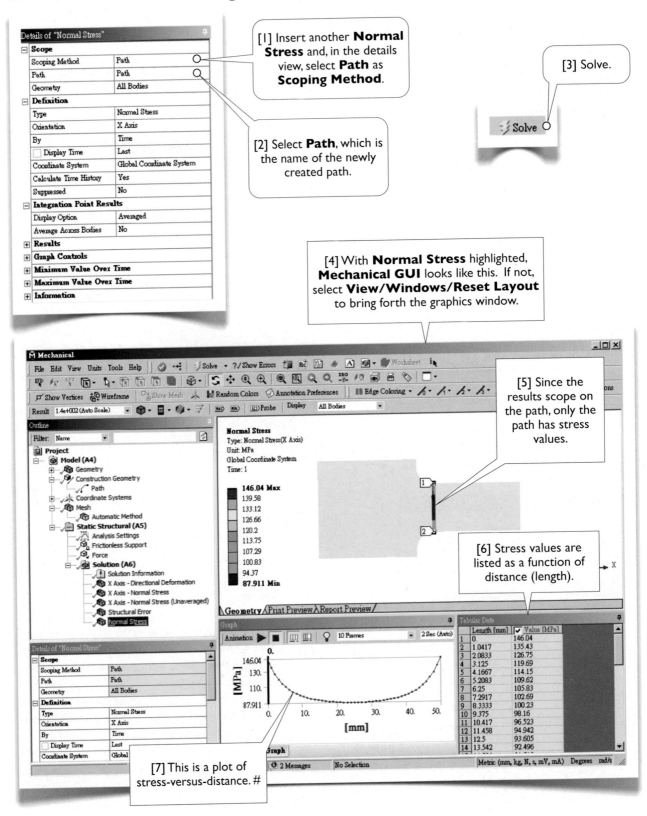

Details of "Normal Stress"

Scope	
Scoping Method	Path
Path	Path
Geometry	All Bodies
Definition	
Type	Normal Stress
Orientation	X Axis
By	Time
☐ Display Time	Last
Coordinate System	Global Coordinate System
Calculate Time History	Yes
Suppressed	No
Integration Point Results	
Display Option	Averaged
Average Across Bodies	No
⊞ **Results**	
⊞ **Graph Controls**	
⊞ **Minimum Value Over Time**	
⊞ **Maximum Value Over Time**	
⊞ **Information**	

[1] Insert another **Normal Stress** and, in the details view, select **Path** as **Scoping Method**.

[2] Select **Path**, which is the name of the newly created path.

[3] Solve.

Solve

[4] With **Normal Stress** highlighted, **Mechanical GUI** looks like this. If not, select **View/Windows/Reset Layout** to bring forth the graphics window.

[5] Since the results scope on the path, only the path has stress values.

[6] Stress values are listed as a function of distance (length).

[7] This is a plot of stress-versus-distance. #

157

PART E. STRESS SINGULARITY

Most of students are aware of the concept of stress concentration: stress is larger at a concave corner than stress at locations away from the corner. To accurately evaluate the concentrated stress, finer mesh needs to be used around the corner. As a general guideline, *an area of higher stress gradient requires finer mesh.*

The degree of stress concentration is related to the radius of the corner: the smaller the radius, the larger the concentrated stress. Naturally, a question raises: what happen if the radius of the concave corner is zero (i.e., a sharp angle corner)? The elasticity theory predicts that, in that case, the stress at that sharp corner is infinity. A stress of infinity is called a *singular stress.*

Fortunately, a corner with zero radius never exists in the real-world. It is difficult to manufacture such a zero-radius fillet. It requires a process of infinite-high precision machining.

However, zero-radius fillets do exist in the virtual-world of simulations. Since many small features such as fillets do not significantly affect the global behavior (e.g., deformation) of a structure, these small features are often not modeled in the simulation model. The bottom line is that zero-radius fillets exist everywhere in a simplified model.

It is important that the engineers be aware of the existence of stress singularities. Novice engineers often mistakenly take the maximum stress as the design stress, while, in fact, that is a singular stress—it doesn't exist in the real-world. Always check if the maximum stress occurs at a singular point. If so, the stress is meaningless. If a concentrated stress is important, alway include the fillet in the model.

3.5-14 Modify the Model in DesignModeler

Close **Mechanical**. Duplicate the **Static Structural** system. Start up DesignModeler [1] and make sure **Auto Constraints** are turned on.

Modify the sketch such that the fillets become sharp corners [2]. Click **Generate**, to update the surface body. Reduce the model to half by taking advantage of symmetry [3] (using **Tools/Symmetry**). We reduce the model because we need to use very fine mesh to study stress singularity. Close DesignModeler and open **Mechanical** again.

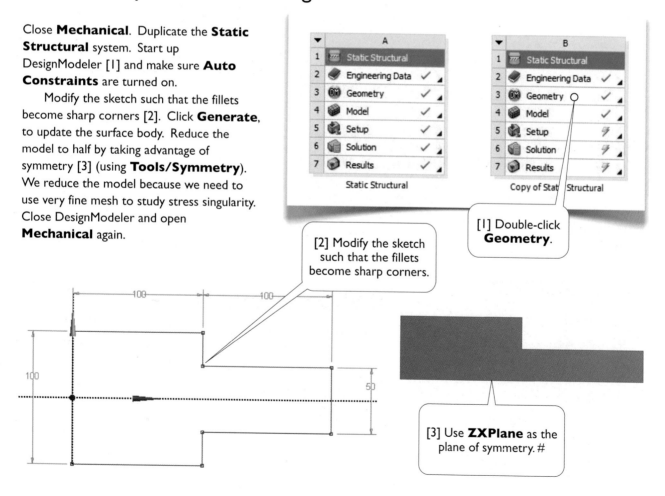

[1] Double-click **Geometry**.

[2] Modify the sketch such that the fillets become sharp corners.

[3] Use **ZXPlane** as the plane of symmetry. #

3.5-15 Modify the Environment Conditions in **Mechanical**

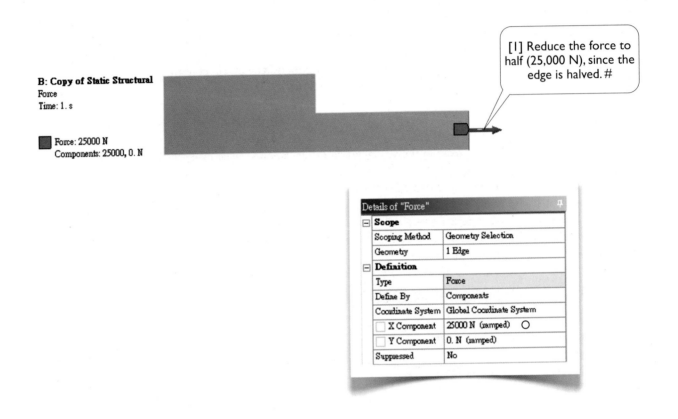

[1] Reduce the force to half (25,000 N), since the edge is halved. #

3.5-16 Set Up Mesh Controls

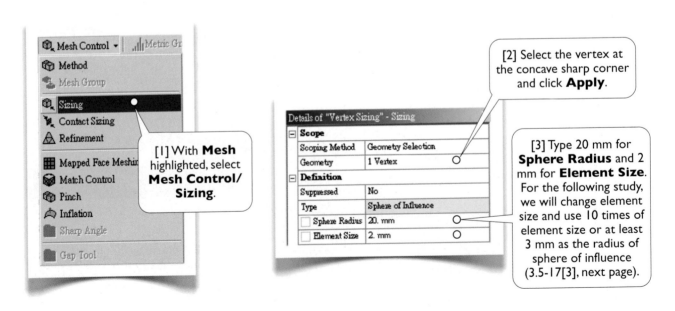

[1] With **Mesh** highlighted, select **Mesh Control/ Sizing**.

[2] Select the vertex at the concave sharp corner and click **Apply**.

[3] Type 20 mm for **Sphere Radius** and 2 mm for **Element Size**. For the following study, we will change element size and use 10 times of element size or at least 3 mm as the radius of sphere of influence (3.5-17[3], next page).

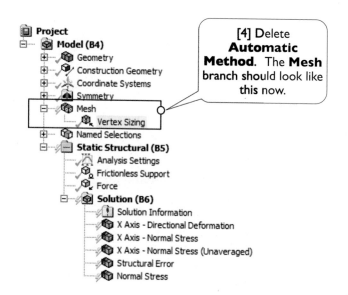

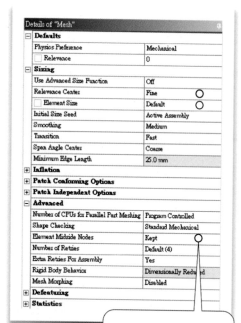

3.5-17 Perform Simulations and Conclusion

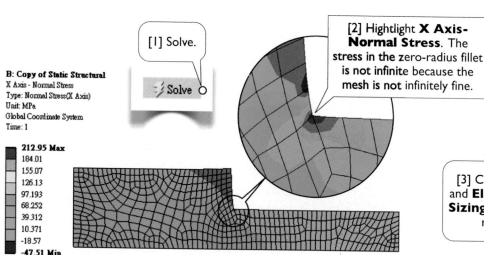

B: Copy of Static Structural
X Axis - Normal Stress
Type: Normal Stress(X Axis)
Unit: MPa
Global Coordinate System
Time: 1

212.95 Max
184.01
155.07
126.13
97.193
68.252
39.312
10.371
-18.57
-47.51 Min

After solving the model, the stress at the zero-radius fillet is not infinite, the theoretical value. This is because our mesh is not fine enough. To achieve an infinite value of stress, you need a zero element size. That is, of course, not possible.

Reduce **Sphere Radius** and **Element Size** and record the maximum stress. Repeat this and you should obtain a table similar to [3].

Interpretation of the data may be easier if we plot them in a graph [4] (next page). It is obvious that, as the mesh gets finer (from right to left), the stress at the sharp corner eventually reaches an infinite value, consistent with the theoretical value.

Sphere Radius (mm)	Element Size (mm)	Max Normal Stress (X Axis) (MPa)
20	2	212.95
10	1	254.70
5	0.5	364.34
3	0.2	604.03
3	0.1	767.93
3	0.05	1220.1

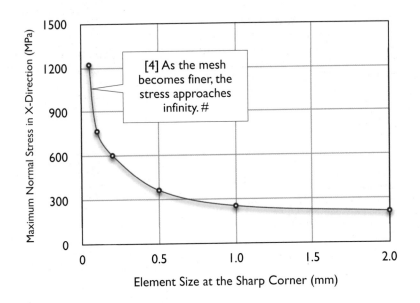

Remark

Stress singularity is not limited to concave sharp corners. Any locations that have stress of infinity are called singular points. For example, a point of concentrated forces is also a singular point, since a point has zero area.

Wrap Up

Save the project. Close **Mechanical** and exit Workbench.

Reference

1. ANSYS Documentation//Mechanical APDL//Theory Reference//17.7.1. Error Approximation Technique for Displacement-Based Problems

Section 3.6

Review

3.6-1 Keywords

Choose a letter for each keyword from the list of descriptions

1. (   ) APDL

2. (   ) Element Convergence Study

3. (   ) Environment Condition

4. (   ) Inconsistent Unit System

5. (   ) Plane Strain Problem

6. (   ) Plane Stress Problem

7. (   ) Project Tree

8. (   ) Stress Discontinuity

9. (   ) Stress Singularity

10. (   ) Structural Error

11. (   ) Weak Spring

Answers:

1. (F)   2. (J)   3. (G)   4. (E)   5. (C)   6. (B)   7. (D)   8. (H)
9. (K)   10.(I)   11.(A)

List of Descriptions

(A) When Workbench detects a structure as unstable, it adds weak springs on the structure to make it capable of withstanding very small external forces.

(B) In a structural simulation problem, if all the stresses in a direction, say Z-direction, vanish, the problem can be reduced to a 2D problem, and is called a plane stress problem.

(C) In a structural simulation problem, if all the strains in a direction, say Z-direction, vanish, the problem can be reduced to a 2D problem, and is called a plane strain problem.

(D) A project tree is a structured representation of an analysis system and displayed on the **Outline** in **Mechanical**. A project tree contains one or more simulation models. A simulation model contains one or more **Environment** branches.

(E) A unit system in which at least one unit is not consistent with other units. For example, **mm-kg-N-s** is inconsistent, since if mm and kg is used, the force must be mN (milli-newton) instead of N.

(F) ANSYS Parametric Design Language. A set of text-based language that is used to drive ANSYS Classic program.

(G) Consist of loads, supports, and inertial effects, which apply on the simulation model.

(H) In the finite element methods, the shape functions are used to interpolate the displacement fields interior to the element. Across the element boundaries, the displacement fields are continuous but not smooth. The strains and stresses, which are calculated by differentiating the displacement fields, become discontinuous across the element boundaries.

(I) For an element, strain energies calculated using averaged stresses and unaveraged stresses are different. The difference between these two energy values is called structural error of the element. The finer the mesh, the smaller the structural error. It is used as an indicator for mesh adequacy.

(J) Study of how the finite element solutions approach to theoretical values as the mesh is getting finer. In 2D, quadrilateral elements converge faster than the triangular. In 3D, the order of convergence speeds are, from faster to slower, hexahedral, prism, pyramid, and tetrahedral.

(K) Any locations that have infinitely large stress are called points of stress singularity. They are often found at concave fillets of zero radius, or at points subject to concentrated forces or displacement constraints.

3.6-2 Additional Workbench Exercises

Stress in a Long Cylinder

Consider the problem described in VM25 of the APDL verification manual[Ref 1]. Find the radial stress and the hoop stress at the inner and outer surfaces and at the middle wall thickness, for both loading cases. Model the problem as (a) axisymmetric problem and (b) plane strain problem.

Reference

1. ANSYS Documentation//Verification Manuals//Mechanical APDL Verification Manual//I. Verification Test Case Description//VM25

Chapter 4
3D Solid Modeling

Creating a 3D geometric model is usually much more elaborate than a 2D model. Most of the techniques you've learned in Chapters 2 and 3 for 2D cases can be applied in constructing 3D models. The types of 3D bodies that Workbench supports are *solid bodies*, *surface bodies*, and *line bodies*, and they may be mixed up in a 3D model. In this chapter, we will mostly focus on models consisting of solid bodies only, except Section 4.3, in which a surface body and a solid body constitute a 3D model. Chapters 6 and 7 will focus on surface bodies and line bodies respectively.

Purpose of This Chapter

This chapter guides students familiarize with 3D solid modeling using DesignModeler. Four mechanical parts are created in this chapter. These models are then used for simulations in the next and later chapters.

About Each Section

We start with a beam bracket model in Section 4.1 to introduce 3D bodies creation basics. Section 4.2 creates a more elaborated model, the cover of a pressure cylinder, to obtain some degree of proficiency at the 3D modeling techniques. Section 4.3 creates a model for a lifting fork, which is a combination of a solid body and a surface body. Section 4.4 overviews 3D solid bodies creation and manipulation tools in a systematic way, intending to fill up what were missed in the first three sections. Section 4.5 provides an additional exercise.

Section 4.1

Beam Bracket

4.1-1 About the Beam Bracket

When constructing a steel beam-column structure, such as high-rise buildings or manufacturing plants, columns are erected before beams can be elevated, positioned, and welded. The functions of beam brackets [1] are to (a) precisely position the beam during the construction, and (b) safely transfer the loads from the beam to the column. The loads are determined by a thorough analysis of the entire structure subject to design loads, such as dead load, live load, earthquake, wind load, etc.

The beam bracket here consists of a seat plate (the flange) [2] and a web plate [3]. The design considerations include: (a) Would the maximum stress excess the allowable stress? (b) Would the web buckle under the load?

In this section, we will create a 3D solid model for the beam bracket. The solid model will be used for a static structural simulation in Section 5.1 and then a buckling analysis in Section 10.3. In Section 6.2, the solid model will be simplified to a surface model and the simulation results are compared with those in Section 5.1.

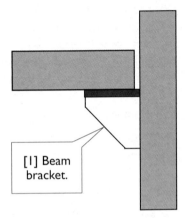

[1] Beam bracket.

Global Coordinate System

Before creating a geometric model in DesignModeler, you need to set up a global coordinate system in mind. This is true both for 2D and 3D. For 2D, it is so trivial that we even didn't mention it. For 3D, you need to pay more attention about the global coordinate system. Many students are easily disoriented when working with 3D models on a 2D computer screen. In this section, the global coordinate system is set up as shown [4-7].

[2] Seat plate.

[3] Web plate.

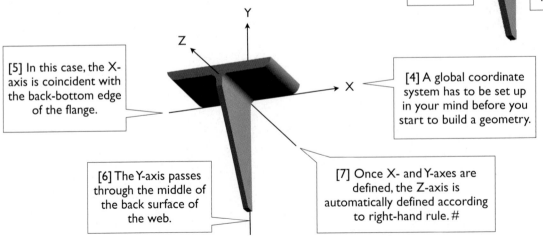

[5] In this case, the X-axis is coincident with the back-bottom edge of the flange.

[4] A global coordinate system has to be set up in your mind before you start to build a geometry.

[6] The Y-axis passes through the middle of the back surface of the web.

[7] Once X- and Y-axes are defined, the Z-axis is automatically defined according to right-hand rule. #

165

4.1-2 Start Up

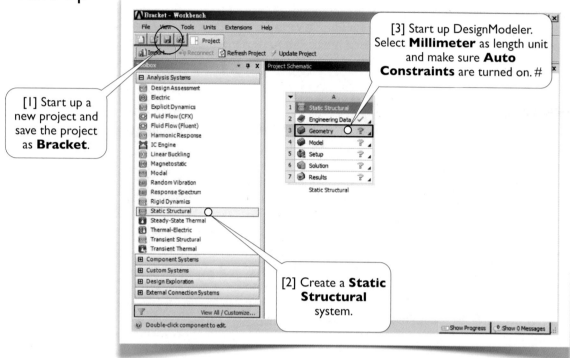

[1] Start up a new project and save the project as **Bracket**.

[3] Start up DesignModeler. Select **Millimeter** as length unit and make sure **Auto Constraints** are turned on. #

[2] Create a **Static Structural** system.

4.1-3 Sketch Seat Plate on ZXPlane

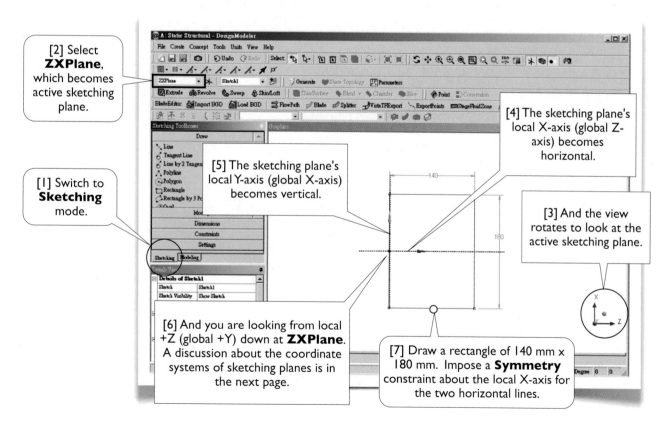

[2] Select **ZXPlane**, which becomes active sketching plane.

[1] Switch to **Sketching** mode.

[5] The sketching plane's local Y-axis (global X-axis) becomes vertical.

[4] The sketching plane's local X-axis (global Z-axis) becomes horizontal.

[3] And the view rotates to look at the active sketching plane.

[6] And you are looking from local +Z (global +Y) down at **ZXPlane**. A discussion about the coordinate systems of sketching planes is in the next page.

[7] Draw a rectangle of 140 mm x 180 mm. Impose a **Symmetry** constraint about the local X-axis for the two horizontal lines.

Local Coordinate Systems of Sketching Planes

We mentioned plane coordinate system in the bottom of page 114. Here is more. Each sketching plane has a local coordinate system. For three pre-defined planes (**XYPlane**, **YZPlane**, and **ZXPlane**), their first two letters, which are the name of the global axes, are the local X-axis and Y-axis respectively; the local Z-axis is then defined according to the right-hand rule. For example, **ZXPlane**'s local-X and local Y-axis coincide with the global Z-axis and global X-axis respectively [8-10].

Plane View vs. 3D View Sketching

When the view rotates to look at the sketching plane [3], the local X-axis and Y-axis of the sketching plane becomes horizontal and vertical respectively [4, 5], and the local Z-axis points toward you. It is a simple rule but may take time to get accustomed to. If you have trouble sketching with plane view [4-6] due to the orientation problems, you always can sketch in 3D view. In our case, you may draw the sketch in 3D space as shown [8-10]. The main problem of sketching in 3D view is that all angles are distorted. For example, a right angle is no longer 90°. If you have no problems with orientation, sketching with plane view is usually more convenient. However, choose whichever is suitable for you. In this book, we will stick to plane view sketching.

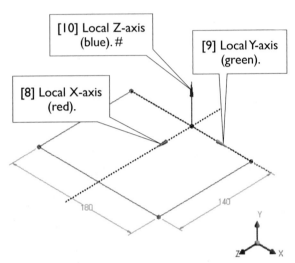

[10] Local Z-axis (blue). #

[9] Local Y-axis (green).

[8] Local X-axis (red).

4.1-4 Extrude the Sketch to Create Seat Plate

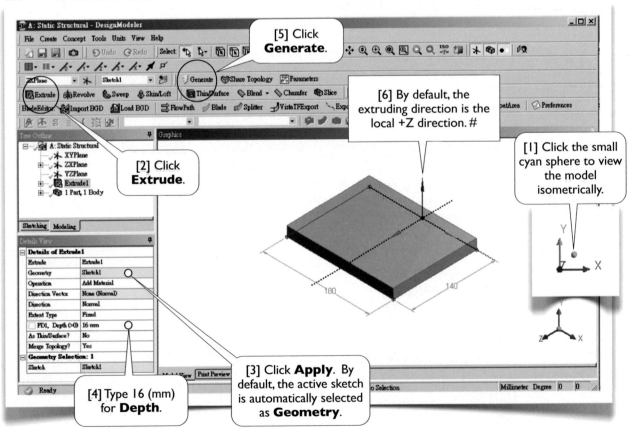

[5] Click **Generate**.

[6] By default, the extruding direction is the local +Z direction. #

[1] Click the small cyan sphere to view the model isometrically.

[2] Click **Extrude**.

[3] Click **Apply**. By default, the active sketch is automatically selected as **Geometry**.

[4] Type 16 (mm) for **Depth**.

4.1-5 Sketch Web Plate on YZPlane

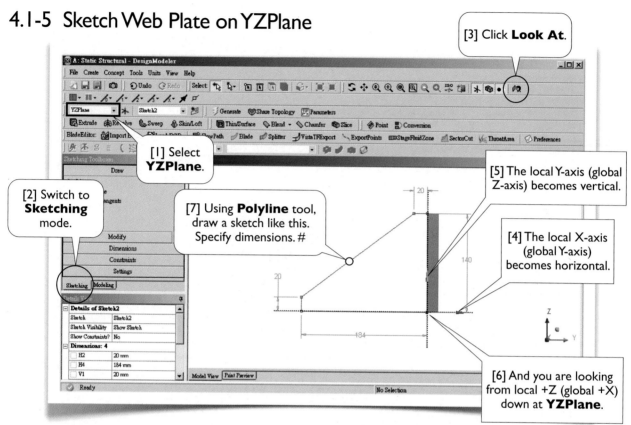

[3] Click **Look At**.

[1] Select **YZPlane**.

[2] Switch to **Sketching** mode.

[7] Using **Polyline** tool, draw a sketch like this. Specify dimensions. #

[5] The local Y-axis (global Z-axis) becomes vertical.

[4] The local X-axis (global Y-axis) becomes horizontal.

[6] And you are looking from local +Z (global +X) down at **YZPlane**.

4.1-6 Extrude the Sketch to Create Web Plate

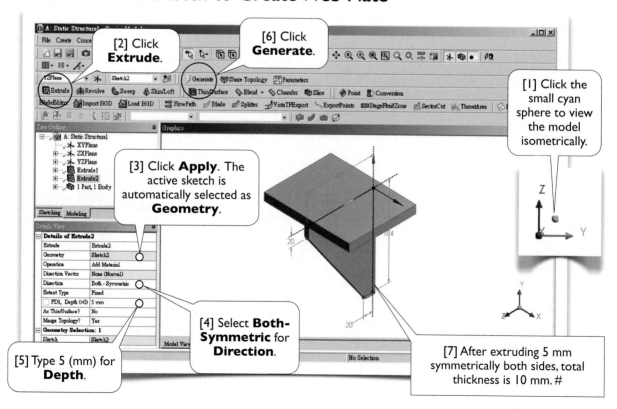

[2] Click **Extrude**.

[6] Click **Generate**.

[1] Click the small cyan sphere to view the model isometrically.

[3] Click **Apply**. The active sketch is automatically selected as **Geometry**.

[4] Select **Both-Symmetric** for **Direction**.

[5] Type 5 (mm) for **Depth**.

[7] After extruding 5 mm symmetrically both sides, total thickness is 10 mm. #

4.1-7 Create Rounds for the Flange

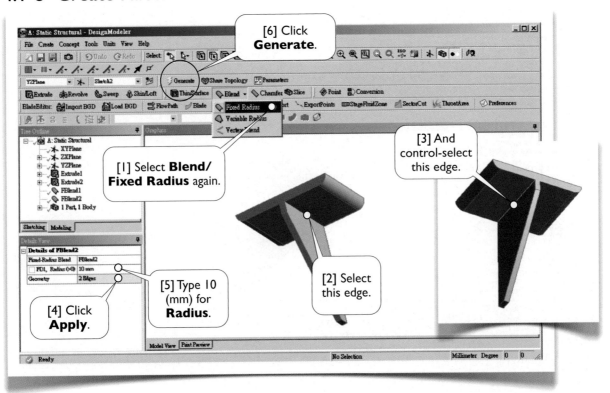

[1] Turn off **Display Plane**.

[7] Click **Generate**. #

[2] Select **Blend/ Fixed Radius**.

[3] Select this edge.

[6] Type 16 (mm) for **Radius**.

[5] Click **Apply**.

[4] And control-select this edge. You may need to rotate the model (using middle-click-drag, see 2.1-7[6], page 59).

4.1-8 Create Fillets

[6] Click **Generate**.

[1] Select **Blend/ Fixed Radius** again.

[3] And control-select this edge.

[2] Select this edge.

[4] Click **Apply**.

[5] Type 10 (mm) for **Radius**.

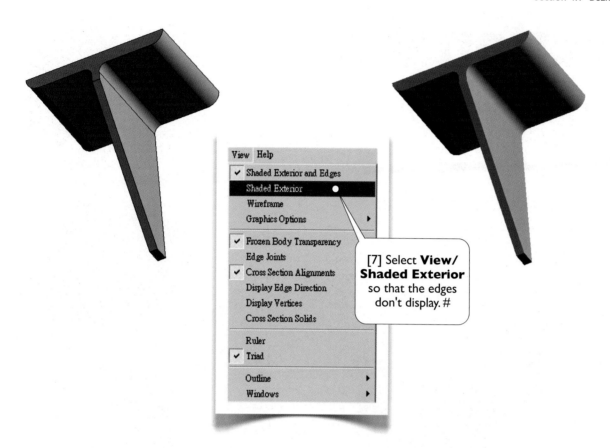

View Help

✓ Shaded Exterior and Edges

   Shaded Exterior   ●

   Wireframe

   Graphics Options   ▶

✓ Frozen Body Transparency

   Edge Joints

✓ Cross Section Alignments

   Display Edge Direction

   Display Vertices

   Cross Section Solids

   Ruler

✓ Triad

   Outline   ▶

   Windows   ▶

[7] Select **View/ Shaded Exterior** so that the edges don't display. #

Wrap Up

Close DesignModeler, save the project, and exit Workbench.

Section 4.2

Cover of Pressure Cylinder

4.2-1 About the Cylinder Cover

The pressure cylinder [1] contains a gas of 0.5 MPa. The cylinder cover [2-4] is made of carbon-fiber reinforced plastic. In this section, we want to create a 3D solid model for the cover; the model will be used for a static structural simulation in Section 5.2 to investigate the deformation under the pressure.

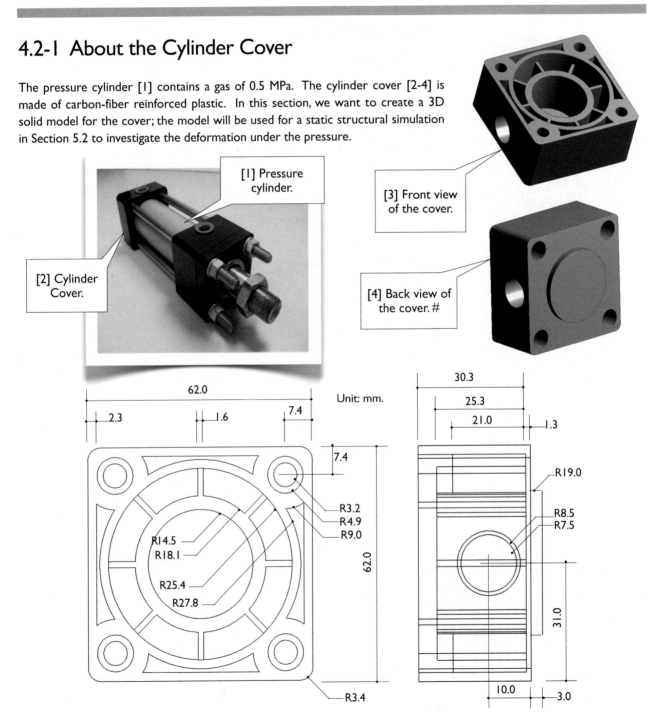

[1] Pressure cylinder.

[2] Cylinder Cover.

[3] Front view of the cover.

[4] Back view of the cover. #

Unit: mm.

62.0

2.3

1.6

7.4

7.4

R3.2
R4.9
R9.0

R14.5
R18.1

R25.4

R27.8

62.0

R3.4

30.3

25.3

21.0

1.3

R19.0

R8.5
R7.5

31.0

10.0

3.0

4.2-2 Start Up

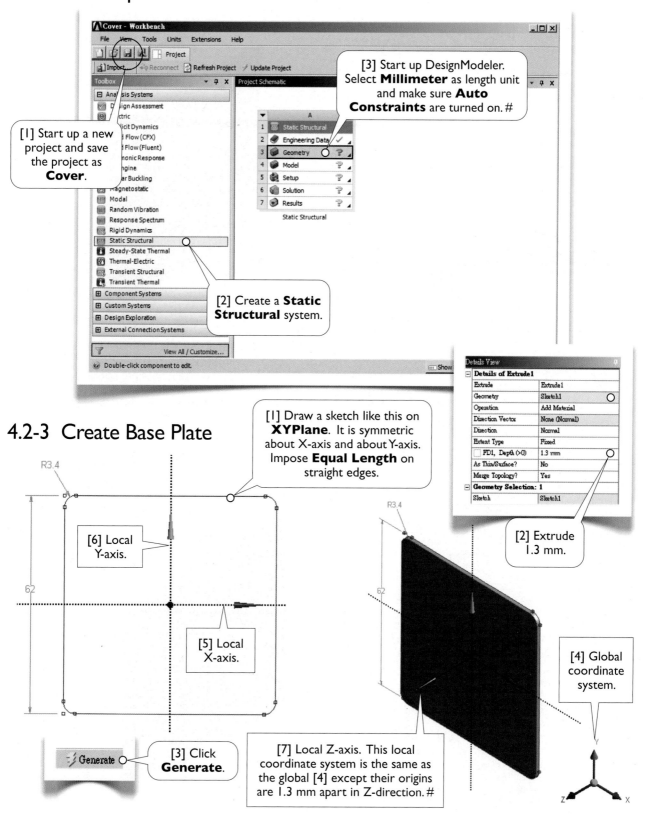

4.2-3 Create Base Plate

[1] Start up a new project and save the project as **Cover**.

[2] Create a **Static Structural** system.

[3] Start up DesignModeler. Select **Millimeter** as length unit and make sure **Auto Constraints** are turned on. #

[1] Draw a sketch like this on **XYPlane**. It is symmetric about X-axis and about Y-axis. Impose **Equal Length** on straight edges.

[2] Extrude 1.3 mm.

[6] Local Y-axis.

[5] Local X-axis.

[4] Global coordinate system.

[3] Click **Generate**.

[7] Local Z-axis. This local coordinate system is the same as the global [4] except their origins are 1.3 mm apart in Z-direction. #

172

4.2-4 Create a New Sketching Plane

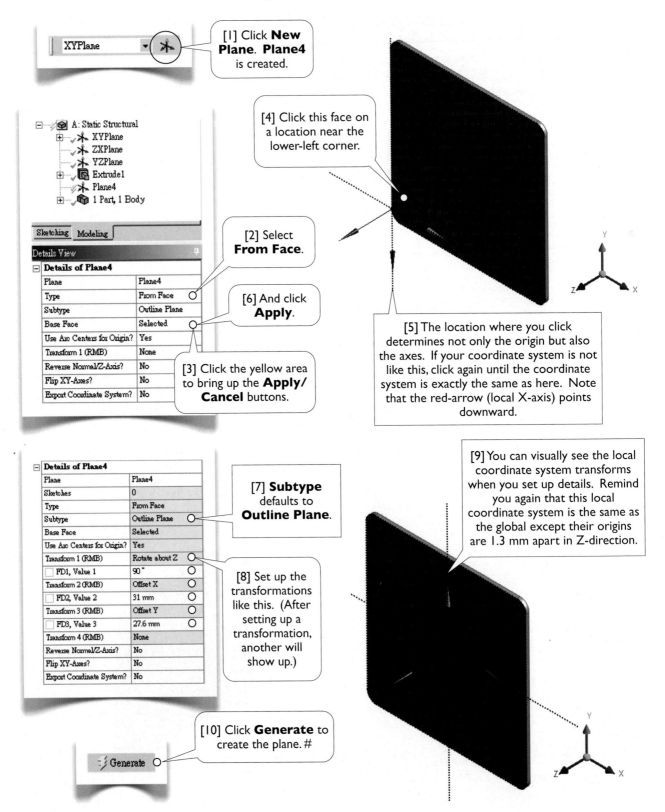

[1] Click **New Plane**. **Plane4** is created.

XYPlane

[4] Click this face on a location near the lower-left corner.

A: Static Structural
XYPlane
ZXPlane
YZPlane
Extrude1
Plane4
1 Part, 1 Body

Sketching Modeling

Details View

Details of Plane4

Plane	Plane4
Type	From Face
Subtype	Outline Plane
Base Face	Selected
Use Arc Centers for Origin?	Yes
Transform 1 (RMB)	None
Reverse Normal/Z-Axis?	No
Flip XY-Axes?	No
Export Coordinate System?	No

[2] Select **From Face**.

[6] And click **Apply**.

[3] Click the yellow area to bring up the **Apply/Cancel** buttons.

[5] The location where you click determines not only the origin but also the axes. If your coordinate system is not like this, click again until the coordinate system is exactly the same as here. Note that the red-arrow (local X-axis) points downward.

Details of Plane4

Plane	Plane4
Sketches	0
Type	From Face
Subtype	Outline Plane
Base Face	Selected
Use Arc Centers for Origin?	Yes
Transform 1 (RMB)	Rotate about Z
FD1, Value 1	90 °
Transform 2 (RMB)	Offset X
FD2, Value 2	31 mm
Transform 3 (RMB)	Offset Y
FD3, Value 3	27.6 mm
Transform 4 (RMB)	None
Reverse Normal/Z-Axis?	No
Flip XY-Axes?	No
Export Coordinate System?	No

[7] **Subtype** defaults to **Outline Plane**.

[8] Set up the transformations like this. (After setting up a transformation, another will show up.)

[9] You can visually see the local coordinate system transforms when you set up details. Remind you again that this local coordinate system is the same as the global except their origins are 1.3 mm apart in Z-direction.

[10] Click **Generate** to create the plane. #

Generate

4.2-5 Duplicate the Plane Outline

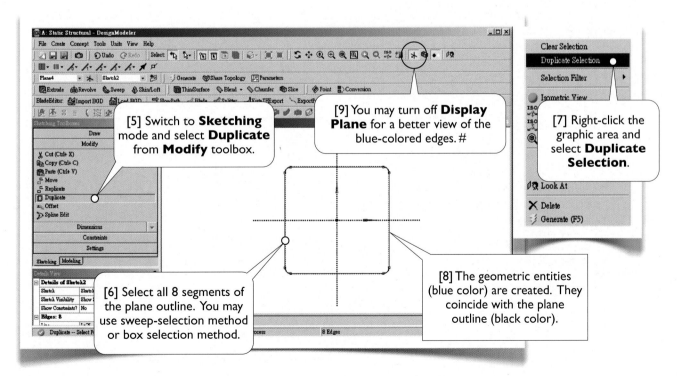

[2] Click **Look At**.

[3] Turn off **Display Model**.

[1] Newly created plane automatically becomes active plane.

[4] Now the active plane is the only entity on the graphics window. The plane includes an outline. We now proceed to create a sketch on this plane.

Outline Plane

When a plane is derived from a face and its **Subtype** is **Outline Plane**, the face boundary becomes part of the plane. The outline is not a geometric entity; you use it mainly as reference, or datum, just like you use local X-axis and Y-axis as datum. However, you may duplicate it to create geometric entities. Here we will demonstrate the duplication of the plane outline.

[5] Switch to **Sketching** mode and select **Duplicate** from **Modify** toolbox.

[6] Select all 8 segments of the plane outline. You may use sweep-selection method or box selection method.

[9] You may turn off **Display Plane** for a better view of the blue-colored edges. #

[7] Right-click the graphic area and select **Duplicate Selection**.

[8] The geometric entities (blue color) are created. They coincide with the plane outline (black color).

174

4.2-6 Draw the Rest of the Sketch

Continue to draw on the same sketch (Sketch2)...

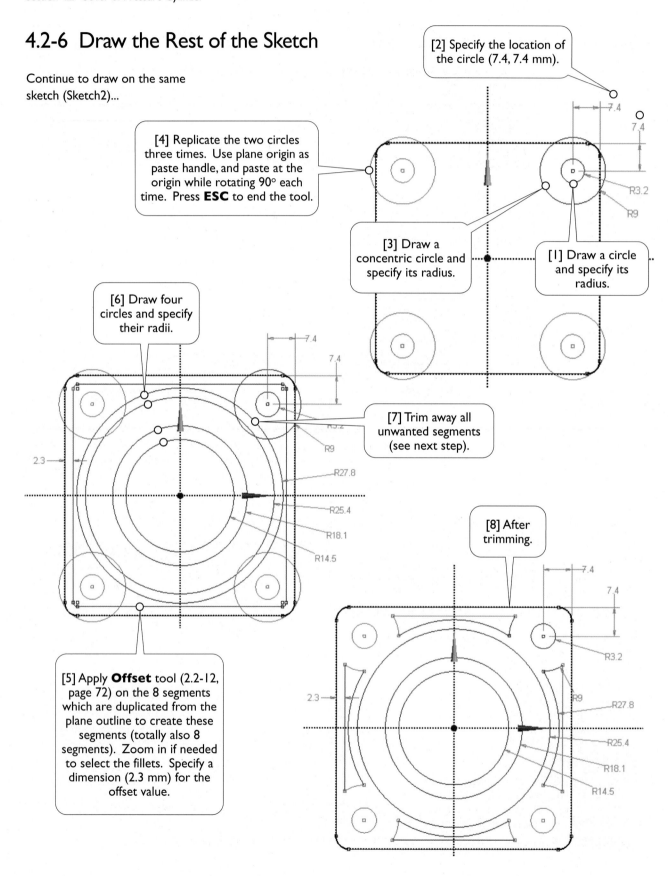

[2] Specify the location of the circle (7.4, 7.4 mm).

[4] Replicate the two circles three times. Use plane origin as paste handle, and paste at the origin while rotating 90° each time. Press **ESC** to end the tool.

[3] Draw a concentric circle and specify its radius.

[1] Draw a circle and specify its radius.

R3.2

R9

[6] Draw four circles and specify their radii.

[7] Trim away all unwanted segments (see next step).

R9

R27.8

R25.4

R18.1

R14.5

2.3

[8] After trimming.

7.4

7.4

R3.2

R9

R27.8

R25.4

R18.1

R14.5

2.3

[5] Apply **Offset** tool (2.2-12, page 72) on the 8 segments which are duplicated from the plane outline to create these segments (totally also 8 segments). Zoom in if needed to select the fillets. Specify a dimension (2.3 mm) for the offset value.

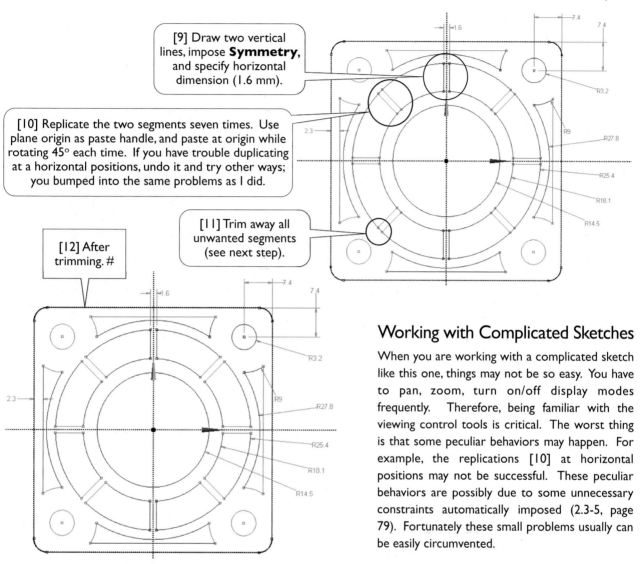

[9] Draw two vertical lines, impose **Symmetry**, and specify horizontal dimension (1.6 mm).

[10] Replicate the two segments seven times. Use plane origin as paste handle, and paste at origin while rotating 45° each time. If you have trouble duplicating at a horizontal positions, undo it and try other ways; you bumped into the same problems as I did.

[11] Trim away all unwanted segments (see next step).

[12] After trimming. #

Working with Complicated Sketches

When you are working with a complicated sketch like this one, things may not be so easy. You have to pan, zoom, turn on/off display modes frequently. Therefore, being familiar with the viewing control tools is critical. The worst thing is that some peculiar behaviors may happen. For example, the replications [10] at horizontal positions may not be successful. These peculiar behaviors are possibly due to some unnecessary constraints automatically imposed (2.3-5, page 79). Fortunately these small problems usually can be easily circumvented.

4.2-7 Extrude the Sketch

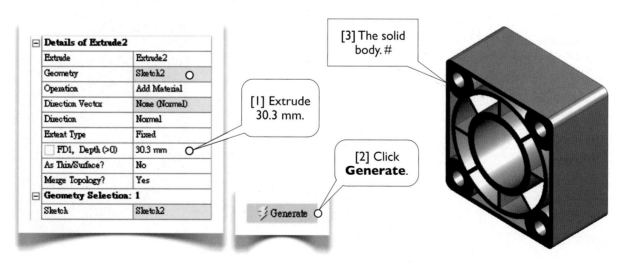

Details of Extrude2	
Extrude	Extrude2
Geometry	Sketch2
Operation	Add Material
Direction Vector	None (Normal)
Direction	Normal
Extent Type	Fixed
FD1, Depth (>0)	30.3 mm
As Thin/Surface?	No
Merge Topology?	Yes
Geometry Selection: 1	
Sketch	Sketch2

[1] Extrude 30.3 mm.

[2] Click **Generate**.

Generate

[3] The solid body. #

4.2-8 Create a New Plane and Draw a Circle on the Plane

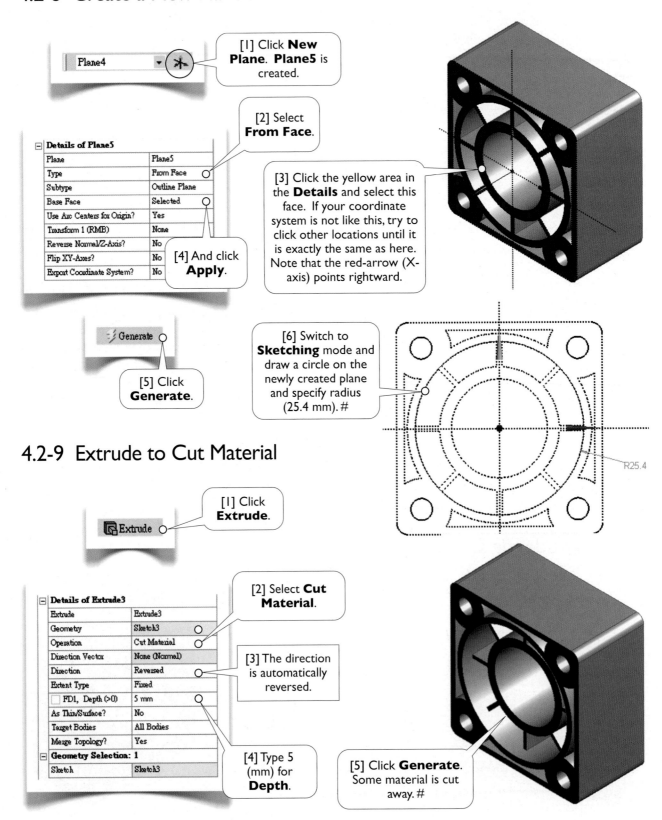

Plane4

[1] Click **New Plane**. **Plane5** is created.

[2] Select **From Face**.

Details of Plane5

Plane	Plane5
Type	From Face
Subtype	Outline Plane
Base Face	Selected
Use Arc Centers for Origin?	Yes
Transform 1 (RMB)	None
Reverse Normal/Z-Axis?	No
Flip XY-Axes?	No
Export Coordinate System?	No

[3] Click the yellow area in the **Details** and select this face. If your coordinate system is not like this, try to click other locations until it is exactly the same as here. Note that the red-arrow (X-axis) points rightward.

[4] And click **Apply**.

Generate

[5] Click **Generate**.

[6] Switch to **Sketching** mode and draw a circle on the newly created plane and specify radius (25.4 mm). #

R25.4

4.2-9 Extrude to Cut Material

Extrude

[1] Click **Extrude**.

Details of Extrude3

Extrude	Extrude3
Geometry	Sketch3
Operation	Cut Material
Direction Vector	None (Normal)
Direction	Reversed
Extent Type	Fixed
FD1, Depth (>0)	5 mm
As Thin/Surface?	No
Target Bodies	All Bodies
Merge Topology?	Yes
Geometry Selection: 1	
Sketch	Sketch3

[2] Select **Cut Material**.

[3] The direction is automatically reversed.

[4] Type 5 (mm) for **Depth**.

[5] Click **Generate**. Some material is cut away. #

4.2-10 Create New Sketch on XYPlane and Draw Circles

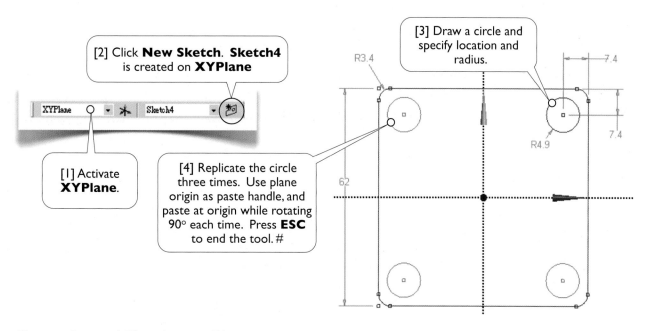

[2] Click **New Sketch**. **Sketch4** is created on **XYPlane**

[1] Activate **XYPlane**.

[4] Replicate the circle three times. Use plane origin as paste handle, and paste at origin while rotating 90° each time. Press **ESC** to end the tool. #

[3] Draw a circle and specify location and radius.

XYPlane   Sketch4

R3.4   7.4

R4.9   7.4

62

Create Second Sketch on a Plane

A plane may contain multiple sketches. When you begin to draw on a fresh plane, a sketch is automatically created; you don't have to explicitly click **New Sketch** tool [2], although there is no harm to do that. When you want to add a second sketch on a plane, you MUST explicitly click **New Sketch** tool, otherwise the entities you draw will become part of the first sketch.

4.2-11 Extrude to Cut Material

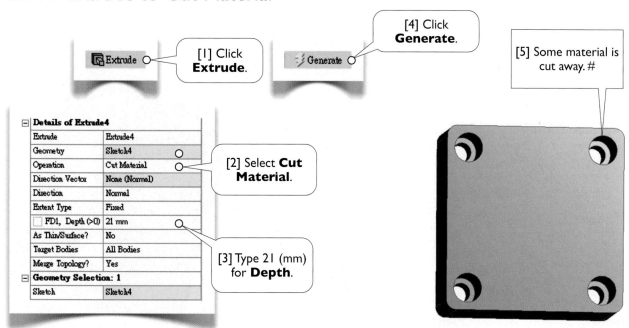

[1] Click **Extrude**.

[4] Click **Generate**.

[5] Some material is cut away. #

Details of Extrude4	
Extrude	Extrude4
Geometry	Sketch4
Operation	Cut Material
Direction Vector	None (Normal)
Direction	Normal
Extent Type	Fixed
FD1, Depth (>0)	21 mm
As Thin/Surface?	No
Target Bodies	All Bodies
Merge Topology?	Yes
Geometry Selection: 1	
Sketch	Sketch4

[2] Select **Cut Material**.

[3] Type 21 (mm) for **Depth**.

4.2-12 Create Another Sketch on XYPlane and Draw a Circle

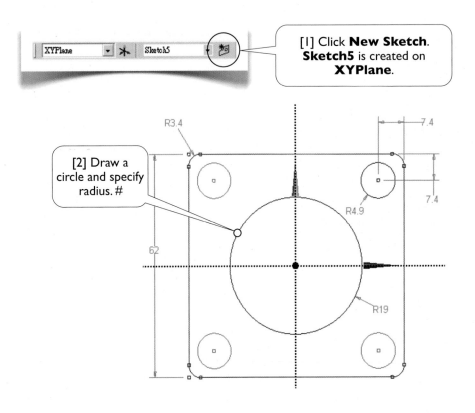

XYPlane | Sketch5

[1] Click **New Sketch**. **Sketch5** is created on **XYPlane**.

[2] Draw a circle and specify radius. #

R3.4

7.4

R4.9

7.4

62

R19

4.2-13 Extrude to Add Material

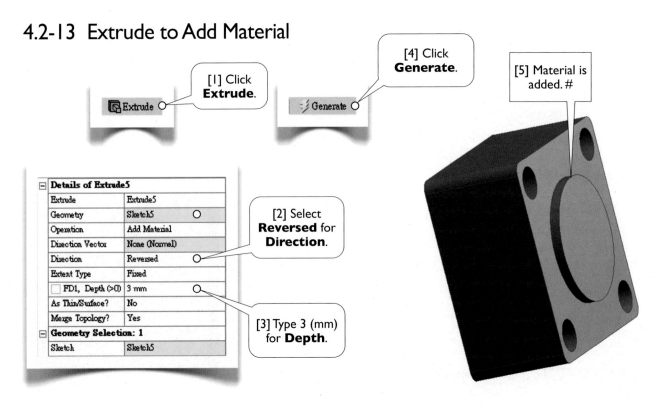

[1] Click **Extrude**.

[4] Click **Generate**.

[5] Material is added. #

Extrude

Generate

Details of Extrude5	
Extrude	Extrude5
Geometry	Sketch5
Operation	Add Material
Direction Vector	None (Normal)
Direction	Reversed
Extent Type	Fixed
FD1, Depth (>0)	3 mm
As Thin/Surface?	No
Merge Topology?	Yes
Geometry Selection: 1	
Sketch	Sketch5

[2] Select **Reversed** for **Direction**.

[3] Type 3 (mm) for **Depth**.

4.2-14 Create a Plane on the Side Wall and Two Sketches on the Plane

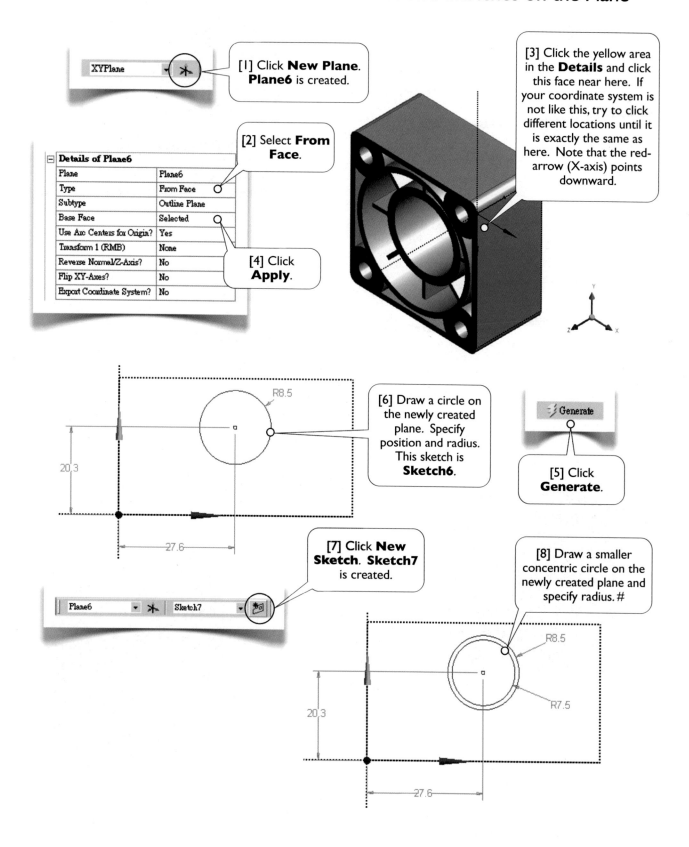

[1] Click **New Plane**. **Plane6** is created.

XYPlane

[2] Select **From Face**.

Details of Plane6

Plane	Plane6
Type	From Face
Subtype	Outline Plane
Base Face	Selected
Use Arc Centers for Origin?	Yes
Transform 1 (RMB)	None
Reverse Normal/Z-Axis?	No
Flip XY-Axes?	No
Export Coordinate System?	No

[4] Click **Apply**.

[3] Click the yellow area in the **Details** and click this face near here. If your coordinate system is not like this, try to click different locations until it is exactly the same as here. Note that the red-arrow (X-axis) points downward.

R8.5

20.3

27.6

[6] Draw a circle on the newly created plane. Specify position and radius. This sketch is **Sketch6**.

Generate

[5] Click **Generate**.

[7] Click **New Sketch**. **Sketch7** is created.

Plane6   Sketch7

[8] Draw a smaller concentric circle on the newly created plane and specify radius. #

R8.5

R7.5

20.3

27.6

4.2-15 Complete the Model

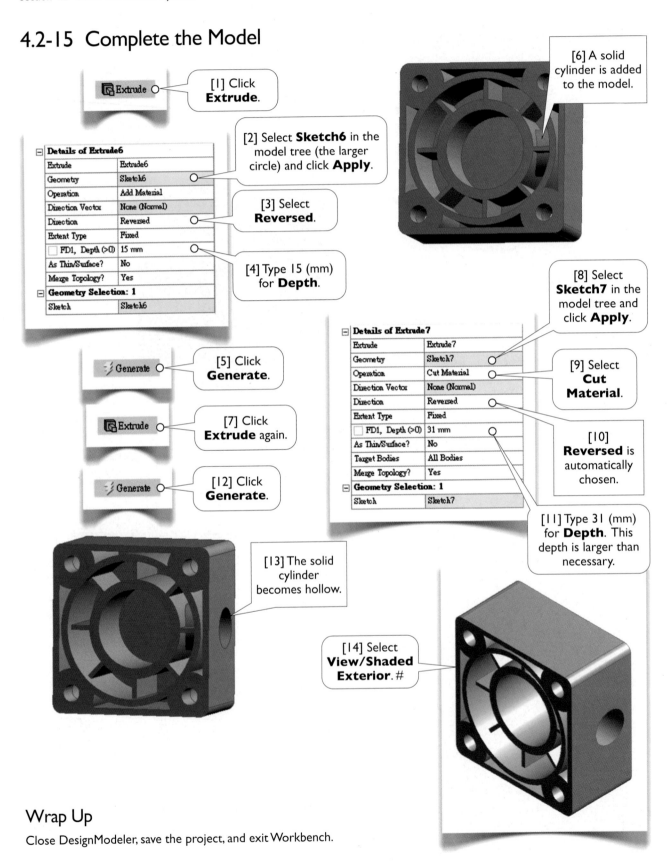

[1] Click **Extrude**.

[2] Select **Sketch6** in the model tree (the larger circle) and click **Apply**.

Details of Extrude6

Extrude	Extrude6
Geometry	Sketch6
Operation	Add Material
Direction Vector	None (Normal)
Direction	Reversed
Extent Type	Fixed
☐ FD1, Depth (>0)	15 mm
As Thin/Surface?	No
Merge Topology?	Yes

Geometry Selection: 1

Sketch	Sketch6

[3] Select **Reversed**.

[4] Type 15 (mm) for **Depth**.

[5] Click **Generate**.

[6] A solid cylinder is added to the model.

[7] Click **Extrude** again.

[8] Select **Sketch7** in the model tree and click **Apply**.

Details of Extrude7

Extrude	Extrude7
Geometry	Sketch7
Operation	Cut Material
Direction Vector	None (Normal)
Direction	Reversed
Extent Type	Fixed
☐ FD1, Depth (>0)	31 mm
As Thin/Surface?	No
Target Bodies	All Bodies
Merge Topology?	Yes

Geometry Selection: 1

Sketch	Sketch7

[9] Select **Cut Material**.

[10] **Reversed** is automatically chosen.

[11] Type 31 (mm) for **Depth**. This depth is larger than necessary.

[12] Click **Generate**.

[13] The solid cylinder becomes hollow.

[14] Select **View/Shaded Exterior**. #

Wrap Up

Close DesignModeler, save the project, and exit Workbench.

Section 4.3

Lifting Fork

4.3-1 About the Lifting Fork

The lifting fork is used in an LCD (liquid crystal display) manufacturing factory to handle glass panels [1-5]. The glass panel is so large (2.5 m x 2.2 m) and thin (1.0 mm) that the engineers are concerned with its deflections during the dynamic handling. The deflection is critical for the design of precision machine handling process. Another process design consideration is the duration that the vibrations settle to a steady state.

In this section, we want to model the fork as a solid body and the glass panel as a surface body. The model will be used for static, modal, and transient simulations in Section 12.2.

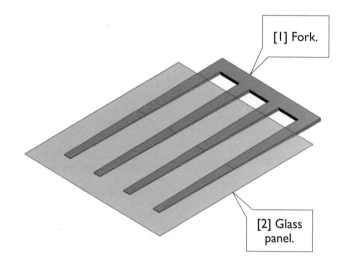

[1] Fork.

[2] Glass panel.

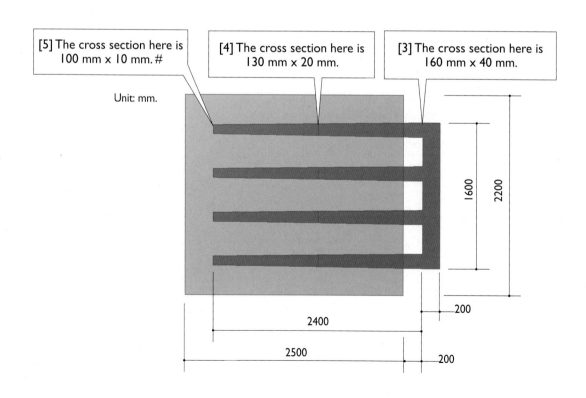

[5] The cross section here is 100 mm x 10 mm. #

[4] The cross section here is 130 mm x 20 mm.

[3] The cross section here is 160 mm x 40 mm.

Unit: mm.

1600

2200

2400

2500

200

200

4.3-2 Start Up

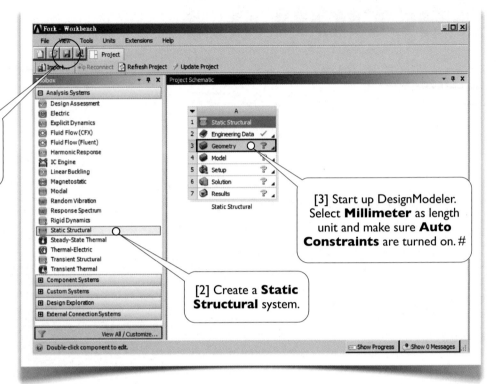

[1] Launch Workbench and save the project as **Fork**.

[2] Create a **Static Structural** system.

[3] Start up DesignModeler. Select **Millimeter** as length unit and make sure **Auto Constraints** are turned on. #

4.3-3 Create the Transversal Beam

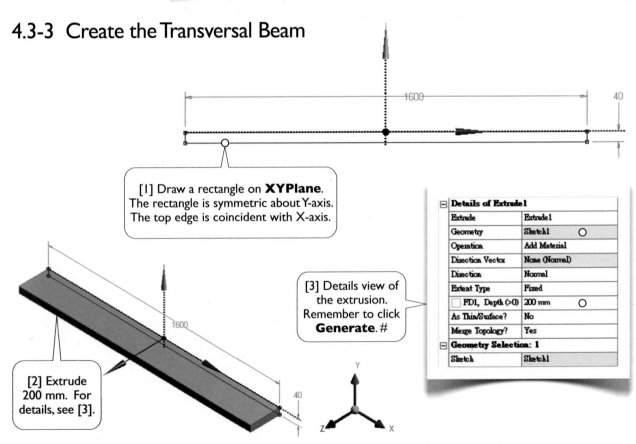

[1] Draw a rectangle on **XYPlane**. The rectangle is symmetric about Y-axis. The top edge is coincident with X-axis.

[2] Extrude 200 mm. For details, see [3].

[3] Details view of the extrusion. Remember to click **Generate**. #

Details of Extrude1	
Extrude	Extrude1
Geometry	Sketch1
Operation	Add Material
Direction Vector	None (Normal)
Direction	Normal
Extent Type	Fixed
FD1, Depth (>0)	200 mm
As Thin/Surface?	No
Merge Topology?	Yes
Geometry Selection: 1	
Sketch	Sketch1

Skin/Loft

Now we want to create a single prong, or finger. The prong is then duplicated to create 3 additional prongs. The prong's cross section is not uniform, and it cannot be created using **Extrude** or **Sweep**. A way to create a solid or surface of different cross sections along its path is using **Skin/Loft**, which takes a series of profiles from different planes and creates a solid or surface that fits through these profiles.

You may view **Sweep** as a specialization of **Skin/Loft**, and **Extrude** as a specialization of **Sweep**.

4.3-4 Create Three Planes Based on a Face of the Beam

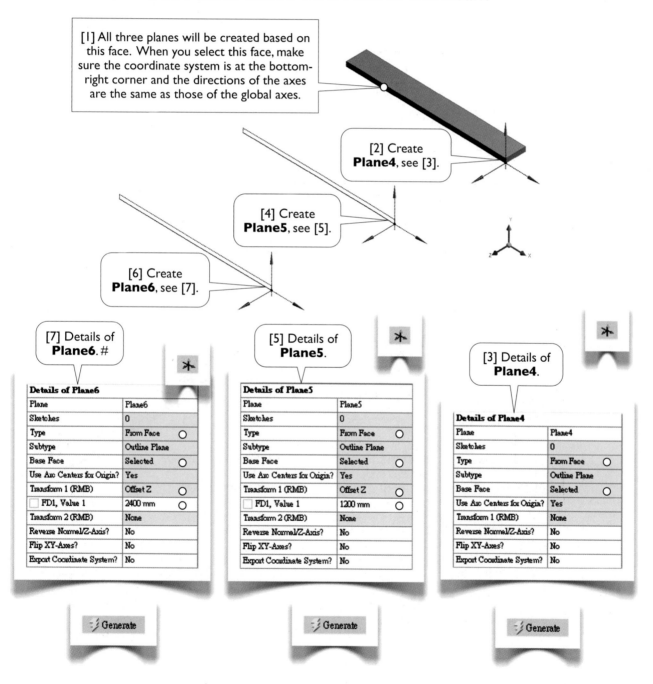

[I] All three planes will be created based on this face. When you select this face, make sure the coordinate system is at the bottom-right corner and the directions of the axes are the same as those of the global axes.

[2] Create **Plane4**, see [3].

[4] Create **Plane5**, see [5].

[6] Create **Plane6**, see [7].

[7] Details of **Plane6**. #

Details of Plane6

Plane	Plane6	
Sketches	0	
Type	From Face	○
Subtype	Outline Plane	
Base Face	Selected	○
Use Arc Centers for Origin?	Yes	
Transform 1 (RMB)	Offset Z	○
☐ FD1, Value 1	2400 mm	○
Transform 2 (RMB)	None	
Reverse Normal/Z-Axis?	No	
Flip XY-Axes?	No	
Export Coordinate System?	No	

[5] Details of **Plane5**.

Details of Plane5

Plane	Plane5	
Sketches	0	
Type	From Face	○
Subtype	Outline Plane	
Base Face	Selected	○
Use Arc Centers for Origin?	Yes	
Transform 1 (RMB)	Offset Z	○
☐ FD1, Value 1	1200 mm	○
Transform 2 (RMB)	None	
Reverse Normal/Z-Axis?	No	
Flip XY-Axes?	No	
Export Coordinate System?	No	

[3] Details of **Plane4**.

Details of Plane4

Plane	Plane4	
Sketches	0	
Type	From Face	○
Subtype	Outline Plane	
Base Face	Selected	○
Use Arc Centers for Origin?	Yes	
Transform 1 (RMB)	None	
Reverse Normal/Z-Axis?	No	
Flip XY-Axes?	No	
Export Coordinate System?	No	

Generate Generate Generate

4.3-5 Create a Sketch on Each Plane

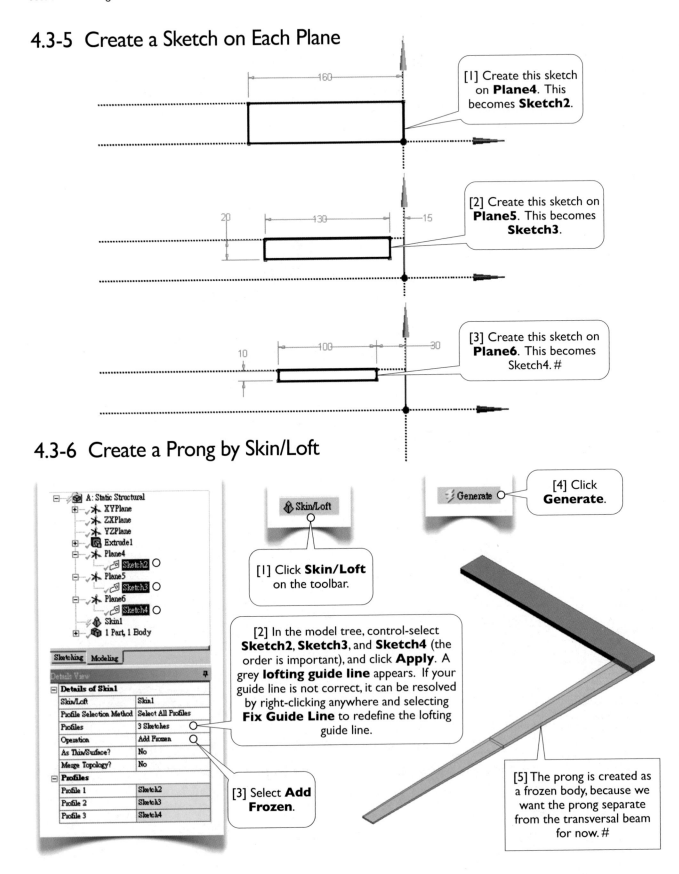

[1] Create this sketch on **Plane4**. This becomes **Sketch2**.

[2] Create this sketch on **Plane5**. This becomes **Sketch3**.

[3] Create this sketch on **Plane6**. This becomes Sketch4. #

4.3-6 Create a Prong by Skin/Loft

[4] Click **Generate**.

[1] Click **Skin/Loft** on the toolbar.

[2] In the model tree, control-select **Sketch2**, **Sketch3**, and **Sketch4** (the order is important), and click **Apply**. A grey **lofting guide line** appears. If your guide line is not correct, it can be resolved by right-clicking anywhere and selecting **Fix Guide Line** to redefine the lofting guide line.

[3] Select **Add Frozen**.

[5] The prong is created as a frozen body, because we want the prong separate from the transversal beam for now. #

4.3-7 Duplicate the Prong Using Pattern

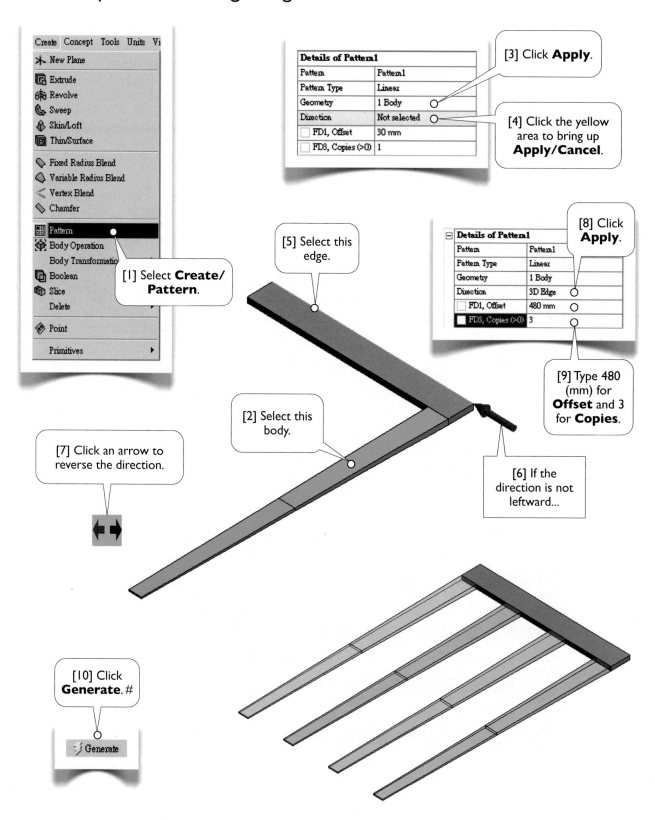

Create Concept Tools Units Vi

New Plane

Extrude
Revolve
Sweep
Skin/Loft
Thin/Surface

Fixed Radius Blend
Variable Radius Blend
Vertex Blend
Chamfer

Pattern
Body Operation
Body Transformatio
Boolean
Slice
Delete

Point

Primitives

[1] Select **Create/ Pattern**.

Details of Pattern1	
Pattern	Pattern1
Pattern Type	Linear
Geometry	1 Body
Direction	Not selected
FD1, Offset	30 mm
FD3, Copies (>0)	1

[3] Click **Apply**.

[4] Click the yellow area to bring up **Apply/Cancel**.

Details of Pattern1	
Pattern	Pattern1
Pattern Type	Linear
Geometry	1 Body
Direction	3D Edge
FD1, Offset	480 mm
FD3, Copies (>0)	3

[8] Click **Apply**.

[9] Type 480 (mm) for **Offset** and 3 for **Copies**.

[5] Select this edge.

[2] Select this body.

[7] Click an arrow to reverse the direction.

[6] If the direction is not leftward...

[10] Click **Generate**. #

Generate

186

4.3-8 Combine the Bodies Using **Boolean**

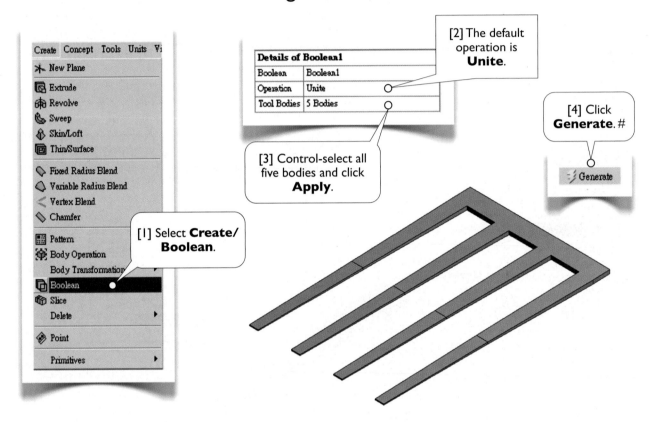

[2] The default operation is **Unite**.

Details of Boolean1

Boolean	Boolean1
Operation	Unite
Tool Bodies	5 Bodies

[3] Control-select all five bodies and click **Apply**.

[4] Click **Generate**. #

[1] Select **Create/ Boolean**.

Create   Concept   Tools   Units   Vi

- New Plane
- Extrude
- Revolve
- Sweep
- Skin/Loft
- Thin/Surface
- Fixed Radius Blend
- Variable Radius Blend
- Vertex Blend
- Chamfer
- Pattern
- Body Operation
- Body Transformation
- Boolean
- Slice
- Delete
- Point
- Primitives

4.3-9 Create a Surface Body for the Glass Panel

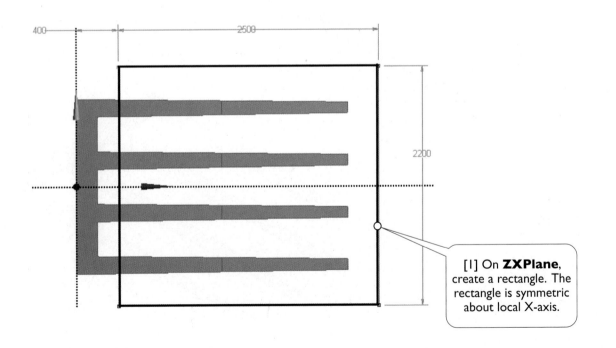

400   2500

2200

[1] On **ZXPlane**, create a rectangle. The rectangle is symmetric about local X-axis.

187

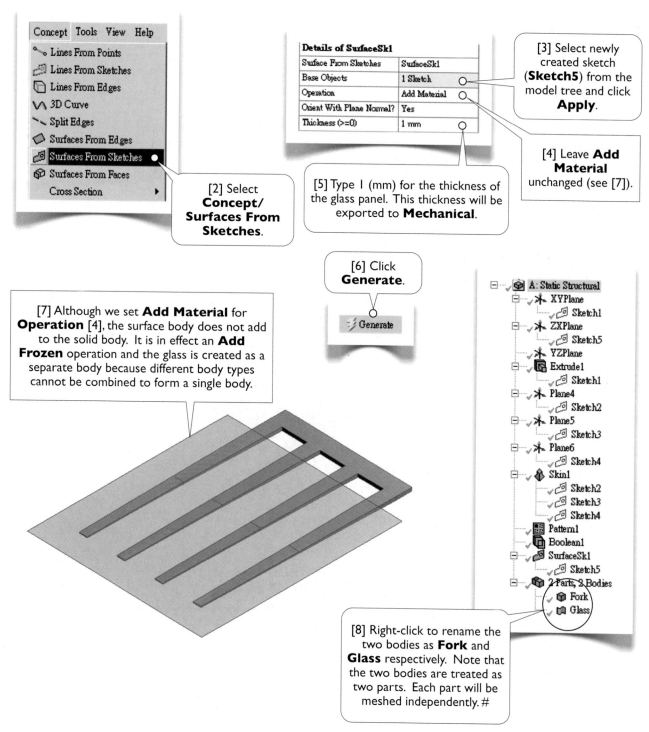

Concept Tools View Help

- Lines From Points
- Lines From Sketches
- Lines From Edges
- 3D Curve
- Split Edges
- Surfaces From Edges
- **Surfaces From Sketches** ●
- Surfaces From Faces
- Cross Section ►

[2] Select **Concept/ Surfaces From Sketches**.

Details of SurfaceSk1

Surface From Sketches	SurfaceSk1
Base Objects	1 Sketch
Operation	Add Material
Orient With Plane Normal?	Yes
Thickness (>=0)	1 mm

[3] Select newly created sketch (**Sketch5**) from the model tree and click **Apply**.

[4] Leave **Add Material** unchanged (see [7]).

[5] Type 1 (mm) for the thickness of the glass panel. This thickness will be exported to **Mechanical**.

[6] Click **Generate**.

Generate

[7] Although we set **Add Material** for **Operation** [4], the surface body does not add to the solid body. It is in effect an **Add Frozen** operation and the glass is created as a separate body because different body types cannot be combined to form a single body.

A: Static Structural
- XYPlane
  - Sketch1
- ZXPlane
  - Sketch5
- YZPlane
- Extrude1
  - Sketch1
- Plane4
  - Sketch2
- Plane5
  - Sketch3
- Plane6
  - Sketch4
- Skin1
  - Sketch2
  - Sketch3
  - Sketch4
- Pattern1
- Boolean1
- SurfaceSk1
  - Sketch5
- 2 Parts, 2 Bodies
  - Fork
  - Glass

[8] Right-click to rename the two bodies as **Fork** and **Glass** respectively. Note that the two bodies are treated as two parts. Each part will be meshed independently. #

Wrap Up

Close DesignModeler, save the project, and exit Workbench.

Section 4.4

More Details

4.4-1 DesignModeler GUI Revisit

In Section 2.3, we overviewed **DesignModeler GUI**, yet skip some 3D tools, such as view orientations [1, 2, 3], and entities selection [4, 5, 6]. Also on the toolbar are tools to create 3D features [7]. These tools are covered in this section.

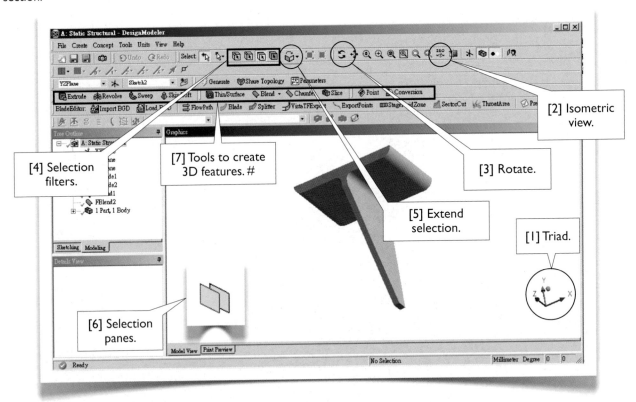

[4] Selection filters.

[7] Tools to create 3D features. #

[2] Isometric view.

[3] Rotate.

[5] Extend selection.

[1] Triad.

[6] Selection panes.

4.4-2 Principal Views and Isometric Views

Triad[Ref 1]

On the bottom right corner of the GUI is a triad (4.4-1[1]). Clicking any of the triad arrows will rotate the view such that it is normal to that arrow [1] (next page). Move the mouse over the negative side of an arrow to display a black arrow [2], which represents the negative direction of that arrow. Clicking black arrows also rotates the view.

Accompanying the triad is a small cyan sphere. When you rotate the view (4.4-3, next page), the triad and the small sphere will rotate accordingly. The sphere represents a point located at an "isometric axis," which is a collection of points having the same magnitude of coordinates (but may be different in signs) in all three axes. Its initial position is (1, 1, 1). Thus, if the sphere coincides with the origin, that means your view is an isometric view [3]. When the sphere does not coincide with the origin, clicking the sphere will rotate to an isometric view [4].

189

Isometric View[Ref 2]

As mentioned, the small cyan ball represents an isometric direction and its initial position is (1, 1, 1). In 3D space, there are totally 8 such directions. For examples: (-1, 1, 1), (1, -1, 1), etc. These are all isometric views. When you click **Isometric View** tool (4.4-1[2], last page), the view will rotate to the isometric view closest to the current view, and the small cyan ball will move to new location accordingly.

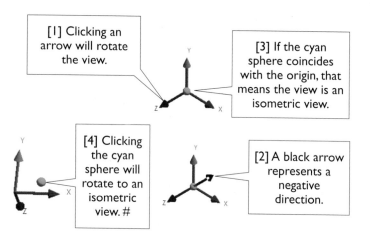

[1] Clicking an arrow will rotate the view.

[3] If the cyan sphere coincides with the origin, that means the view is an isometric view.

[4] Clicking the cyan sphere will rotate to an isometric view. #

[2] A black arrow represents a negative direction.

4.4-3 View Rotations[Ref 3]

Rotate with Mouse Wheel

Hold the middle mouse button down while moving around the graphic area, you can rotate the view to any orientations [1]. It is simple and convenient.

Rotate tool

Rotate tool (4.4-1[3], last page) gives you more controls for rotating the model. After activating **Rotate** tool by clicking it, the mouse cursor becomes one of the four shapes [2-5], and the type of rotation depends on the location of your mouse cursor [6]: free rotation [2] when the cursor is near the center of the graphics window; roll [3] when the cursor outside the center; yaw [4] when the cursor near the vertical edges; pitch [5] when the cursor near the horizontal edges.

By default, the model center is the center of rotation. You can set the center of rotation (a red sphere) by clicking on the model. The red sphere will stay in the middle of the graphics window.

To restore the center of rotation to the model center, click anywhere in the graphics window away from the model. This will re-center the model in the middle of the graphics window.

[1] Hold the middle mouse button down while moving around the graphic area, you can rotate the view.

[6] The type of rotation depends on the location of the cursor. #

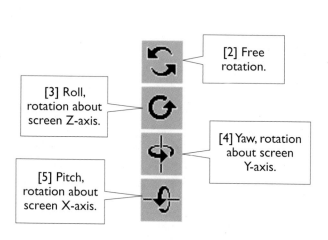

[2] Free rotation.

[3] Roll, rotation about screen Z-axis.

[4] Yaw, rotation about screen Y-axis.

[5] Pitch, rotation about screen X-axis.

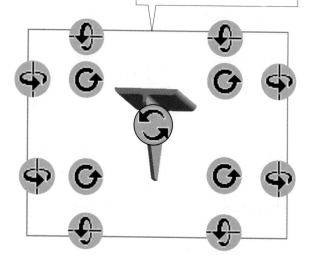

4.4-4 Mouse Cursor

Various mouse cursors are used to indicate the current operation [1].

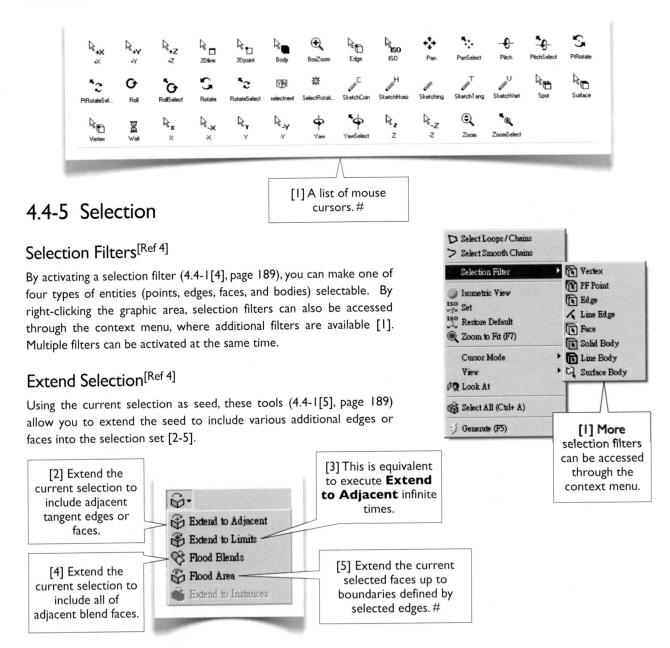

[1] A list of mouse cursors. #

4.4-5 Selection

Selection Filters[Ref 4]

By activating a selection filter (4.4-1[4], page 189), you can make one of four types of entities (points, edges, faces, and bodies) selectable. By right-clicking the graphic area, selection filters can also be accessed through the context menu, where additional filters are available [1]. Multiple filters can be activated at the same time.

Extend Selection[Ref 4]

Using the current selection as seed, these tools (4.4-1[5], page 189) allow you to extend the seed to include various additional edges or faces into the selection set [2-5].

[2] Extend the current selection to include adjacent tangent edges or faces.

[3] This is equivalent to execute **Extend to Adjacent** infinite times.

[4] Extend the current selection to include all of adjacent blend faces.

[5] Extend the current selected faces up to boundaries defined by selected edges. #

[1] **More** selection filters can be accessed through the context menu.

Selection Panes[Ref 5]

When you select an entity by clicking your mouse on the model, and if more than one entity lie under the mouse cursor, the graphics window displays a stack of rectangles in the lower-left corner (4.4-1[6], page 189). The rectangles are orderly stacked with the topmost rectangle representing the most visible entity and subsequent rectangles representing entities underneath the mouse cursor, front to back. These rectangles are aliases of selectable entities, that is, highlighting and picking these rectangles are identical to that for the entities. When you move the mouse over these rectangles, mouse cursor will change to show the type of the entity.

191

4.4-6 Edge Display Controls[Ref 9]

When the model is displayed with edges (or wireframe) [1], each type of edge display can be turned on/off through **Edge Display Controls** tools. When the button **/n** is turned on, the edges which are shared by **n** faces will be displayed [2]. For example, **/2** means that the edge is shared by two faces, and **/0** means that the edge is not shared by any faces.

[1] When the model is displayed with edges, each type of edge display can be turned on/off.

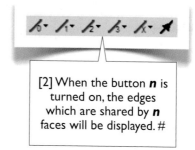

[2] When the button **n** is turned on, the edges which are shared by **n** faces will be displayed. #

4.4-7 Bodies and Parts[Ref 6]

The last branch of the model tree contains the bodies and parts of the model [1], which are the only geometric entities that will be exported to **Mechanical** for simulations.

A body is a continuum and made of one kind of material. A 3D body is either a solid body, a surface body, or a line body.

A part consists of a single body or multiple bodies of the same type. If multiple, then all bodies are assumed to be bonded together; i.e., they form a single continuum (but multiple materials are allowed). In **Mechanical**, each part is meshed independently—this is the most important concept about parts. Within a part, the boundary nodes are shared between contacting bodies.

A model may consist of one or more parts. Since each part is meshed independently, mesh at the boundaries between parts is not compatible. In **Mechanical**, connections[Ref 7] (e.g., contacts, joints) among parts must be established to complete a model.

[1] Bodies and parts. #

4.4-8 Feature-Based 3D Modeling

A geometric model consists of features; an object in a model tree is called a feature of the model. Features include *planes*, *base features*, *placed features*, etc.

Base Features

Base features are also called *sketched features* because they are created by first drawing one or more sketches, and then "growing" to 3D features by means of **Extrude**, **Revolve**, **Sweep**, or **Skin/Loft**. A newly created base feature can add material to or subtract material from the existing bodies.

Placed Features

Some features have predefined shapes and behaviors. To add these features to existing bodies, all we have to do is to specify where we want to place these features, along with a few other settings. Therefore, these features are called *placed features*, Including **Blend**, **Chamfer**, **Thin/Surface**, **Slice**.

4.4-9 Base Features[Ref 8]

Extrude

The tool is used to extrude a sketch along its normal direction to create a 3D body. The extrusion may be symmetric or asymmetric to the sketching plane. The extrusion depth may be a fixed value, through all bodies (used only for cutting the material), up to a face, or up to a surface. A *face* is a bounded region and has a finite area while a *surface* is an unbounded region and has infinite area. A surface is often the extension of a face.

Revolve

The tool is used to revolve a sketch about an axis to create a 3D body. An angle of revolution can be specified.

Sweep

Sweep can be thought of a generalization of **Extrude**. The tool is used to sweep a profile along a path to create a 3D body. Both the profile and the path must be defined using sketches.

Skin/Loft

Skin/Loft can be thought of a generalization of **Sweep**. It takes a series of profiles to create a 3D body by fitting through them. The profiles must be defined using sketches.

4.4-10 Placed Features[Ref 8]

Thin/Surface

The tool is used to convert a solid into a thin solid body or a surface body. Typically, you will select one or more faces to remove, and then specify a thickness. If the thickness is a positive value, then a thin solid body is created. If the thickness is zero, then a surface body is created.

Blend

The tool is used to create *rounds* or *fillets* on edges, or on vertices. The radius of the rounds or fillets may be fixed or variable.

Chamfer

The tool is used to create chamfer faces on edges.

193

References

1. ANSYS Documentation//DesignModeler User's Guide//Viewing//Model Appearance Controls//Triad
2. ANSYS Documentation//DesignModeler User's Guide//Viewing//Rotation Modes Toolbar//Isometric View
3. ANSYS Documentation//Mechanical Applications// Mechanical User's Guide//Application Interface//Working with Graphics//Controlling the Viewing Orientation
4. ANSYS Documentation//DesignModeler User's Guide//Selection//Selection Toolbar
5. ANSYS Documentation//DesignModeler User's Guide//Selection//Graphical Selection//Depth Picking
6. ANSYS Documentation//DesignModeler User's Guide//3D Modeling//Bodies and Parts
7. ANSYS Documentation//Mechanical Applications// Mechanical User's Guide//Objects Reference//Connections
8. ANSYS Documentation//DesignModeler User's Guide//3D Modeling//3D Features
9. ANSYS Documentation//DesignModeler User's Guide//Viewing//Model Appearance Controls//Graphics Options//Edge Coloring

Section 4.5

LCD Display Support

4.5-1 About the LCD Display Support

The LCD Display support is made of an ABS (acrylonitrile-butadiene-styrene) plastic. The thickness is 3 mm [1]. Details of the hinge [2] are not shown in the figure but will be illustrated in 4.5-4 (pages 197-198).

This model will be used in Section 5.4 for a static structural simulation to assess the deformation and stress under a design load.

[1] The thickness of the plastic is 3 mm.

[2] Details of the hinge design will be illustrated in 4.5-4. #

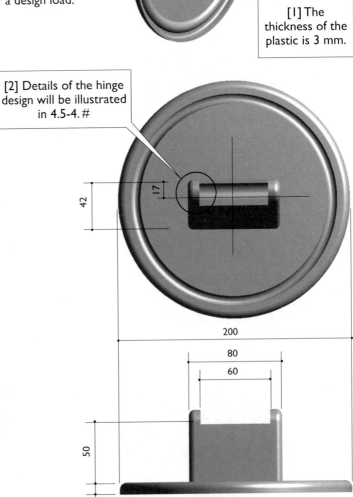

4.5-2 Create the Base

Launch Workbench, create a **Static Structural** system, and save the project as **Support**. Start up DesignModeler.
Select **Millimeter** as length unit and make sure **Auto Constraints** are turned on.

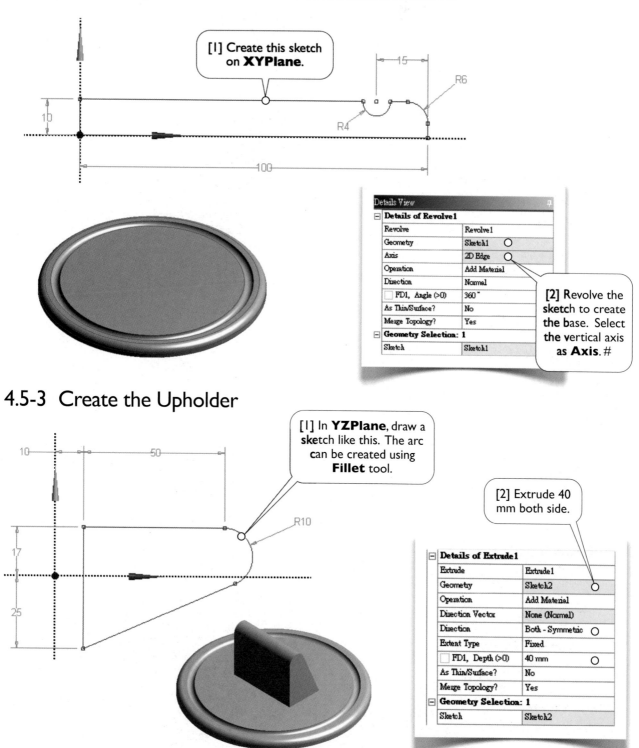

[1] Create this sketch on **XYPlane**.

[2] Revolve the sketch to create the base. Select the vertical axis as **Axis**. #

4.5-3 Create the Upholder

[1] In **YZPlane**, draw a sketch like this. The arc can be created using **Fillet** tool.

[2] Extrude 40 mm both side.

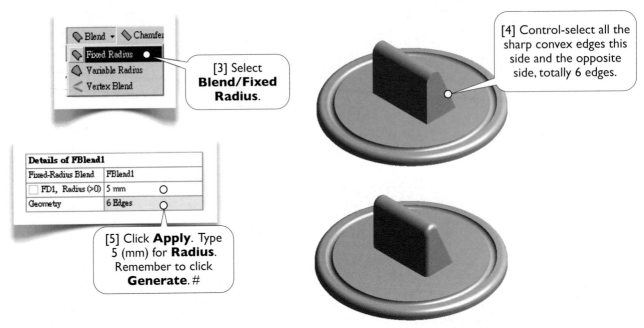

[3] Select **Blend/Fixed Radius**.

Details of FBlend1

Fixed-Radius Blend	FBlend1	
☐ FD1, Radius (>0)	5 mm	○
Geometry	6 Edges	○

[5] Click **Apply**. Type 5 (mm) for **Radius**. Remember to click **Generate**. #

[4] Control-select all the sharp convex edges this side and the opposite side, totally 6 edges.

4.5-4 Create Hinge

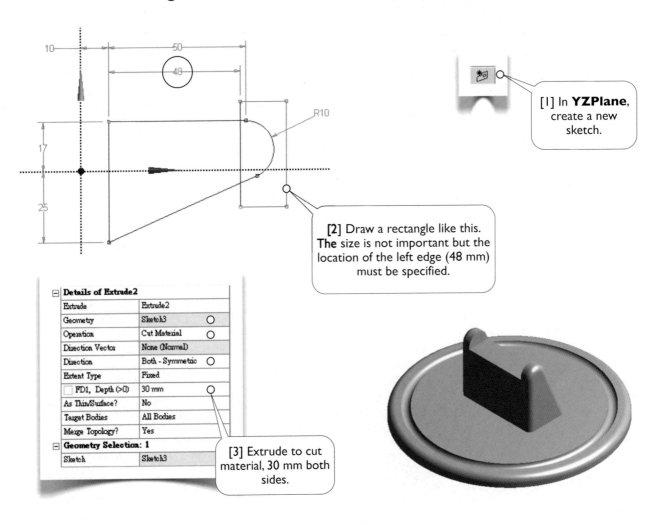

[1] In **YZPlane**, create a new sketch.

[2] Draw a rectangle like this. **The** size is not important but the location of the left edge (48 mm) must be specified.

Details of Extrude2

Extrude	Extrude2	
Geometry	Sketch3	○
Operation	Cut Material	○
Direction Vector	None (Normal)	
Direction	Both - Symmetric	○
Extent Type	Fixed	
☐ FD1, Depth (>0)	30 mm	○
As Thin/Surface?	No	
Target Bodies	All Bodies	
Merge Topology?	Yes	
Geometry Selection: 1		
Sketch	Sketch3	

[3] Extrude to cut material, 30 mm both sides.

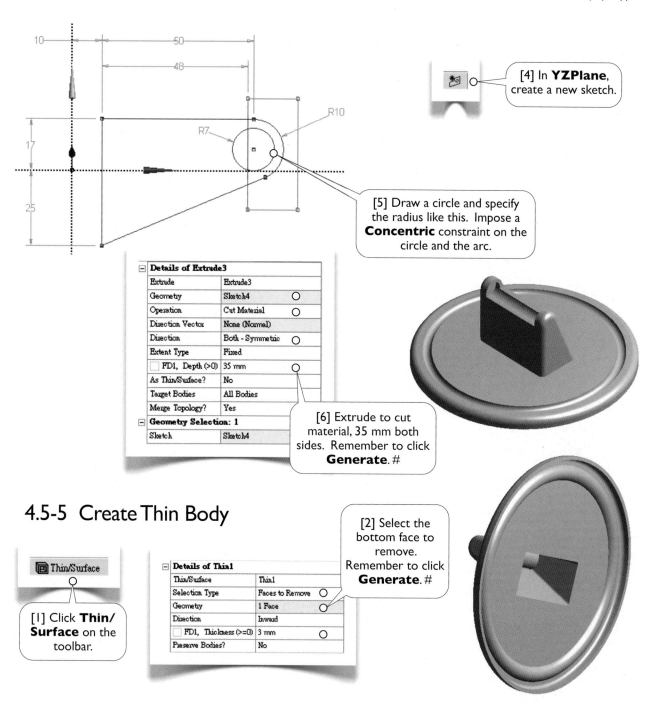

[4] In **YZPlane**, create a new sketch.

[5] Draw a circle and specify the radius like this. Impose a **Concentric** constraint on the circle and the arc.

Details of Extrude3	
Extrude	Extrude3
Geometry	Sketch4
Operation	Cut Material
Direction Vector	None (Normal)
Direction	Both - Symmetric
Extent Type	Fixed
FD1, Depth (>0)	35 mm
As Thin/Surface?	No
Target Bodies	All Bodies
Merge Topology?	Yes
Geometry Selection: 1	
Sketch	Sketch4

[6] Extrude to cut material, 35 mm both sides. Remember to click **Generate**. #

4.5-5 Create Thin Body

[2] Select the bottom face to remove. Remember to click **Generate**. #

Thin/Surface

[1] Click **Thin/Surface** on the toolbar.

Details of Thin1	
Thin/Surface	Thin1
Selection Type	Faces to Remove
Geometry	1 Face
Direction	Inward
FD1, Thickness (>=0)	3 mm
Preserve Bodies?	No

Wrap Up

Close DesignModeler, save the project, and exit Workbench.

Section 4.6

Review

4.6-1 Keywords

Choose a letter for each keyword from the list of descriptions

1. (   ) Base Features
2. (   ) Body
3. (   ) Connections
4. (   ) Extend Selection
5. (   ) Feature
6. (   ) Isotropic View
7. (   ) Part
8. (   ) Pitch

9. (   ) Placed Feature
10. (   ) Plane Outline
11. (   ) Plane Coordinate System
12. (   ) Roll
13. (   ) Selection Filters
14. (   ) Selection Panes
15. (   ) Yaw

Answers:

1. (N)   2. (J)   3. (L)   4. (H)   5. (M)   6. (C)   7. (K)   8. (F)
9. (O)   10.(B)   11.(A)   12.(D)   13.(G)   14.(I)   15.(E)

List of Descriptions

(A) Each plane is attached a local coordinate system, called the plane coordinate system. Its Z-axis always points out of the plane.

(B) A plane may be created by deriving from a face. In such cases, the plane may be designated as **Outline Plane** (i.e., a plane with boundary). You can draw entities beyond the boundary. The boundary itself is not a geometric entity. The plane boundary is used as datum edges, or as source of replication/duplication.

(C) A model view in which the view direction follows an isometric axis, which forms the same angles with the three principal axes, is called an isometric view.

(D) Rotate the model about the screen Z-axis.

(E) Rotate the model about the screen Y-axis.

(F) Rotate the model about the screen X-axis.

(G) A selection filter is a tool used to make a specific type of entities selectable. Multiple filters can be activated at the same time.

(H) Using the current selection as seed, these tools allow you to extend the seed to include additional edges or faces into the selection set.

(I) Rectangles appear in the lower-left corner of the graphics window. These rectangles are aliases of selectable entities, that is, highlighting and picking these rectangles are identical and synchronized for the selectable entities.

(J) A body is a continuum and made of one kind of material. A 3D body is either a solid body, a surface body, or a line body.

(K) A part is a collection of same type of bodies. If multiple, then all bodies are assumed to be bonded together i.e., they form a single continuum (but multiple materials are allowed). In **Mechanical** parts are meshed independently. Within a part, the boundary nodes are shared between contacting bodies.

(L) In **Mechanical**, connections are kinematic relations between parts. Typical connections are contacts, joints, etc.

(M) A 3D body is created by combining various features. The features include planes, base features, and placed features.

(N) Base features are also called sketched features; they are created by first drawing one or more sketches, and then "growing" to 3D features by means of extrusion, revolution, sweeping, or lofting. A newly created base feature can add to or subtract material from the existing bodies.

(O) Placed features have predefined shapes and behaviors. To add these features to existing bodies, all we have to do is to specify where we want to place these features, along with a few other settings.

4.6-2 Additional Workbench Exercises

Create a 3D geometry for the Bolt-Nut-Plate Assembly

Create a model as shown in 3.2-1[4], page 118. You can download the finished project file from the Companion Webpage (see Preface). Or, if you have a DVD that comes with the copy of the book, you can find the finished project file in the DVD.

Chapter 5
3D Simulations

Many concepts and techniques in 2D simulations (Chapter 3) are applicable to 3D simulations. Comparing with 2D cases, problem sizes (number of nodes and elements) in 3D cases are usually larger and the geometries are usually more complicated; that implies more computing and engineering time. On the other hand, because we've been accustomed to the 3D world, the 3D simulations are more natural to us. As a result, newcomers often stick at 3D ways of thinking and forget that a problem often can be modeled as 2D. Remember that, if a problem can be reduced to a 2D, you have no reasons to go for a 3D simulation.

Purpose of This Chapter

This chapter focuses on 3D simulations with solid models. Problems involving surface models and line models are discussed in Chapters 6 and 7 respectively. Like the 2D simulations in Chapter 3, this chapter introduces linear static structural simulations. Dynamic and nonlinear simulations will be discussed in latter chapters.

About Each Section

As usual, we will conduct two step-by-step examples in the first two sections. Section 5.1 is a simple example, serving as an introductory tutorial. Section 5.2 is a more involved example. Section 5.3 provides a systematic discussion trying to complement what have been missed in the first two sections. Section 5.4 is an additional exercise. All exercises in this chapter use models created in Chapter 4.

Section 5.1

Beam Bracket

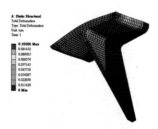

5.1-1 About the Beam Bracket

In Section 4.1, we created a 3D solid model for a beam bracket. In this section, we will use the model for a static structural simulation. Besides the geometry, we need more information for a simulation, which can be summarized into two categories: material properties and the environment conditions (loads and supports).

The bracket is made of structural steel [1], with a Young's modulus of 200 GPa and a Poisson's ratio of 0.3, Its yield strength 250 MPa. The yield strength is needed for assessing safety factors.

The load, determined by a global analysis of the entire structure, is 27 kN uniformly distributed over the seat plate [2].

As to the support conditions, we assume the beam bracket's back face is rigidly welded on a steel column; i.e., a fixed support [3].

[2] The bracket is designed to withstand a load of 27 kN uniformly distributed over the seat plate.

[3] Fixed support at the back face. #

[1] The bracket is made of structural steel.

5.1-2 Open the Project **Bracket**

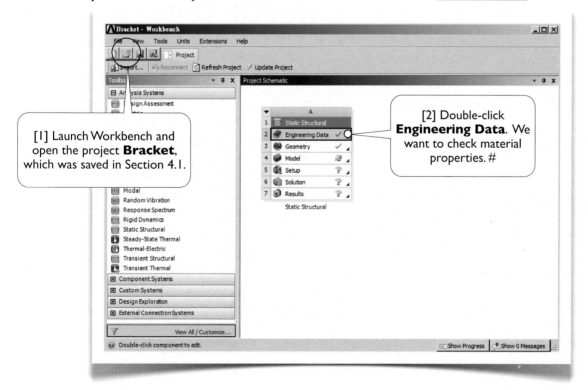

[1] Launch Workbench and open the project **Bracket**, which was saved in Section 4.1.

[2] Double-click **Engineering Data**. We want to check material properties. #

202

5.1-3 Check Material in **Engineering Data**

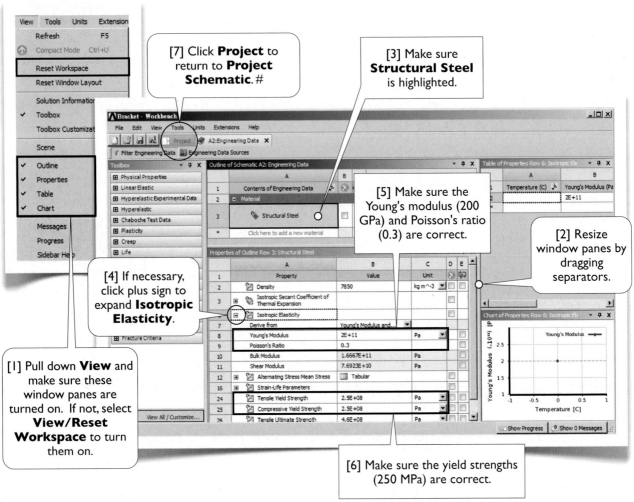

[7] Click **Project** to return to **Project Schematic**. #

[3] Make sure **Structural Steel** is highlighted.

[5] Make sure the Young's modulus (200 GPa) and Poisson's ratio (0.3) are correct.

[2] Resize window panes by dragging separators.

[4] If necessary, click plus sign to expand **Isotropic Elasticity**.

[1] Pull down **View** and make sure these window panes are turned on. If not, select **View/Reset Workspace** to turn them on.

[6] Make sure the yield strengths (250 MPa) are correct.

5.1-4 Start Up **Mechanical**

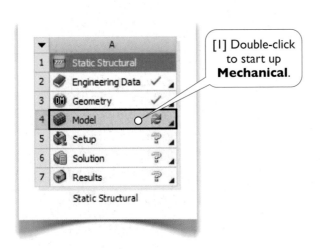

[1] Double-click to start up **Mechanical**.

Static Structural

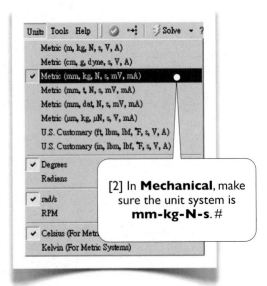

[2] In **Mechanical**, make sure the unit system is **mm-kg-N-s**. #

5.1-5 Check Material Assignment

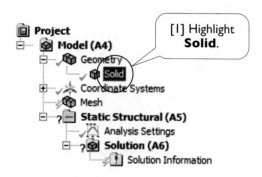

[1] Highlight **Solid**.

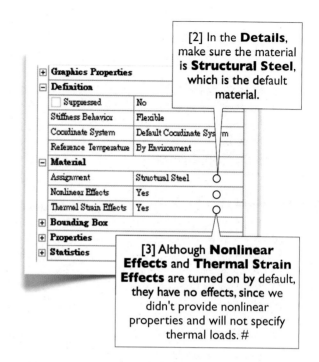

[2] In the **Details**, make sure the material is **Structural Steel**, which is the default material.

[3] Although **Nonlinear Effects** and **Thermal Strain Effects** are turned on by default, they have no effects, since we didn't provide nonlinear properties and will not specify thermal loads. #

Material Assignment

By default, **Mechanical** assigns **Structural Steel** for each body; therefore, we don't need to do anything about the material assignment in this case.

5.1-6 Specify Support

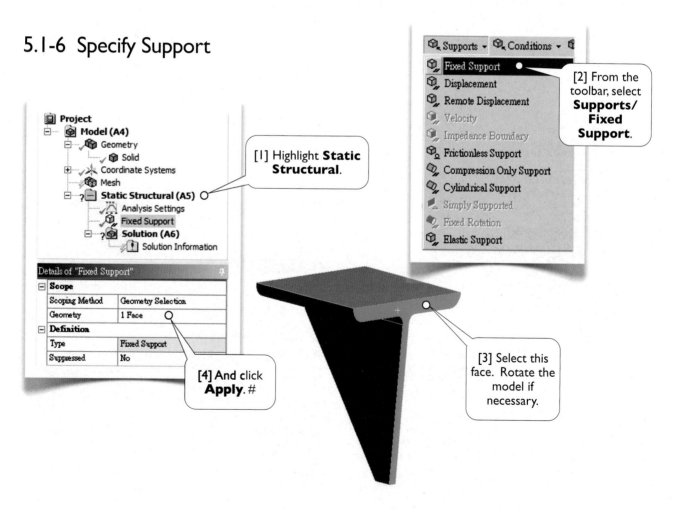

[1] Highlight **Static Structural**.

[2] From the toolbar, select **Supports/ Fixed Support**.

[3] Select this face. Rotate the model if necessary.

[4] And click **Apply**. #

5.1-7 Apply Force

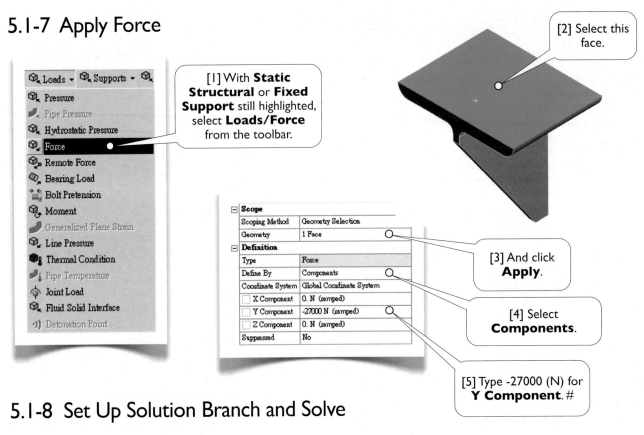

[1] With **Static Structural** or **Fixed Support** still highlighted, select **Loads/Force** from the toolbar.

[2] Select this face.

[3] And click **Apply**.

[4] Select **Components**.

[5] Type -27000 (N) for **Y Component**. #

5.1-8 Set Up Solution Branch and Solve

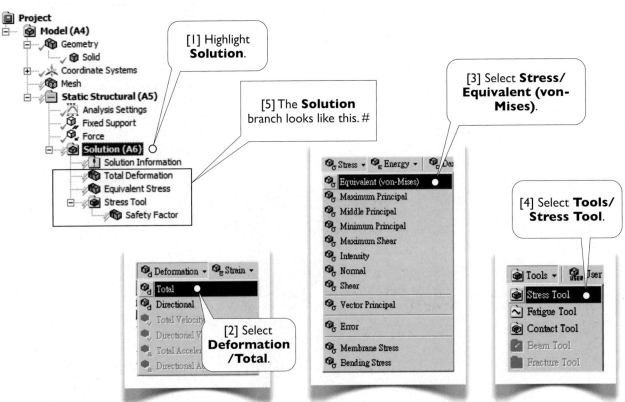

[1] Highlight **Solution**.

[5] The **Solution** branch looks like this. #

[3] Select **Stress/Equivalent (von-Mises)**.

[4] Select **Tools/Stress Tool**.

[2] Select **Deformation/Total**.

5.1-9 Generate Mesh

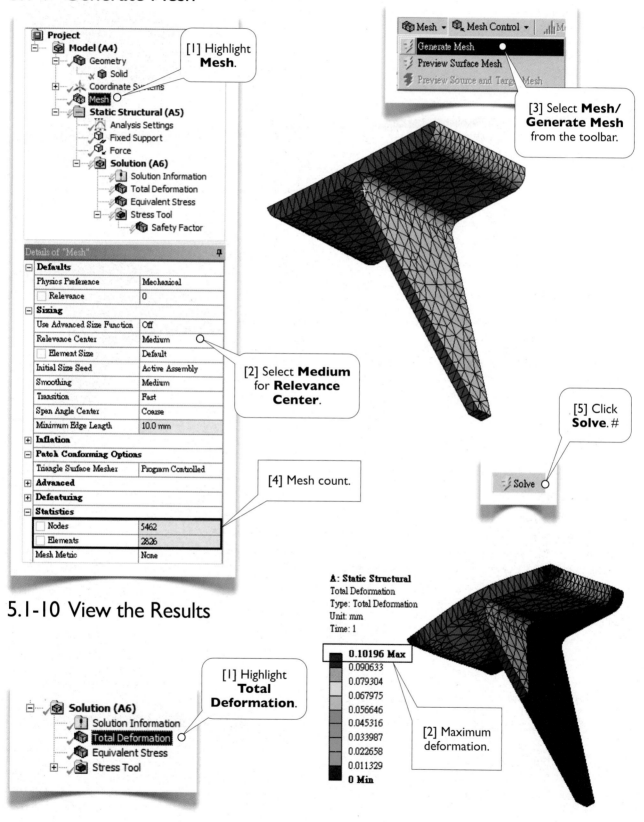

[1] Highlight **Mesh**.

[3] Select **Mesh/ Generate Mesh** from the toolbar.

[2] Select **Medium** for **Relevance Center**.

[4] Mesh count.

[5] Click **Solve**. #

5.1-10 View the Results

[1] Highlight **Total Deformation**.

A: Static Structural
Total Deformation
Type: Total Deformation
Unit: mm
Time: 1

[2] Maximum deformation.

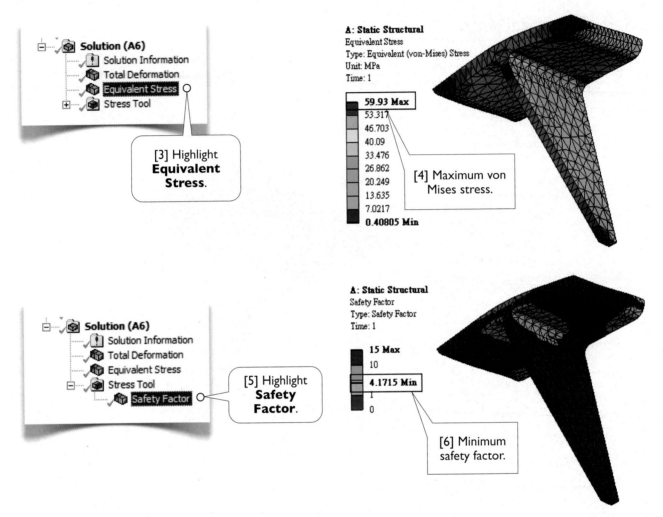

[3] Highlight **Equivalent Stress**.

[4] Maximum von Mises stress.

[5] Highlight **Safety Factor**.

[6] Minimum safety factor.

Safety Factors

Highlight **Stress Tool** in the project tree [7], and the details view shows how the safety factors are calculated [8]. In this case, the safety factor is the ratio between the yield stress (250 MPa, see 5.1-3[6], page 203) and the calculated equivalent stress [3-4]. For example, at the most critical region, the equivalent stress is 59.93 MPa [4], therefore the safety factor is 4.1715 (250/59.93) [6].

You can change the way safety factors are calculated by changing the settings in the details view [8].

[7] Highlight **Stress Tool**.

[8] Details of **Stress Tool** shows how safety factors are calculated. #

5.1-11 Animate the Results

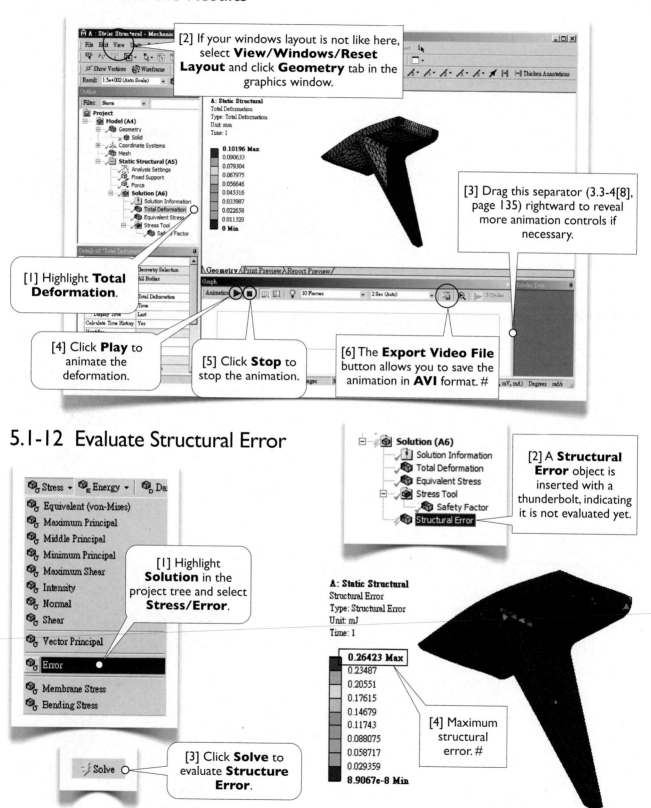

[2] If your windows layout is not like here, select **View/Windows/Reset Layout** and click **Geometry** tab in the graphics window.

[3] Drag this separator (3.3-4[8], page 135) rightward to reveal more animation controls if necessary.

[1] Highlight **Total Deformation**.

[4] Click **Play** to animate the deformation.

[5] Click **Stop** to stop the animation.

[6] The **Export Video File** button allows you to save the animation in **AVI** format. #

5.1-12 Evaluate Structural Error

[1] Highlight **Solution** in the project tree and select **Stress/Error**.

[2] A **Structural Error** object is inserted with a thunderbolt, indicating it is not evaluated yet.

[3] Click **Solve** to evaluate **Structure Error**.

[4] Maximum structural error. #

5.1-13 Improve Mesh Quality

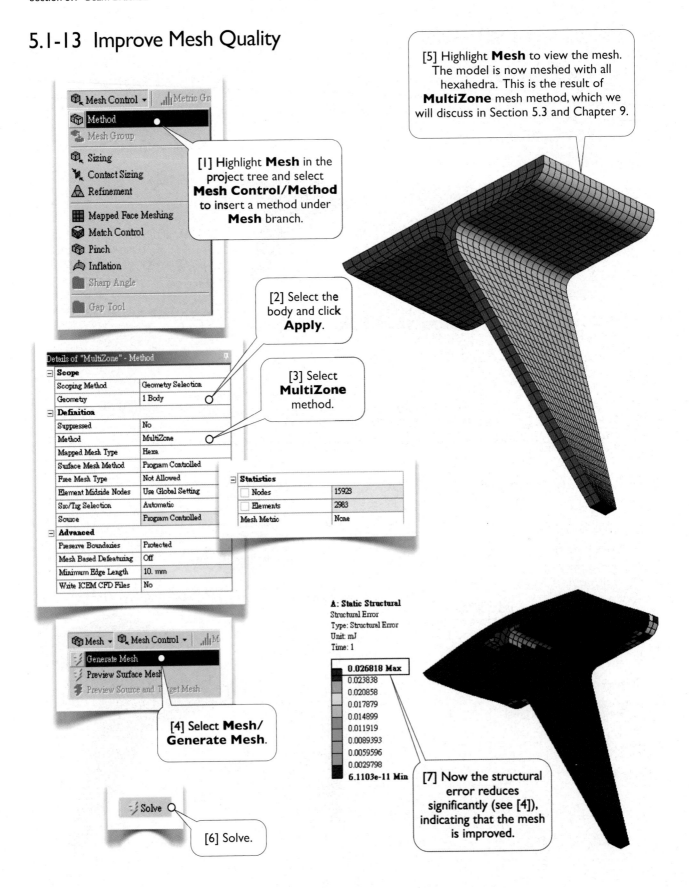

[5] Highlight **Mesh** to view the mesh. The model is now meshed with all hexahedra. This is the result of **MultiZone** mesh method, which we will discuss in Section 5.3 and Chapter 9.

[1] Highlight **Mesh** in the project tree and select **Mesh Control/Method** to insert a method under **Mesh** branch.

[2] Select the body and click **Apply**.

[3] Select **MultiZone** method.

[4] Select **Mesh/ Generate Mesh**.

[6] Solve.

A: Static Structural
Structural Error
Type: Structural Error
Unit: mJ
Time: 1

0.026818 Max
0.023838
0.020858
0.017879
0.014899
0.011919
0.0089393
0.0059596
0.0029798
6.1103e-11 Min

[7] Now the structural error reduces significantly (see [4]), indicating that the mesh is improved.

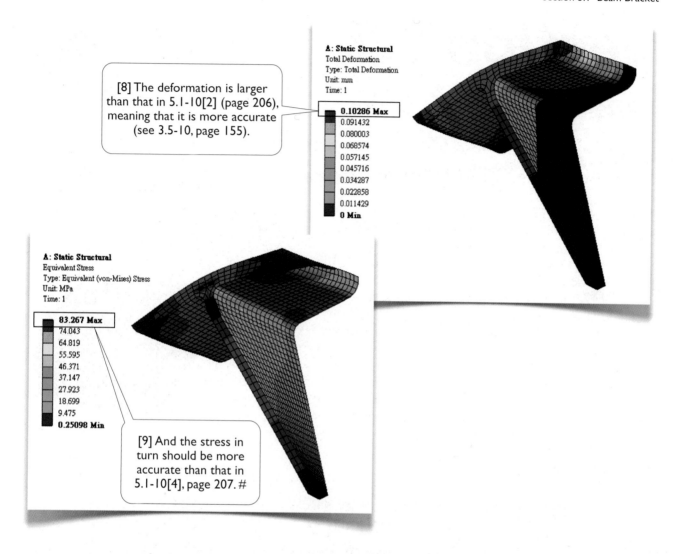

A: Static Structural
Total Deformation
Type: Total Deformation
Unit: mm
Time: 1

0.10286 Max
0.091432
0.080003
0.068574
0.057145
0.045716
0.034287
0.022858
0.011429
0 Min

[8] The deformation is larger than that in 5.1-10[2] (page 206), meaning that it is more accurate (see 3.5-10, page 155).

A: Static Structural
Equivalent Stress
Type: Equivalent (von-Mises) Stress
Unit: MPa
Time: 1

83.267 Max
74.043
64.819
55.595
46.371
37.147
27.923
18.699
9.475
0.25098 Min

[9] And the stress in turn should be more accurate than that in 5.1-10[4], page 207. #

Meshing 3D Solid Bodies

Solution accuracy depends not only on mesh density but also on mesh quality. In nonlinear problems, poorer mesh quality often leads to more computing time or even fails to find a solution.

Achieving a high mesh quality is, however, not trivial. In this section, we've use a new meshing technology, namely multi-zone method. It will be conceptually introduced in Section 5.3. Various meshing methods will be introduced in Chapter 9.

Wrap Up

Close **Mechanical**, save the project, and exit Workbench.

Section 5.2

Cover of Pressure Cylinder

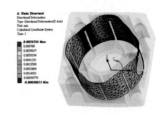

5.2-1 About the Cylinder Cover

In Section 4.2, we created a 3D solid model for the cover of a pressure cylinder, which is designed to hold a pressure of 0.5 MPa. In this section, we will use this model for a static structural simulation.

The cover was originally made of aluminum alloy. The purpose of the simulation is to assess the possibility of replacing the aluminum alloy by a new type of engineering plastic. The engineering plastic has a Young's modulus of 22 GPa and a Poisson's ratio of 0.3.

What concerns the engineers is the deformation, specifically the circularity (more accurately, cylindricity) of the internal surface that encompasses the pressure cylinder [1]. The circularity of a cylindrical surface can be defined as the difference between the radii of its circumscribed circle and its inscribed circle.

It is required that the circularity should be less than 10 micrometers, excess of circularity beyond this value may impair the tightness and cause a leakage of gas.

The unit system used in this exercise is **mm-kg-N-s**.

[1] We want to investigate the circularity of this internal surface. #

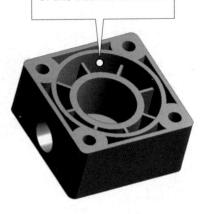

5.2-2 Open the Project Cover

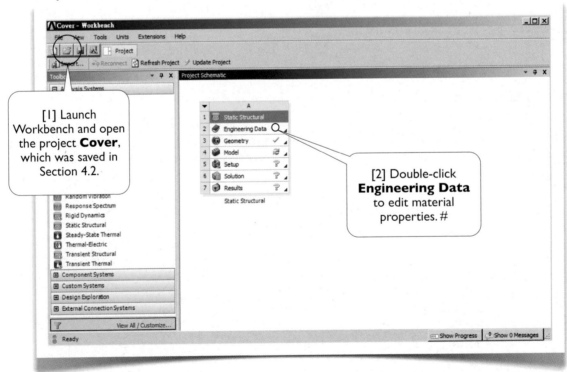

[1] Launch Workbench and open the project **Cover**, which was saved in Section 4.2.

[2] Double-click **Engineering Data** to edit material properties. #

5.2-3 Prepare Material Properties

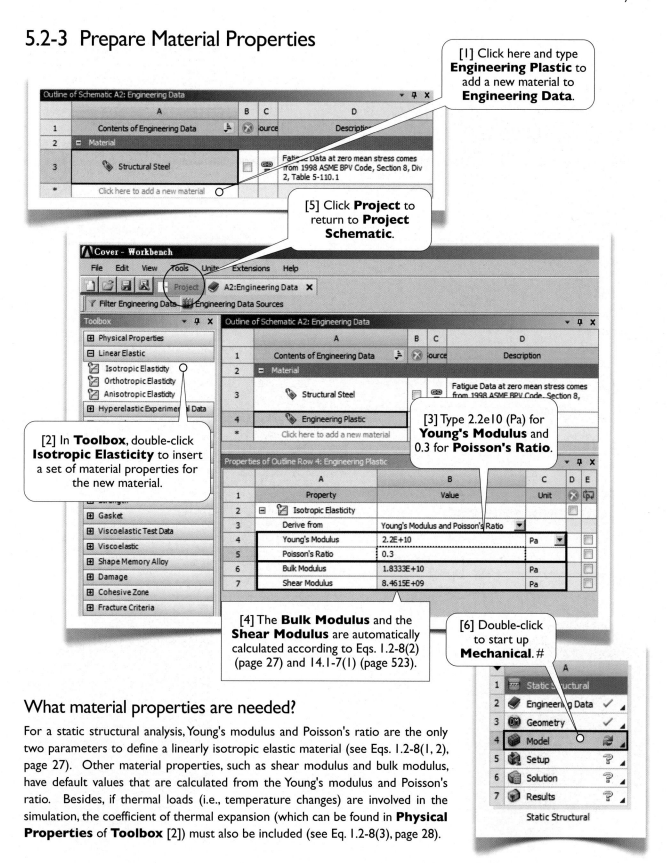

[1] Click here and type **Engineering Plastic** to add a new material to **Engineering Data**.

[5] Click **Project** to return to **Project Schematic**.

[2] In **Toolbox**, double-click **Isotropic Elasticity** to insert a set of material properties for the new material.

[3] Type 2.2e10 (Pa) for **Young's Modulus** and 0.3 for **Poisson's Ratio**.

[4] The **Bulk Modulus** and the **Shear Modulus** are automatically calculated according to Eqs. 1.2-8(2) (page 27) and 14.1-7(1) (page 523).

[6] Double-click to start up **Mechanical**. #

What material properties are needed?

For a static structural analysis, Young's modulus and Poisson's ratio are the only two parameters to define a linearly isotropic elastic material (see Eqs. 1.2-8(1, 2), page 27).   Other material properties, such as shear modulus and bulk modulus, have default values that are calculated from the Young's modulus and Poisson's ratio.   Besides, if thermal loads (i.e., temperature changes) are involved in the simulation, the coefficient of thermal expansion (which can be found in **Physical Properties** of **Toolbox** [2]) must also be included (see Eq. 1.2-8(3), page 28).

5.2-4 Assign Material to the Body

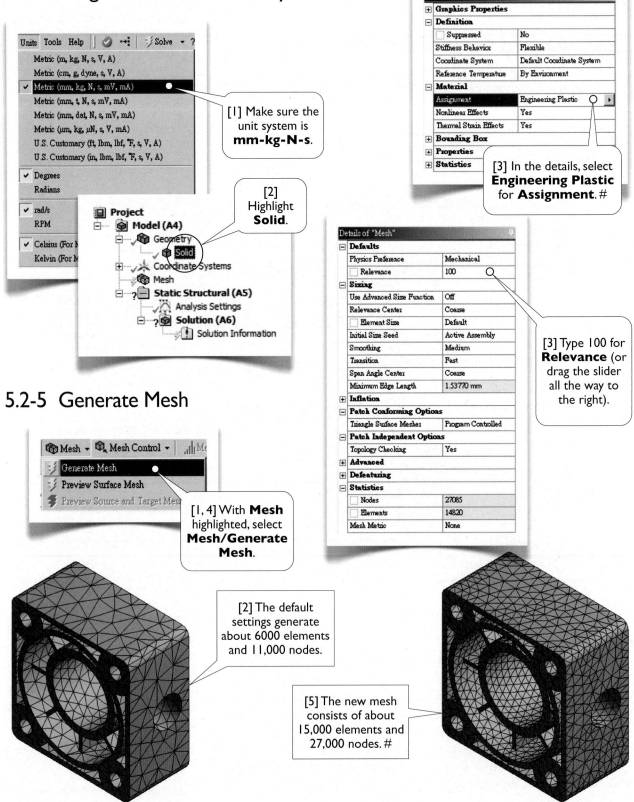

[1] Make sure the unit system is **mm-kg-N-s**.

[2] Highlight **Solid**.

[3] In the details, select **Engineering Plastic** for **Assignment**. #

5.2-5 Generate Mesh

[1, 4] With **Mesh** highlighted, select **Mesh/Generate Mesh**.

[3] Type 100 for **Relevance** (or drag the slider all the way to the right).

[2] The default settings generate about 6000 elements and 11,000 nodes.

[5] The new mesh consists of about 15,000 elements and 27,000 nodes. #

5.2-6 Specify Supports

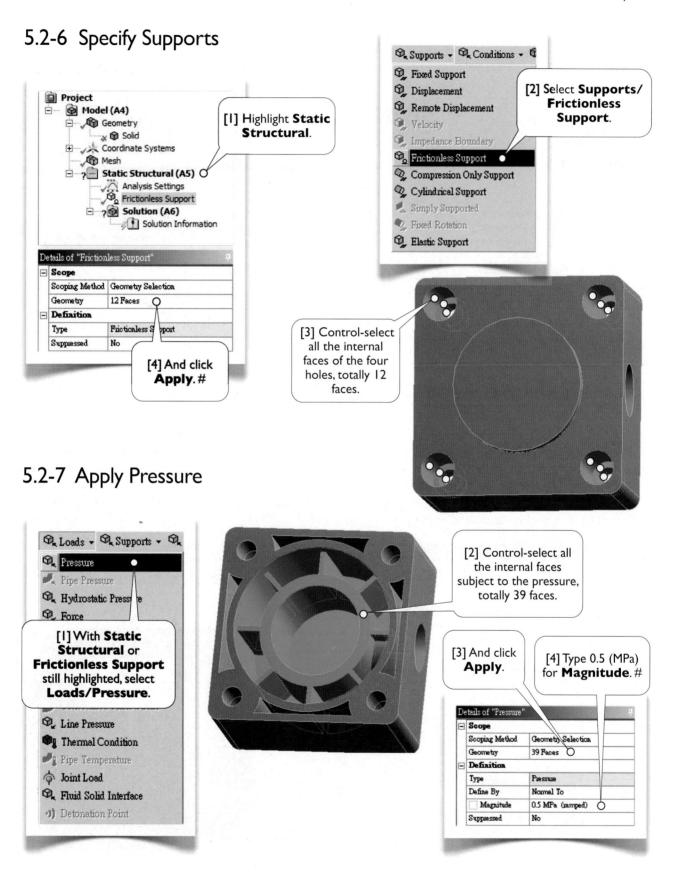

[1] Highlight **Static Structural**.

[2] Select **Supports/ Frictionless Support**.

[4] And click **Apply**. #

[3] Control-select all the internal faces of the four holes, totally 12 faces.

5.2-7 Apply Pressure

[1] With **Static Structural** or **Frictionless Support** still highlighted, select **Loads/Pressure**.

[2] Control-select all the internal faces subject to the pressure, totally 39 faces.

[3] And click **Apply**.

[4] Type 0.5 (MPa) for **Magnitude**. #

5.2-8 Set Up Solution Branch and Solve

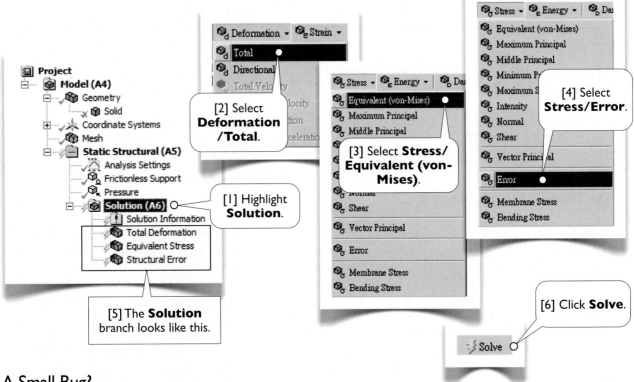

[1] Highlight **Solution**.

[2] Select **Deformation/Total**.

[3] Select **Stress/Equivalent (von-Mises)**.

[4] Select **Stress/Error**.

[5] The **Solution** branch looks like this.

[6] Click **Solve**.

A Small Bug?

After solving the model, you may notice that a warning message saying, **One or more bodies may be underconstrained and experiencing rigid body motion. Weak springs have been added to attain a solution** [7]. If you carefully examine the support conditions, the rigid body motions in all directions have been prevented by the supports. Ignore the warning message; it could be a small bug.

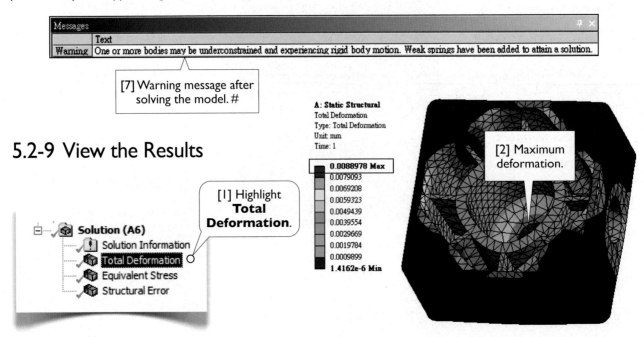

	Text
Warning	One or more bodies may be underconstrained and experiencing rigid body motion. Weak springs have been added to attain a solution.

[7] Warning message after solving the model. #

5.2-9 View the Results

[1] Highlight **Total Deformation**.

[2] Maximum deformation.

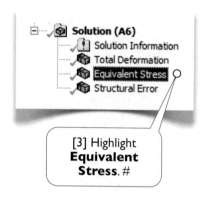

[3] Highlight **Equivalent Stress**. #

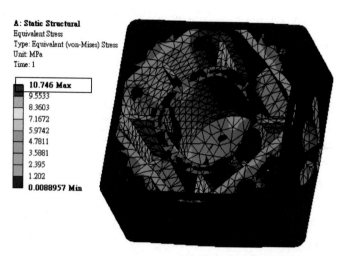

A: Static Structural
Equivalent Stress
Type: Equivalent (von-Mises) Stress
Unit: MPa
Time: 1

10.746 Max
9.5533
8.3603
7.1672
5.9742
4.7811
3.5881
2.395
1.202
0.0088957 Min

5.2-10 Create a Cylindrical Coordinate System

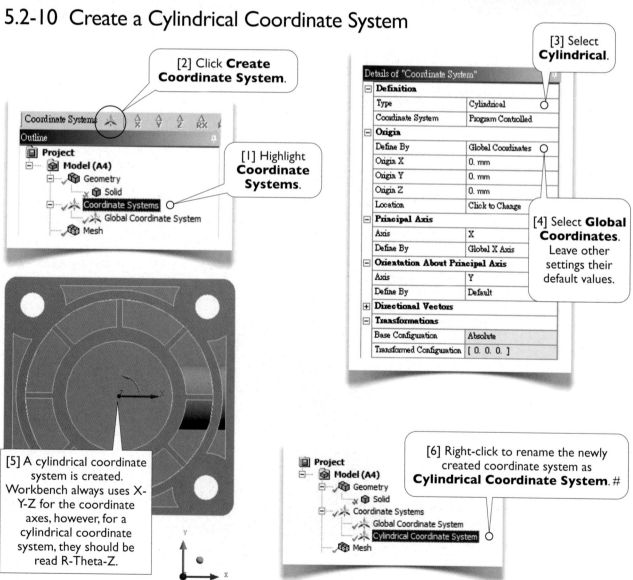

[2] Click **Create Coordinate System**.

[1] Highlight **Coordinate Systems**.

Coordinate Systems

Outline

- **Project**
  - **Model (A4)**
    - Geometry
      - Solid
    - Coordinate Systems
      - Global Coordinate System
    - Mesh

[3] Select **Cylindrical**.

Details of "Coordinate System"

Definition	
Type	Cylindrical
Coordinate System	Program Controlled
Origin	
Define By	Global Coordinates
Origin X	0. mm
Origin Y	0. mm
Origin Z	0. mm
Location	Click to Change
Principal Axis	
Axis	X
Define By	Global X Axis
Orientation About Principal Axis	
Axis	Y
Define By	Default
Directional Vectors	
Transformations	
Base Configuration	Absolute
Transformed Configuration	[0. 0. 0.]

[4] Select **Global Coordinates**. Leave other settings their default values.

[5] A cylindrical coordinate system is created. Workbench always uses X-Y-Z for the coordinate axes, however, for a cylindrical coordinate system, they should be read R-Theta-Z.

[6] Right-click to rename the newly created coordinate system as **Cylindrical Coordinate System**. #

- **Project**
  - **Model (A4)**
    - Geometry
      - Solid
    - Coordinate Systems
      - Global Coordinate System
      - Cylindrical Coordinate System
    - Mesh

216

5.2-11 Assess Circularity

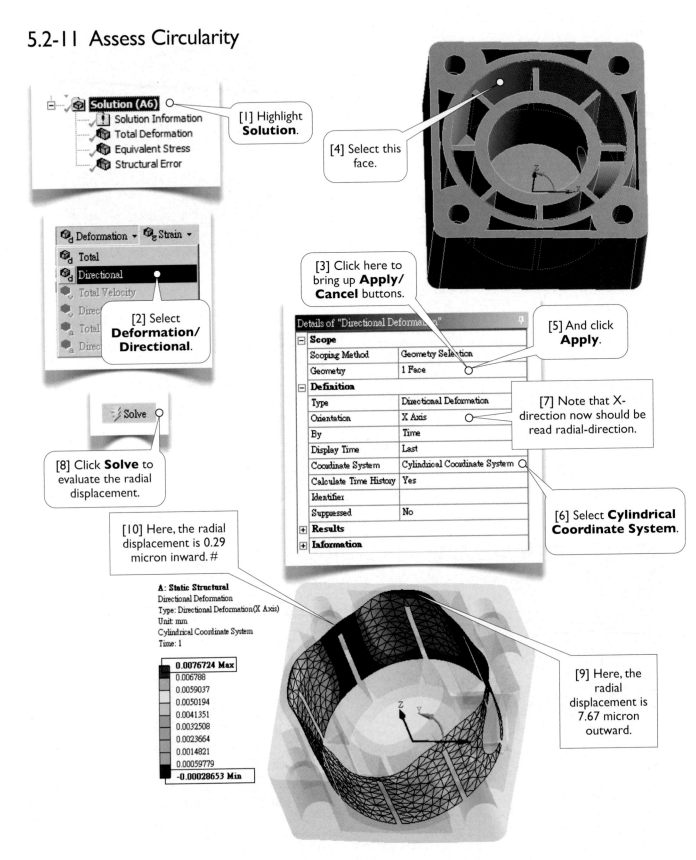

[1] Highlight **Solution**.

[4] Select this face.

[2] Select **Deformation/Directional**.

[3] Click here to bring up **Apply/Cancel** buttons.

[5] And click **Apply**.

Details of "Directional Deformation"

Scope	
Scoping Method	Geometry Selection
Geometry	1 Face
Definition	
Type	Directional Deformation
Orientation	X Axis
By	Time
Display Time	Last
Coordinate System	Cylindrical Coordinate System
Calculate Time History	Yes
Identifier	
Suppressed	No
Results	
Information	

[7] Note that X-direction now should be read radial-direction.

[6] Select **Cylindrical Coordinate System**.

[8] Click **Solve** to evaluate the radial displacement.

[10] Here, the radial displacement is 0.29 micron inward. #

A: Static Structural
Directional Deformation
Type: Directional Deformation(X Axis)
Unit: mm
Cylindrical Coordinate System
Time: 1

0.0076724 Max
0.006788
0.0059037
0.0050194
0.0041351
0.0032508
0.0023664
0.0014821
0.00059779
-0.00028653 Min

[9] Here, the radial displacement is 7.67 micron outward.

Conclusions

The circularity of the cover under the pressure, according to the definition, is 7.96 micrometer (7.67 + 0.29). It is within the spec requirement (10 micrometers).

We don't need to worry about the stress, since the stress (10.746 MPa, 5.2-9[3], page 216) is well below the fracture stress (which is about 50+ MPa).

Wrap Up

Close **Mechanical**, save the project, and exit Workbench.

Section 5.3

More Details

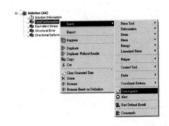

5.3-1 Global Mesh Controls[Ref 1]

By highlighting **Mesh** in a project tree, you can access the details of **Mesh** [1-4].

After the model meshed, a statistics of the mesh shows up on the bottom of the details view [1]. The mesh count provides an estimation of the problem size. In 3D cases, as mentioned, the total degrees of freedom equal three times of the number of nodes (1.3-1[1], page 31). In this case [1], Workbench was solving a system of equations of degrees of freedom 81255 (27085x3). The matrix [K] has a size of 81255x81255!

Mesh Metric [2] provides ways of measuring the mesh quality. This topic will be covered in Chapter 9.

Relevance Center [3] and **Relevance** [4] together provide a way of global mesh control. **Relevance Center** has three options: **Coarse**, **Medium**, or **Fine**. **Relevance** ranges from -100 to 100, the larger the finer. These two values are related roughly as follows: (Coarse, 0) = (Medium, -100), (Coarse, 100) = (Medium, 0) = (Fine, -100), and (Medium, 100) = (Fine, 0).

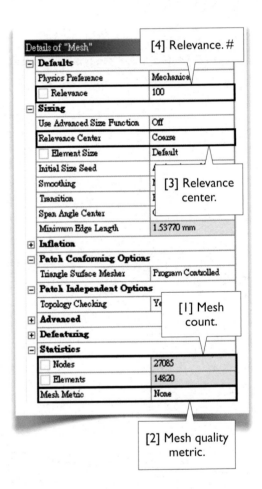

[4] Relevance. #

[3] Relevance center.

[1] Mesh count.

[2] Mesh quality metric.

5.3-2 Mesh with MultiZone Method

Generally, hexahedral elements are more desirable than other shapes, such as tetrahedral. Similarly, in 2D cases, quadrilateral elements are more desirable than triangular. The main reason is that hexahedral (or quadrilateral in 2D) has better convergence behavior (3.5-10 (page 155), 9.3-13, and 9.3-14 (page 353)). That implies, with the same problem size, hexahedral or quadrilateral gives more accurate results. Besides the shapes, mesh quality (5.3-1[2]) is also a key factor affecting convergence behavior. A mesh of hexahedral elements with poor mesh quality might be less desirable than tetrahedral with good mesh quality. Mesh metric will be discussed in Chapter 9.

For 2D models, Workbench usually does good jobs and meshes them with all-quadrilateral elements, or at least quadrilateral-dominated. For examples, in Sections 3.1, 3.2, and 3.4, all models are meshed with quadrilateral elements without further mesh controls.

For 3D models, meshing is much more challenging. In 5.2-5 (page 213), the model is relatively complicated that Workbench chooses to mesh with all-tetrahedral.

A simple idea of creating hexahedral elements is to mesh faces of a body with quadrilaterals and then "sweep" along a depth up to other end faces of the body. The starting faces are called *source faces* and the ending faces are target faces. The source or target faces can be either manually or automatically selected. Not all bodies are sweepable. In Section 5.1, the body (the bracket) is not sweepable without further decomposition.

The idea of **MultiZone** method is to decompose a non-sweepable body into several sweepable bodies, and then apply **Sweep** method on each of bodies. This is what we had done in 5.1-13 (page 209), where we inserted a **MultiZone** method and the Workbench decomposed the body into several sweepable bodies and easily "swept" each body with hexahedral elements. The result is an all-hexa mesh.

5.3-3 Coordinate Systems[Ref 2]

When defining an environment condition or a solution object by **Components**, you need to refer to a coordinate system. By default, **Global Coordinate System** is used, which is a Cartesian coordinate system. Sometimes this coordinate system is not convenient. In such cases, we may define additional coordinate systems for further use.

To define a coordinate system, you need to define the type of the coordinate system [1], the origin [2], and the axes [3].

Currently, workbench supports only two types of coordinate systems: Cartesian and Cylindrical [1].

Defining the origin is straightforward. You can click a location or type the coordinates [2].

There are many ways to define axes. Basically, you need to define two of the three axes and the third axis is automatically defined according to right-hand rule.

The first axis you define is called the **Principal Axis** and the second axis is defined by **Orientation About Principal Axis**.

For cylindrical coordinate system, you may be confused by such terminology. Fortunately, a triad always appears on the graphics window and you should be visually aware if you make mistakes.

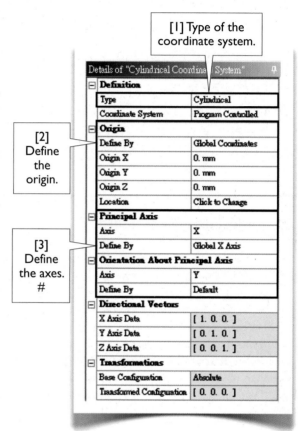

[1] Type of the coordinate system.

[2] Define the origin.

[3] Define the axes. #

5.3-4 Results View Controls[Ref 3]

With a results object highlighted, the toolbar displays tools that can be used to control how the results are presented [1-6].

[5] Vector display.

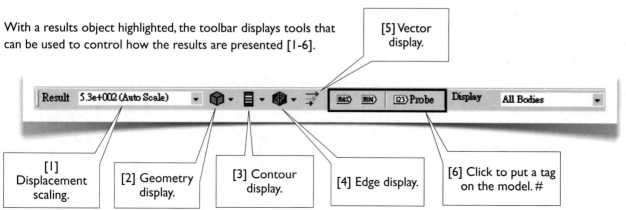

[1] Displacement scaling.

[2] Geometry display.

[3] Contour display.

[4] Edge display.

[6] Click to put a tag on the model. #

5.3-5 Legend Controls

Right-clicking the legend in the graphics window allows the user to modify the legend [1-7].

Independent bands [6] allow neutral colors to represent regions of the model above or below the specified legend limits.

Whenever needed, Select **Reset All** to return to the default settings [7].

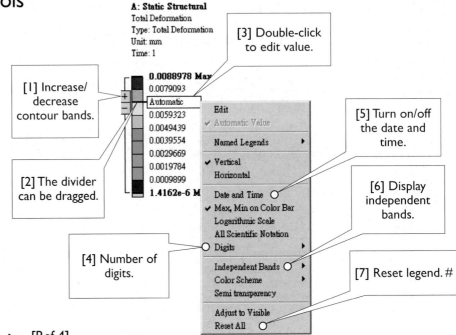

5.3-6 Adaptive Meshing[Ref 4]

We demonstrated the procedure of finite element convergence study in 3.5-8 through 3.5-10 (pages 154-155). Given an error, say 5%, we may refine mesh until the accuracy reaches this level. Performing these tasks manually is cumbersome. Workbench provides a tool to automate the mesh refinement until a user-specified level of accuracy is reached. This idea is termed *adaptive meshing*. Internally, Workbench uses structural errors (3.5-7, page 153) to help adjust the mesh, that is, it refines the mesh size in the area of large structural errors, and vice versa.

To use this tool, right-click a results object and select **Insert/Convergence** [1]. In the details view, specify the accuracy, or **Allowable Change** [2]. Also, highlight **Solution** in the project tree and, in the details view, specify the maximum number of mesh refinement loops [3]. When solving the model, Workbench will iterate to refine the mesh until the difference between two iterations is less than the **Allowable Change** or the **Max Refinement Loops** is reached.

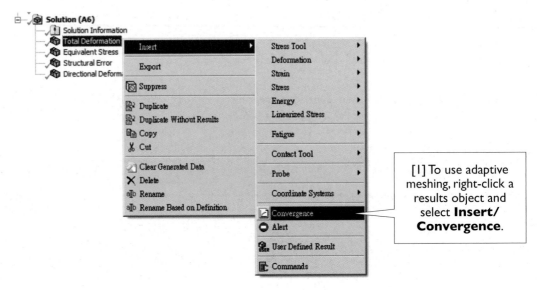

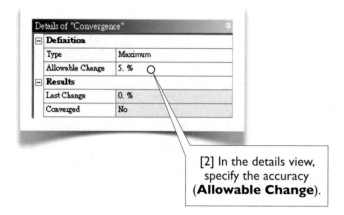

[2] In the details view, specify the accuracy (**Allowable Change**).

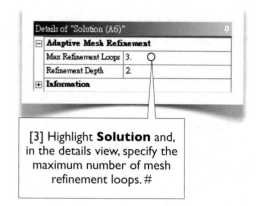

[3] Highlight **Solution** and, in the details view, specify the maximum number of mesh refinement loops. #

References

1. ANSYS Documentation//Meshing User's Guide//Global Mesh Controls
2. ANSYS Documentation//Mechanical Applications//Mechanical User's Guide//Objects Reference//Coordinate System
3. ANSYS Documentation //Mechanical Applications//Mechanical User's Guide//Approach//Steps for Using the Application//Review Results
4. ANSYS Documentation //Mechanical Applications//Mechanical User's Guide//Objects Reference//Convergence

Section 5.4

LCD Display Support

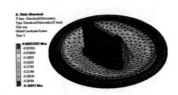

5.4-1 About the LCD Display Support

The geometry created in Section 4.5 is used in this section for a static structural simulation to assess the deformation and stress under a design load [1].

The display support is made of an ABS (acrylonitrile-butadiene-styrene) plastic, its Young's modulus 2.62 GPa and Poisson's ratio 0.34. In a tensile test, the material starts to develop fine cracks at 37 MPa, and fracture at 54 MPa.

The support is designed for a 17" LCD display, which weighs 40 N and is used as the design load.

[1] The load applies on the trough. #

5.4-2 Prepare Material Properties

Open the project **Support**, which was saved in Section 4.5. Double-click **Engineering Data** to edit material properties. Add a material to **Engineering Data** [1] and specify the Young's modulus and Poisson's ratio [2-3].

Return to **Project Schematic** [4] and double-click **Model** to start up **Mechanical**. In **Mechanical**, make sure the unit system is **mm-kg-N-s**, and assign the new material to the solid body [5].

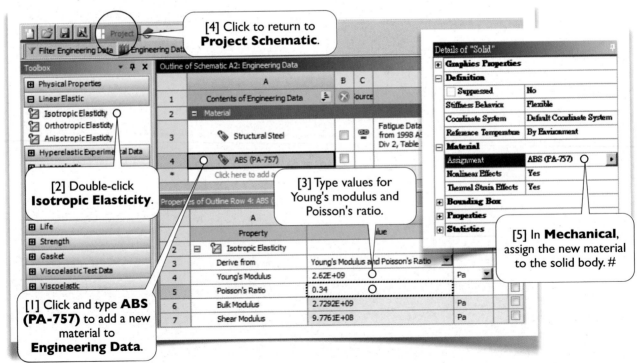

[4] Click to return to **Project Schematic**.

[2] Double-click **Isotropic Elasticity**.

[3] Type values for Young's modulus and Poisson's ratio.

[1] Click and type **ABS (PA-757)** to add a new material to **Engineering Data**.

[5] In **Mechanical**, assign the new material to the solid body. #

5.4-3 Specify Support

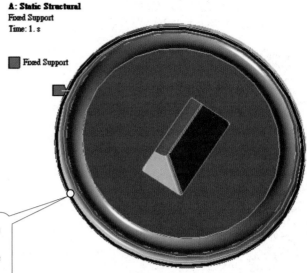

A: **Static Structural**
Fixed Support
Time: 1. s

■ Fixed Support

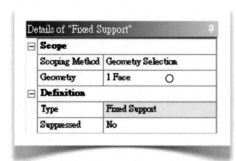

Details of "Fixed Support"	📌
□ **Scope**	
Scoping Method	Geometry Selection
Geometry	1 Face ○
□ **Definition**	
Type	Fixed Support
Suppressed	No

[1] Specify **Fixed Support** for the bottom rim face. Note that it would be more realistic if we specify **Compression Only Support** for the bottom rim face. For this case, however, the calculated results would be the same, since the rim face will not separate from the surface that supports the model. A **Compression Only Support** would introduce contact nonlinearity into the problem; that's why we avoid it. #

5.4-4 Specify Load

[1] Insert a **Bearing Load**.

A: **Static Structural**
Bearing Load
Time: 1. s

■ Bearing Load: 40. N
Components: 0., -40., 0. N

[2] Select the cylindrical surface of the hinge (see [3]).

Details of "Bearing Load"	
□ **Scope**	
Scoping Method	Geometry Selection
Geometry	1 Face ○
□ **Definition**	
Type	Bearing Load
Define By	Components ○
Coordinate System	Global Coordinate System
☐ X Component	0. N
☐ Y Component	-40. N ○
☐ Z Component	0. N
Suppressed	No

[4] The magnitude is 40 (N) in negative Y-direction. #

[3] The bearing load is applied on this cylindrical surface.

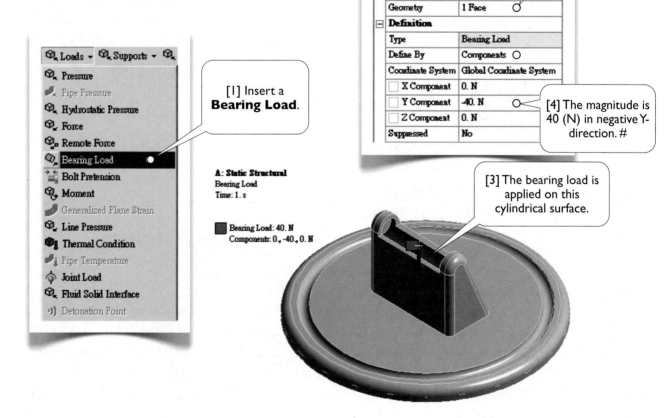

5.4-5 Generate Mesh

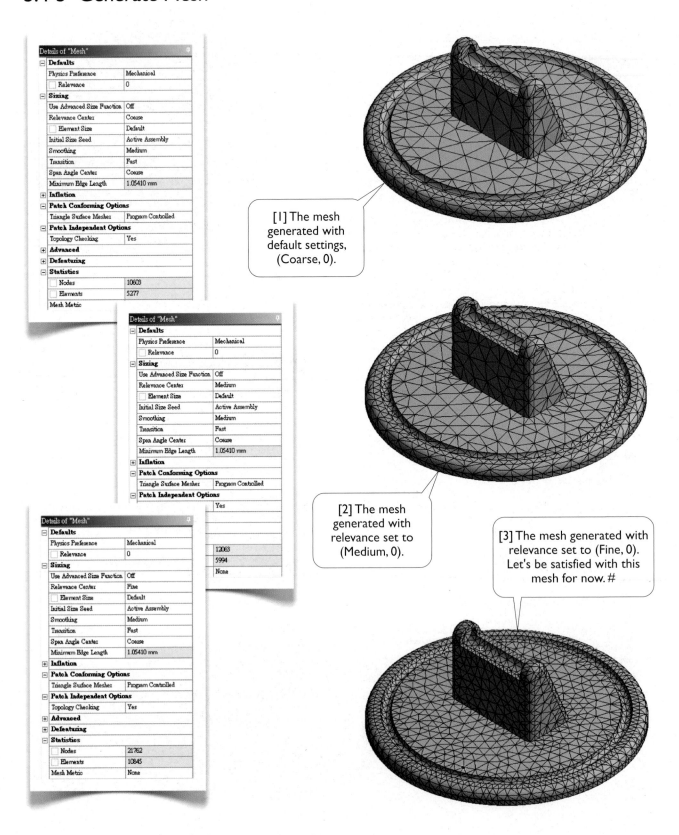

[1] The mesh generated with default settings, (Coarse, 0).

[2] The mesh generated with relevance set to (Medium, 0).

[3] The mesh generated with relevance set to (Fine, 0). Let's be satisfied with this mesh for now. #

5.4-6 Set Up Solution Branch and Solve

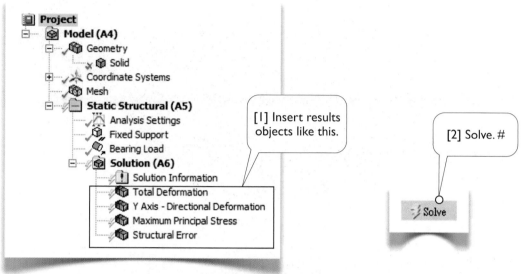

[1] Insert results objects like this.

[2] Solve. #

5.4-7 View the Results

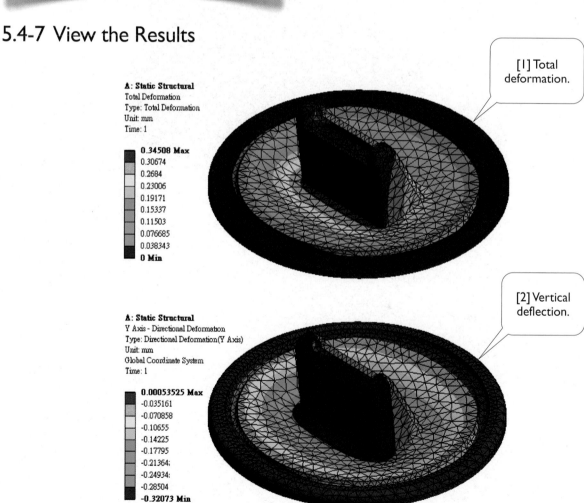

[1] Total deformation.

A: Static Structural
Total Deformation
Type: Total Deformation
Unit: mm
Time: 1

0.34508 Max
0.30674
0.2684
0.23006
0.19171
0.15337
0.11503
0.076685
0.038343
0 Min

[2] Vertical deflection.

A: Static Structural
Y Axis - Directional Deformation
Type: Directional Deformation(Y Axis)
Unit: mm
Global Coordinate System
Time: 1

0.00053525 Max
-0.035161
-0.070858
-0.10655
-0.14225
-0.17795
-0.21364;
-0.24934;
-0.28504
-0.32073 Min

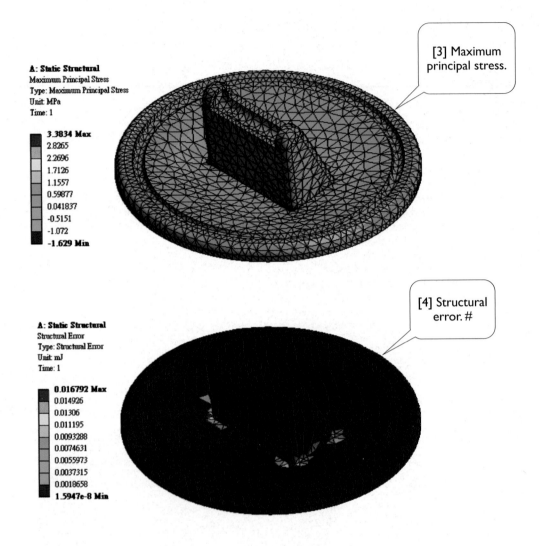

[3] Maximum principal stress.

[4] Structural error. #

Wrap Up

Close **Mechanical**, save the project, and exit Workbench.

Section 5.5

Review

5.5-1 Keywords

Choose a letter for each keyword from the list of descriptions

1. (   ) Adaptive Meshing
2. (   ) Bearing Load
3. (   ) Coordinate System
4. (   ) MultiZone Method
5. (   ) Sweep Method

Answers:

1. (E)   2. (A)   3. (B)   4. (D)   5. (C)

List of Descriptions

(A)   In 3D simulations, a bearing load applies on cylindrical faces. The total force is distributed on the compressive side of the cylindrical faces.

(B)   To define a coordinate system, we need to specify the type of coordinate system, the location of origin, and the axes. The types of coordinate systems may be **Cartesian** or **Cylindrical** in the current version of the Workbench.

(C)   A meshing method, in which the source faces, selected manually or automatically, are meshed, and the 2D mesh grows to become 3D elements by sweeping along a path up to target faces. Not all 3D solid bodies are sweepable. A sweepable 3D solid body often can be meshed with hexahedral elements.

(D)   A meshing method. For a non-sweepable 3D solid body, the method tries to decompose the body into several sweepable bodies and then uses Sweep method to mesh each body.

(E)   An automatic and iterative solution process to meet a user-specified solution accuracy. Basic idea is to refine the mesh size in the area of large structural errors until the specified accuracy is satisfied.

5.5-2 Questions

Symmetries

We didn't take any advantages of symmetries in all three simulation cases of this chapter. Point out the planes of symmetry in each model.

5.5-3 Additional Workbench Exercises

Simulation with Symmetric Models

It is a good exercise to redo the simulations in this chapter, taking advantages of the symmetries.

Additional Exercises

Additional Exercises of 3D simulations can be found in the Verification Manual for Workbench[Ref 1].

Reference

1. ANSYS Documentation//Verification Manuals//Workbench Verification Manual

---

Chapter 6
Surface Models

Many real-world objects can be modeled as surface bodies. In Section 4.3, we modeled the glass as a surface body. The 2D bodies in Chapter 3, although created as surfaces, are called 2D solid bodies in **Mechanical**. Remember that 2D solid bodies do not have out-of-plane bending.

When a real-world body is thin enough, it is usually a good candidate for a 3D surface body. Workbench will mesh surface bodies with shell elements (1.3-3[7, 8], page 34). There are many advantages of using surface models over 3D solid models. First, creating surface models is usually easier. Second, the problem size is much smaller; that implies a much less computing time. Third, it often results a more accurate solution due to the efficiency of shell elements. Therefore, engineers should consider using surface models instead of solid models whenever possible. In the old days, surface models are visually awkward since they have zero thickness and occupy zero volume in the space. Workbench avoids this awkwardness by allowing a rendering of thickness, so that the surface bodies can be visually the same as solid bodies.

Purpose of This Chapter

The main purpose of this chapter is to practice the use of shell elements (1.3-3[7, 8], page 34). This chapter guides the students to create surface models and perform simulations using surface models. The chapter uses three examples, two of them are purely surface models while the other one is a model mixed up with solid bodies and surface bodies.

About Each Section

Section 6.1 creates a bellows joint model and performs a simulation using the model. In Section 6.2, the bracket, which we introduced in Sections 4.1 and 5.1, is used again as an example to demonstrate how an existing solid model can be transformed to a surface model using the tool **Mid-Surface**. The solution is numerically comparable with that obtained using the solid model in Section 5.1, while using much less computing time. Section 6.3 provides an additional exercise and demonstrates the mix-up of solid bodies and surface bodies.

Section 6.1

Bellows Joints

6.1-1 About Bellows Joints

The bellows joints [1-2] are used as expansion joints, absorbing thermal or vibrational movement in a pipeline system that transports high pressure gases, designed to sustain internal pressure as well as external pressure. The external pressure is considered when the piping system is used under the ocean. With the internal pressure, the engineers are concerned about its radial deformation (due to an engineering tolerance consideration) and hoop stress (due to a safety consideration). Under the external pressure, buckling is the main concern.

In this section, we will create a full 3D surface model for the bellows joint and perform a static structural simulation under the internal pressure of 0.5 MPa. A buckling simulation under the external pressure is left as an exercise in 10.4-2, page 380.

Note that the problem is axisymmetric both in geometry and loading. We could take advantage of this property and model the problem as 2D solid body or 2D line body (both as axisymmetric models). The latter, 2D line model, is not supported in the current version of **Mechanical** (however, it is supported through **APDL**). The former, 2D solid model, usually results a poorer solution than 3D surface body, for this particular case, because the bending dominates the structural behavior.

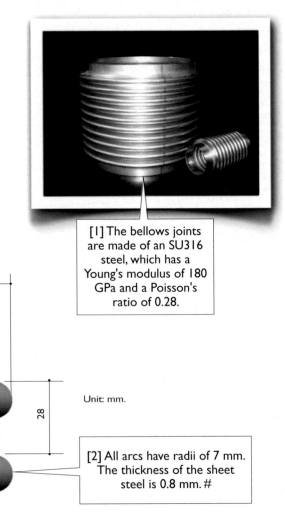

[1] The bellows joints are made of an SU316 steel, which has a Young's modulus of 180 GPa and a Poisson's ratio of 0.28.

Unit: mm.

[2] All arcs have radii of 7 mm. The thickness of the sheet steel is 0.8 mm. #

R315   R315   28

20

28

PART A. GEOMETRIC MODELING

6.1-2 Start Up

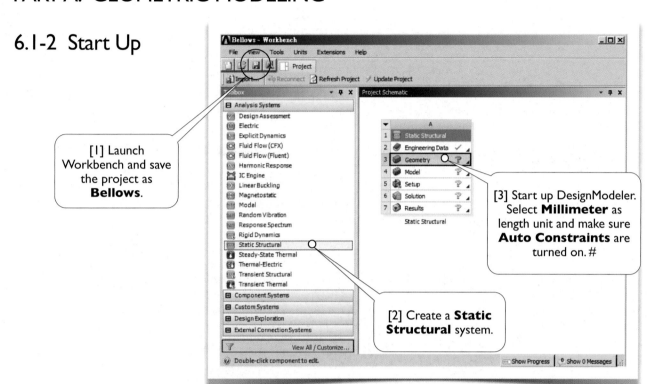

[1] Launch Workbench and save the project as **Bellows**.

[2] Create a **Static Structural** system.

[3] Start up DesignModeler. Select **Millimeter** as length unit and make sure **Auto Constraints** are turned on. #

6.1-3 Draw a Pitch of the Bellows

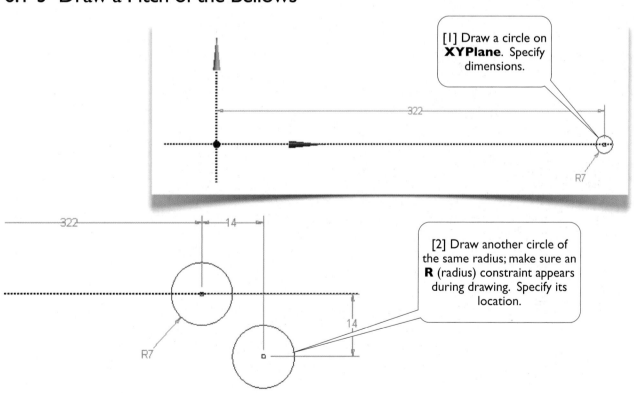

[1] Draw a circle on **XYPlane**. Specify dimensions.

[2] Draw another circle of the same radius; make sure an **R** (radius) constraint appears during drawing. Specify its location.

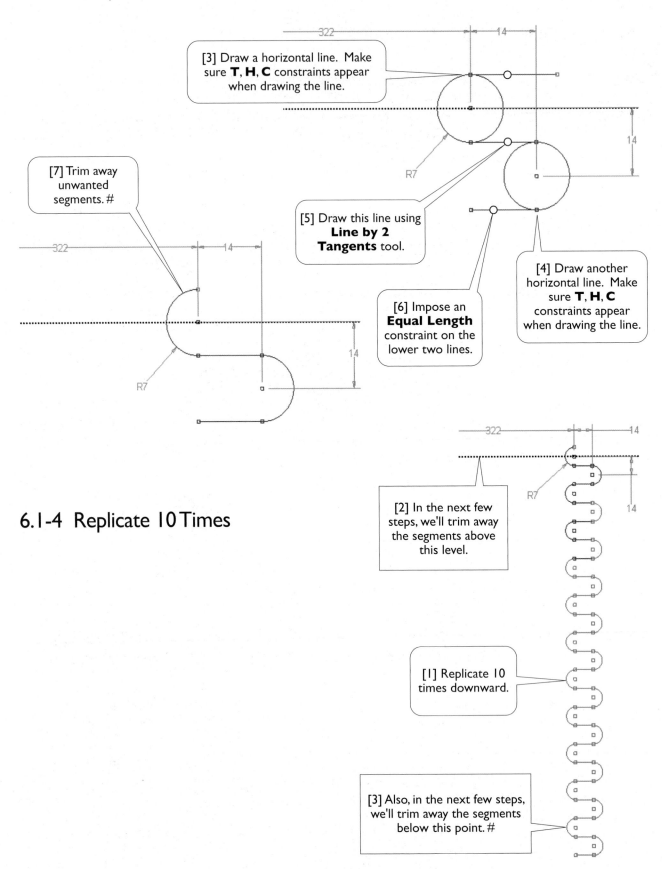

[3] Draw a horizontal line. Make sure **T**, **H**, **C** constraints appear when drawing the line.

[7] Trim away unwanted segments. #

[5] Draw this line using **Line by 2 Tangents** tool.

[6] Impose an **Equal Length** constraint on the lower two lines.

[4] Draw another horizontal line. Make sure **T**, **H**, **C** constraints appear when drawing the line.

6.1-4 Replicate 10 Times

[2] In the next few steps, we'll trim away the segments above this level.

[1] Replicate 10 times downward.

[3] Also, in the next few steps, we'll trim away the segments below this point. #

233

6.1-5 Finish Both Ends

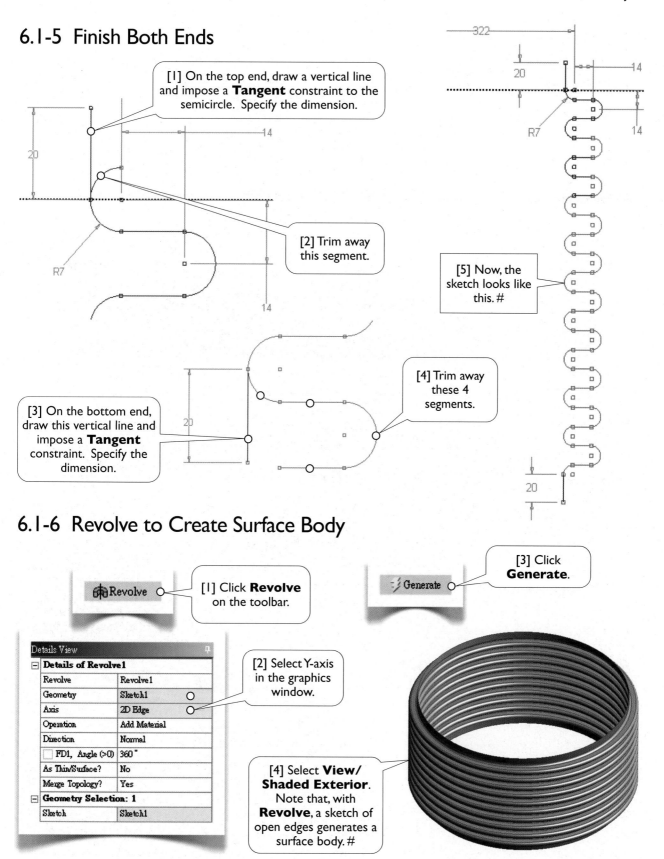

[1] On the top end, draw a vertical line and impose a **Tangent** constraint to the semicircle. Specify the dimension.

[2] Trim away this segment.

[3] On the bottom end, draw this vertical line and impose a **Tangent** constraint. Specify the dimension.

[4] Trim away these 4 segments.

[5] Now, the sketch looks like this. #

6.1-6 Revolve to Create Surface Body

[1] Click **Revolve** on the toolbar.

[3] Click **Generate**.

Details of Revolve1	
Revolve	Revolve1
Geometry	Sketch1
Axis	2D Edge
Operation	Add Material
Direction	Normal
FD1, Angle (>0)	360 °
As Thin/Surface?	No
Merge Topology?	Yes
Geometry Selection: 1	
Sketch	Sketch1

[2] Select Y-axis in the graphics window.

[4] Select **View/Shaded Exterior**. Note that, with **Revolve**, a sketch of open edges generates a surface body. #

PART B. SIMULATION

6.1-7 Prepare Material Properties

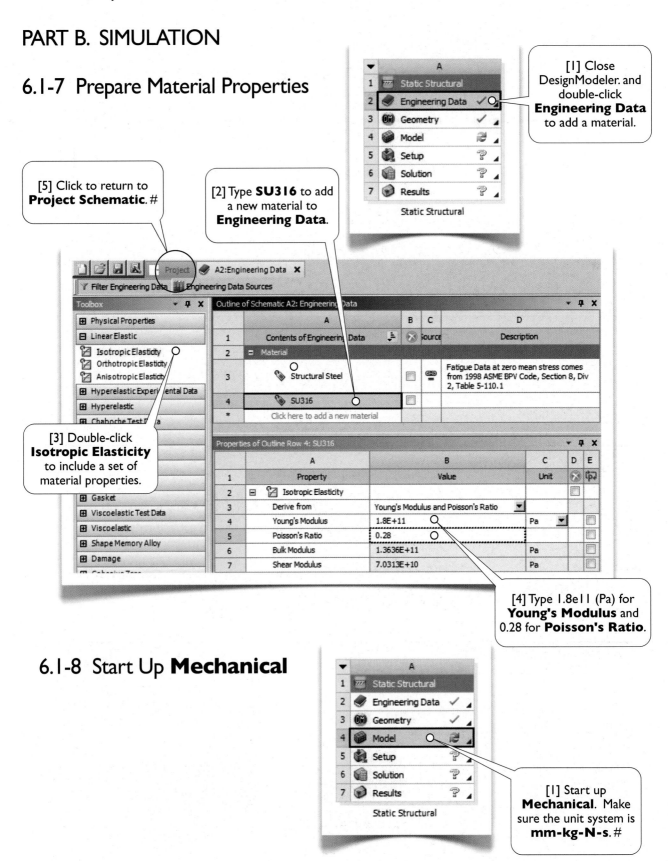

[1] Close DesignModeler. and double-click **Engineering Data** to add a material.

[5] Click to return to **Project Schematic**. #

[2] Type **SU316** to add a new material to **Engineering Data**.

[3] Double-click **Isotropic Elasticity** to include a set of material properties.

[4] Type 1.8e11 (Pa) for **Young's Modulus** and 0.28 for **Poisson's Ratio**.

6.1-8 Start Up **Mechanical**

[1] Start up **Mechanical**. Make sure the unit system is **mm-kg-N-s**. #

6.1-9 Assign Material and Thickness

[1] Highlight **Surface Body**.

[2] Type 0.8 (mm) for **Thickness**.

[3] Select **SU316**. #

6.1-10 Generate Mesh

[1] Highlight **Mesh** in the project tree, turn off **Use Advanced Size Function**.

[2] Select **Mesh/ Generate Mesh**.

[3] The mesh consists of about 12,000 shell elements. #

6.1-11 Specify Supports

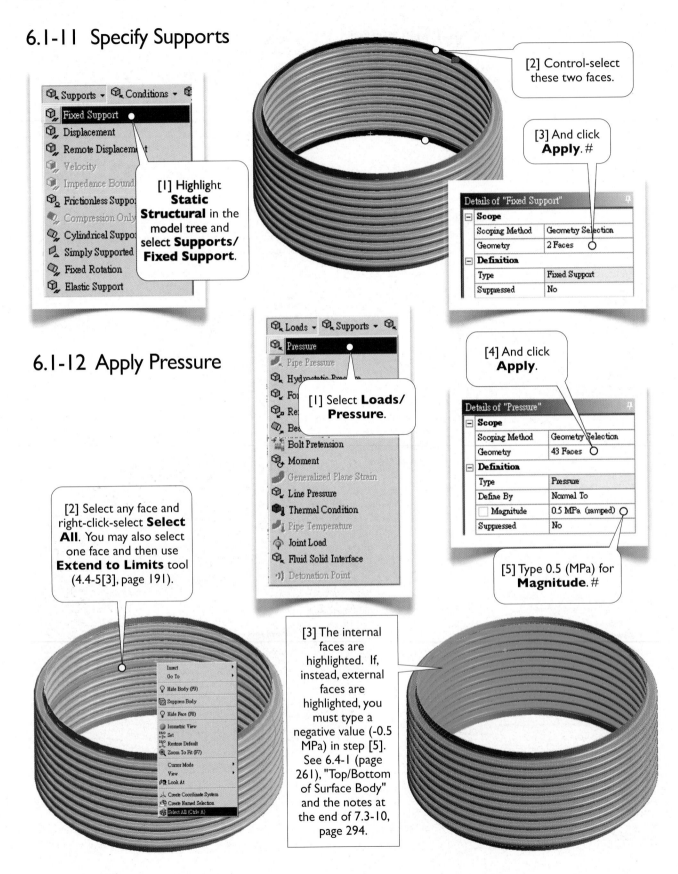

[2] Control-select these two faces.

[3] And click **Apply**. #

[1] Highlight **Static Structural** in the model tree and select **Supports/ Fixed Support**.

Details of "Fixed Support"

Scope	
Scoping Method	Geometry Selection
Geometry	2 Faces
Definition	
Type	Fixed Support
Suppressed	No

6.1-12 Apply Pressure

[1] Select **Loads/ Pressure**.

[4] And click **Apply**.

Details of "Pressure"

Scope	
Scoping Method	Geometry Selection
Geometry	43 Faces
Definition	
Type	Pressure
Define By	Normal To
Magnitude	0.5 MPa (ramped)
Suppressed	No

[5] Type 0.5 (MPa) for **Magnitude**. #

[2] Select any face and right-click-select **Select All**. You may also select one face and then use **Extend to Limits** tool (4.4-5[3], page 191).

[3] The internal faces are highlighted. If, instead, external faces are highlighted, you must type a negative value (-0.5 MPa) in step [5]. See 6.4-1 (page 261), "Top/Bottom of Surface Body" and the notes at the end of 7.3-10, page 294.

6.1-13 Create a Cylindrical Coordinate System

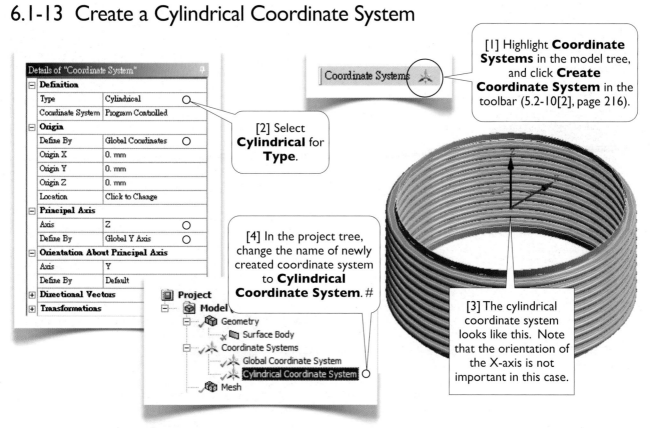

[1] Highlight **Coordinate Systems** in the model tree, and click **Create Coordinate System** in the toolbar (5.2-10[2], page 216).

[2] Select **Cylindrical** for **Type**.

[4] In the project tree, change the name of newly created coordinate system to **Cylindrical Coordinate System**. #

[3] The cylindrical coordinate system looks like this. Note that the orientation of the X-axis is not important in this case.

6.1-14 Set Up Solution Branch and Solve the Model

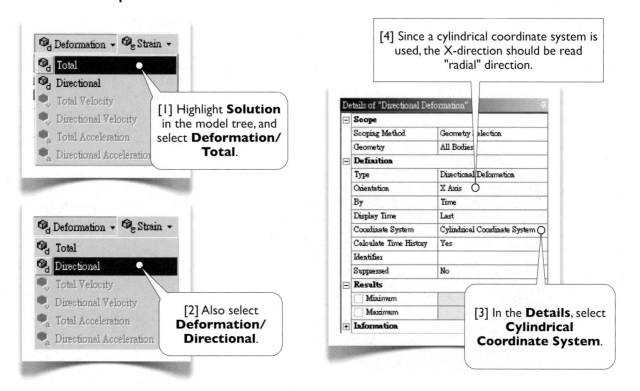

[1] Highlight **Solution** in the model tree, and select **Deformation/ Total**.

[2] Also select **Deformation/ Directional**.

[4] Since a cylindrical coordinate system is used, the X-direction should be read "radial" direction.

[3] In the **Details**, select **Cylindrical Coordinate System**.

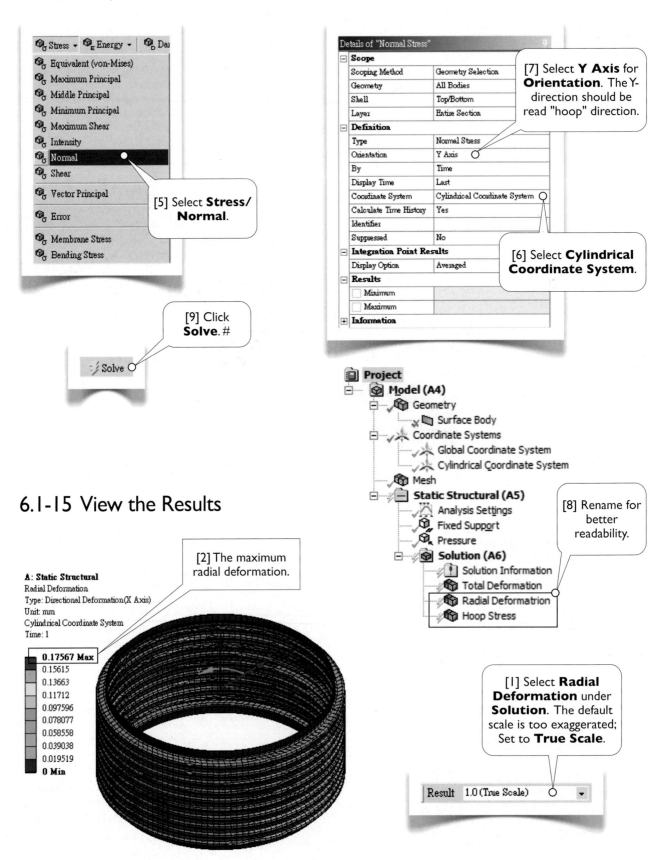

[5] Select **Stress/ Normal**.

[9] Click **Solve**. #

[7] Select **Y Axis** for **Orientation**. The Y-direction should be read "hoop" direction.

[6] Select **Cylindrical Coordinate System**.

[8] Rename for better readability.

6.1-15 View the Results

[2] The maximum radial deformation.

[1] Select **Radial Deformation** under **Solution**. The default scale is too exaggerated; Set to **True Scale**.

A: Static Structural
Radial Deformation
Type: Directional Deformation(X Axis)
Unit: mm
Cylindrical Coordinate System
Time: 1

0.17567 Max
0.15615
0.13663
0.11712
0.097596
0.078077
0.058558
0.039038
0.019519
0 Min

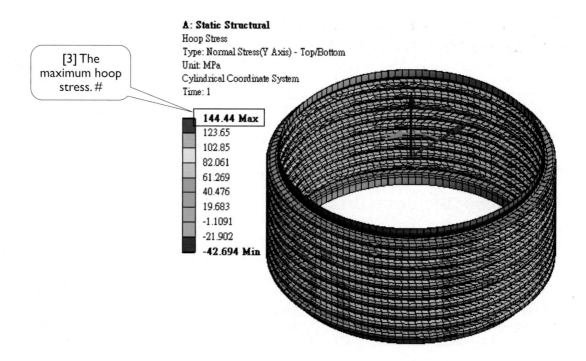

Wrap Up

Save the project and exit Workbench.

Remark: Symmetry of the Model

As mentioned in 6.1-1, the model is axisymmetric (page 231). We could take advantage of this property and model the problem as 2D line body and mesh the body using axisymmetric shell elements (such as **SHELL208**[Ref 1]).   For the current version of Workbench, that is possible only by using **APDL** commands.

In the foregoing simulation, we didn't take the advantage of axisymmetry. There are, however, at least three planes of symmetry we may used. This leaves for the students as an exercise (6.4-3, page 262). One reason we model the bellows without taking the advantage of any symmetries is that we want to perform a buckling simulation in 10.4-2, as an exercise (page 380).

While the geometry and the loading are symmetric, the buckling mode may be asymmetric. If we model a problem as symmetric, then we would miss all asymmetric buckling modes. This is also true for a free vibration simulation (modal analysis), which if we model as symmetric, then we would miss all asymmetric vibration modes.

Reference

1.   ANSYS Documentation//Mechanical APDL//Element Reference//I. Element Library//SHELL208

Section 6.2

Beam Bracket

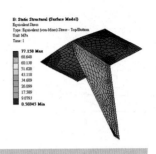

6.2-1 About the Beam Bracket

In Section 4.1, we created a 3D solid model for the beam bracket, the model simulated in Section 5.1. Since the seat plate (flange) and the web plate are relatively thin and have uniform thicknesses, it raises a curiosity that if it is possible to model the beam bracket as a surface model and obtain a comparable result.

To create a surface model for the beam bracket, we don't have to start from scratch; we can use a tool in DesignModeler, called **Mid-Surface**. In fact, surface models are often created this way. CAD models are usually created as 3D solids, since they are created for many purposes, and simulation is only one of them. When a surface model is needed for simulation, **Mid-Surface** is a powerful tool to extract a surface model from a solid model.

PART A. GEOMETRIC MODELING

6.2-2 Start Up

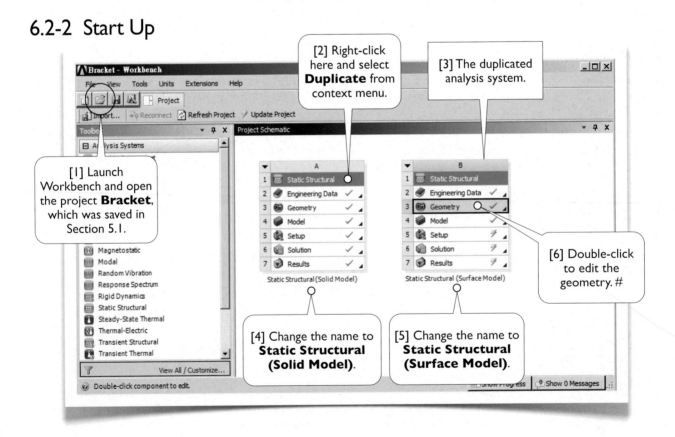

[1] Launch Workbench and open the project **Bracket**, which was saved in Section 5.1.

[2] Right-click here and select **Duplicate** from context menu.

[3] The duplicated analysis system.

[4] Change the name to **Static Structural (Solid Model)**.

[5] Change the name to **Static Structural (Surface Model)**.

[6] Double-click to edit the geometry. #

6.2-3 Automatically Find Face Pairs

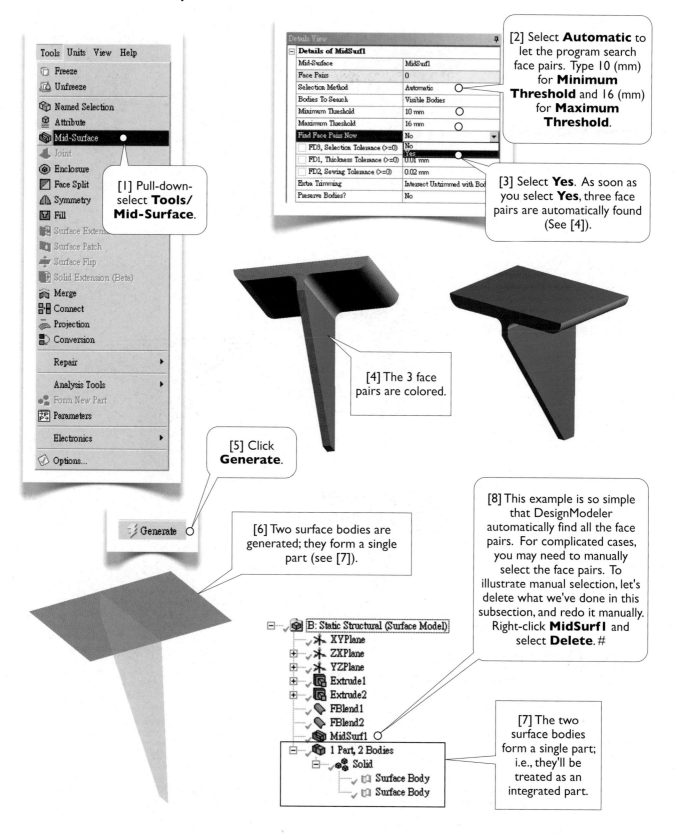

Details View

Details of MidSurf1	
Mid-Surface	MidSurf1
Face Pairs	0
Selection Method	Automatic
Bodies To Search	Visible Bodies
Minimum Threshold	10 mm
Maximum Threshold	16 mm
Find Face Pairs Now	No
FD3, Selection Tolerance (>=0)	0.01 mm
FD1, Thickness Tolerance (>=0)	0.01 mm
FD2, Sewing Tolerance (>=0)	0.02 mm
Extra Trimming	Intersect Untrimmed with Bod
Preserve Bodies?	No

Tools menu:
Freeze
Unfreeze
Named Selection
Attribute
Mid-Surface
Joint
Enclosure
Face Split
Symmetry
Fill
Surface Extension
Surface Patch
Surface Flip
Solid Extension (Beta)
Merge
Connect
Projection
Conversion
Repair
Analysis Tools
Form New Part
Parameters
Electronics
Options...

[1] Pull-down-select **Tools/Mid-Surface**.

[2] Select **Automatic** to let the program search face pairs. Type 10 (mm) for **Minimum Threshold** and 16 (mm) for **Maximum Threshold**.

[3] Select **Yes**. As soon as you select **Yes**, three face pairs are automatically found (See [4]).

[4] The 3 face pairs are colored.

[5] Click **Generate**.

Generate

[6] Two surface bodies are generated; they form a single part (see [7]).

[8] This example is so simple that DesignModeler automatically find all the face pairs. For complicated cases, you may need to manually select the face pairs. To illustrate manual selection, let's delete what we've done in this subsection, and redo it manually. Right-click **MidSurf1** and select **Delete**. #

Tree:
B: Static Structural (Surface Model)
XYPlane
ZXPlane
YZPlane
Extrude1
Extrude2
FBlend1
FBlend2
MidSurf1
1 Part, 2 Bodies
Solid
Surface Body
Surface Body

[7] The two surface bodies form a single part; i.e., they'll be treated as an integrated part.

6.2-4 Manually Select Face Pairs

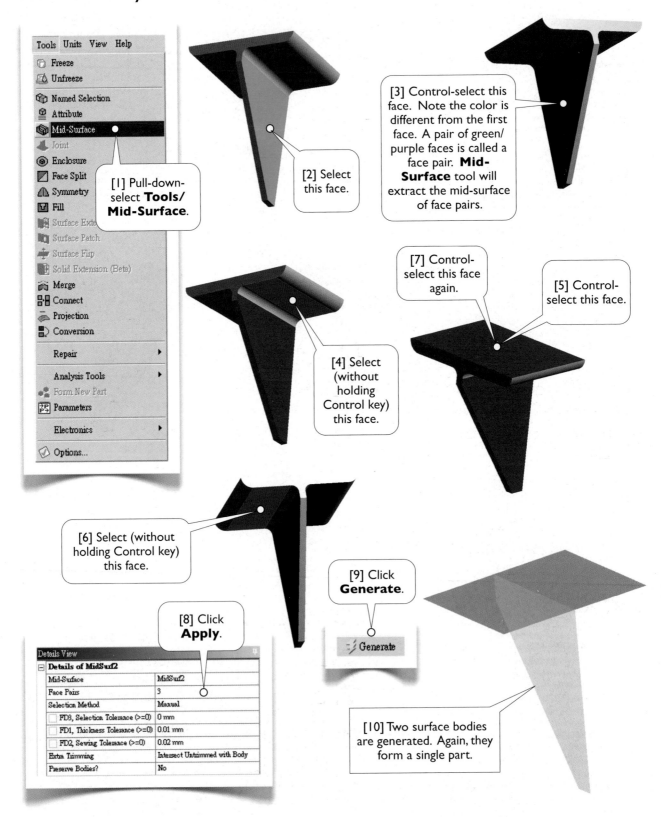

[1] Pull-down-select **Tools/ Mid-Surface**.

[2] Select this face.

[3] Control-select this face. Note the color is different from the first face. A pair of green/ purple faces is called a face pair. **Mid-Surface** tool will extract the mid-surface of face pairs.

[4] Select (without holding Control key) this face.

[7] Control-select this face again.

[5] Control-select this face.

[6] Select (without holding Control key) this face.

[8] Click **Apply**.

[9] Click **Generate**.

[10] Two surface bodies are generated. Again, they form a single part.

Details View

Details of MidSurf2	
Mid-Surface	MidSurf2
Face Pairs	3
Selection Method	Manual
FD3, Selection Tolerance (>=0)	0 mm
FD1, Thickness Tolerance (>=0)	0.01 mm
FD2, Sewing Tolerance (>=0)	0.02 mm
Extra Trimming	Intersect Untrimmed with Body
Preserve Bodies?	No

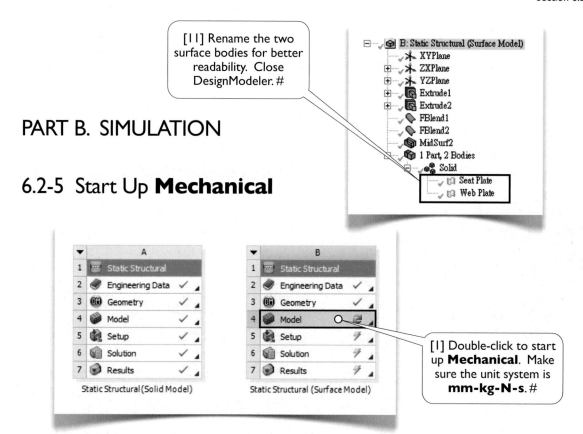

[11] Rename the two surface bodies for better readability. Close DesignModeler. #

PART B. SIMULATION

6.2-5 Start Up **Mechanical**

[1] Double-click to start up **Mechanical**. Make sure the unit system is **mm-kg-N-s**. #

Static Structural (Solid Model)

Static Structural (Surface Model)

6.2-6 Check Material and Thickness

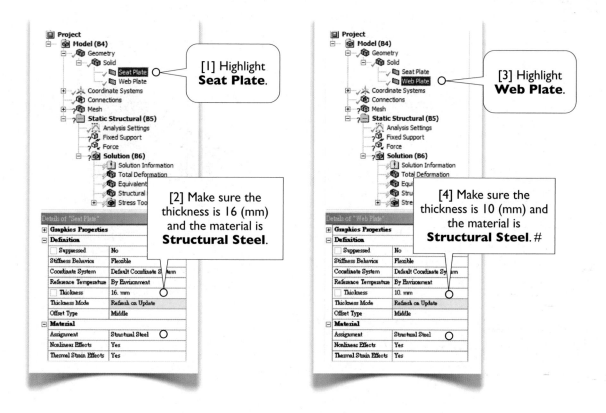

[1] Highlight **Seat Plate**.

[2] Make sure the thickness is 16 (mm) and the material is **Structural Steel**.

[3] Highlight **Web Plate**.

[4] Make sure the thickness is 10 (mm) and the material is **Structural Steel**. #

244

Thickness Assignment

The surface bodies are created using **Mid-Surface** tool. In most cases the thicknesses of bodies are correctly calculated and transferred to **Mechanical**. For some complicated cases, the thicknesses may not be correctly calculated and transferred. It is a good practice to always check material and thickness assignments in **Mechanical**.

6.2-7 Redefine Fixed Support

6.2-8 Redefine Force

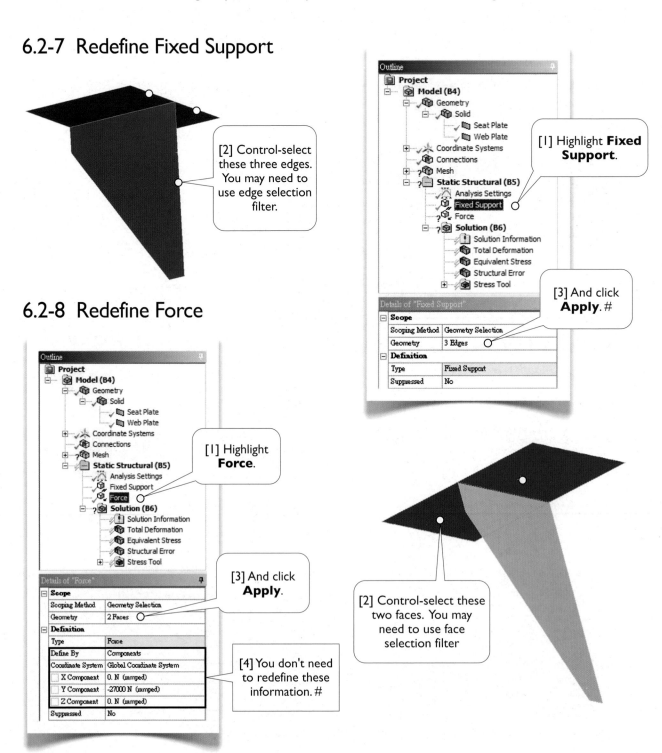

[2] Control-select these three edges. You may need to use edge selection filter.

[1] Highlight **Fixed Support**.

[3] And click **Apply**. #

[1] Highlight **Force**.

[3] And click **Apply**.

[4] You don't need to redefine these information. #

[2] Control-select these two faces. You may need to use face selection filter

6.2-9 Generate Mesh

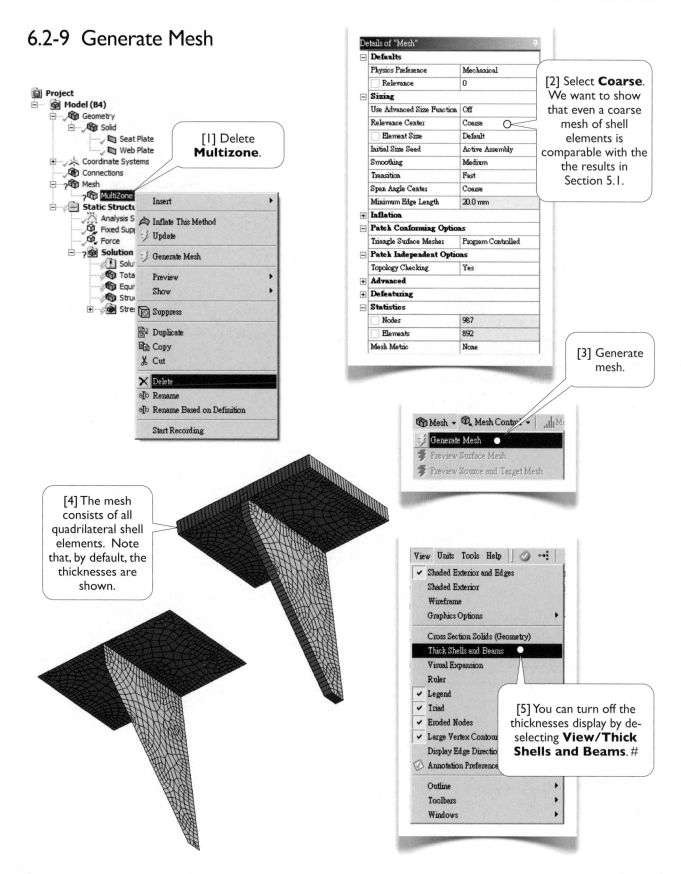

Project
Model (B4)
  Geometry
    Solid
      Seat Plate
      Web Plate
  Coordinate Systems
  Connections
  Mesh
    MultiZone

[1] Delete **Multizone**.

Static Structu
  Analysis S
  Fixed Sup
  Force
  Solution
    Solu
    Tota
    Equi
    Stru
    Stre

Insert
Inflate This Method
Update
Generate Mesh
Preview
Show
Suppress
Duplicate
Copy
Cut
Delete
Rename
Rename Based on Definition
Start Recording

Details of "Mesh"

Defaults	
Physics Preference	Mechanical
Relevance	0
Sizing	
Use Advanced Size Function	Off
Relevance Center	Coarse
Element Size	Default
Initial Size Seed	Active Assembly
Smoothing	Medium
Transition	Fast
Span Angle Center	Coarse
Minimum Edge Length	20.0 mm
Inflation	
Patch Conforming Options	
Triangle Surface Mesher	Program Controlled
Patch Independent Options	
Topology Checking	Yes
Advanced	
Defeaturing	
Statistics	
Nodes	987
Elements	892
Mesh Metric	None

[2] Select **Coarse**. We want to show that even a coarse mesh of shell elements is comparable with the the results in Section 5.1.

[3] Generate mesh.

Mesh ▾   Mesh Control ▾
Generate Mesh
Preview Surface Mesh
Preview Source and Target Mesh

[4] The mesh consists of all quadrilateral shell elements. Note that, by default, the thicknesses are shown.

View Units Tools Help
✓ Shaded Exterior and Edges
  Shaded Exterior
  Wireframe
  Graphics Options
  Cross Section Solids (Geometry)
  Thick Shells and Beams
  Visual Expansion
  Ruler
✓ Legend
✓ Triad
✓ Eroded Nodes
✓ Large Vertex Contour
  Display Edge Direction
  Annotation Preference
  Outline
  Toolbars
  Windows

[5] You can turn off the thicknesses display by de-selecting **View/Thick Shells and Beams**. #

6.2-10 Solve the Model

[1] Solve. #

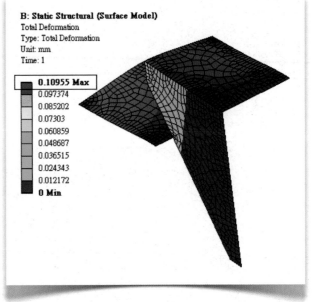

B: Static Structural (Surface Model)
Total Deformation
Type: Total Deformation
Unit: mm
Time: 1

0.10955 Max
0.097374
0.085202
0.07303
0.060859
0.048687
0.036515
0.024343
0.012172
0 Min

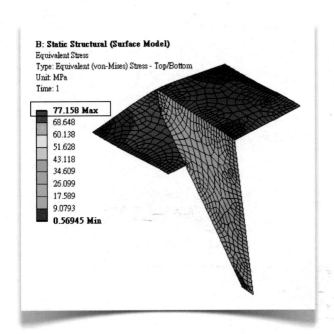

B: Static Structural (Surface Model)
Equivalent Stress
Type: Equivalent (von-Mises) Stress - Top/Bottom
Unit: MPa
Time: 1

77.158 Max
68.648
60.138
51.628
43.118
34.609
26.099
17.589
9.0793
0.56945 Min

Surface Model versus Solid Model

The results are quite comparable to the results obtained in 5.1-13[8, 9] (page 210). The surface model in this section consists of less than 1000 nodes (6.2-9[2], last page) while the solid model consists of about 16,000 nodes (page 209)! In general, before resorting to a solid modeling, consider a surface/line modeling first.

Wrap Up

Save the project and quit Workbench.

Section 6.3

Gearbox

In DesignModeler, a surface body is usually created using one of four techniques: (a) Creating a 2D sketch and applying **Concept/Surface From Sketches** to create a planar surface body; this technique has been demonstrated in Section 4.3 to create the planar glass. (b) Creating an open sketch and then using **Extrude**, **Revolve**, **Sweep**, or **Skin/Loft** to create planar or non-planar surfaces; this technique has been demonstrated in Section 6.1 to create the bellows. (c) Applying **Tool/Mid-Surface** to extract mid-surface from a solid body; this technique has been demonstrated in Section 6.2 to create a surface model for the beam bracket. (d) Using **Thin/Surface** tool; the idea is to extract the exterior "shell" of a solid body. We will demonstrate this technique in this section.

6.3-1 About the Gearbox[Ref 1]

Gears are used to transmit mechanical power. During power transmission, gears exert forces on each other. Through the gear shafts, the forces eventually reach the bearing supports. A gearbox is designed to provide structural functions to withstand the bearing forces.

In this case, what the engineers' concern is the deformation of the gearbox, for it may cause a displacement of the rotating axes, which may reach a point where the transmission becomes defective.

In this section, we will create a 3D model for the gearbox and conduct a simulation. The model will consist of two solid bodies for the flange and the base, made of gray cast iron [1-2], and a surface body for the housing, made of stainless steel [3].

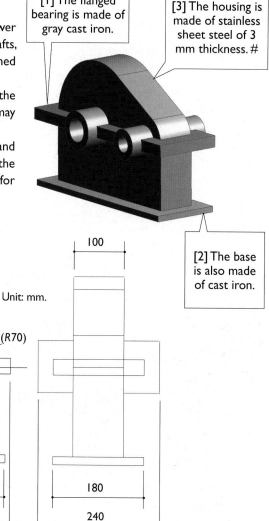

[1] The flanged bearing is made of gray cast iron.

[3] The housing is made of stainless sheet steel of 3 mm thickness. #

[2] The base is also made of cast iron.

Unit: mm.

PART A. GEOMETRIC MODELING

6.3-2 Start Up

[1] Launch Workbench. Create a **Static Structural** system. Save the project as **Gearbox**.

[2] Start up DesignModeler. Select **Millimeter** as length unit and make sure **Auto Constraints** are turne on. #

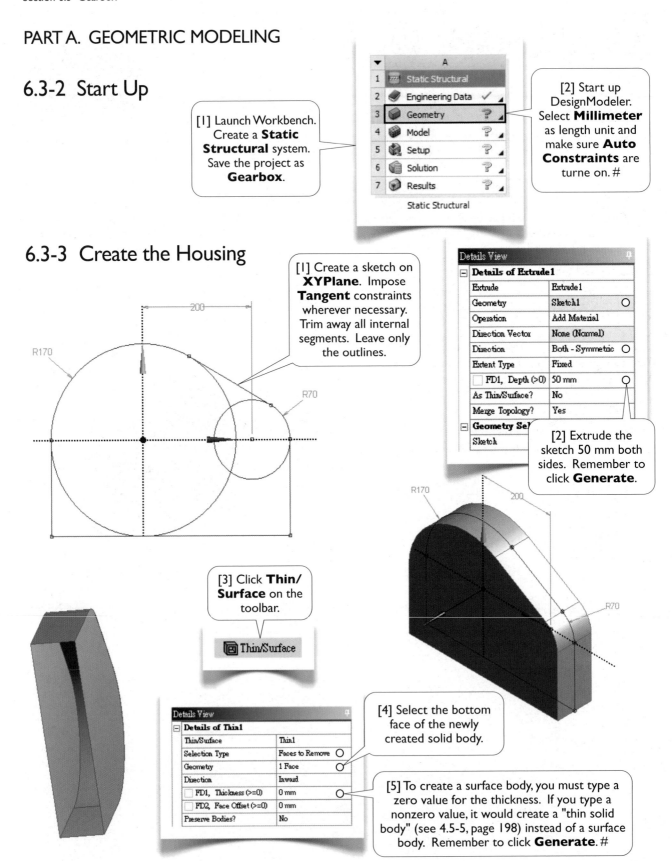

	A	
1	Static Structural	
2	Engineering Data	✓
3	Geometry	?
4	Model	?
5	Setup	?
6	Solution	?
7	Results	?

Static Structural

6.3-3 Create the Housing

[1] Create a sketch on **XYPlane**. Impose **Tangent** constraints wherever necessary. Trim away all internal segments. Leave only the outlines.

200

R170

R70

Details of Extrude1

Extrude	Extrude1	
Geometry	Sketch1	○
Operation	Add Material	
Direction Vector	None (Normal)	
Direction	Both - Symmetric	○
Extent Type	Fixed	
FD1, Depth (>0)	50 mm	○
As Thin/Surface?	No	
Merge Topology?	Yes	

Geometry Se...

Sketch

[2] Extrude the sketch 50 mm both sides. Remember to click **Generate**.

R170
200
R70

[3] Click **Thin/Surface** on the toolbar.

Thin/Surface

[4] Select the bottom face of the newly created solid body.

Details View

Details of Thin1

Thin/Surface	Thin1	
Selection Type	Faces to Remove	○
Geometry	1 Face	○
Direction	Inward	
FD1, Thickness (>=0)	0 mm	○
FD2, Face Offset (>=0)	0 mm	
Preserve Bodies?	No	

[5] To create a surface body, you must type a zero value for the thickness. If you type a nonzero value, it would create a "thin solid body" (see 4.5-5, page 198) instead of a surface body. Remember to click **Generate**. #

6.3-4 Create the Bearings

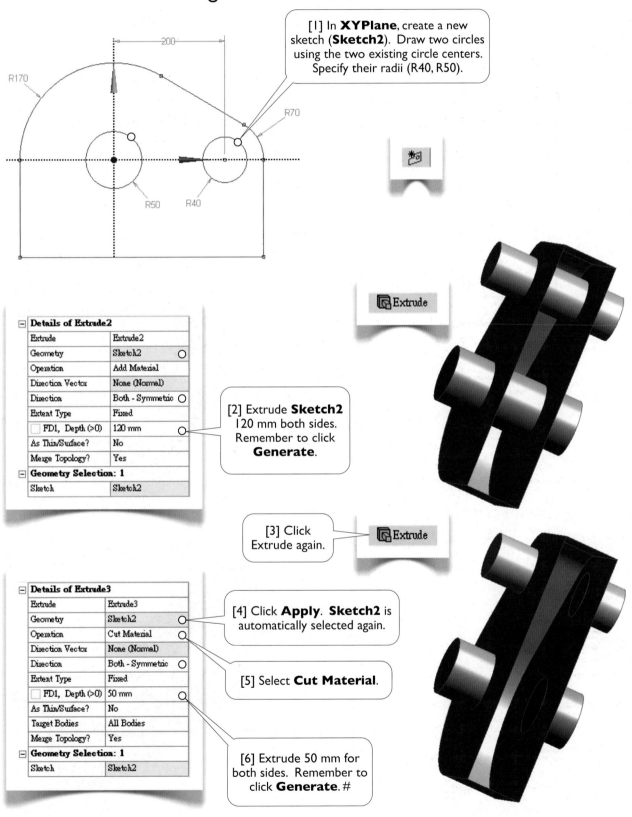

[1] In **XYPlane**, create a new sketch (**Sketch2**). Draw two circles using the two existing circle centers. Specify their radii (R40, R50).

Details of Extrude2

Extrude	Extrude2
Geometry	Sketch2
Operation	Add Material
Direction Vector	None (Normal)
Direction	Both - Symmetric
Extent Type	Fixed
☐ FD1, Depth (>0)	120 mm
As Thin/Surface?	No
Merge Topology?	Yes
Geometry Selection: 1	
Sketch	Sketch2

[2] Extrude **Sketch2** 120 mm both sides. Remember to click **Generate**.

[3] Click Extrude again.

Details of Extrude3

Extrude	Extrude3
Geometry	Sketch2
Operation	Cut Material
Direction Vector	None (Normal)
Direction	Both - Symmetric
Extent Type	Fixed
☐ FD1, Depth (>0)	50 mm
As Thin/Surface?	No
Target Bodies	All Bodies
Merge Topology?	Yes
Geometry Selection: 1	
Sketch	Sketch2

[4] Click **Apply**. **Sketch2** is automatically selected again.

[5] Select **Cut Material**.

[6] Extrude 50 mm for both sides. Remember to click **Generate**. #

6.3-5 Create the Flange

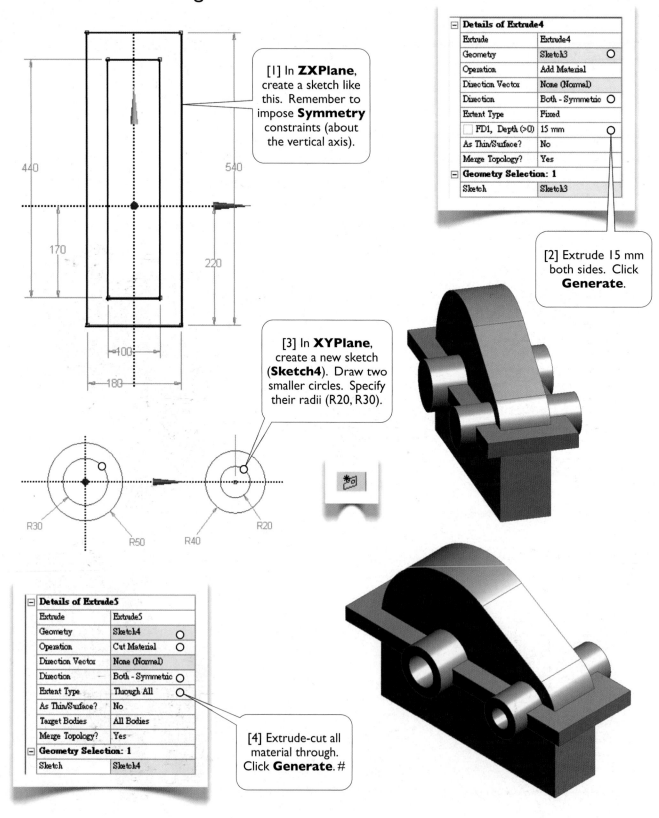

[1] In **ZXPlane**, create a sketch like this. Remember to impose **Symmetry** constraints (about the vertical axis).

Details of Extrude4		
Extrude	Extrude4	
Geometry	Sketch3	O
Operation	Add Material	
Direction Vector	None (Normal)	
Direction	Both - Symmetric	O
Extent Type	Fixed	
☐ FD1, Depth (>0)	15 mm	O
As Thin/Surface?	No	
Merge Topology?	Yes	
Geometry Selection: 1		
Sketch	Sketch3	

[2] Extrude 15 mm both sides. Click **Generate**.

[3] In **XYPlane**, create a new sketch (**Sketch4**). Draw two smaller circles. Specify their radii (R20, R30).

R30 R50 R40 R20

Details of Extrude5		
Extrude	Extrude5	
Geometry	Sketch4	O
Operation	Cut Material	O
Direction Vector	None (Normal)	
Direction	Both - Symmetric	O
Extent Type	Through All	O
As Thin/Surface?	No	
Target Bodies	All Bodies	
Merge Topology?	Yes	
Geometry Selection: 1		
Sketch	Sketch4	

[4] Extrude-cut all material through. Click **Generate**. #

6.3-6 Create the Base

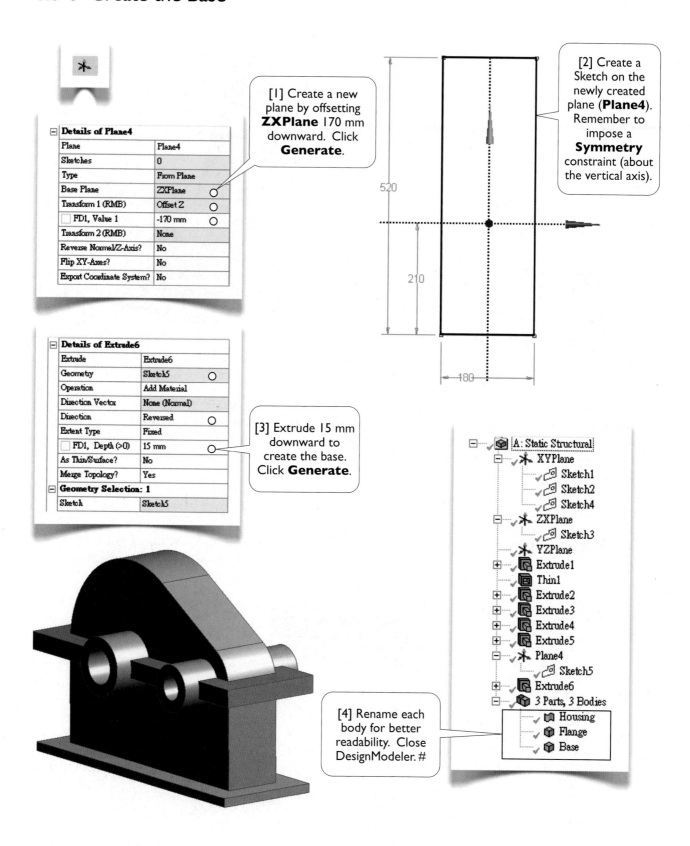

Details of Plane4

Plane	Plane4
Sketches	0
Type	From Plane
Base Plane	ZXPlane
Transform 1 (RMB)	Offset Z
FD1, Value 1	-170 mm
Transform 2 (RMB)	None
Reverse Normal/Z-Axis?	No
Flip XY-Axes?	No
Export Coordinate System?	No

[1] Create a new plane by offsetting **ZXPlane** 170 mm downward. Click **Generate**.

[2] Create a Sketch on the newly created plane (**Plane4**). Remember to impose a **Symmetry** constraint (about the vertical axis).

520

210

180

Details of Extrude6

Extrude	Extrude6
Geometry	Sketch5
Operation	Add Material
Direction Vector	None (Normal)
Direction	Reversed
Extent Type	Fixed
FD1, Depth (>0)	15 mm
As Thin/Surface?	No
Merge Topology?	Yes
Geometry Selection: 1	
Sketch	Sketch5

[3] Extrude 15 mm downward to create the base. Click **Generate**.

A: Static Structural
  XYPlane
    Sketch1
    Sketch2
    Sketch4
  ZXPlane
    Sketch3
  YZPlane
  Extrude1
  Thin1
  Extrude2
  Extrude3
  Extrude4
  Extrude5
  Plane4
    Sketch5
  Extrude6
  3 Parts, 3 Bodies
    Housing
    Flange
    Base

[4] Rename each body for better readability. Close DesignModeler. #

PART B. SIMULATION

6.3-7 Bearing Loads

The bearing loads can be calculated according to transmitted power, which is 175 hp in this case. Skipping the calculation details (they are not the emphasis here; if you are interested in the calculation details, please refer to Zahavi's book[Ref 1]), we summarize the bearing loads as follows. Note that these forces sum up to zero.

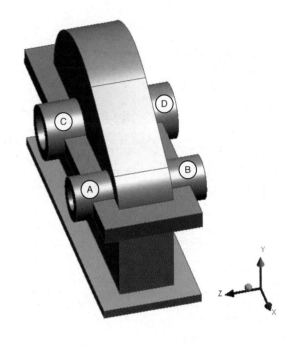

Bearing	F_X (N)	F_Y (N)	F_Z (N)
A	16000	30000	16000
B	6000	30000	0
C	3000	-30000	0
D	-25000	-30000	-16000

Bearing Loads

In **Mechanical**, a bearing load can be specified on a cylindrical surface. It is important to note that the bearing load is distributed on compressive side using projected area. This implies that axial components (here, the Z-components) will be zeros. In our case, the X- and Y-components (please refer to the global coordinates) of the bearing loads can be specified as bearing loads components, but the Z-components cannot be specified as bearing load components. They will be specified separately as surface forces.

6.3-8 Prepare Material Properties

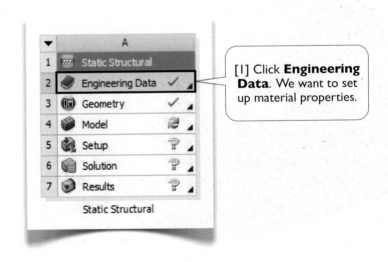

[1] Click **Engineering Data**. We want to set up material properties.

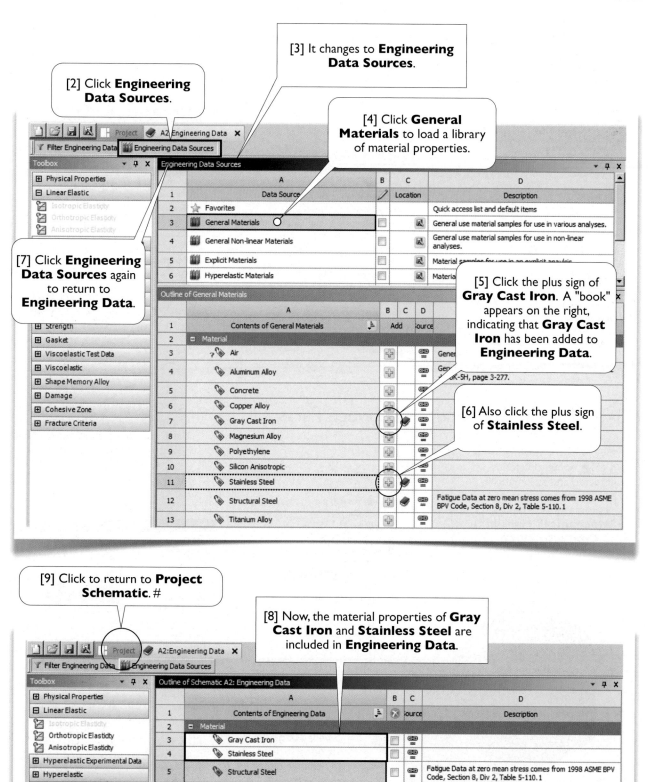

[3] It changes to **Engineering Data Sources**.

[2] Click **Engineering Data Sources**.

[4] Click **General Materials** to load a library of material properties.

[7] Click **Engineering Data Sources** again to return to **Engineering Data**.

[5] Click the plus sign of **Gray Cast Iron**. A "book" appears on the right, indicating that **Gray Cast Iron** has been added to **Engineering Data**.

[6] Also click the plus sign of **Stainless Steel**.

[9] Click to return to **Project Schematic**. #

[8] Now, the material properties of **Gray Cast Iron** and **Stainless Steel** are included in **Engineering Data**.

6.3-9 Assign Material and Thickness

Start up **Mechanical**. Make sure the unit system is **mm-kg-N-s**. Assign material and thickness for each body [1-4].

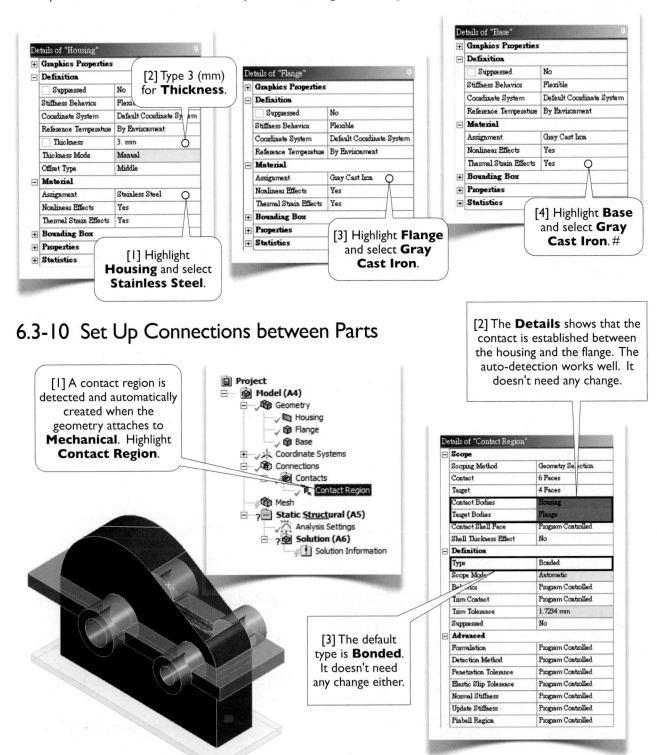

[2] Type 3 (mm) for **Thickness**.

[1] Highlight **Housing** and select **Stainless Steel**.

[3] Highlight **Flange** and select **Gray Cast Iron**.

[4] Highlight **Base** and select **Gray Cast Iron**. #

6.3-10 Set Up Connections between Parts

[1] A contact region is detected and automatically created when the geometry attaches to **Mechanical**. Highlight **Contact Region**.

[2] The **Details** shows that the contact is established between the housing and the flange. The auto-detection works well. It doesn't need any change.

[3] The default type is **Bonded**. It doesn't need any change either.

255

[4] Contact between the housing and the base, however, is not detected. This is because **Mechanical** doesn't detect edge-to-surface contact automatically. We need to manually establish contact between the housing and the base. Select **Contact/Bonded**.

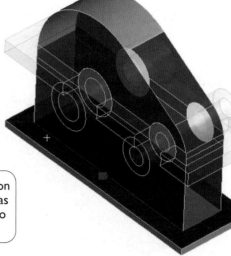

Details of "Bonded - Housing To Base"

Scope	
Scoping Method	Geometry Selection
Contact	4 Edges
Target	1 Face
Contact Bodies	Housing
Target Bodies	Base
Shell Thickness Effect	No
Definition	
Type	Bonded
Scope Mode	Manual
Trim Contact	Program Controlled
Suppressed	No
Advanced	
Formulation	Program Controlled
Penetration Tolerance	Program Controlled
Elastic Slip Tolerance	Program Controlled
Normal Stiffness	Program Controlled
Update Stiffness	Program Controlled
Pinball Region	Program Controlled

[5] Control-select 4 edges on the bottom of the housing as **Contact**. You may need to use edge filter.

[6] Select top face of the base as **Target**. Use face filter. #

Do contacts always imply nonlinearity?

It depends on the type of contacts. In this case, the **Bonded** contacts do not introduce any nonlinearities. For details, see 13.1-8, page 468.

6.3-11 Generate Mesh

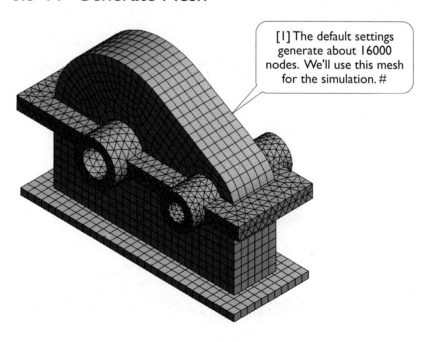

[1] The default settings generate about 16000 nodes. We'll use this mesh for the simulation. #

Details of "Mesh"

Defaults	
Physics Preference	Mechanical
Relevance	0
Sizing	
Use Advanced Size Function	On: Curvature
Relevance Center	Coarse
Initial Size Seed	Active Assembly
Smoothing	Medium
Transition	Fast
Span Angle Center	Coarse
Curvature Normal Angle	Default (30.0°)
Min Size	Default (3.36540 mm)
Max Face Size	Default (16.8270 mm)
Max Size	Default (16.8270 mm)
Growth Rate	Default
Minimum Edge Length	15.0 mm
Inflation	
Patch Conforming Options	
Triangle Surface Mesher	Program Controlled
Patch Independent Options	
Topology Checking	Yes
Advanced	
Defeaturing	
Statistics	
Nodes	15876
Elements	8286
Mesh Metric	None

6.3-12 Specify Support

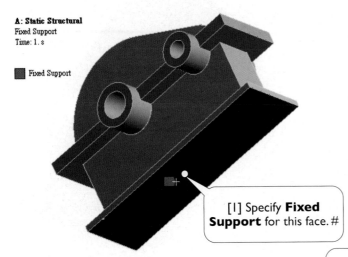

A: Static Structural
Fixed Support
Time: 1. s

■ Fixed Support

[1] Specify **Fixed Support** for this face. #

Support Conditions

In this case, a **Compression Only Support** or a **Frictionless Support** would be more realistic than a **Fixed Supported**. However, the results will be the same for these three support conditions. Since the external forces sum up to zero, the support reaction is zero. **Fixed Support** has an advantage that it doesn't introduce nonlinearities or weak springs (3.1-8, page 111). **Compression Only Support** would introduce nonlinearities, and **Frictionless Support** would introduce weak springs.

6.3-13 Apply Bearing Loads

[4] Apply a bearing load on the internal cylindrical surface of bearing B.

🔻 Loads ▾ 🔻 Supports ▾ 🔻
🔻 Pressure
🔻 Pipe Pressure
🔻 Hydrostatic Pressure
🔻 Force
🔻 Remote Force
🔻 Bearing Load ●
🔻 Bolt Pretension
🔻 Moment
🔻 Generalized Plane rain
🔻 L
🔻 T
🔻 E
🔻 J
🔻 F
·)) Detonation Point

[1, 3, 5, 7] Select **Loads/Bearing Load**.

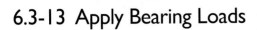

Details of "Bearing Load A"

Scope	
Scoping Method	Geometry Selection
Geometry	1 Face
Definition	
Type	Bearing Load
Define By	Components
Coordinate System	Global Coordinate System
X Component	16000 N
Y Component	30000 N
Z Component	0. N
Suppressed	No

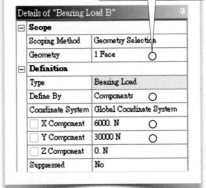

Details of "Bearing Load B"

Scope	
Scoping Method	Geometry Selection
Geometry	1 Face
Definition	
Type	Bearing Load
Define By	Components
Coordinate System	Global Coordinate System
X Component	6000. N
Y Component	30000 N
Z Component	0. N
Suppressed	No

■ Bearing Load A: 34000 N
Components: 16000, 30000, 0. N

■ Bearing Load B: 30594 N
Components: 6000., 30000, 0. N

[2] Apply a bearing load on the internal cylindrical surface of bearing A.

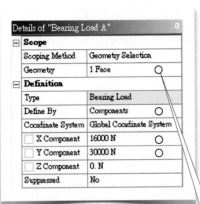

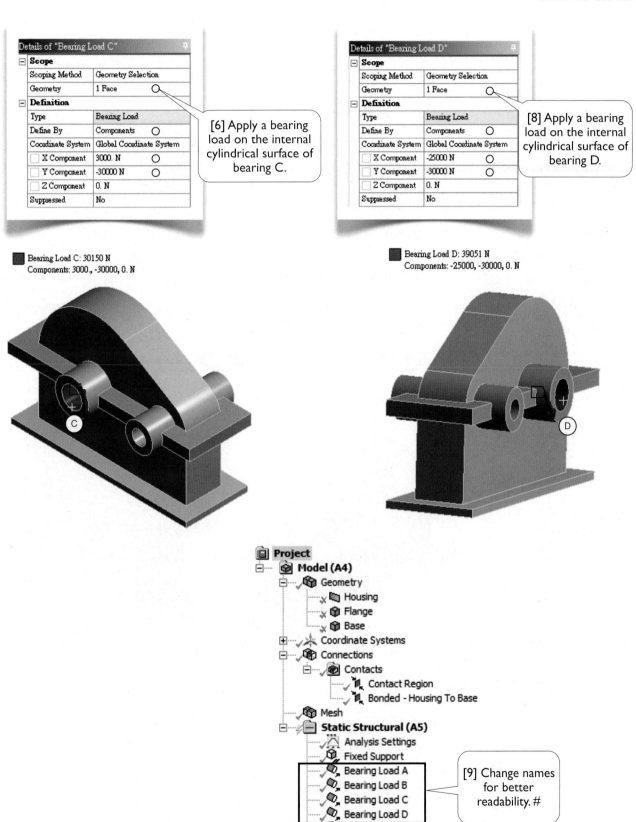

Details of "Bearing Load C"

Scope	
Scoping Method	Geometry Selection
Geometry	1 Face

Definition	
Type	Bearing Load
Define By	Components
Coordinate System	Global Coordinate System
☐ X Component	3000. N
☐ Y Component	-30000 N
☐ Z Component	0. N
Suppressed	No

[6] Apply a bearing load on the internal cylindrical surface of bearing C.

Details of "Bearing Load D"

Scope	
Scoping Method	Geometry Selection
Geometry	1 Face

Definition	
Type	Bearing Load
Define By	Components
Coordinate System	Global Coordinate System
☐ X Component	-25000 N
☐ Y Component	-30000 N
☐ Z Component	0. N
Suppressed	No

[8] Apply a bearing load on the internal cylindrical surface of bearing D.

Bearing Load C: 30150 N
Components: 3000., -30000, 0. N

Bearing Load D: 39051 N
Components: -25000, -30000, 0. N

Project
Model (A4)
  Geometry
    Housing
    Flange
    Base
  Coordinate Systems
  Connections
    Contacts
      Contact Region
      Bonded - Housing To Base
  Mesh
  Static Structural (A5)
    Analysis Settings
    Fixed Support
    Bearing Load A
    Bearing Load B
    Bearing Load C
    Bearing Load D
  Solution (A6)
    Solution Information

[9] Change names for better readability. #

258

6.3-14 Apply Axial Loads

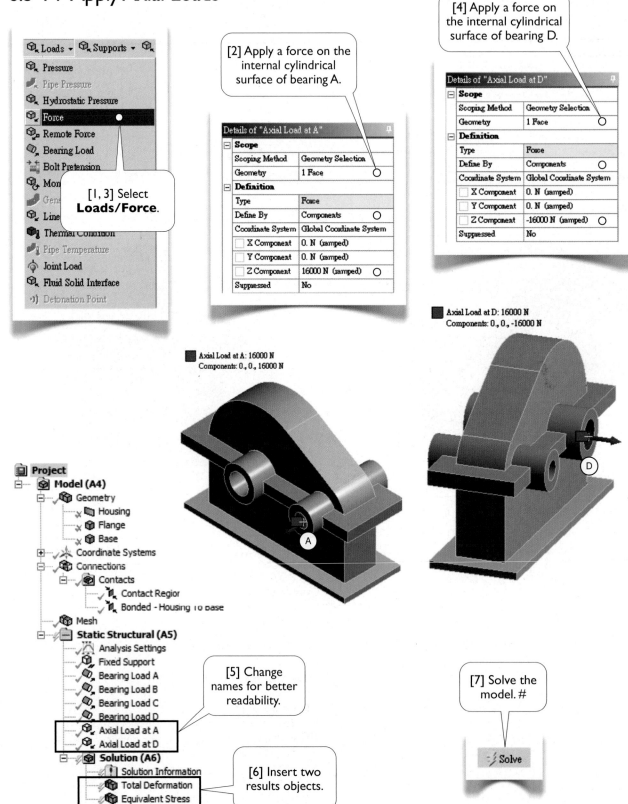

[1, 3] Select **Loads/Force**.

[2] Apply a force on the internal cylindrical surface of bearing A.

[4] Apply a force on the internal cylindrical surface of bearing D.

Details of "Axial Load at A"

Scope	
Scoping Method	Geometry Selection
Geometry	1 Face
Definition	
Type	Force
Define By	Components
Coordinate System	Global Coordinate System
X Component	0. N (ramped)
Y Component	0. N (ramped)
Z Component	16000 N (ramped)
Suppressed	No

Details of "Axial Load at D"

Scope	
Scoping Method	Geometry Selection
Geometry	1 Face
Definition	
Type	Force
Define By	Components
Coordinate System	Global Coordinate System
X Component	0. N (ramped)
Y Component	0. N (ramped)
Z Component	-16000 N (ramped)
Suppressed	No

Axial Load at A: 16000 N
Components: 0., 0., 16000 N

Axial Load at D: 16000 N
Components: 0., 0., -16000 N

[5] Change names for better readability.

[6] Insert two results objects.

[7] Solve the model. #

259

6.3-15 View the Results

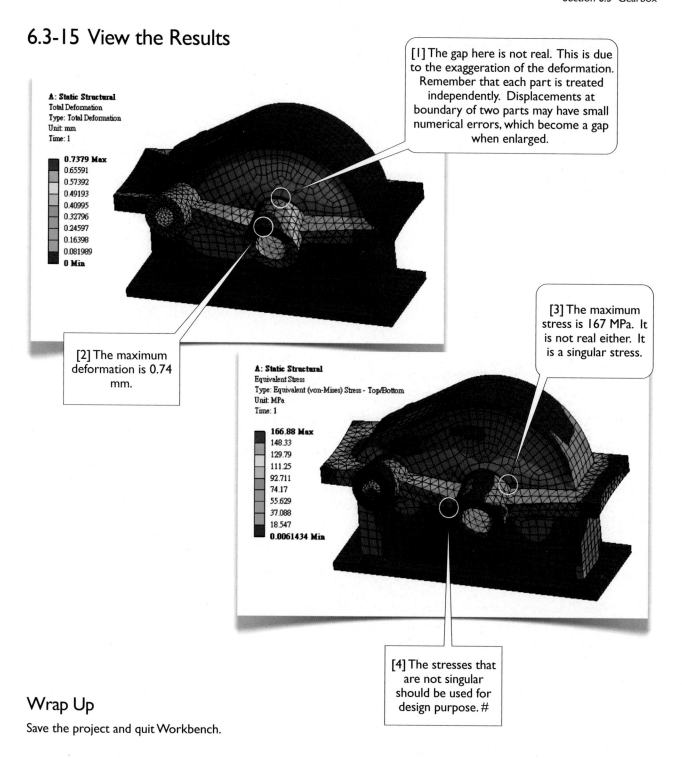

A: Static Structural
Total Deformation
Type: Total Deformation
Unit: mm
Time: 1

0.7379 Max
0.65591
0.57392
0.49193
0.40995
0.32796
0.24597
0.16398
0.081989
0 Min

[1] The gap here is not real. This is due to the exaggeration of the deformation. Remember that each part is treated independently. Displacements at boundary of two parts may have small numerical errors, which become a gap when enlarged.

[2] The maximum deformation is 0.74 mm.

[3] The maximum stress is 167 MPa. It is not real either. It is a singular stress.

A: Static Structural
Equivalent Stress
Type: Equivalent (von-Mises) Stress - Top/Bottom
Unit: MPa
Time: 1

166.88 Max
148.33
129.79
111.25
92.711
74.17
55.629
37.088
18.547
0.0061434 Min

[4] The stresses that are not singular should be used for design purpose. #

Wrap Up

Save the project and quit Workbench.

Reference

1. Zahavi, E., *The Finite Element Method in Machine Design*, Prentice-Hall, 1992; Chapter 10. Gear Box.

Section 6.4

Review

6.4-1 Keywords

Choose a letter for each keyword from the list of descriptions

1. (   ) Auto-Detection of Contact
2. (   ) Shell Elements
3. (   ) Top/Bottom of Surface Body

Answers:

1. (C)   2. (A)   3. (B)

List of Descriptions

(A) A shell element is a planar (2D) element that can be arranged in the 3D space. It is used to mesh a body when one of its dimensions is much smaller than the other two dimensions. Each node has 6 degrees of freedom: 3 translational and 3 rotational. Due to the presence of rotational degrees of freedom, it is very efficient to model the problems dominated by the out-of-plane bending modes, contrasting to a solid element, which does not have rotational degrees of freedom.

(B) Each surface body has a top side and a bottom side. When you select a surface body, only the top side is highlighted. Loads are applied on the top side. By default, results are reported on both sides.

(C) When a geometry attaches to **Mechanical**, it automatically detects and establishes possible contacts between parts, wherever the gaps between parts are less than a tolerance. The contact type is **Bonded** by default. The result of auto-detection may not be accurate and needs to be manually modified.

6.4-2 Questions

Surface Body versus Thin Solid Body

Q: In 6.3-3[5] (page 249), you type a zero value for the thickness to create a surface body. If you provide a nonzero value, it would create a "thin" solid body instead of a surface body. What is an essential difference between a surface body and a thin solid body?

A: A surface body will be meshed with shell elements (1.3-3[7, 8], page 34), while a thin solid body will be meshed with solid elements (1.3-3[1-4], page 34).

Shell Elements versus Solid Elements

Q: What is the essential difference between shell elements and solid elements?

A: Besides the geometry, the shell elements have rotational degrees of freedom (1.3-3[7, 8], page 34), while solid elements do not (1.3-3[1-4], page 34).

Triangular Plate

Q: Can we perform a simulation with a 3D surface model for the triangular plate in Section 3.1? If positive, is there any advantages of doing that over the 2D solid model?

A: Yes, we could, but there is no advantages over the 2D solid mode. Remember that 3D surface bodies are meshed with shell elements. The essential difference between shell elements and 2D solid elements is that shell elements can have out-of-plane deformation (warpage) while 2D solid elements cannot. In the case of triangular plate, there is no out-of-plane deformation. There is no needs to use shell elements. The above discussions also apply on the spur gear of Section 3.4 and the filleted bar of Section 3.5.

Axisymmetric Bodies

In the beginning and the end of Section 6.1, we mentioned that we could model the bellows using axisymmetric 2D solid body or 2D line body. Make sure that you do understand the meaning of the statement.

6.4-3 Additional Workbench Exercises

Exploring the Symmetries of the Bellows[Ref 1]

In the end of 6.1-15 (page 240), we mentioned that there are at least three planes of symmetry we may exploit for a static structural simulation. In fact, there are infinite number of planes of symmetry: any planes passing through Z-axis are planes of symmetry. In 3D, however, we may use at most three planes of symmetry. In this exercise, you are asked to redo the static structural simulation, exploiting the symmetries to reduce the model. You can let Workbench automatically set up the boundary conditions on the symmetric planes (3.1-11[6], page 114), or manually set them up yourself. How do you manually set up the rotational degrees of freedom on a plane of symmetry (see 7.1-11, page 270)? Also, after reducing the model by using symmetries, refine the mesh and expect to obtain more accurate results.

Edge Joints[Ref 2]

In 6.2-7 (page 245), we formed a single part by grouping three surface bodies to ensure the meshing continuity. Another way is using the **Edge Joints** feature. Edge joints are the glue that holds together bodies where a continuous mesh is desired. Edge joints can be created manually by pull-down-selecting **Tools/Joint**, and can be viewed by turning on **View/Edge Joints**. Explore this feature yourself.

Cover of Pressure Cylinder

Using **Mid-Surface** tool, transform the cover of pressure cylinder (Sections 4.2 and 5.2) to a surface model. Redo the simulation using the surface model.

References

1. ANSYS Documentation//Mechanical Applications//Mechanical User's Guide//Specify Geometry//Symmetry//Symetry in the Mechanical Application
2. ANSYS Documentation//DesignModeler User's Guide//3D Modeling//Advanced Features and Tools//Joint

Chapter 7
Line Models

Many real-world objects can be modeled as line bodies. When a body has small lateral dimensions and has a uniform cross-section, it is often modeled as a line body. The most important applications of line models are frame, beam, and truss structures. Workbench meshes a line body with beam elements (1.3-3[9], page 35). Advantages of using line models over surface models or solid models include: (a) creating line models is usually easier, (b) the problem size is much smaller, and (c) the solution can be more accurate. Therefore, engineers should consider using line models instead of surface or solid models whenever possible. Besides, Workbench stores many built-in cross sections in the database to be chosen from by the users. Workbench allows the rendering of cross-sections when displaying the model, so that the line bodies visually look like solid bodies.

Purpose of This Chapter

The main purpose of this chapter is to practice the use of beam elements (1.3-3[9], page 35). This chapter guides the students to create line models and perform simulations using these models. The chapter provides three examples, two of them are entirely line models while the other one is a model mixed up with line bodies and surface bodies.

About Each Section

Section 7.1 creates a flexible gripper model and performs a simulation using the model. Section 7.2 demonstrates the creation of a truss structure and the simulation. Section 7.3 uses a two-story building as an example to demonstrate the creation of a building structure and the simulation. Another purpose of Section 7.3 is to demonstrate how surface bodies and line bodies can be mixed-up in a simulation model.

Section 7.1

Flexible Gripper[Ref 1]

7.1-1 About the Flexible Gripper

The gripper [1-4] is made of POM (polyoxymethylene, a plastic polymer), which has a Young's modulus of 2 GPa and a Poisson's ratio of 0.35. It has a rectangular cross section of 1x5 mm².

One of concerns in designing this gripper is its geometric advantage, GA. The GA is defined as the ratio of the horizontal output displacement [4] to the input actuation [3]. The GA value is used to assess the efficiency of the gripper, the larger the better. The main purpose of this simulation is to assess the GA value of the current design as shown in the figure [5]. Note that only half of the gripper is modeled due to the symmetry.

The profile consists of two smooth spline curves defined by 7 key points, whose numbering and coordinates are shown in the figure [5, 6].

We could model the problem as 2D. With 2D simulations, however, current version of **Mechanical** supports only solid models (doesn't support 2D surface or line model). Comparing with a 3D line model, 2D solid model is not efficient (in terms of accuracy and CPU time.) We decide to go for a 3D line model, although the geometry and the motion are entirely on a plane.

We will create a line model in the first part of this section. The model will be used in the second part of this section to simulate the motion of gripping and assess the GA value. The model will be used in Section 8.1 to demonstrate an optimization capability of Workbench; that is, we want to re-configure the positions of the key points to achieve an optimum GA value.

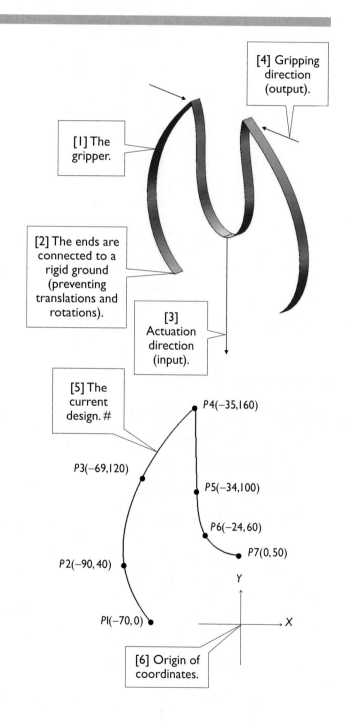

[4] Gripping direction (output).

[1] The gripper.

[2] The ends are connected to a rigid ground (preventing translations and rotations).

[3] Actuation direction (input).

[5] The current design. #

P4(−35,160)

P3(−69,120)

P5(−34,100)

P6(−24,60)

P7(0,50)

P2(−90,40)

P1(−70,0)

[6] Origin of coordinates.

PART A. GEOMETRIC MODELING

7.1-2 Start Up

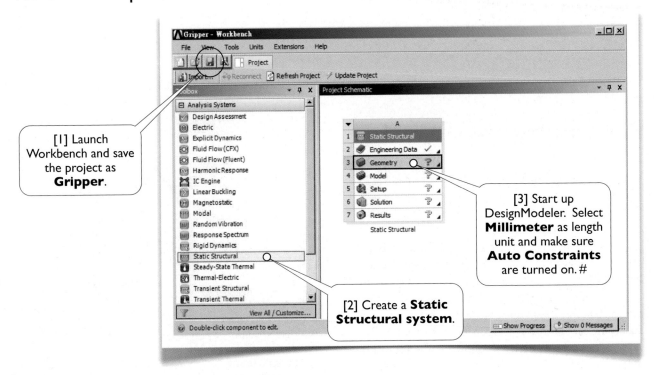

[1] Launch Workbench and save the project as **Gripper**.

[2] Create a **Static Structural system**.

[3] Start up DesignModeler. Select **Millimeter** as length unit and make sure **Auto Constraints** are turned on. #

7.1-3 Create Sketch on **XYPlane**

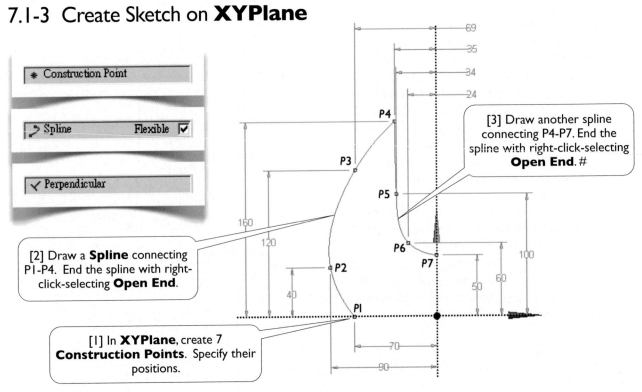

[2] Draw a **Spline** connecting P1-P4. End the spline with right-click-selecting **Open End**.

[3] Draw another spline connecting P4-P7. End the spline with right-click-selecting **Open End**. #

[1] In **XYPlane**, create 7 **Construction Points**. Specify their positions.

7.1-4 Create Line Body

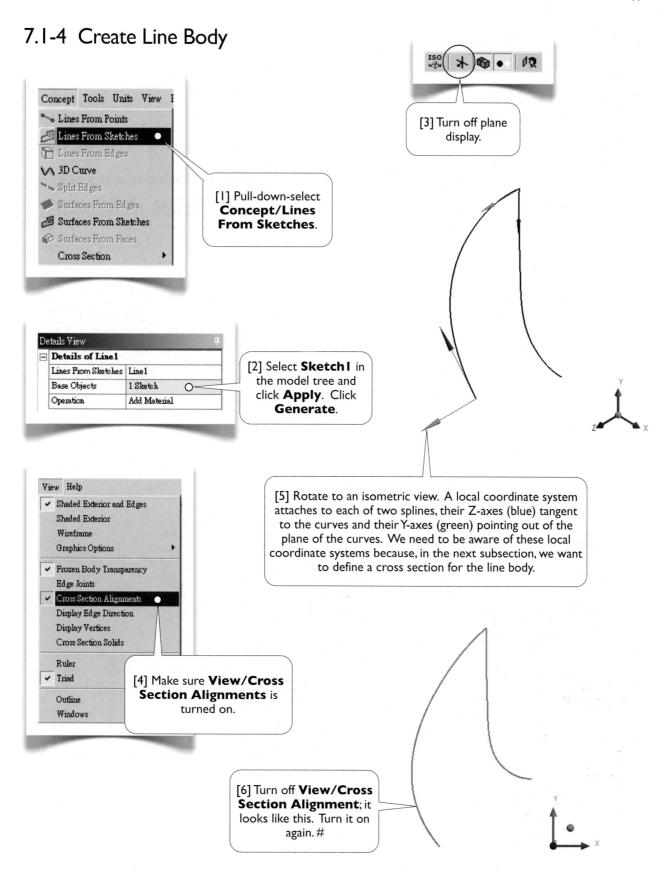

[1] Pull-down-select **Concept/Lines From Sketches**.

[2] Select **Sketch1** in the model tree and click **Apply**. Click **Generate**.

[3] Turn off plane display.

[4] Make sure **View/Cross Section Alignments** is turned on.

[5] Rotate to an isometric view. A local coordinate system attaches to each of two splines, their Z-axes (blue) tangent to the curves and their Y-axes (green) pointing out of the plane of the curves. We need to be aware of these local coordinate systems because, in the next subsection, we want to define a cross section for the line body.

[6] Turn off **View/Cross Section Alignment**; it looks like this. Turn it on again. #

7.1-5 Create a Rectangular Cross Section

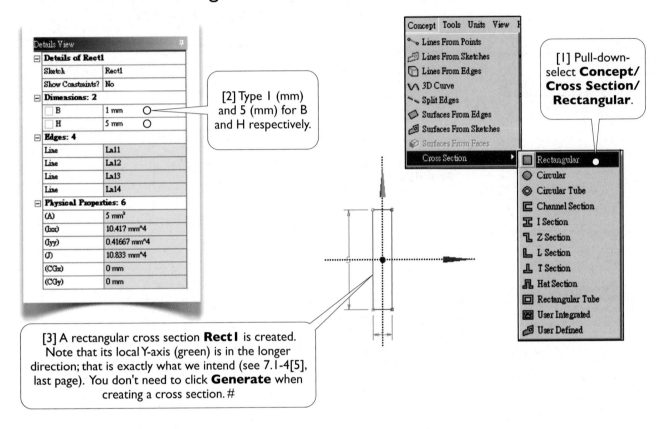

[2] Type 1 (mm) and 5 (mm) for B and H respectively.

[1] Pull-down-select **Concept/ Cross Section/ Rectangular**.

[3] A rectangular cross section **Rect1** is created. Note that its local Y-axis (green) is in the longer direction; that is exactly what we intend (see 7.1-4[5], last page). You don't need to click **Generate** when creating a cross section. #

7.1-6 Assign the Cross Section to the Line Body

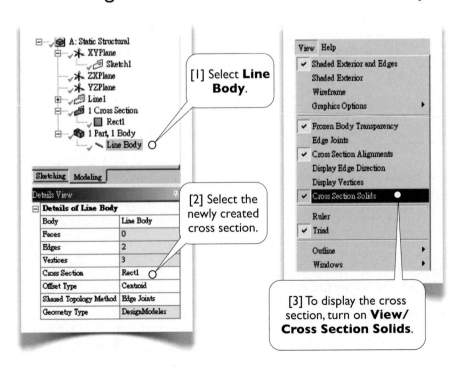

[4] The model displays its rectangular cross section. Note that this is for visual effect only. The model is still a line body, not a solid body.

[1] Select **Line Body**.

[2] Select the newly created cross section.

[3] To display the cross section, turn on **View/ Cross Section Solids**.

Adjusting Cross Section Alignments

When creating line models, make sure the cross section alignments are correct. In our case, the default alignments are exactly what we intend, so we don't need to make any adjustment. In other cases, if adjustment of cross section alignments is needed, you can turn on the edge selection filter, select the edges of the line bodies, and type the rotation angle [5]. We will demonstrate this procedure in 7.3-9, page 292.

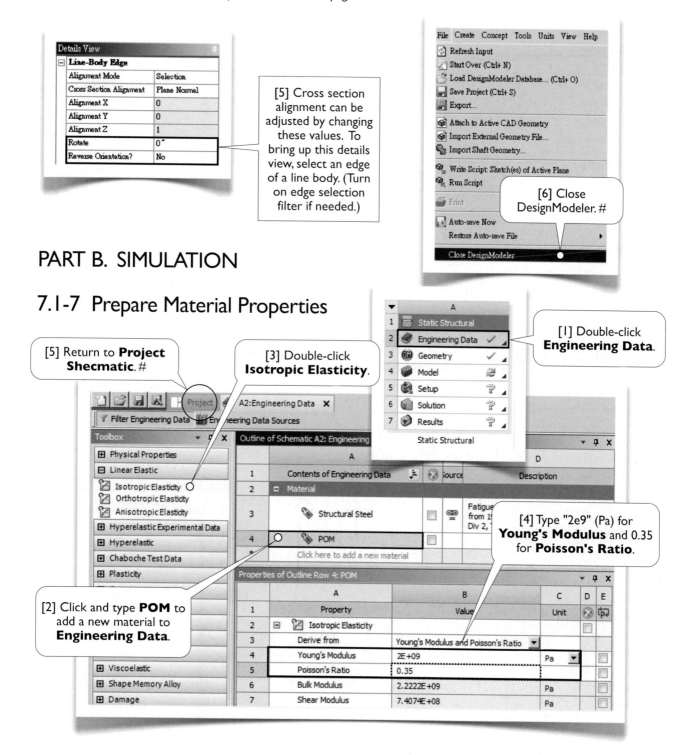

[5] Cross section alignment can be adjusted by changing these values. To bring up this details view, select an edge of a line body. (Turn on edge selection filter if needed.)

[6] Close DesignModeler. #

PART B. SIMULATION

7.1-7 Prepare Material Properties

[1] Double-click **Engineering Data**.

[5] Return to **Project Shecmatic**. #

[3] Double-click **Isotropic Elasticity**.

[4] Type "2e9" (Pa) for **Young's Modulus** and 0.35 for **Poisson's Ratio**.

[2] Click and type **POM** to add a new material to **Engineering Data**.

7.1-8 Start Up **Mechanical** and Assign Material

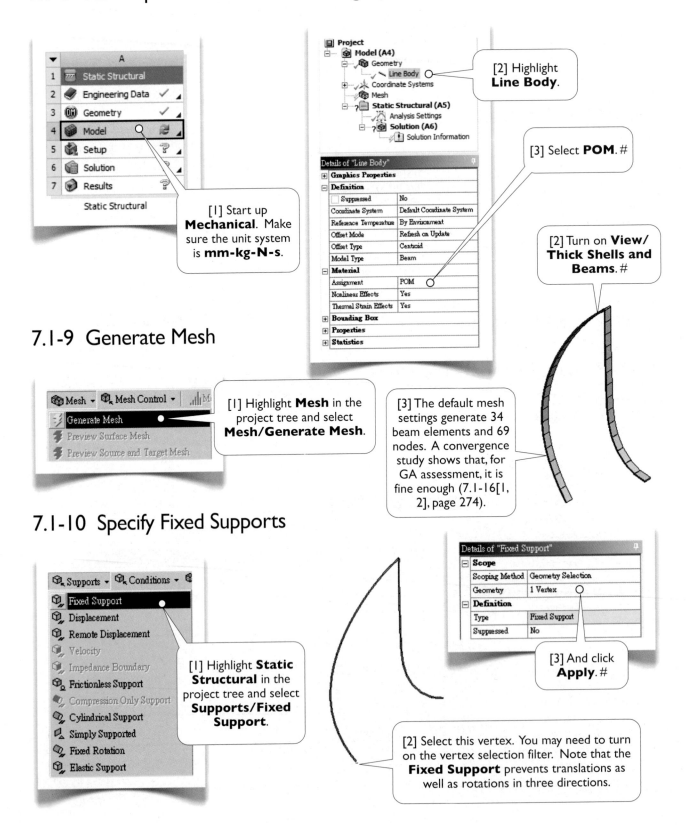

[1] Start up **Mechanical**. Make sure the unit system is **mm-kg-N-s**.

[2] Highlight **Line Body**.

[3] Select **POM**. #

[2] Turn on **View/Thick Shells and Beams**. #

7.1-9 Generate Mesh

[1] Highlight **Mesh** in the project tree and select **Mesh/Generate Mesh**.

[3] The default mesh settings generate 34 beam elements and 69 nodes. A convergence study shows that, for GA assessment, it is fine enough (7.1-16[1, 2], page 274).

7.1-10 Specify Fixed Supports

[1] Highlight **Static Structural** in the project tree and select **Supports/Fixed Support**.

[3] And click **Apply**. #

[2] Select this vertex. You may need to turn on the vertex selection filter. Note that the **Fixed Support** prevents translations as well as rotations in three directions.

7.1-11 Specify Symmetry Condition and the Actuation

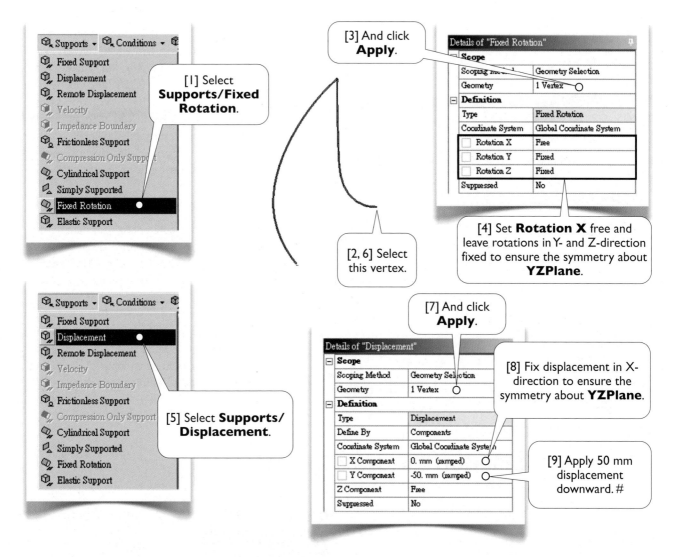

Symmetry Conditions for Shell and Beam Elements

The purpose of steps [1-8] is to set up conditions of symmetry about **YZPlane**, while step [9] is to set up a downward actuation.

Recall that, in the first exercise of 6.4-3 (page 262), you are asked how to set up the rotational degrees of freedom on a plane of symmetry. Beam elements, like shell elements, have rotational degrees of freedom (1.3-3[9], page 35). The rule of symmetry conditions for shell and beam elements is: fixing (zero values) out-of-plane translations and in-plane rotations.

Consider the boundary conditions at the vertex [2, 6]. In this case, **YZPlane** is the plane of symmetry, X-displacement is the out-of-plane translation [8], Y-rotation and Z-rotation are the in-plane rotations [4].

You could let Workbench automatically set up the symmetry boundary conditions (3.1-11[6], page 114), but make sure you know how Workbench sets up these symmetry boundary conditions. Workbench will set up the symmetry boundary conditions as we described above: fixing **out-of-plane translations** and **in-plane rotations**. As a good engineer, use software only when you know how they work.

7.1-12 Set Up **Analysis Settings**

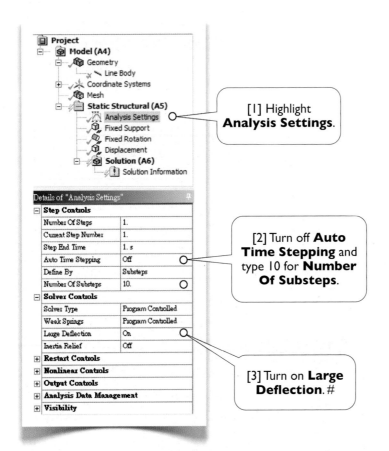

[1] Highlight **Analysis Settings**.

[2] Turn off **Auto Time Stepping** and type 10 for **Number Of Substeps**.

[3] Turn on **Large Deflection**. #

Why Large Deflection?

Turning on large deflection is to include geometry nonlinearity, which always gives more accurate solutions but takes more computing time. To justify the inclusion of geometry nonlinearity, turn off **Large Deflection** and rerun this case at the end of the section. A substantial difference of the results is an indication that the inclusion of geometry nonlinearity is necessary.

Auto Time Stepping

This model is simple enough that we actually don't need to change any default settings other than just turn on **Large Deflection**. The solution would be complete in just 3 substeps, with a program controlled auto time stepping.

The reason we turn off **Auto Time Stepping** is because we want to gather more data for plotting an input-displacement-versus-output-displacement chart (7.1-15[6], page 273).

7.1-13 Set Up **Solution** Branch and Solve the Model

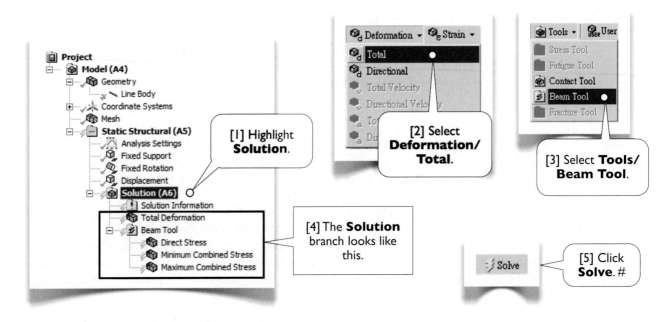

[1] Highlight **Solution**.

[2] Select **Deformation/Total**.

[3] Select **Tools/Beam Tool**.

[4] The **Solution** branch looks like this.

[5] Click **Solve**. #

7.1-14 View Results

Messages
Mechanical Wizard
Graphics Annotations
Section Planes
Selection Information
• Reset Layout

[1] Close the warning message and If your window layout is different from what we have here, pull-down-select **View/Windows/Reset Layout** and click **Geometry** tab.

M A : Static Structural - Mechanical [ANSYS Academic Research]

File Edit View Units Tools Help   Solve ▾ ?/Show Errors   Worksheet

Show Vertices   Wireframe   Show Mesh   Random Colors   Annotation Pref   Thicken Annotations
Result 1.0 (True Scale)   Probe   Display

[3] The maximum total displacement is 77.153 mm.

Outline
Filter: Name
Project
  Model (A4)
    Geometry
      Line Body
    Coordinat
    Mesh
    Static St
      Anal
      Fixed
      Fixed
      Displacement
    Solution (A6)
      Solution Information
      Total Deformation
      Beam Tool

A: Static Structural
Total Deformation
Type: Total Deformation
Unit: mm
Time: 1.

77.153 Max
68.581
60.008
51.436
42.863
34.29
25.718
17.145
8.5726
0 Min

[2] Highlight **Total Deformation**.

[4] Click **Result Sets**.

[7] Numerical data for each substep are available. You can copy/paste to your spread sheet for further processing.

Geometry \ Print Preview \ Report Preview /

[5] Click **Play**. Click **Stop** to stop the animation.

Graph
Animation ▶ ■   ║│ ║║   10 Frames   2 Sec (Auto)

Details of "Total De
Scope
  Scoping Method   Geometry Selection

[6] A graph of "time" versus maximum response versus is displayed here. The graph is replicated below.

77.153

[mm]

30.

0.

0.   0.125   0.25   0.375   0.5   0.625   0.75   0.875   1.

[s]
1

	Time [s]	☑ Minimum [mm]	☑ Maximum [m
1	0.1	0.	8.0845
2	0.2	0.	16.164
3	0.3	0.	24.189
4	0.4	0.	32.126
5	0.5	0.	39.954
6	0.6	0.	47.66
7	0.7	0.	55.236
8	0.8	0.	62.68
9	0.9	0.	69.986
10	1.	0.	77.153

Tabular Data

1 Message   No Selection   Metric (mm, kg, N, s, mV, mA)   Degrees   rad/s

[9] The input displacement versus total displacement curve is almost a straight line. #

[8] The total input displacement (50 mm) is divided into 10 substeps, therefore each substep is 5 mm of input displacement.

7.1-15 Assess GA Value

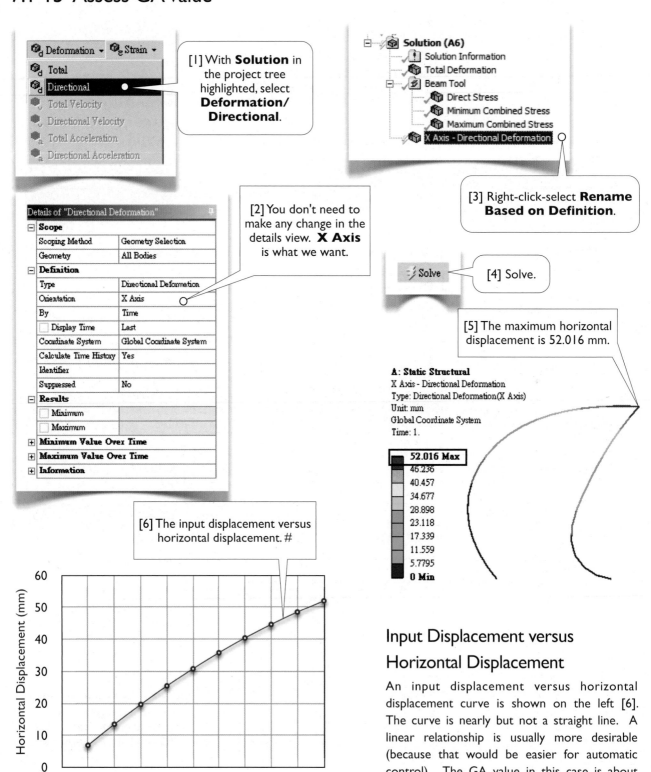

[1] With **Solution** in the project tree highlighted, select **Deformation/ Directional**.

[2] You don't need to make any change in the details view. **X Axis** is what we want.

[3] Right-click-select **Rename Based on Definition**.

[4] Solve.

[5] The maximum horizontal displacement is 52.016 mm.

A: Static Structural
X Axis - Directional Deformation
Type: Directional Deformation(X Axis)
Unit: mm
Global Coordinate System
Time: 1.

52.016 Max
46.236
40.457
34.677
28.898
23.118
17.339
11.559
5.7795
0 Min

[6] The input displacement versus horizontal displacement. #

Input Displacement versus Horizontal Displacement

An input displacement versus horizontal displacement curve is shown on the left [6]. The curve is nearly but not a straight line. A linear relationship is usually more desirable (because that would be easier for automatic control). The GA value in this case is about 1.04 (52/50).

7.1-16 Convergence Study of Beam Elements

A convergence study of the gripper model is summarized in the figure below. The figure shows that the displacement converges to 52.035 mm [1]. In the foregoing simulation, we meshed with 34 elements under the default settings and obtained a displacement of 52.016 mm [2]. That is accurate enough for GA assessment, but may not be adequate for other purposes (e.g., stress assessment).

Generally, models meshed with beam elements or shell elements converge very fast. The meshing consideration is mostly geometric. In our case, since it is a curved structure, the major concern is to mesh the structure so that the meshed finite element model doesn't deviate the original geometry too much.

For a structure composed by straight beams, it is possible to obtain a solution equal to theoretical values by meshing each straight beam a single element! This will be demonstrated in Section 7.2.

Element Size (mm)	Number of Elements	Output Displacement (mm)
10	32	52.010
6	53	52.028
4	79	52.032
3	104	52.033
2	157	52.034
1	312	52.035
0.5	624	52.035

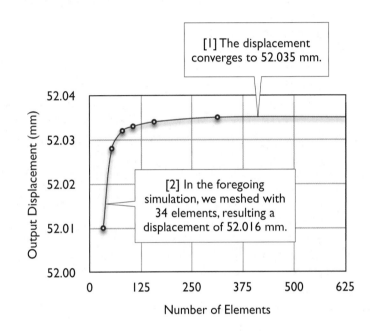

[1] The displacement converges to 52.035 mm.

[2] In the foregoing simulation, we meshed with 34 elements, resulting a displacement of 52.016 mm.

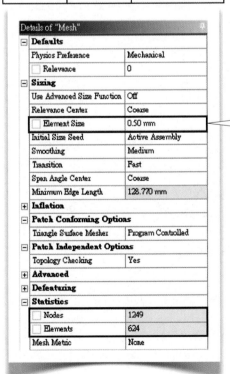

[3] At the end of this exercise, **Details of Mesh** looks like this. We will use this mesh for another exercise in Section 8.1. #

Wrap Up

Save the project and quit Workbench.

Reference

1. Chao-Chieh Lan and Yung-Jen Cheng, 2008, "Distributed Shape Optimization of Compliant Mechanisms Using Intrinsic Functions," *ASME Journal of Mechanical Design*, Vol. 130, 072304.

Section 7.2

3D Truss

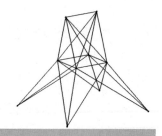

7.2-1 About the 3D Truss

Traditionally, a truss is defined as a structure consisting of two-force members; i.e., the members are pin-jointed at ends and loads are applied on joints so that the members are either stretched or compressed but not bent. Two members connected by a pin-joint can rotate about the joint independently. In reality, structural members are rarely connected each other by pin-joints. Modern structures are constructed using either welds or multiple bolt-and-nuts; the members are rigid jointed, not pin-jointed. Even in the old days, pin-jointed structures are not common. Main reason of pin-joint assumption is to ease the computational difficulty, in the days when computers were not widespread, if existing. Note that, due to the neglect of joint rigidity, pin-joint assumption leads to a conservative design: safer, but over-designed.

How much is the error caused by the pin-joint assumption? This is a good exercise problem for engineering students (7.4-2, page 302). The amount of error depends on the slenderness of the structural member. If the members are slender enough, there are no essential errors caused by the pin-joint assumption. On the other hand, if the members are not slender enough, then the pin-joints assumption may induce substantial errors.

The beam element **BEAM188**[Ref 1] (1.3-3[9], page 35) is the only element supported in **Mechanical** for line bodies. The so called "truss elements" (such as LINK180[Ref 2]) are not supported. To model a pin-jointed structure in **Mechanical**, you need to either specify revolute joints between the members or utilize **End Release** feature (see exercises in 7.4-2, page 302).

In this section, we will create a line model for a power transmission tower as shown below. All members are made of structural steel angle of $1\frac{1}{2} \times 1\frac{1}{2} \times \frac{1}{4}$ cross section. Note that we've assigned a number for each joint (P1-P10) and each member (1-25). The design loads are also tabulated below.

Design Loads

Joint	F_X (lb)	F_Y (lb)	F_Z (lb)
P1	1,000	-10,000	-10,000
P2	0	-10,000	-10,000
P3	500	0	0
P6	600	0	0

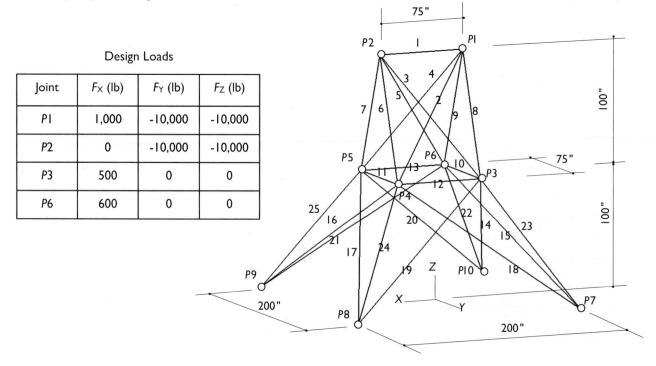

PART A. GEOMETRIC MODELING

7.2-2 Start Up

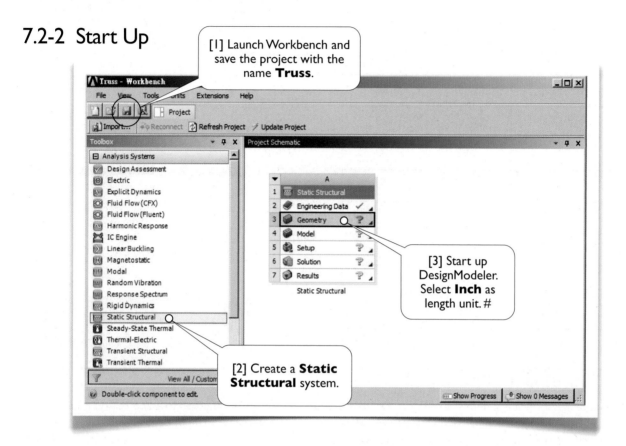

[1] Launch Workbench and save the project with the name **Truss**.

[2] Create a **Static Structural** system.

[3] Start up DesignModeler. Select **Inch** as length unit. #

7.2-3 Create 10 Construction Points in the 3D Space

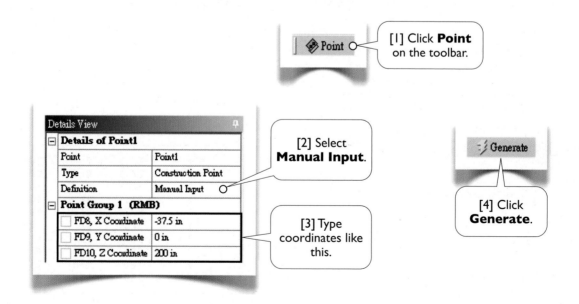

[1] Click **Point** on the toolbar.

[2] Select **Manual Input**.

[3] Type coordinates like this.

[4] Click **Generate**.

Point	X Coordinate (in)	Y Coordinate (in)	Z Coordinate (in)
1	-37.5	0	200
2	37.5	0	200
3	-37.5	37.5	100
4	37.5	37.5	100
5	37.5	-37.5	100
6	-37.5	-37.5	100
7	-100	100	0
8	100	100	0
9	100	-100	0
10	-100	-100	0

[5] Repeat steps [1-4] for additional nine points (Points 2-10). Their respective coordinates are like this.

[6] The 10 points. (The numbers are added by the author for clarity.) Note that the model has been rotated such that Z-axis directs upward.

Using **Coordinates File** to Define Points

An alternative way of defining points is to select **From Coordinates File** for **Definition** and read the coordinates from a file [7, 8]. When the number of points is large, this is obviously a better way to input coordinates.

A **Coordinates file** [8] is a text file describing the coordinates of points. The file has 5 columns, or fields: (a) group number, (b) ID number, (c) X-coordinate, (d) Y-coordinate, and (e) Z-coordinate. The group number and the ID number can be arbitrarily chosen and they together uniquely identify a point from others.

Fields of numbers can be separated by spaces or TABs. You may prepare the text file using a text editor; another way is using a spread-sheet program such as Microsoft Excel and saving as a text file.

[7] An alternative way of defining coordinates is to select **From Coordinates File** for **Definition** and read the coordinates from a file.

[8] The format of a **Coordinate File**. #

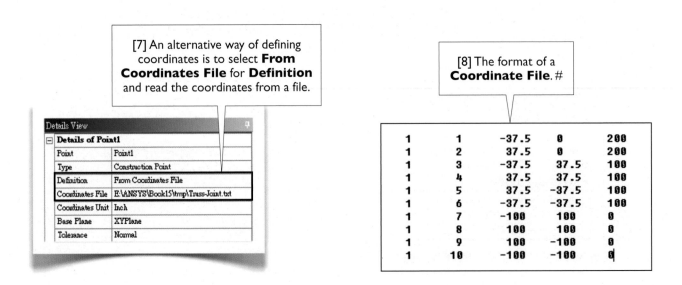

1	1	-37.5	0	200
1	2	37.5	0	200
1	3	-37.5	37.5	100
1	4	37.5	37.5	100
1	5	37.5	-37.5	100
1	6	-37.5	-37.5	100
1	7	-100	100	0
1	8	100	100	0
1	9	100	-100	0
1	10	-100	-100	0

7.2-4 Create Line Body

[2] Click **Point 1** and control-click **Point 2** to create a line segment.

[1] Pull-down-select **Concept/Lines From Points**.

Concept Tools Units View

- Lines From Points ●
- Lines From Sketches
- Lines From Edges
- 3D Curve
- Split Edges
- Surfaces From Edges
- Surfaces From Sketches
- Surfaces From Faces
- Cross Section ▶

Member	Start Point	End Point
1	P1	P2
2	P1	P4
3	P2	P3
4	P1	P5
5	P2	P6
6	P2	P4
7	P2	P5
8	P1	P3
9	P1	P6
10	P3	P6
11	P4	P5
12	P3	P4
13	P5	P6
14	P3	P10
15	P6	P7
16	P4	P9
17	P5	P8
18	P4	P7
19	P3	P8
20	P5	P10
21	P6	P9
22	P6	P10
23	P3	P7
24	P4	P8
25	P5	P9

[3] Repeat step [2] for additional 24 line segments (totally 25 line segments). Each line segment is created by clicking the starting point and then control-clicking the ending point. If you make a mistake, you can remove a line segment by clicking its starting point and then control-clicking ending point again. Note that, when creating or removing a line segment, order of starting or ending points is not relevant.

Details View

Details of Line1	
Lines From Points	Line1
Point Segments	25 ○
Operation	Add Material

[4] Click **Apply**.

View Help
- ✓ Shaded Exterior and Edges
- Shaded Exterior
- Wireframe
- Graphics Options ▶
- ✓ Frozen Body Transparency
- Edge Joints
- Cross Section Alignments ●
- Display Edge Direction
- Display Vertices
- Cross Section Solids
- Ruler
- ✓ Triad
- Outline
- Window

[6] Turn off **View/ Cross Section Alignments** for clarity.

Generate

[5] Click **Generate**.

[7] A line body of 25 members is created. (The numbers are added by the author for clarity.) #

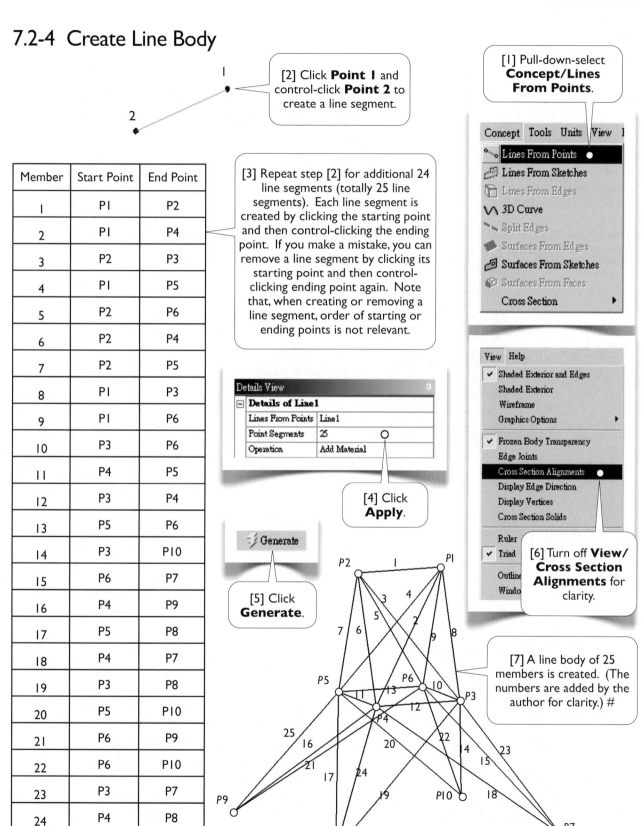

278

7.2-5 Create Cross Section

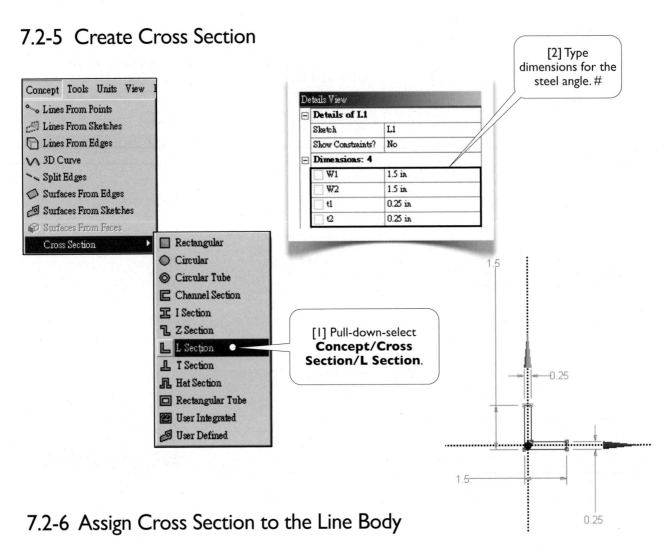

[2] Type dimensions for the steel angle. #

Details View

Details of L1	
Sketch	L1
Show Constraints?	No
Dimensions: 4	
W1	1.5 in
W2	1.5 in
t1	0.25 in
t2	0.25 in

[1] Pull-down-select **Concept/Cross Section/L Section**.

7.2-6 Assign Cross Section to the Line Body

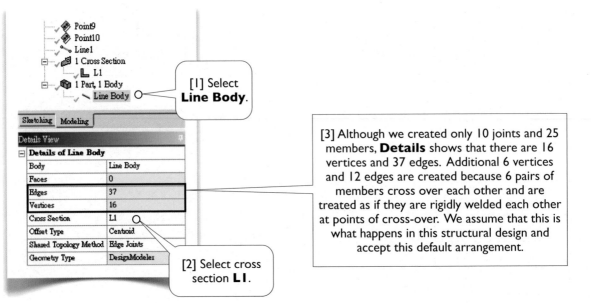

[1] Select **Line Body**.

[2] Select cross section **LI**.

Details View

Details of Line Body	
Body	Line Body
Faces	0
Edges	37
Vertices	16
Cross Section	L1
Offset Type	Centroid
Shared Topology Method	Edge Joints
Geometry Type	DesignModeler

[3] Although we created only 10 joints and 25 members, **Details** shows that there are 16 vertices and 37 edges. Additional 6 vertices and 12 edges are created because 6 pairs of members cross over each other and are treated as if they are rigidly welded each other at points of cross-over. We assume that this is what happens in this structural design and accept this default arrangement.

279

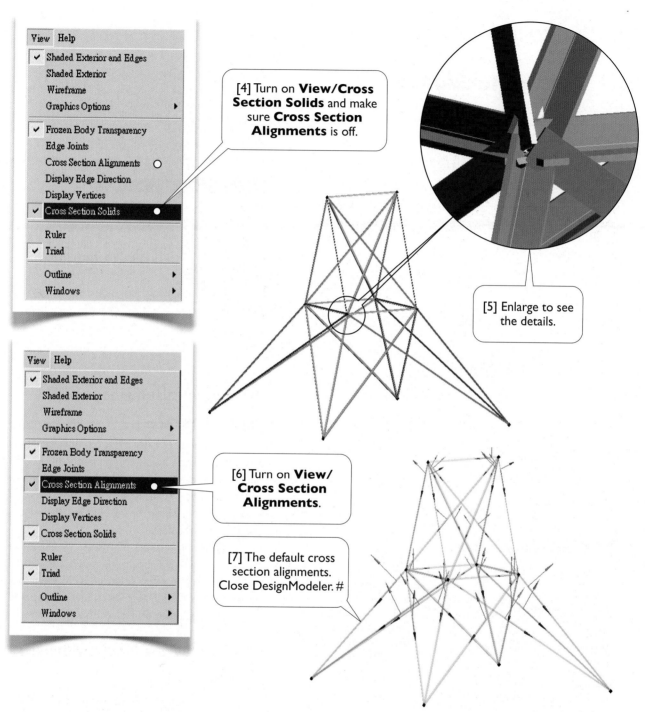

[4] Turn on **View/Cross Section Solids** and make sure **Cross Section Alignments** is off.

[5] Enlarge to see the details.

[6] Turn on **View/ Cross Section Alignments**.

[7] The default cross section alignments. Close DesignModeler. #

Cross Section Alignments

The default cross section alignments [7] are not entirely consistent with the reality in this case. However, we decided to neglect the difference between the reality and the default alignments. Since the structural members are quite slender, the behaviors are essentially two-force members, therefore alignments should not be critical. In other words, cross section alignments have little effects on the structural response in this case. We will demonstrate the adjustment of cross section alignments in 7.3-9 (page 292), in which the alignments must be adjusted, otherwise it would deviate from the reality too much.

PART B. SIMULATION

7.2-7 Start Up **Mechanical**

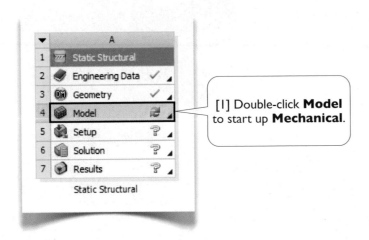

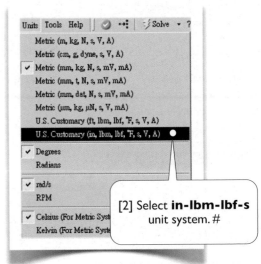

[1] Double-click **Model** to start up **Mechanical**.

[2] Select **in-lbm-lbf-s** unit system. #

7.2-8 Generate Mesh

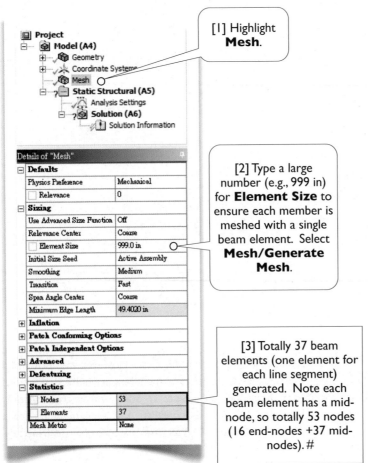

[1] Highlight **Mesh**.

[2] Type a large number (e.g., 999 in) for **Element Size** to ensure each member is meshed with a single beam element. Select **Mesh/Generate Mesh**.

[3] Totally 37 beam elements (one element for each line segment) generated. Note each beam element has a mid-node, so totally 53 nodes (16 end-nodes +37 mid-nodes). #

Why mesh each member with single element?

We mentioned (7.1-16, page 274) that it is possible to obtain a solution equal to theoretical values by meshing each straight beam with a single element. The reason we meshed each member with a single beam element here is to demonstrate this fact.

The default settings of **Mesh** would mesh the model with 205 beam elements and would result exactly the same solution as 37 elements. As an exercise (7.4-2, page 302), verify it yourself after the completion of this section.

Use Surface/Line Models Whenever Possible

Since the solution of a model meshed with beam elements or shell elements converges very fast (i.e., very accurate solution can be obtained with only a few elements), we should consider a line model or surface model whenever possible. This is particularly true for those problems requiring many number of iterations or substeps, such as nonlinear problems, dynamic problems, optimization problems, etc.

7.2-9 Specify Supports

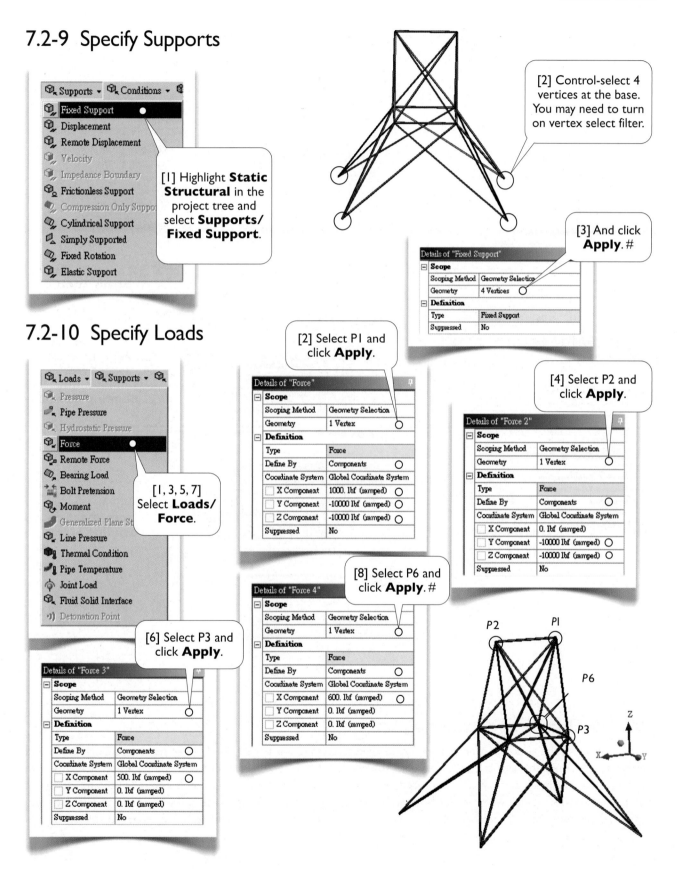

Supports ▾ Conditions ▾

- **Fixed Support** ●
- Displacement
- Remote Displacement
- Velocity
- Impedance Boundary
- Frictionless Support
- Compression Only Support
- Cylindrical Support
- Simply Supported
- Fixed Rotation
- Elastic Support

[1] Highlight **Static Structural** in the project tree and select **Supports/ Fixed Support**.

[2] Control-select 4 vertices at the base. You may need to turn on vertex select filter.

[3] And click **Apply**. #

Details of "Fixed Support"

Scope	
Scoping Method	Geometry Selection
Geometry	4 Vertices
Definition	
Type	Fixed Support
Suppressed	No

7.2-10 Specify Loads

Loads ▾ Supports ▾

- Pressure
- Pipe Pressure
- Hydrostatic Pressure
- **Force** ●
- Remote Force
- Bearing Load
- Bolt Pretension
- Moment
- Generalized Plane St
- Line Pressure
- Thermal Condition
- Pipe Temperature
- Joint Load
- Fluid Solid Interface
- Detonation Point

[1, 3, 5, 7] Select **Loads/ Force**.

[2] Select P1 and click **Apply**.

Details of "Force"

Scope	
Scoping Method	Geometry Selection
Geometry	1 Vertex
Definition	
Type	Force
Define By	Components
Coordinate System	Global Coordinate System
☐ X Component	1000. lbf (ramped)
☐ Y Component	-10000 lbf (ramped)
☐ Z Component	-10000 lbf (ramped)
Suppressed	No

[4] Select P2 and click **Apply**.

Details of "Force 2"

Scope	
Scoping Method	Geometry Selection
Geometry	1 Vertex
Definition	
Type	Force
Define By	Components
Coordinate System	Global Coordinate System
☐ X Component	0. lbf (ramped)
☐ Y Component	-10000 lbf (ramped)
☐ Z Component	-10000 lbf (ramped)
Suppressed	No

[8] Select P6 and click **Apply**. #

Details of "Force 4"

Scope	
Scoping Method	Geometry Selection
Geometry	1 Vertex
Definition	
Type	Force
Define By	Components
Coordinate System	Global Coordinate System
☐ X Component	600. lbf (ramped)
☐ Y Component	0. lbf (ramped)
☐ Z Component	0. lbf (ramped)
Suppressed	No

[6] Select P3 and click **Apply**.

Details of "Force 3"

Scope	
Scoping Method	Geometry Selection
Geometry	1 Vertex
Definition	
Type	Force
Define By	Components
Coordinate System	Global Coordinate System
☐ X Component	500. lbf (ramped)
☐ Y Component	0. lbf (ramped)
☐ Z Component	0. lbf (ramped)
Suppressed	No

P2 P1 P6 P3

7.2-11 Set Up Solution Branch and Solve the Model

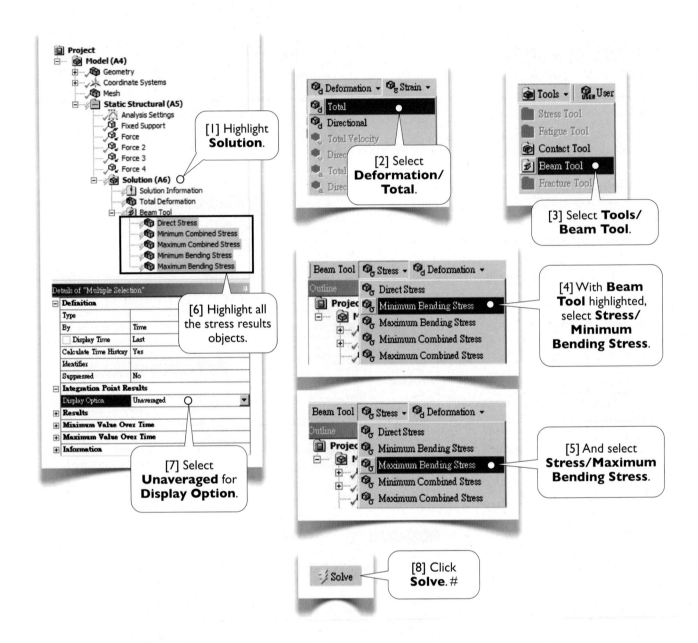

Why Use **Unaveraged** Stresses?

We've introduced the notion of averaged and unaveraged stresses (3.5-6, page 152). Unaveraged stresses are also called *element stresses*, since they are calculated at points (usually geometric center or integration points) inside elements. On the other hand, averaged stresses are also called *nodal stresses*, since they are calculated at nodes, which are located at element boundaries. Since we mesh each member with a single beam element, and if we select to display averaged stresses, every two adjacent members' stresses would have been averaged and reported. The averaged stresses in this case would not have any meaning.

7.2-12 View Results

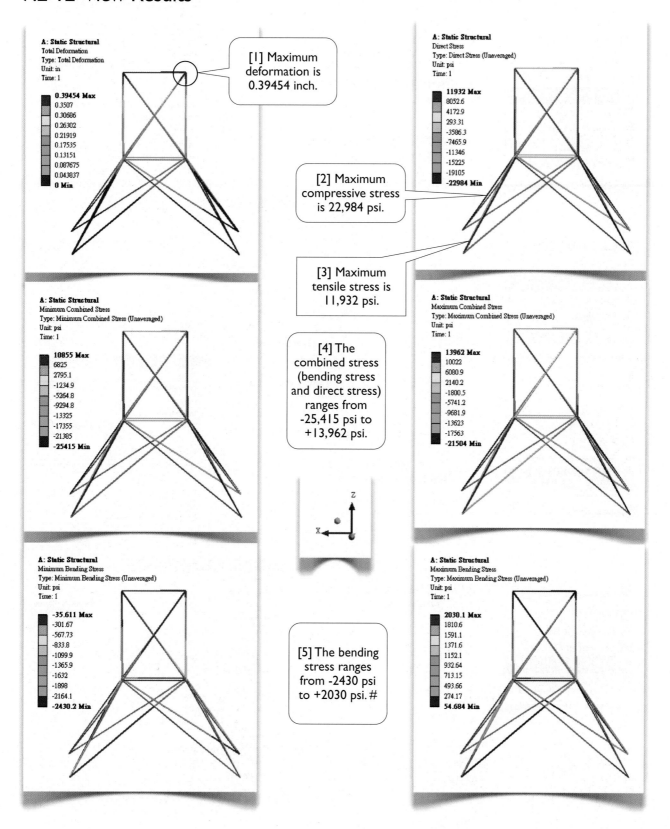

A: Static Structural
Total Deformation
Type: Total Deformation
Unit: in
Time: 1

0.39454 Max
0.3507
0.30686
0.26302
0.21919
0.17535
0.13151
0.087675
0.043837
0 Min

[1] Maximum deformation is 0.39454 inch.

A: Static Structural
Direct Stress
Type: Direct Stress (Unaveraged)
Unit: psi
Time: 1

11932 Max
8052.6
4172.9
293.31
-3586.3
-7465.9
-11346
-15225
-19105
-22984 Min

[2] Maximum compressive stress is 22,984 psi.

[3] Maximum tensile stress is 11,932 psi.

A: Static Structural
Minimum Combined Stress
Type: Minimum Combined Stress (Unaveraged)
Unit: psi
Time: 1

10855 Max
6825
2795.1
-1234.9
-5264.8
-9294.8
-13325
-17355
-21385
-25415 Min

[4] The combined stress (bending stress and direct stress) ranges from -25,415 psi to +13,962 psi.

A: Static Structural
Maximum Combined Stress
Type: Maximum Combined Stress (Unaveraged)
Unit: psi
Time: 1

13962 Max
10022
6080.9
2140.2
-1800.5
-5741.2
-9681.9
-13623
-17563
-21504 Min

A: Static Structural
Minimum Bending Stress
Type: Minimum Bending Stress (Unaveraged)
Unit: psi
Time: 1

-35.611 Max
-301.67
-567.73
-833.8
-1099.9
-1365.9
-1632
-1898
-2164.1
-2430.2 Min

[5] The bending stress ranges from -2430 psi to +2030 psi. #

A: Static Structural
Maximum Bending Stress
Type: Maximum Bending Stress (Unaveraged)
Unit: psi
Time: 1

2030.1 Max
1810.6
1591.1
1371.6
1152.1
932.64
713.15
493.66
274.17
54.684 Min

7.2-13 View Member Forces/Moments

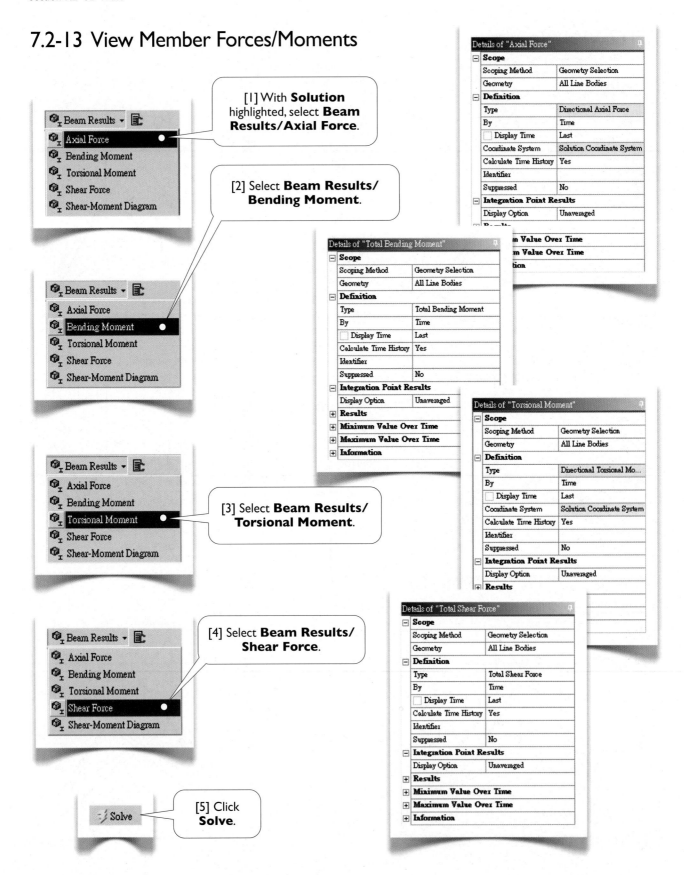

[1] With **Solution** highlighted, select **Beam Results/Axial Force**.

[2] Select **Beam Results/ Bending Moment**.

[3] Select **Beam Results/ Torsional Moment**.

[4] Select **Beam Results/ Shear Force**.

[5] Click **Solve**.

Beam Results ▾
- Axial Force
- Bending Moment
- Torsional Moment
- Shear Force
- Shear-Moment Diagram

Details of "Axial Force"

Scope	
Scoping Method	Geometry Selection
Geometry	All Line Bodies
Definition	
Type	Directional Axial Force
By	Time
Display Time	Last
Coordinate System	Solution Coordinate System
Calculate Time History	Yes
Identifier	
Suppressed	No
Integration Point Results	
Display Option	Unaveraged

Details of "Total Bending Moment"

Scope	
Scoping Method	Geometry Selection
Geometry	All Line Bodies
Definition	
Type	Total Bending Moment
By	Time
Display Time	Last
Calculate Time History	Yes
Identifier	
Suppressed	No
Integration Point Results	
Display Option	Unaveraged
+ **Results**	
+ **Minimum Value Over Time**	
+ **Maximum Value Over Time**	
+ **Information**	

Details of "Torsional Moment"

Scope	
Scoping Method	Geometry Selection
Geometry	All Line Bodies
Definition	
Type	Directional Torsional Mo...
By	Time
Display Time	Last
Coordinate System	Solution Coordinate System
Calculate Time History	Yes
Identifier	
Suppressed	No
Integration Point Results	
Display Option	Unaveraged
+ **Results**	

Details of "Total Shear Force"

Scope	
Scoping Method	Geometry Selection
Geometry	All Line Bodies
Definition	
Type	Total Shear Force
By	Time
Display Time	Last
Calculate Time History	Yes
Identifier	
Suppressed	No
Integration Point Results	
Display Option	Unaveraged
+ **Results**	
+ **Minimum Value Over Time**	
+ **Maximum Value Over Time**	
+ **Information**	

285

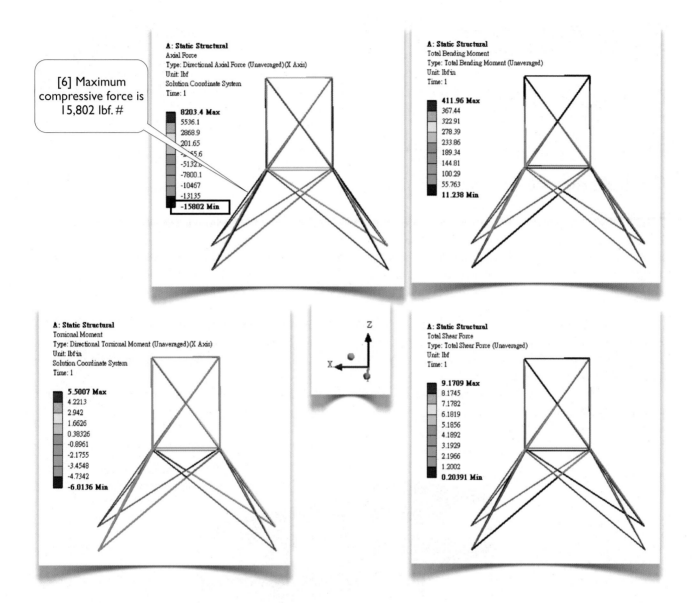

Wrap Up

Save the project and quit Workbench.

References

1. ANSYS Documentation//Mechanical APDL//Element Reference//I. Element Library//BEAM188
2. ANSYS Documentation//Mechanical APDL//Element Reference//I. Element Library//LINK180

Section 7.3

Two-Story Building

7.3-1 About the Two-Story Building

A two-story building [1-6] is constructed for residential usage. The local building code requires that a live load of 50 lb/ft^2 be considered, along with its own weight. Since the building is in an earthquake zone, an earthquake load must be considered. For a low-rise building like this, the building code allows an equivalent static analysis instead of a dynamic analysis. Here we consider a static earthquake load, which is equivalent to 0.2 times of gravitational acceleration, applying horizontally in the shorter direction of the building. In a practical design project, earthquake load applying in other directions should also be simulated.

The beams and columns will be modeled as line bodies and the floor slabs as surface bodies. All bodies will be combined to form a single part to ensure a perfect bonding between bodies. The geometric model will be used for a static structural simulation in this section. The model will be used again in Section 11.2 for a modal analysis and Section 12.3 for a harmonic response analysis.

In this exercise, we will use **in-lbm-lbf-s** unit system most of time. In some occasions, we will change the unit system to **ft-lbm-lbf-s** unit system. One feature of Workbench is the flexibility of unit systems. When you switch to another unit system, Workbench will transform all the units nicely for you.

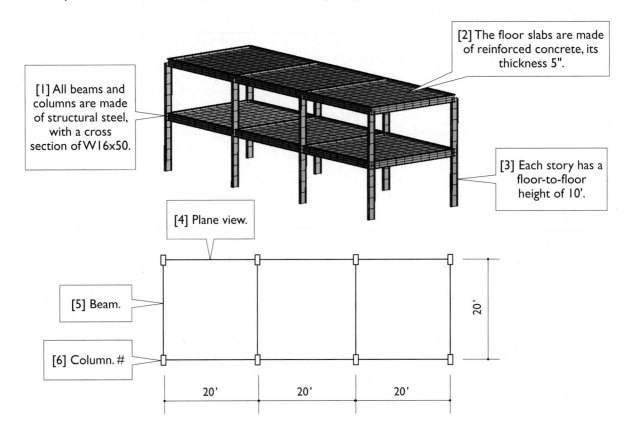

[1] All beams and columns are made of structural steel, with a cross section of W16x50.

[2] The floor slabs are made of reinforced concrete, its thickness 5".

[3] Each story has a floor-to-floor height of 10'.

[4] Plane view.

[5] Beam.

[6] Column. #

20'

20' 20' 20'

PART A. GEOMETRIC MODELING

7.3-2 Start Up

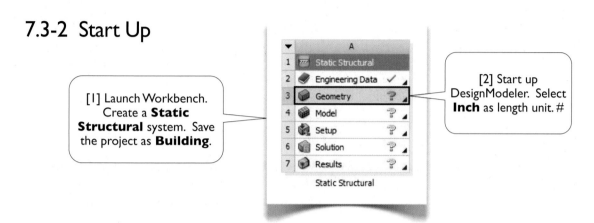

[1] Launch Workbench. Create a **Static Structural** system. Save the project as **Building**.

[2] Start up DesignModeler. Select **Inch** as length unit. #

7.3-3 Create 16 Points in the Space

Point	X Coordinate (in)	Y Coordinate (in)	Z Coordinate (in)
1	0	0	0
2	240	0	0
3	480	0	0
4	720	0	0
5	0	0	240
6	240	0	240
7	480	0	240
8	720	0	240
9	0	120	0
10	240	120	0
11	480	120	0
12	720	120	0
13	0	120	240
14	240	120	240
15	480	120	240
16	720	120	240

[1] Click **Point** on the toolbar.

[2] Select **Manual Input**.

Details View

Details of Point1

Point	Point1
Type	Construction Point
Definition	Manual Input

Point Group 1 (RMB)

FD8, X Coordinate	0 in
FD9, Y Coordinate	0 in
FD10, Z Coordinate	0 in

[3] Type coordinates for **Point1**.

Generate

[4] Click **Generate**.

[5] Repeat steps [1-4] for additional 15 points (Point 2-16) and type their respective coordinates shown in this table.

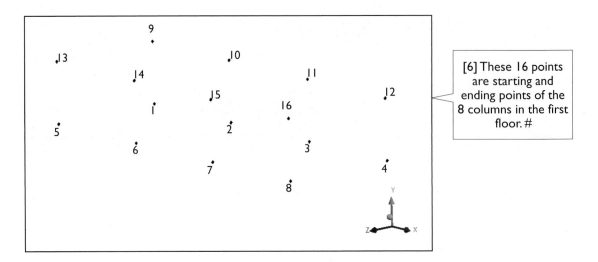

[6] These 16 points are starting and ending points of the 8 columns in the first floor. #

7.3-4 Create a Line Body of 10 Beams

[1] Pull-down-select **Concept/Lines From Points**.

[2] Define 6 line segments. Remember that each line segment is defined by first clicking its starting point then control-clicking its ending point.

Member	Start Point	End Point
1	P9	P12
2	P13	P16
3	P9	P13
4	P10	P14
5	P11	P15
6	P12	P16

[3] Click **Apply**.

[4] Click **Generate**.

[5] Turn off **View/ Cross Section Alignment**. #

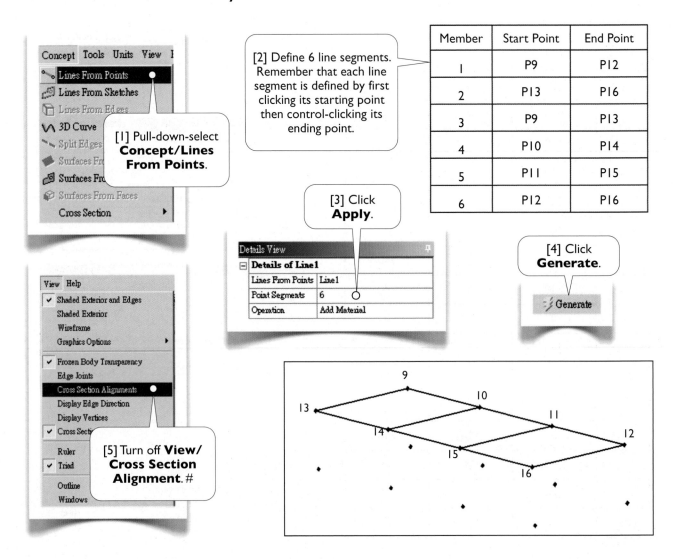

7.3-5 Copy the Line Body

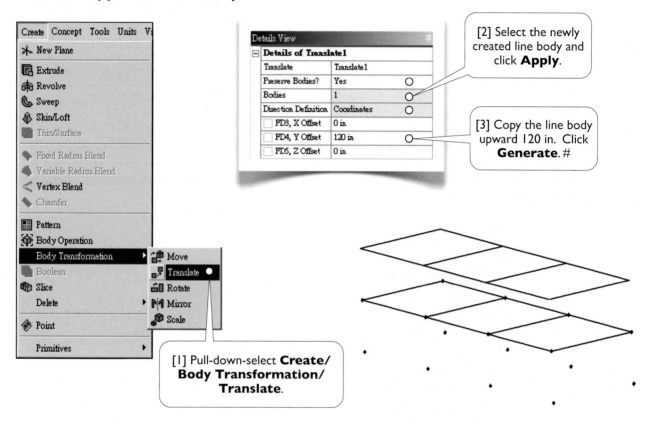

[2] Select the newly created line body and click **Apply**.

[3] Copy the line body upward 120 in. Click **Generate**. #

[1] Pull-down-select **Create/ Body Transformation/ Translate**.

Details of Translate1	
Translate	Translate1
Preserve Bodies?	Yes
Bodies	1
Direction Definition	Coordinates
FD3, X Offset	0 in
FD4, Y Offset	120 in
FD5, Z Offset	0 in

7.3-6 Create Line Body for Columns

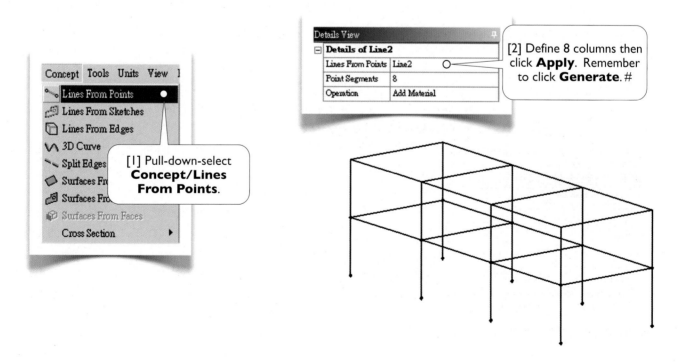

[1] Pull-down-select **Concept/Lines From Points**.

[2] Define 8 columns then click **Apply**. Remember to click **Generate**. #

Details of Line2	
Lines From Points	Line2
Point Segments	8
Operation	Add Material

7.3-7 Create a Cross Section

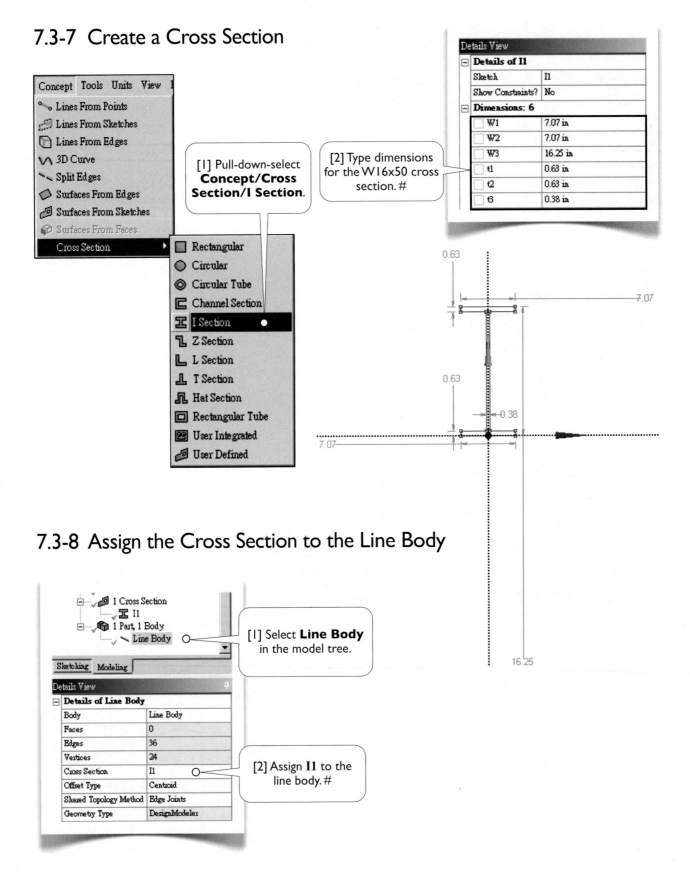

Details View		
Details of I1		
Sketch	I1	
Show Constraints?	No	
Dimensions: 6		
W1	7.07 in	
W2	7.07 in	
W3	16.25 in	
t1	0.63 in	
t2	0.63 in	
t3	0.38 in	

Concept Tools Units View

- Lines From Points
- Lines From Sketches
- Lines From Edges
- 3D Curve
- Split Edges
- Surfaces From Edges
- Surfaces From Sketches
- Surfaces From Faces
- Cross Section

- Rectangular
- Circular
- Circular Tube
- Channel Section
- I Section
- Z Section
- L Section
- T Section
- Hat Section
- Rectangular Tube
- User Integrated
- User Defined

[1] Pull-down-select **Concept/Cross Section/I Section**.

[2] Type dimensions for the W16x50 cross section. #

0.63

7.07

0.63

−0.38

7.07

16.25

7.3-8 Assign the Cross Section to the Line Body

- 1 Cross Section
  - I1
- 1 Part, 1 Body
  - Line Body

[1] Select **Line Body** in the model tree.

Sketching Modeling

Details View	
Details of Line Body	
Body	Line Body
Faces	0
Edges	36
Vertices	24
Cross Section	I1
Offset Type	Centroid
Shared Topology Method	Edge Joints
Geometry Type	DesignModeler

[2] Assign **I1** to the line body. #

291

7.3-9 Adjust Cross Section Alignments

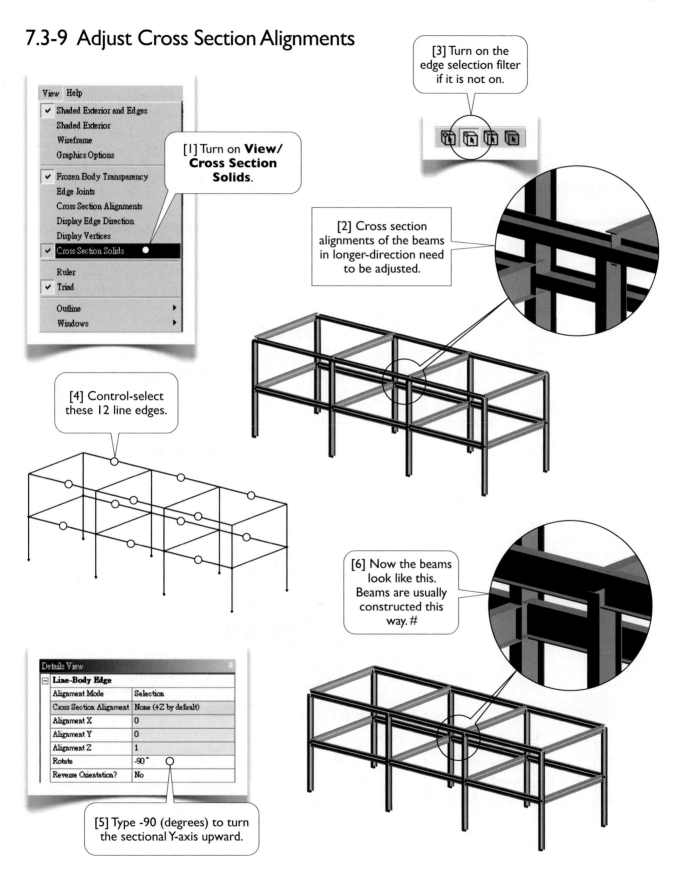

[3] Turn on the edge selection filter if it is not on.

[1] Turn on **View/ Cross Section Solids**.

[2] Cross section alignments of the beams in longer-direction need to be adjusted.

[4] Control-select these 12 line edges.

[6] Now the beams look like this. Beams are usually constructed this way. #

[5] Type -90 (degrees) to turn the sectional Y-axis upward.

7.3-10 Create Surface Bodies

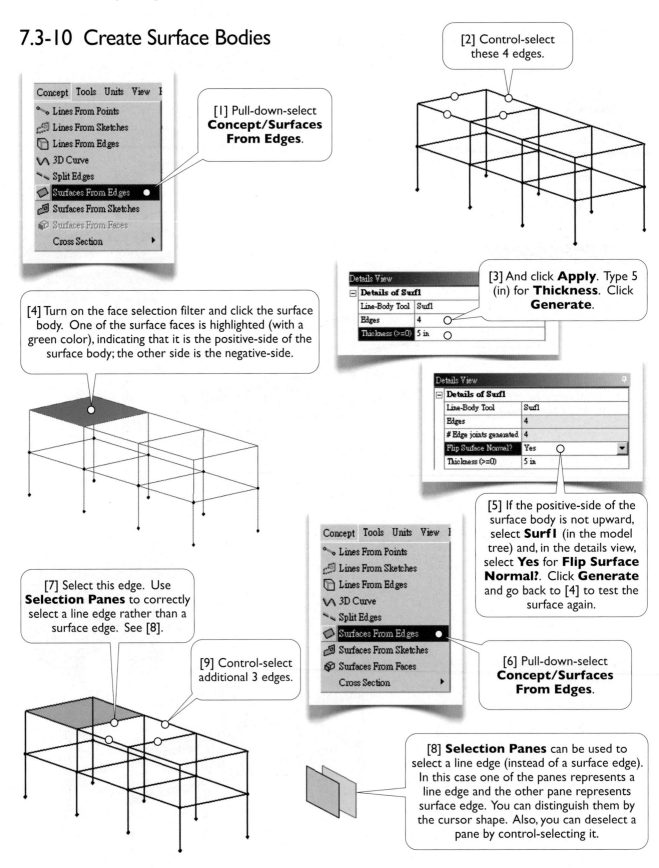

[1] Pull-down-select **Concept/Surfaces From Edges**.

[2] Control-select these 4 edges.

[3] And click **Apply**. Type 5 (in) for **Thickness**. Click **Generate**.

[4] Turn on the face selection filter and click the surface body. One of the surface faces is highlighted (with a green color), indicating that it is the positive-side of the surface body; the other side is the negative-side.

[5] If the positive-side of the surface body is not upward, select **Surf1** (in the model tree) and, in the details view, select **Yes** for **Flip Surface Normal?**. Click **Generate** and go back to [4] to test the surface again.

[7] Select this edge. Use **Selection Panes** to correctly select a line edge rather than a surface edge. See [8].

[9] Control-select additional 3 edges.

[6] Pull-down-select **Concept/Surfaces From Edges**.

[8] **Selection Panes** can be used to select a line edge (instead of a surface edge). In this case one of the panes represents a line edge and the other pane represents surface edge. You can distinguish them by the cursor shape. Also, you can deselect a pane by control-selecting it.

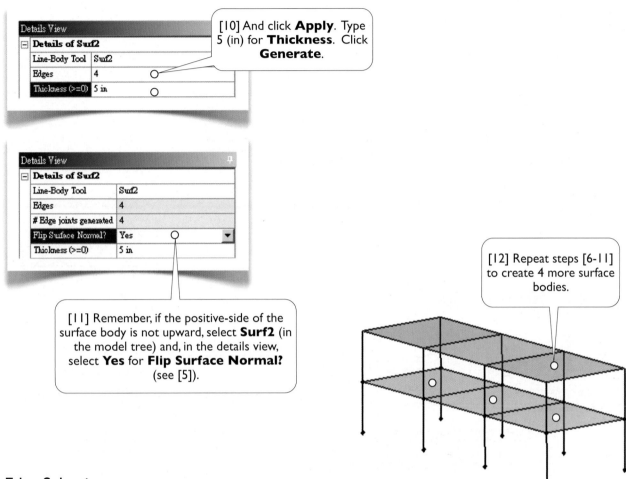

[10] And click **Apply**. Type 5 (in) for **Thickness**. Click **Generate**.

[11] Remember, if the positive-side of the surface body is not upward, select **Surf2** (in the model tree) and, in the details view, select **Yes** for **Flip Surface Normal?** (see [5]).

[12] Repeat steps [6-11] to create 4 more surface bodies.

Edge Selection

When you select edges, you may need to use Selection Panes to correctly select the line edge rather than a surface edge [7, 8]. You always can distinguish them by the cursor shape. If you select a surface edge instead of a line edge, the slab would not connect to the line body, and would have a gap after deformation [13].

Direction of Surface Bodies

Each surface body has a positive-side and a negative-side. In **Mechanical**, the pressure always applies on the positive-side. Therefore it is important to know which is the positive-side: turn on the face selection filter and click the surface body, and the positive-side will be highlighted with a green color. Make sure the positive-side of each surface body is upward in this case. If not, select **Yes** for **Flip Surface Normal?** in the details view [5, 11].

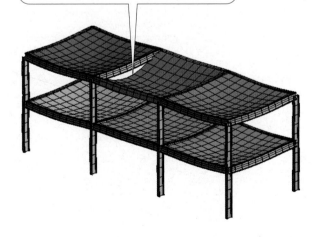

[13] If you select a surface edge instead of a line edge, the slab would not connect to the line body and would have a gap after deformation. #

7.3-11 Form a Single Part

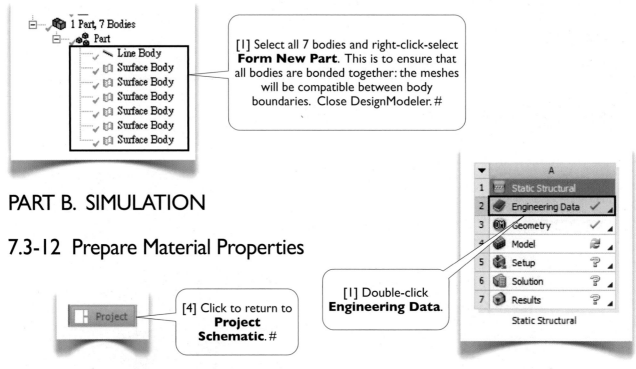

[1] Select all 7 bodies and right-click-select **Form New Part**. This is to ensure that all bodies are bonded together: the meshes will be compatible between body boundaries. Close DesignModeler. #

PART B. SIMULATION

7.3-12 Prepare Material Properties

[4] Click to return to **Project Schematic**. #

[1] Double-click **Engineering Data**.

Static Structural

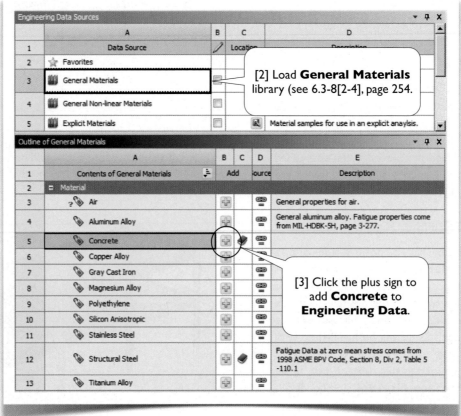

[2] Load **General Materials** library (see 6.3-8[2-4], page 254.

[3] Click the plus sign to add **Concrete** to **Engineering Data**.

7.3-13 Start Up **Mechanical**

[1] Start up **Mechanical**. Make sure the unit system is **in-lbm-lbf-s**. #

7.3-14 Assign Materials

[1] Highlight all 6 surface bodies.

[2] Select **Concrete**. #

7.3-15 Generate Mesh

[1] Highlight **Mesh** and select **Mesh/Generate Mesh**.

[3] To display the model like this, turn on **View/Thick Shells and Beams**. #

[2] The default mesh settings generate 608 elements (including shell elements and beam elements).

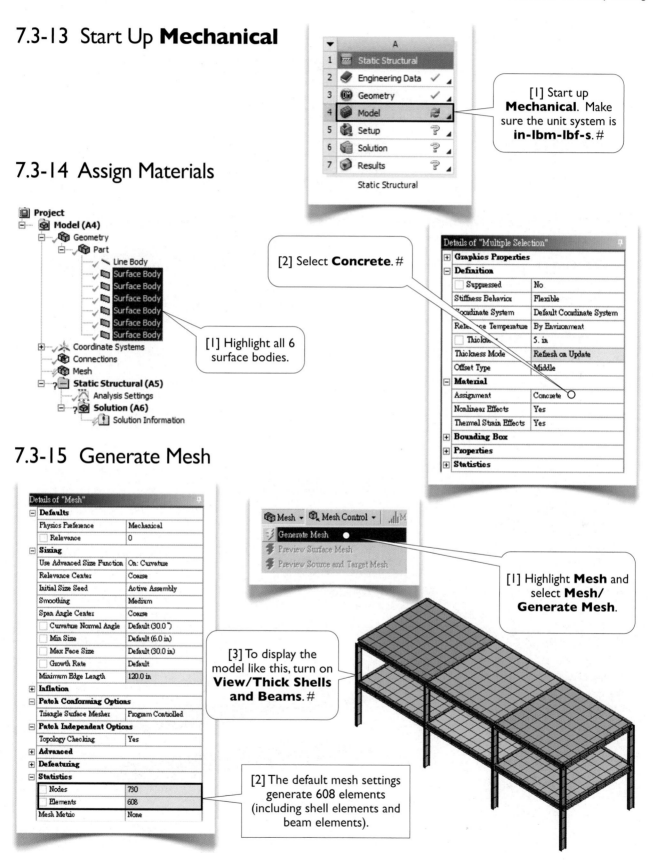

7.3-16 Specify Supports

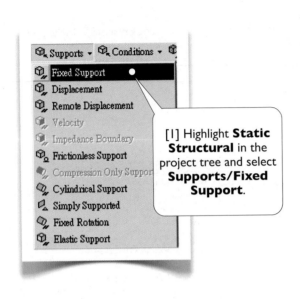

[1] Highlight **Static Structural** in the project tree and select **Supports/Fixed Support**.

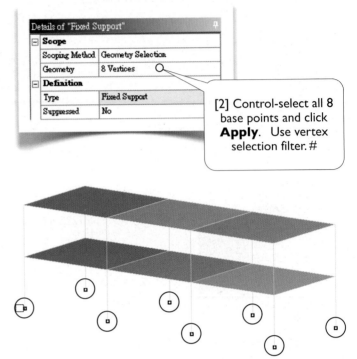

[2] Control-select all 8 base points and click **Apply**. Use vertex selection filter. #

7.3-17 Apply the Live Load

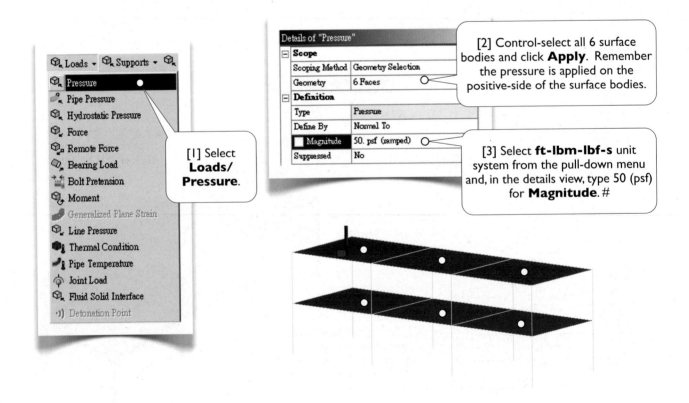

[1] Select **Loads/ Pressure**.

[2] Control-select all 6 surface bodies and click **Apply**. Remember the pressure is applied on the positive-side of the surface bodies.

[3] Select **ft-lbm-lbf-s** unit system from the pull-down menu and, in the details view, type 50 (psf) for **Magnitude**. #

7.3-18 Apply Earth Gravity

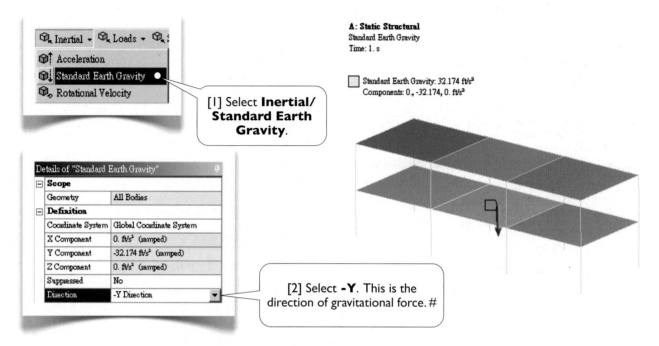

[1] Select **Inertial/ Standard Earth Gravity**.

[2] Select **-Y**. This is the direction of gravitational force. #

7.3-19 Apply the Earthquake Load

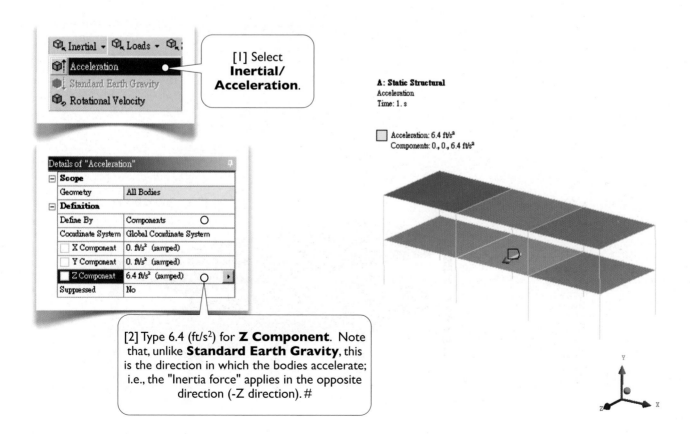

[1] Select **Inertial/ Acceleration**.

[2] Type 6.4 (ft/s²) for **Z Component**. Note that, unlike **Standard Earth Gravity**, this is the direction in which the bodies accelerate; i.e., the "Inertia force" applies in the opposite direction (-Z direction). #

7.3-20 Solve the Model and View the Results

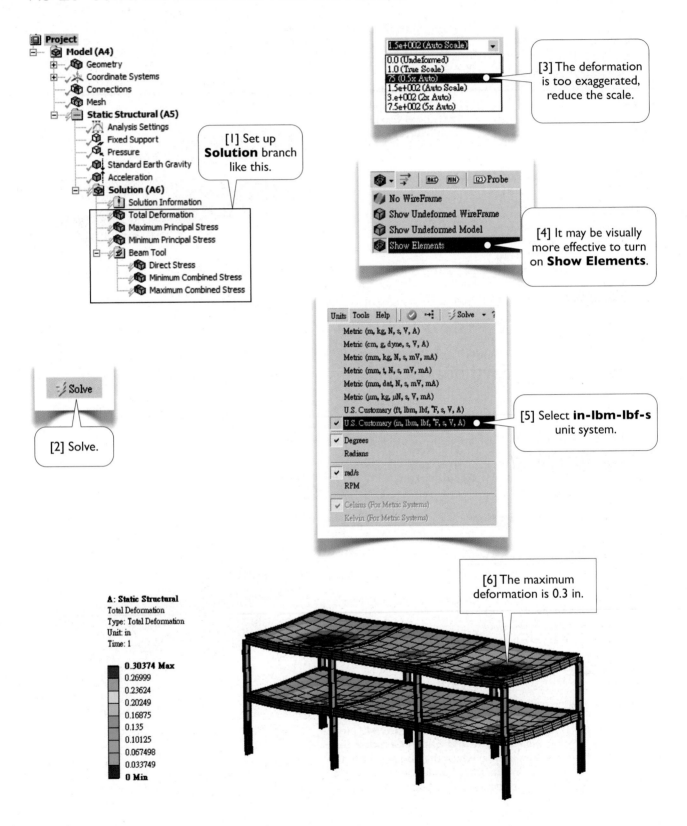

Project
- **Model (A4)**
  - Geometry
  - Coordinate Systems
  - Connections
  - Mesh
  - **Static Structural (A5)**
    - Analysis Settings
    - Fixed Support
    - Pressure
    - Standard Earth Gravity
    - Acceleration
    - **Solution (A6)**
      - Solution Information
      - Total Deformation
      - Maximum Principal Stress
      - Minimum Principal Stress
      - Beam Tool
        - Direct Stress
        - Minimum Combined Stress
        - Maximum Combined Stress

[1] Set up **Solution** branch like this.

1.5e+002 (Auto Scale)
0.0 (Undeformed)
1.0 (True Scale)
75 (0.5x Auto)
1.5e+002 (Auto Scale)
3.e+002 (2x Auto)
7.5e+002 (5x Auto)

[3] The deformation is too exaggerated, reduce the scale.

MAX   MIN   [123] Probe
No WireFrame
Show Undeformed WireFrame
Show Undeformed Model
Show Elements

[4] It may be visually more effective to turn on **Show Elements**.

Solve

[2] Solve.

Units   Tools   Help   Solve
Metric (m, kg, N, s, V, A)
Metric (cm, g, dyne, s, V, A)
Metric (mm, kg, N, s, mV, mA)
Metric (mm, t, N, s, mV, mA)
Metric (mm, dat, N, s, mV, mA)
Metric (μm, kg, μN, s, V, mA)
U.S. Customary (ft, lbm, lbf, °F, s, V, A)
✓ U.S. Customary (in, lbm, lbf, °F, s, V, A)
✓ Degrees
  Radians
✓ rad/s
  RPM
✓ Celsius (For Metric Systems)
  Kelvin (For Metric Systems)

[5] Select **in-lbm-lbf-s** unit system.

[6] The maximum deformation is 0.3 in.

A: Static Structural
Total Deformation
Type: Total Deformation
Unit: in
Time: 1

0.30374 Max
0.26999
0.23624
0.20249
0.16875
0.135
0.10125
0.067498
0.033749
0 Min

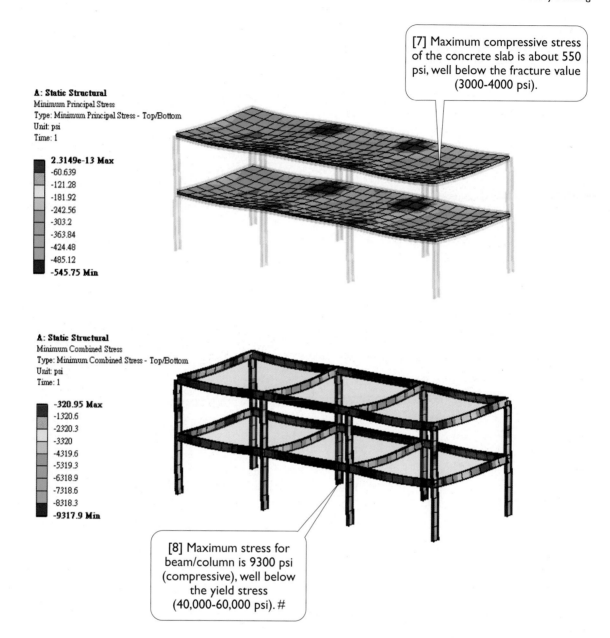

A: Static Structural
Minimum Principal Stress
Type: Minimum Principal Stress - Top/Bottom
Unit: psi
Time: 1

2.3149e-13 Max
-60.639
-121.28
-181.92
-242.56
-303.2
-363.84
-424.48
-485.12
-545.75 Min

[7] Maximum compressive stress of the concrete slab is about 550 psi, well below the fracture value (3000-4000 psi).

A: Static Structural
Minimum Combined Stress
Type: Minimum Combined Stress - Top/Bottom
Unit: psi
Time: 1

-320.95 Max
-1320.6
-2320.3
-3320
-4319.6
-5319.3
-6318.9
-7318.6
-8318.3
-9317.9 Min

[8] Maximum stress for beam/column is 9300 psi (compressive), well below the yield stress (40,000-60,000 psi). #

Wrap Up

Save the project and exit Workbench.

Section 7.4

Review

7.4-1 Keywords

Choose a letter for each keyword from the list of descriptions

1. (   ) Beam Elements
2. (   ) Coordinates File
3. (   ) Direct Stress and Bending Stress
4. (   ) Maximum/Minimum Bending Stress
5. (   ) Maximum/Minimum Combined Stress
6. (   ) Truss

Answers:

1. (A)   2. (F)   3. (C)   4. (D)   5. (E)   6. (B)

List of Descriptions

(A) A beam element is a line (1D) element that can be arranged in the 3D space. It is used to mesh a body when two of its dimensions are much smaller than the third dimension. Each node has 6 degrees of freedom: 3 translational and 3 rotational. Due to the presence of rotational degrees of freedom, it is very efficient to model the problems dominated by bending modes, contrasting to a solid element, which does not have rotational degrees of freedom.

(B) A truss is defined as a structure consisting of two-force members. By two-force member, we mean that the members are pin-jointed at the ends, and the loads apply on the joints so that the members are either stretched or compressed but not bent.

(C) The resultant forces acting on a beam cross section can be summarized into three components: direct force F, bending moment M, and shear force V. Only the direct force and the bending moment contribute to the axial stress. The axial stress caused by the direct force is called the direct stress $\sigma = F/A$, where A is the cross-sectional area. The direct stress can be tensile or compressive. The axial stress caused by the bending moment is called the bending stress $\sigma_b = My/I$, where y is the distance from the neutral axis to the point of concern, and I is the moment of inertia of the cross section.

(D) The maximum or minimum bending stress occurs at the either top or bottom beam edge, depending on the direction of the bending moment. The maximum bending stress is always tensile while the minimum bending stress is always compressive.

(E) The maximum combined stress is the superposition of the direct stress plus the maximum bending stress. The minimum combined stress is the superposition of the direct stress plus the minimum bending stress.

(F) A text file describing the coordinates of points. The file has 5 columns, or fields: (a) group number, (b) ID number, (c) X-coordinate, (d) Y-coordinate, and (e) Z-coordinate. The group number and the ID number are arbitrary and they together uniquely identify a point from others.

7.4-2 Additional Workbench Exercises

Convergence Study of Beam Elements

Generate a convergence curve like the one in 7.1-16[1] (page 274) for the flexible gripper.

Mesh Each Member with More Elements

In 7.2-8 (page 281), we mentioned that the default settings of **Mesh** would mesh the model into 205 beam elements and would result exactly the same solution as 37 elements. Verify this. Type 0 for **Element Size** to set to the default element size.

Pin-Jointed or Rigid-Jointed?

In 7.2-1 (page 275), we mentioned that if the members of a truss structure are slender enough, there is no essential difference between pin-jointed model and rigid-jointed model. Verify this. You may use the 3D truss example (Section 7.2) or select a simple truss structure from any of your Engineering Mechanics textbook. Solve the problem with the pin-joint assumption, and then solve the problem again using a rigid-jointed model. Compare the results and draw your conclusions.

To model a pin-jointed structure in Workbench, you need to either specify revolute joints between the members or utilize **End Release**[Ref 1] feature in **Mechanical** [1].

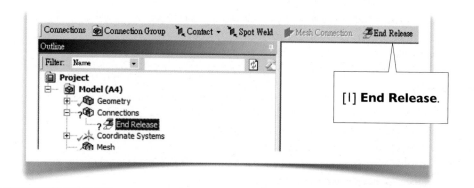

[1] **End Release**.

References

1. ANSYS Documentation//Mechanical Applications//Mechanical User's Guide//Setting Connections//End Releases

Chapter 8
Optimization

Design Process

A typical engineering design process involves several steps as shown [1-6]. First, the engineer sets up an initial design [1]. The design is simulated [2] and the performance is evaluated [3]. If the performance is satisfied, then the design is accepted [6], otherwise the design must be improved somehow [5]. It is an iterative process. After several iterations, the design hopefully converges to an optimal design.

Now, let's look at the way of improving a design [5]. Conventionally, the way of improving a design relies on engineers' experience. The consequence is that the process is costly and often does not converge to an optimal design.

This chapter focuses on a capability of Workbench, which automates the entire design process [1-6]. Workbench provides several algorithms for improving a design [5].

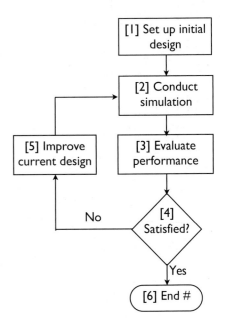

Purpose of This Chapter

This chapter is strategically arranged in the middle of the book because it is a suitable time to learn optimization techniques before we look into more advanced topics. The idea is that the students should be aware as early as possible of the optimization capabilities provided by Workbench so that they can use the features when doing their exercises.

This chapter provides two hands-on examples, which cover main ideas of optimization. After these exercises, the students should be able to use these capabilities on their own.

About Each Section

Section 8.1 revisits the flexible gripper, introduced in Section 7.1. This time, we want to improve the shape of the gripper to achieve an optimal geometric advantage. Section 8.2 provides an additional exercise, in which the triangular plate, introduced in Section 2.2 and simulated in Section 3.1, is to be improved to achieve a minimal weight design.

Section 8.1

Flexible Gripper[Ref 1]

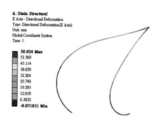

8.1-1 About the Flexible Gripper

In Section 7.1, we created a model for the flexible gripper [1], and performed a static structural simulation to assess the GA (geometric advantage, defined as the ratio of the horizontal output displacement to the vertical input displacement) of the design. For that particular design, the GA value is 1.04 (7.1-15, page 273). In this section, we want to improve the GA value by adjusting the shape of the flexible gripper.

The shape of the gripper is defined by 7 key points [2]. Positions of the P1, P4, P5, and P7 are fixed (cannot be changed) due to constraints imposed by some functional requirements. The positions of the P2, P3, and P6 are free to be adjusted.

The idea is to fix the X-coordinates of these three points and adjust their Y-coordinates to achieve a better GA value. The allowable adjustment ranges for the Y-coordinates are ±10 mm for P2, ±20 mm for P3, and ±5 mm for P6. There is a limitation on the maximum stress, either tensile or compresive, which should not exceed 15 MPa for a reliability consideration.

In Workbench, **DesignXplorer** is used to complete the task.

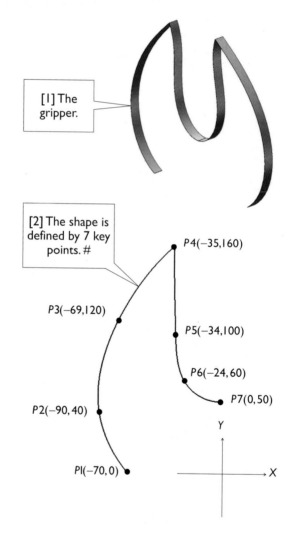

[1] The gripper.

[2] The shape is defined by 7 key points. #

P4(−35,160)

P3(−69,120)

P5(−34,100)

P6(−24,60)

P7(0,50)

P2(−90,40)

P1(−70,0)

Y

X

Keep the Number of Design Variables Minimum

The design parameters that can be changed to improve a design are called *design variables* (in this case, Y-coordinates of P2, P3, and P6). The overall computing time of an optimization process will increase dramatically as the number of design variables increases. Therefore it is important to keep the number of design variables as few as possible.

In this case, we could have chosen both X-coordinates and Y-coordinates of P2, P3, P6 as design variables. That would increase the number of design variables from 3 to 6, without any advantages.

8.1-2 Resume the Project **Gripper**

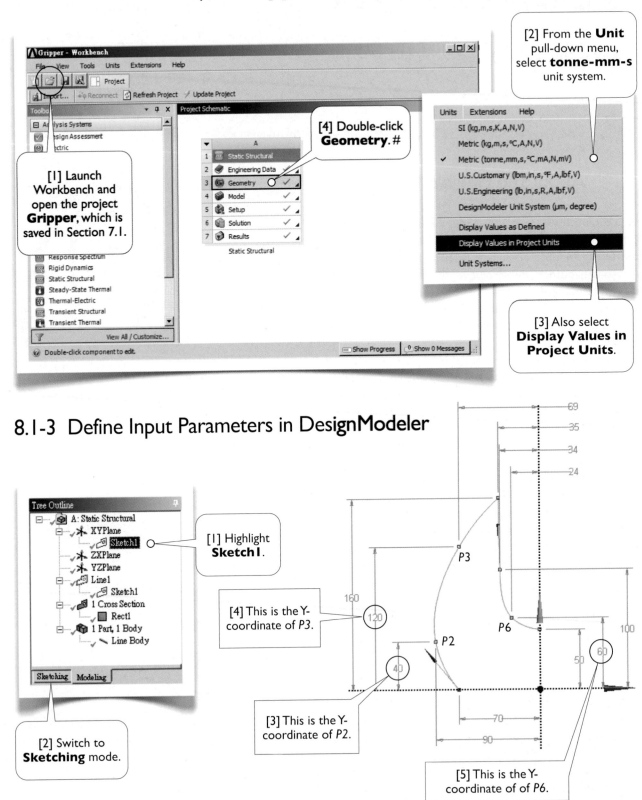

[2] From the Unit pull-down menu, select **tonne-mm-s** unit system.

[4] Double-click Geometry. #

[1] Launch Workbench and open the project Gripper, which is saved in Section 7.1.

[3] Also select Display Values in Project Units.

8.1-3 Define Input Parameters in DesignModeler

[1] Highlight Sketch1.

[4] This is the Y-coordinate of P3.

[2] Switch to Sketching mode.

[3] This is the Y-coordinate of P2.

[5] This is the Y-coordinate of of P6.

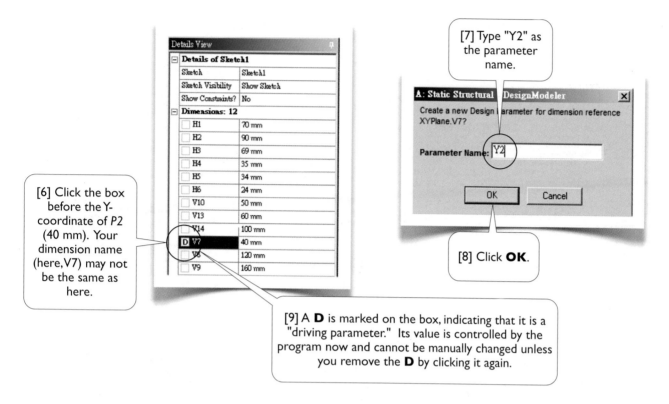

[7] Type "Y2" as the parameter name.

[6] Click the box before the Y-coordinate of *P2* (40 mm). Your dimension name (here, V7) may not be the same as here.

[8] Click **OK**.

[9] A **D** is marked on the box, indicating that it is a "driving parameter." Its value is controlled by the program now and cannot be manually changed unless you remove the **D** by clicking it again.

Repeat steps [6-8] for the Y-coordinates of *P3* and *P6* [10, 11]. Remember that your dimension names may not be the same as they appear here.

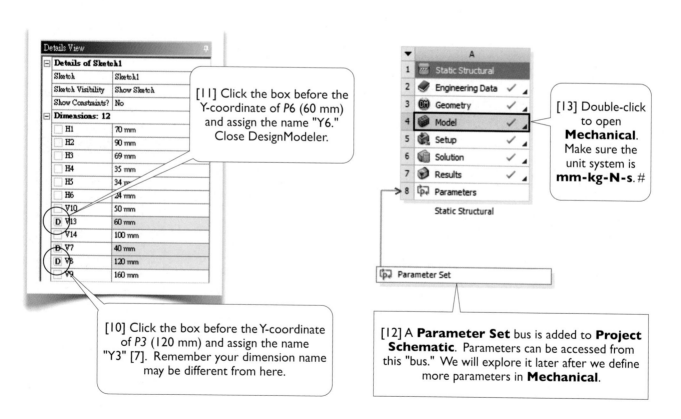

[11] Click the box before the Y-coordinate of *P6* (60 mm) and assign the name "Y6." Close DesignModeler.

[13] Double-click to open **Mechanical**. Make sure the unit system is **mm-kg-N-s**. #

[10] Click the box before the Y-coordinate of *P3* (120 mm) and assign the name "Y3" [7]. Remember your dimension name may be different from here.

[12] A **Parameter Set** bus is added to **Project Schematic**. Parameters can be accessed from this "bus." We will explore it later after we define more parameters in **Mechanical**.

8.1-4 Define Output Parameters in **Mechanical**

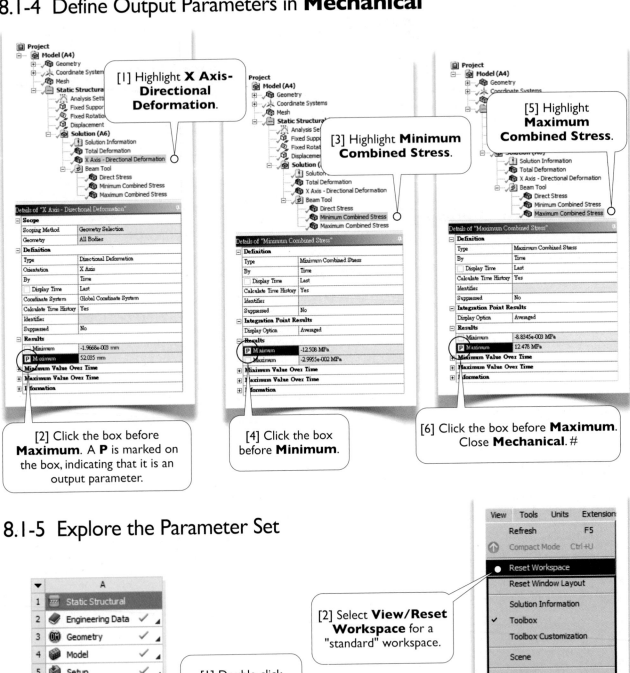

[1] Highlight **X Axis-Directional Deformation**.

[3] Highlight **Minimum Combined Stress**.

[5] Highlight **Maximum Combined Stress**.

[2] Click the box before **Maximum**. A **P** is marked on the box, indicating that it is an output parameter.

[4] Click the box before **Minimum**.

[6] Click the box before **Maximum**. Close **Mechanical**. #

8.1-5 Explore the Parameter Set

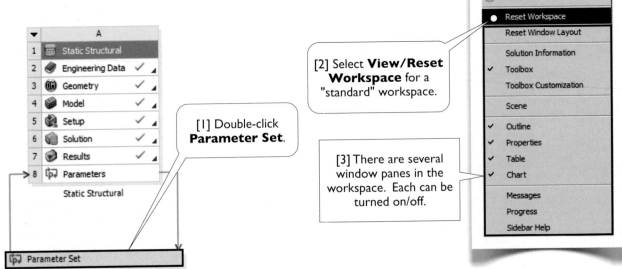

[1] Double-click **Parameter Set**.

[2] Select **View/Reset Workspace** for a "standard" workspace.

[3] There are several window panes in the workspace. Each can be turned on/off.

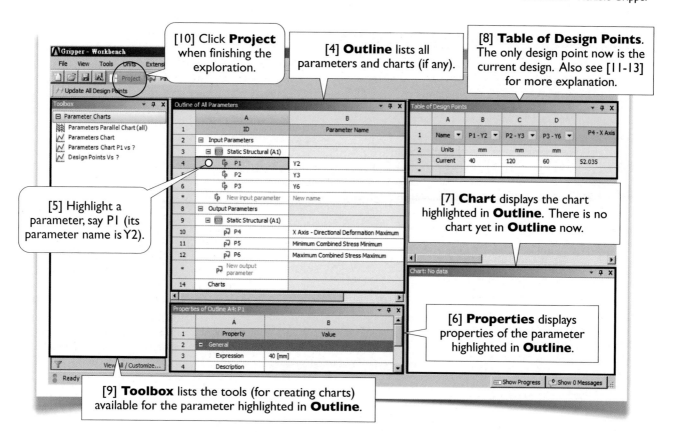

[10] Click **Project** when finishing the exploration.

[4] **Outline** lists all parameters and charts (if any).

[8] **Table of Design Points**. The only design point now is the current design. Also see [11-13] for more explanation.

[5] Highlight a parameter, say P1 (its parameter name is Y2).

[7] **Chart** displays the chart highlighted in **Outline**. There is no chart yet in **Outline** now.

[6] **Properties** displays properties of the parameter highlighted in **Outline**.

[9] **Toolbox** lists the tools (for creating charts) available for the parameter highlighted in **Outline**.

Notations Used in This Section

Workbench refers the parameters as P1, P2, P3 (non-italic), etc. [4]. In our case of flexible gripper, we use *P2*, *P3*, and *P6* (italic) for three control points of the shape. Be careful not to confuse with these two sets of notations.

Input Parameters and Output Parameters

In optimization jargon, input parameters are also called *design variables*, and output parameters are also called *state variables*. Initial values of input parameters are set up by the user, and Workbench subsequently changes these values to improve the design. Values of output parameters are calculated from the simulation applications (e.g., **Mechanical**). In our case, the input parameters are Y2 (P1), Y3 (P2), and Y6 (P3), and the output parameters are the maximum horizontal displacement (P4), the minimum combined stress (P5), and the maximum combined stress (P6). Note that the input parameters are not necessarily defined in DesignModeler. For example, the loads defined in **Mechanical** could be defined as input parameters.

Design Space

The space spanned by the input parameters is called the *design space*. In our case, the space spanned by parameters Y2, Y3, and Y6 is the design space, which is a three-dimensional space. The optimization process can be thought of a process of searching an optimal point in the design space.

Design Points

Any point in the design space is called a *design point*. **Table of Design Points** [8] lists a series of design points with which the simulations have been carried out. In our case, we have only one design point: Y2 = 40, Y3 = 120, Y6 = 60. This design point is now called *current design* [8]. **Current Design** is a design point of which the data are stored in **Mechanical** database.

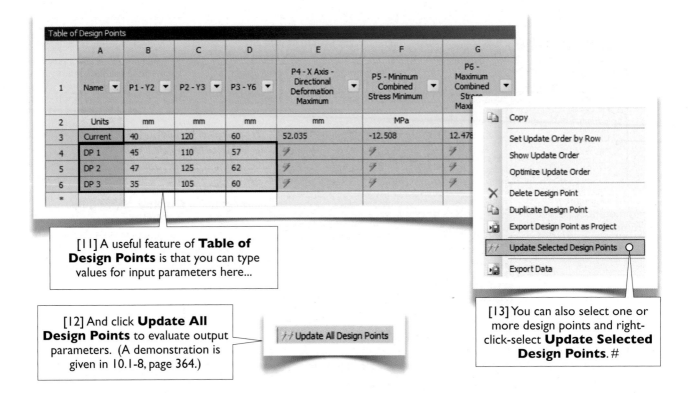

	A	B	C	D	E	F	G
1	Name ▾	P1 - Y2 ▾	P2 - Y3 ▾	P3 - Y6 ▾	P4 - X Axis - Directional Deformation Maximum ▾	P5 - Minimum Combined Stress Minimum ▾	P6 - Maximum Combined Stress Maxi... ▾
2	Units	mm	mm	mm	mm	MPa	
3	Current	40	120	60	52.035	-12.508	12.478
4	DP 1	45	110	57	⚡	⚡	⚡
5	DP 2	47	125	62	⚡	⚡	⚡
6	DP 3	35	105	60	⚡	⚡	⚡
*							

Right-click menu:
- Copy
- Set Update Order by Row
- Show Update Order
- Optimize Update Order
- ✕ Delete Design Point
- Duplicate Design Point
- Export Design Point as Project
- ⚡ **Update Selected Design Points** ○
- Export Data

[11] A useful feature of **Table of Design Points** is that you can type values for input parameters here...

[12] And click **Update All Design Points** to evaluate output parameters. (A demonstration is given in 10.1-8, page 364.)

⚡ Update All Design Points

[13] You can also select one or more design points and right-click-select **Update Selected Design Points**. #

8.1-6 Create a **Direct Optimization** System

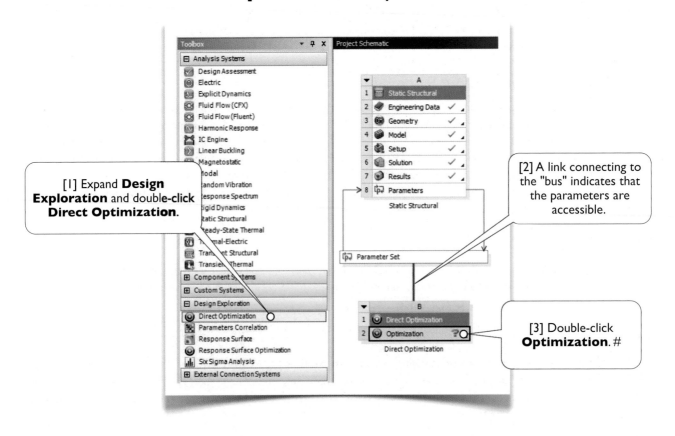

[1] Expand **Design Exploration** and double-click **Direct Optimization**.

[2] A link connecting to the "bus" indicates that the parameters are accessible.

[3] Double-click **Optimization**. #

309

8.1-7 Set Up Optimization Method[Ref 2]

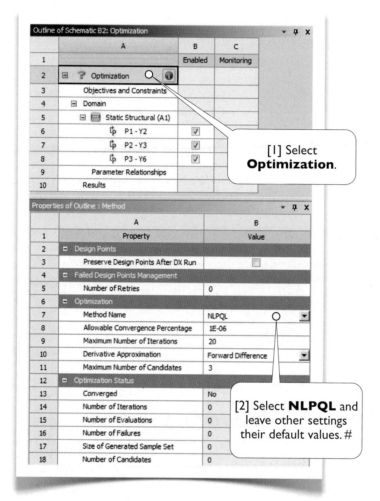

[1] Select **Optimization**.

[2] Select **NLPQL** and leave other settings their default values. #

NLPQL

The method (Nonlinear Programming by Quadratic Lagrangian) is a gradient-base algorithm to provide a local optimization result. It supports single objective and is limited to continuous input parameters.

Here, we select this method simply because it is one of classical optimization methods. After completing this exercise, you are encouraged to try other methods. Other methods available in Workbench are briefly described as follows.

Adaptive Single-Objective

It is also a gradient-based algorithm, providing a global optimization result. It supports single objective and is limited to continuous and manufacturable input parameters.

Adaptive Multiple-Objective

It supports multiple objectives and aims at finding the global optimum. It is also limited to continuous and manufacturable input parameters.

MISQP

The method (Mixed-Integer Sequential Qudratic Programing) solves mixed-integer nonlinear programming problems by a modified SQP method. Under the assumption that integer variables have a smooth influence on the model functions.

Screening

It uses a simple approach based on sampling and sorting. It supports mutiple objectives as well as all type of input parameters. It is usually used for preliminary design; e.g., global behavior study.

MOGA

The method (Multi-Objective Genetic Algorithm) supports multiple objectives and aims at finding global optimum.

8.1-8 Set Up Objective and Constraints

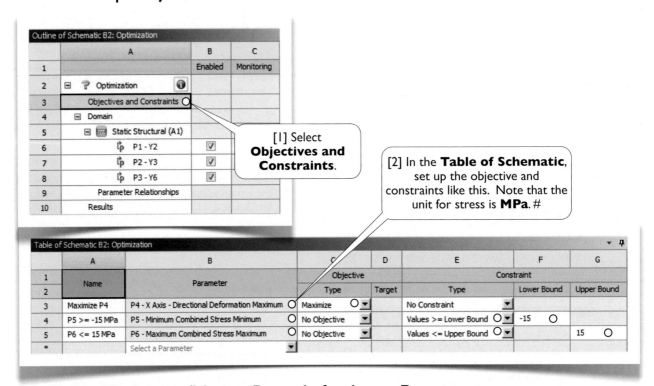

[1] Select **Objectives and Constraints**.

[2] In the **Table of Schematic**, set up the objective and constraints like this. Note that the unit for stress is **MPa**. #

8.1-9 Set Up Lower/Upper Bounds for Input Parameters

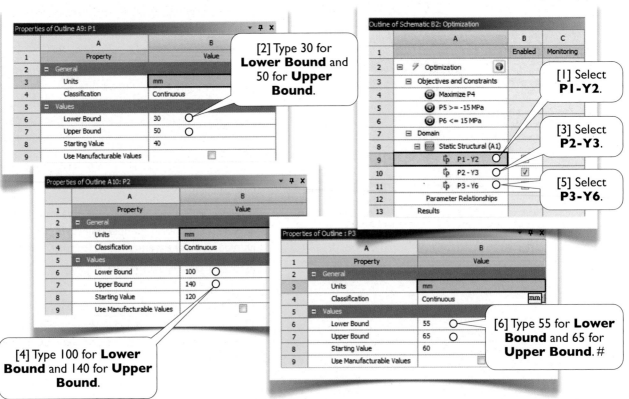

[2] Type 30 for **Lower Bound** and 50 for **Upper Bound**.

[1] Select **P1-Y2**.

[3] Select **P2-Y3**.

[5] Select **P3-Y6**.

[6] Type 55 for **Lower Bound** and 65 for **Upper Bound**. #

[4] Type 100 for **Lower Bound** and 140 for **Upper Bound**.

8.1-10 Run the Optimization

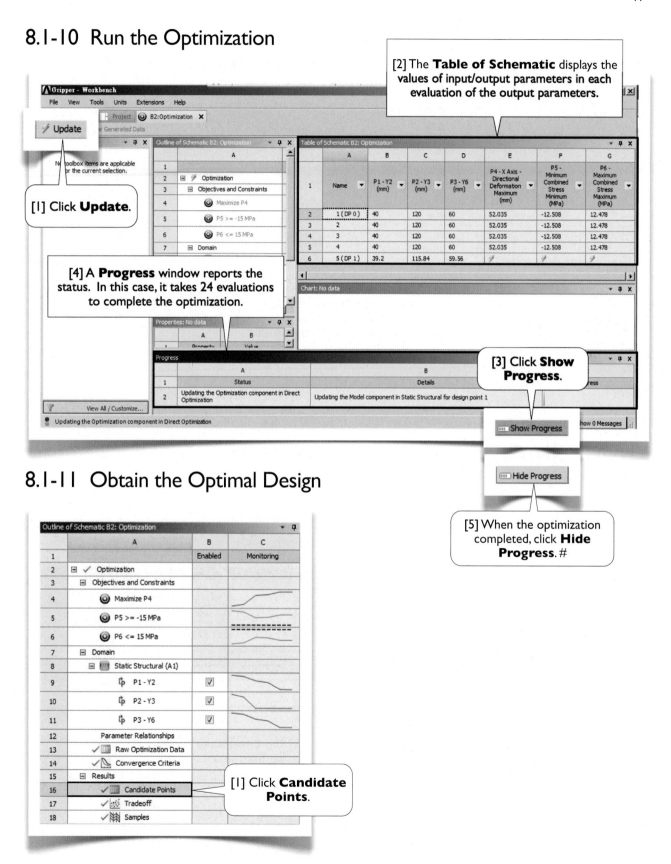

[2] The **Table of Schematic** displays the values of input/output parameters in each evaluation of the output parameters.

[1] Click **Update**.

[4] A **Progress** window reports the status. In this case, it takes 24 evaluations to complete the optimization.

[3] Click **Show Progress**.

8.1-11 Obtain the Optimal Design

[5] When the optimization completed, click **Hide Progress**. #

[1] Click **Candidate Points**.

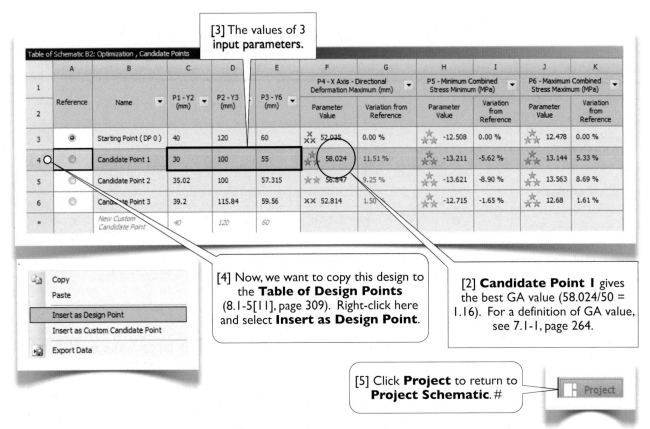

[3] The values of 3 input parameters.

[4] Now, we want to copy this design to the **Table of Design Points** (8.1-5[11], page 309). Right-click here and select **Insert as Design Point**.

[2] **Candidate Point 1** gives the best GA value (58.024/50 = 1.16). For a definition of GA value, see 7.1-1, page 264.

[5] Click **Project** to return to **Project Schematic**. #

8.1-12 Set the Optimal Design as Current Design

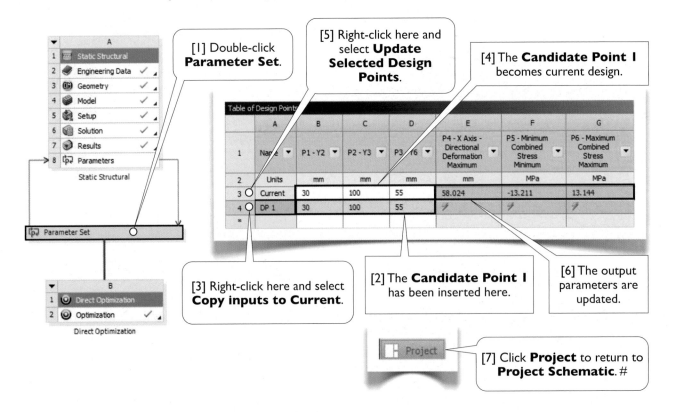

[1] Double-click **Parameter Set**.

[5] Right-click here and select **Update Selected Design Points**.

[4] The **Candidate Point 1** becomes current design.

[3] Right-click here and select **Copy inputs to Current**.

[2] The **Candidate Point 1** has been inserted here.

[6] The output parameters are updated.

[7] Click **Project** to return to **Project Schematic**. #

313

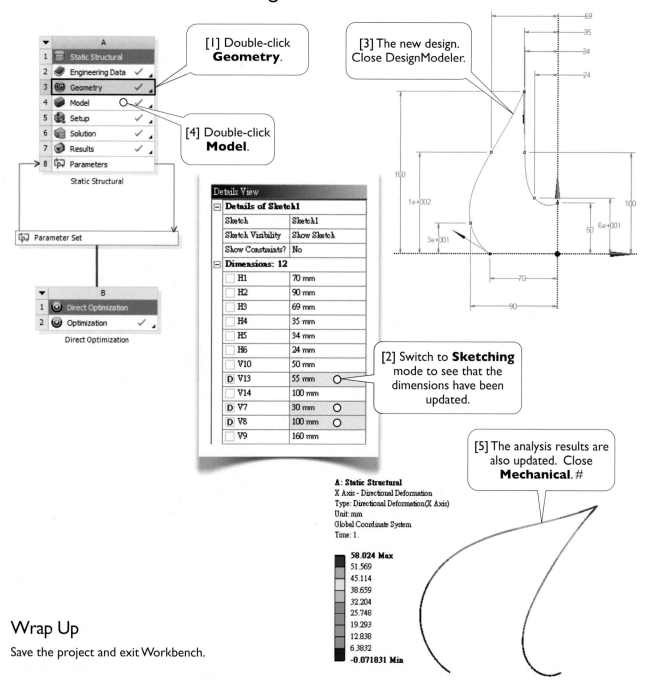

8.1-13 View the Current Design

[1] Double-click **Geometry**.

[3] The new design. Close DesignModeler.

[4] Double-click **Model**.

Static Structural

Parameter Set

Direct Optimization

Details View

Details of Sketch1	
Sketch	Sketch1
Sketch Visibility	Show Sketch
Show Constraints?	No
Dimensions: 12	
H1	70 mm
H2	90 mm
H3	69 mm
H4	35 mm
H5	34 mm
H6	24 mm
V10	50 mm
D V13	55 mm
V14	100 mm
D V7	30 mm
D V8	100 mm
V9	160 mm

[2] Switch to **Sketching** mode to see that the dimensions have been updated.

[5] The analysis results are also updated. Close **Mechanical**. #

A: Static Structural
X Axis - Directional Deformation
Type: Directional Deformation(X Axis)
Unit: mm
Global Coordinate System
Time: 1.

58.024 Max
51.569
45.114
38.659
32.204
25.748
19.293
12.838
6.3832
-0.071831 Min

Wrap Up

Save the project and exit Workbench.

Reference

1. Chao-Chieh Lan and Yung-Jen Cheng, 2008, "Distributed Shape Optimization of Compliant Mechanisms Using Intrinsic Functions," *ASME Journal of Mechanical Design*, Vol. 130, 072304.
2. ANSYS Documentation//Workbench//Design Exploration User's Guide//Using Goal Driven Optimization Methods

Section 8.2

Triangular Plate

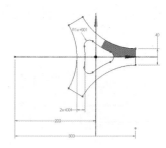

8.2-1 About the Triangular Plate

In Section 3.1, we generated a 2D solid model and performed a 2D static simulation [1-4] for the triangular plate introduced in Section 2.2.

The plate is made of steel with an allowable stress of 100 MPa. According to the simulation results in Section 3.1, the initial design gives a maximum stress of about 57 MPa (3.1-13[2], page 117), much less than the allowable stress. That means the initial design is over-designed: the material can be cut down somehow. In this section, we want to redesign the triangular plate to reduce the amount of material.

The design variables (input parameters) are the width of the bridges [3] and the radius of the fillets [4]. For the width, the initial design is 30 mm and its allowable range is 20-30 mm. For the radius, the initial design is 10 mm and its allowable range is 5-15 mm. The project unit system used in this exercise is **tonne-mm-s**.

8.2-2 Resume Project **Triplate**

Launch Workbench, open the project **Triplate**, which was saved in Section 3.1. Select **tonne-mm-s** as the project units.

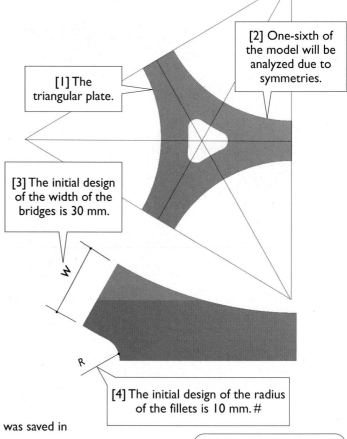

[1] The triangular plate.

[2] One-sixth of the model will be analyzed due to symmetries.

[3] The initial design of the width of the bridges is 30 mm.

[4] The initial design of the radius of the fillets is 10 mm. #

[1] Double-click to start up DesignModeler. #

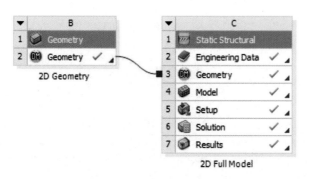

3D Geometry

2D Geometry

2D Full Model

2D Symmetric Model

8.2-3 Define Input Parameters in DesignModeler

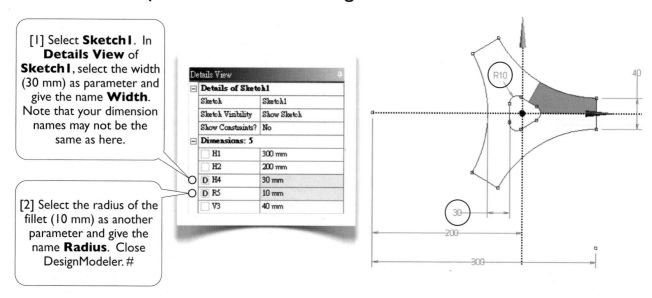

[1] Select **Sketch1**. In **Details View** of **Sketch1**, select the width (30 mm) as parameter and give the name **Width**. Note that your dimension names may not be the same as here.

[2] Select the radius of the fillet (10 mm) as another parameter and give the name **Radius**. Close DesignModeler. #

8.2-4 Define Output Parameters in **Mechanical**

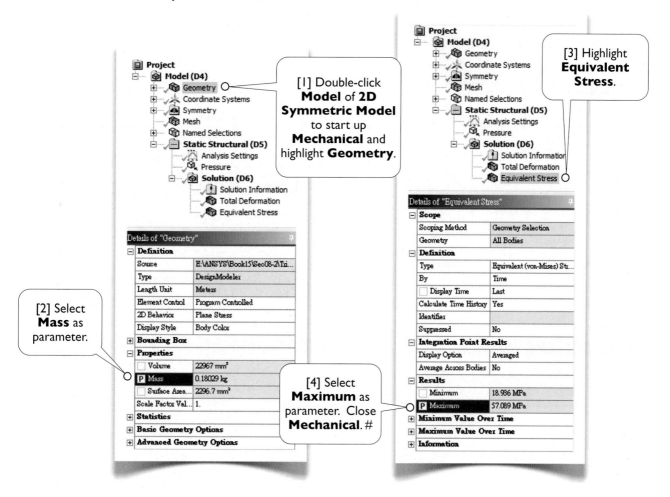

[1] Double-click **Model** of **2D Symmetric Model** to start up **Mechanical** and highlight **Geometry**.

[2] Select **Mass** as parameter.

[3] Highlight **Equivalent Stress**.

[4] Select **Maximum** as parameter. Close **Mechanical**. #

8.2-5 Create a Direct Optimization System

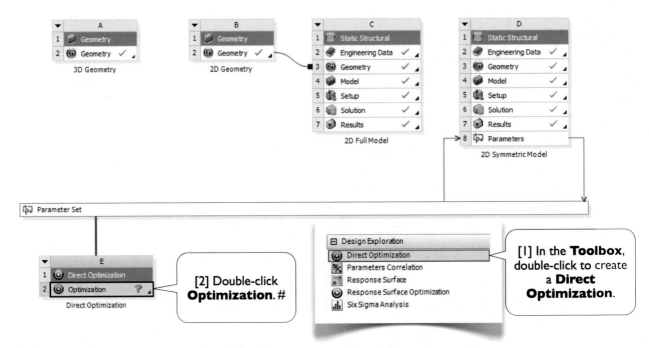

[1] In the **Toolbox**, double-click to create a **Direct Optimization**.

[2] Double-click **Optimization**. #

8.2-6 Define and Run the Optimization Problem

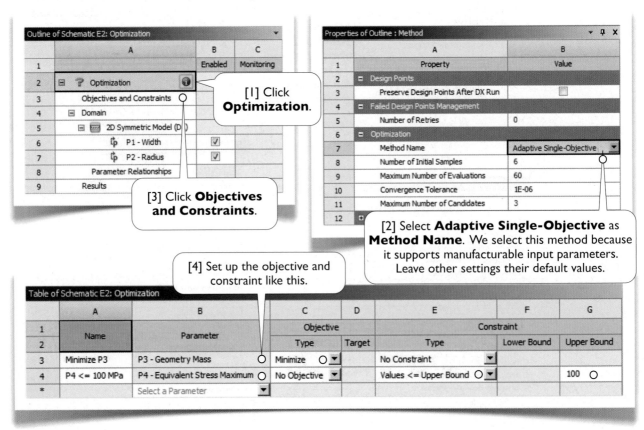

[1] Click **Optimization**.

[3] Click **Objectives and Constraints**.

[2] Select **Adaptive Single-Objective** as **Method Name**. We select this method because it supports manufacturable input parameters. Leave other settings their default values.

[4] Set up the objective and constraint like this.

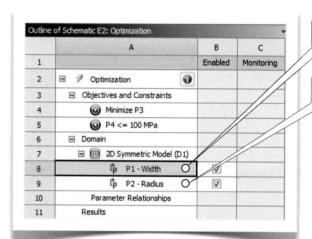

[5] Click **P1-Width**.

[8] Click **P2-Radius**.

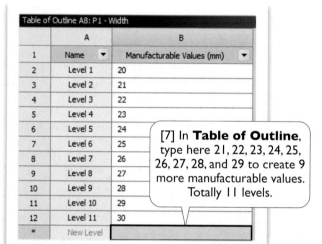

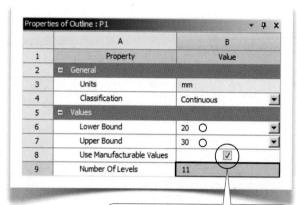

[6] Type 20 (mm) for **Lower Bound** and 30 (mm) for **Upper Bound**. Turn on **Use Manufacturable Values**.

[7] In **Table of Outline**, type here 21, 22, 23, 24, 25, 26, 27, 28, and 29 to create 9 more manufacturable values. Totally 11 levels.

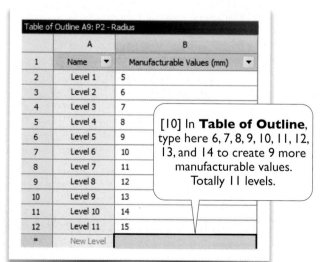

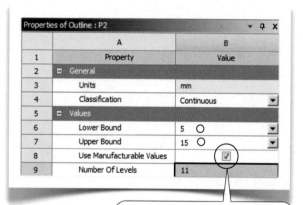

[9] Type 5 (mm) as **Lower Bound** and 15 (mm) as **Upper Bound**. Turn on **Use Manufacturable Values**.

[10] In **Table of Outline**, type here 6, 7, 8, 9, 10, 11, 12, 13, and 14 to create 9 more manufacturable values. Totally 11 levels.

[11] Click **Update**. It takes 10 evaluations to complete the optimization. #

8.2-7 View the Optimal Design

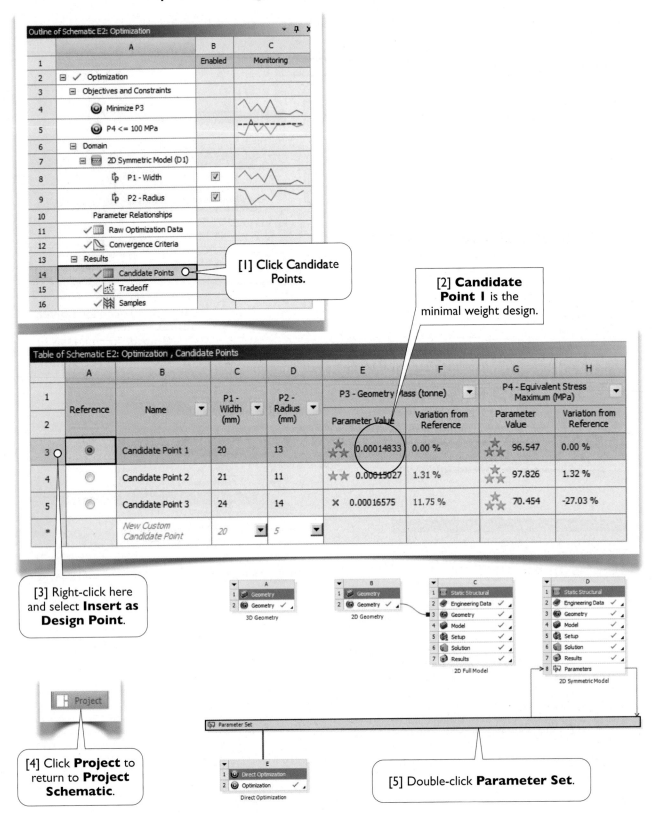

[7] Right-click here and select **Update Selected Design Points**.

[6] Right-click here and select **Copy inputs to Current**.

	A	B	C	D	E
1	Name ▼	P1 - Width ▼	P2 - Radius ▼	P3 - Geometry Mass ▼	P4 - Equivalent Stress Maximum ▼
2	Units	mm	mm	tonne	MPa
3	Current	20	13	0.00014833	96.547
4	DP 1	20	13	⚡	⚡
*					

Table of Design Points

Project

[8] Click **Project** to return to **Project Schematic**.

[9] Double-click **Geometry**.

Details View
Details of Sketch1	
Sketch	Sketch1
Sketch Visibility	Show Sketch
Show Constraints?	No
Dimensions: 5	
□ H1	300 mm
□ H2	200 mm
D H4	20 mm ○
D R5	13 mm ○
□ V3	40 mm

[10] Switch to **Sketching** mode to see the dimensions updated.

	D
1	Static Structural
2	Engineering Data ✓
3	Geometry ○ ✓
4	Model ○ ✓
5	Setup ✓
6	Solution ✓
7	Results ✓
8	Parameters

2D Symmetric Model

[12] Double-click **Model**.

[11] The new design. Close DesignModeler.

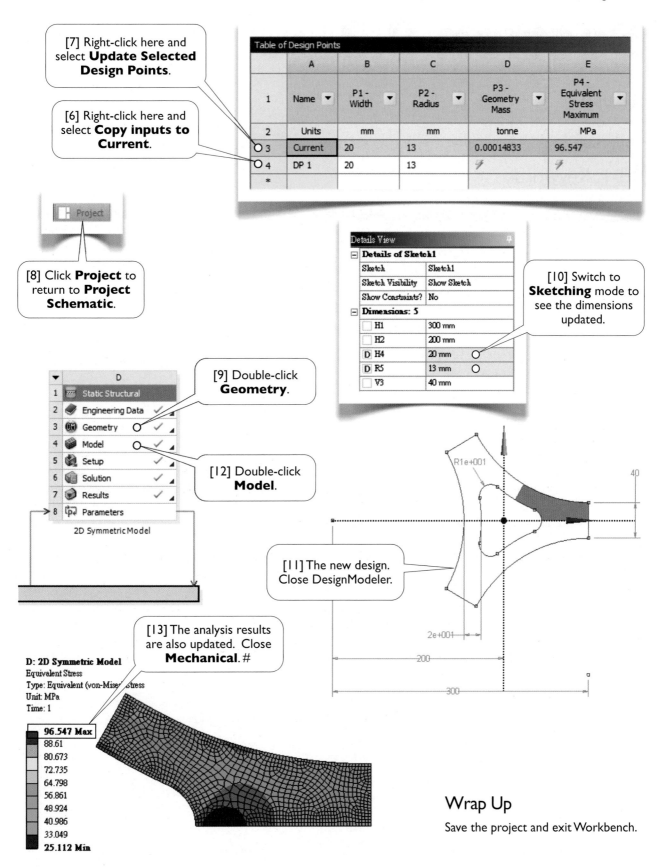

R1e+001

40

2e+001

200

300

[13] The analysis results are also updated. Close **Mechanical**. #

D: 2D Symmetric Model
Equivalent Stress
Type: Equivalent (von-Mises) Stress
Unit: MPa
Time: 1

96.547 Max
88.61
80.673
72.735
64.798
56.861
48.924
40.986
33.049
25.112 Min

Wrap Up

Save the project and exit Workbench.

Section 8.3

Review

8.3-1 Keywords

Choose a letter for each keyword from the list of descriptions

1. (   ) Design Points
2. (   ) Design Space
3. (   ) Input Parameters and Output Parameters
4. (   ) NLPQL

Answers:

1. (C)   2. (B)   3. (A)   4. (D)

List of Descriptions

(A)   Initial values of input parameters, also called design variables, are set up by the user and subsequently updated by Workbench. Values of output parameters, also called state variables, are calculated from the simulation applications.

(B)   The space spanned by the input parameters, or design variables.

(C)   In a broader sense, any point in the design space is called a design point. In **DesignXplorer**, design points are those with which the simulations have been carried out.

(D)   Short for nonlinear programming by quadratic Lagrangian. A method of finding an optimal design. In each iteration, it approaches the problem using a quadratic polynomial and finds the optimal design in the subproblem described by the quadratic polynomial. The process is repeated until the optimal design is found. The method deals with only single-objective problems.

Chapter 9
Meshing

So far, we haven't emphasized much on meshing because, for linear static problems, we usually can solve them with global mesh controls (e.g., **Relevance Center**, **Relevance**, and **Element Size**) and obtain solutions which are acceptable both in computing time and accuracy. For the rest of the book, we will be dealing with dynamic and nonlinear problems. Solutions of dynamic and nonlinear problems are sensitive to meshing quality. With poor mesh quality, a simulation may end up with solutions of poor accuracy or even run into convergence problems. Dynamic and nonlinear problems require much of computational resource, and poor mesh quality may aggravate the situation and result a lengthy computing time. For nonlinear problems, it is possible to reduce much of runtime by improving the mesh quality (because they converge easier). On contrast, it is possible that a nonlinear solution fails to converge just because of the poor mesh quality.

Purpose of This Chapter

This chapter introduces meshing methods provided by Workbench and demonstrates how to use them with hands-on examples. One of the mesh quality metrics, called *skewness*, is also introduced and used as a measure of mesh quality in this chapter.

About Each Section

Using the pneumatic finger example, Section 9.1 provides a step-by-step exercise to introduce some important concepts of meshing technologies. Section 9.2 uses a more involved model, the cover of pressure cylinder, to provide more exercises on meshing technologies. Section 9.3, a sequel of Section 3.5, studies 3D elements convergence behaviors. We postpone the study of 3D elements convergence until now because we need more meshing techniques to control the meshing density.

Section 9.1

Pneumatic Fingers[1]

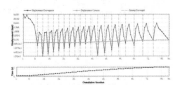

9.1-1 About the Pneumatic Fingers

Simulation of a pneumatic finger has been previewed in Section 1.1. In this section, we will walk through each step. Besides the information described in 1.1-1 (pages 7-8), additional geometric details are shown in the figure below. Due to the symmetry [1], we will model only half of the finger. The **mm-kg-N-s** unit system is used in this section.

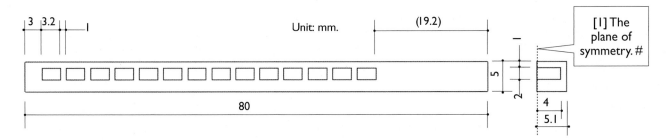

9.1-2 Start Up and Prepare Material Properties

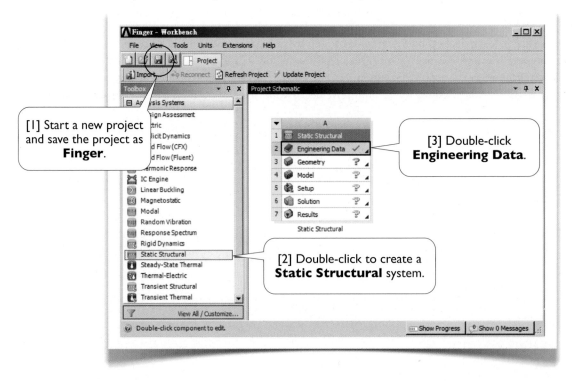

[1] Start a new project and save the project as **Finger**.

[2] Double-click to create a **Static Structural** system.

[3] Double-click **Engineering Data**.

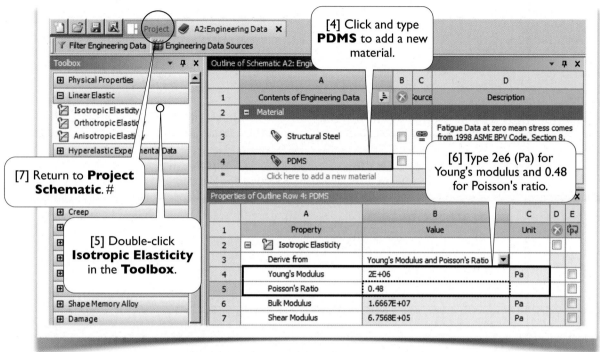

[4] Click and type **PDMS** to add a new material.

[6] Type 2e6 (Pa) for Young's modulus and 0.48 for Poisson's ratio.

[7] Return to **Project Schematic**. #

[5] Double-click **Isotropic Elasticity** in the **Toolbox**.

9.1-3 Create Geometry

Start up DesignModeler; select **Millimeter** as length unit and make sure **Auto Constraints** are turned on.

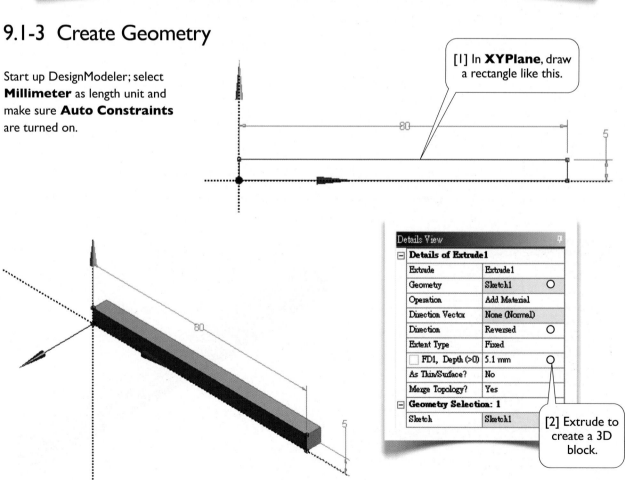

[1] In **XYPlane**, draw a rectangle like this.

[2] Extrude to create a 3D block.

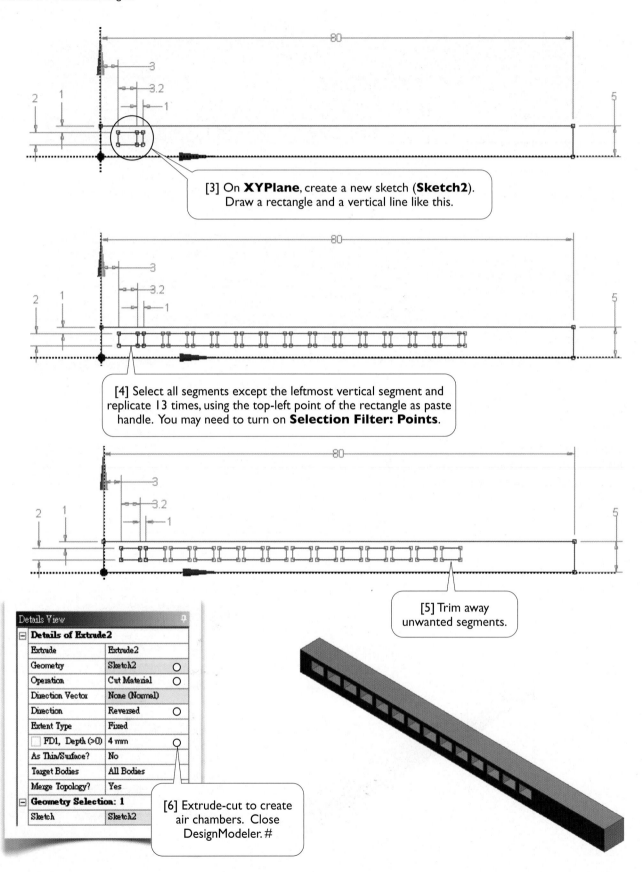

[3] On **XYPlane**, create a new sketch (**Sketch2**). Draw a rectangle and a vertical line like this.

[4] Select all segments except the leftmost vertical segment and replicate 13 times, using the top-left point of the rectangle as paste handle. You may need to turn on **Selection Filter: Points**.

[5] Trim away unwanted segments.

Details of Extrude2	
Extrude	Extrude2
Geometry	Sketch2
Operation	Cut Material
Direction Vector	None (Normal)
Direction	Reversed
Extent Type	Fixed
FD1, Depth (>0)	4 mm
As Thin/Surface?	No
Target Bodies	All Bodies
Merge Topology?	Yes
Geometry Selection: 1	
Sketch	Sketch2

[6] Extrude-cut to create air chambers. Close DesignModeler. #

9.1-4 Assign Material

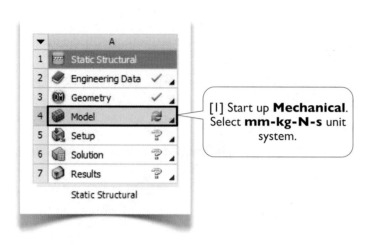

[1] Start up **Mechanical**. Select **mm-kg-N-s** unit system.

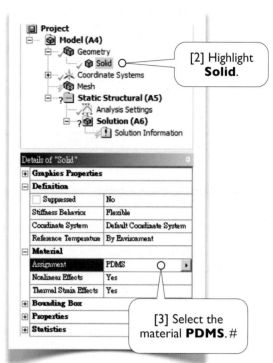

[2] Highlight **Solid**.

[3] Select the material **PDMS**. #

9.1-5 Set Up Environment Conditions

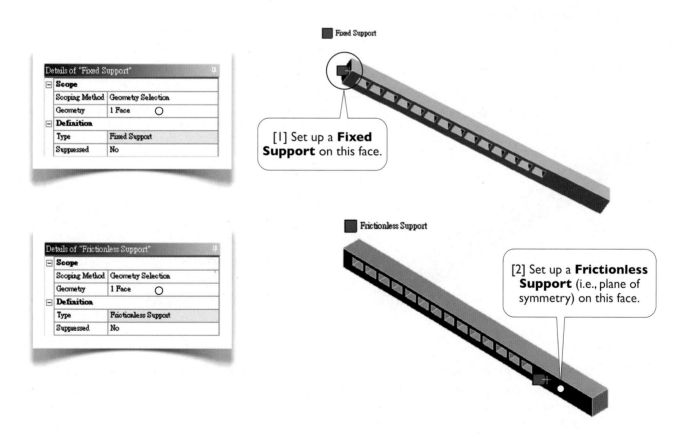

Fixed Support

[1] Set up a **Fixed Support** on this face.

Frictionless Support

[2] Set up a **Frictionless Support** (i.e., plane of symmetry) on this face.

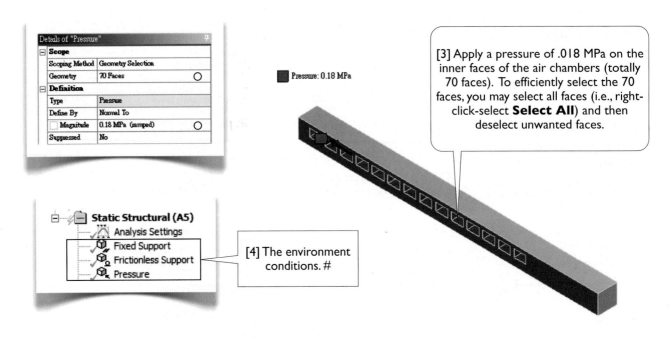

[3] Apply a pressure of .018 MPa on the inner faces of the air chambers (totally 70 faces). To efficiently select the 70 faces, you may select all faces (i.e., right-click-select **Select All**) and then deselect unwanted faces.

[4] The environment conditions. #

9.1-6 Mesh with Default Settings

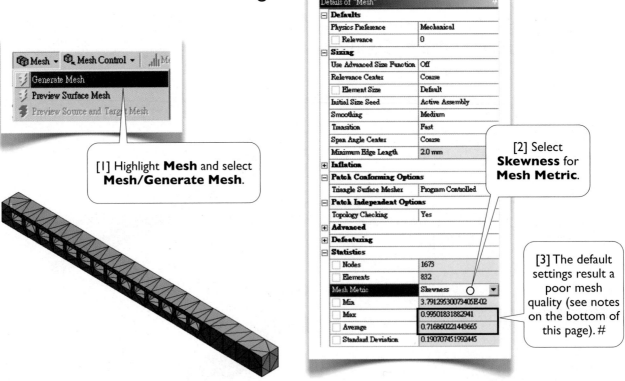

[1] Highlight **Mesh** and select **Mesh/Generate Mesh**.

[2] Select **Skewness** for **Mesh Metric**.

[3] The default settings result a poor mesh quality (see notes on the bottom of this page). #

Skewness[Ref 2, 3]

Skewness, a measure of mesh quality, can be calculated for each element according to its geometry. Definition of skewness can be found in the on-line documentation[Ref 2]. For now, all you need to know is that it is a value ranging from 0 to 1, the smaller the better, and, as a guideline, elements of skewness of more than 0.95 are considered unacceptable.

9.1-7 Improve Mesh Quality

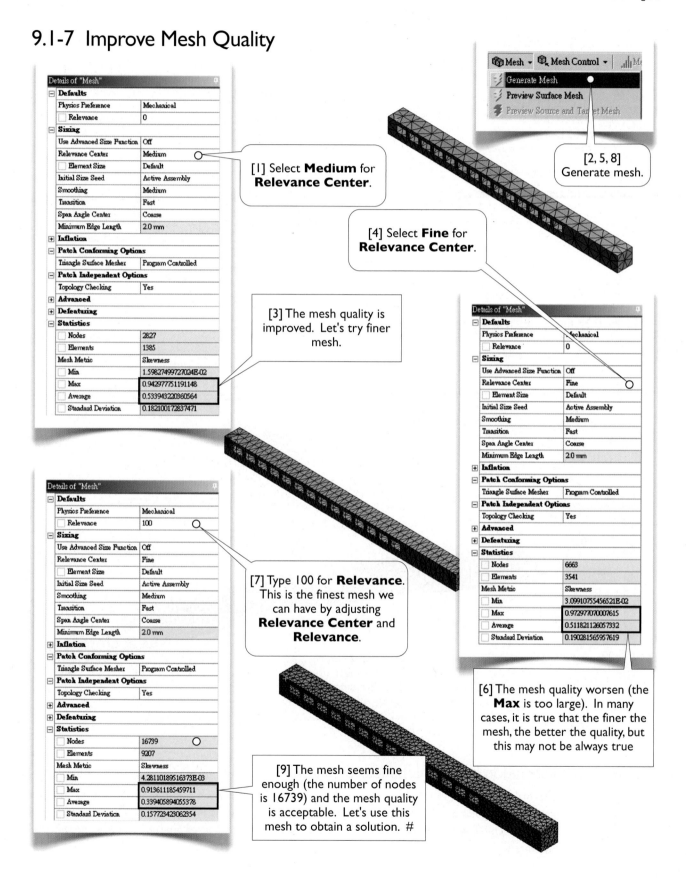

[1] Select **Medium** for **Relevance Center**.

[2, 5, 8] Generate mesh.

[4] Select **Fine** for **Relevance Center**.

[3] The mesh quality is improved. Let's try finer mesh.

[7] Type 100 for **Relevance**. This is the finest mesh we can have by adjusting **Relevance Center** and **Relevance**.

[6] The mesh quality worsen (the **Max** is too large). In many cases, it is true that the finer the mesh, the better the quality, but this may not be always true

[9] The mesh seems fine enough (the number of nodes is 16739) and the mesh quality is acceptable. Let's use this mesh to obtain a solution. #

328

9.1-8 Set Up Solution Branch

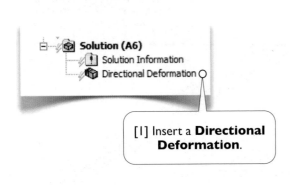

[1] Insert a **Directional Deformation**.

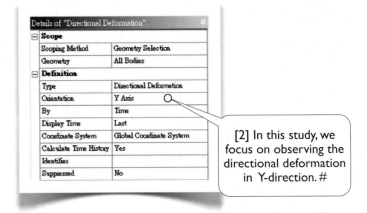

[2] In this study, we focus on observing the directional deformation in Y-direction. #

9.1-9 Obtain a Linear Solution

[1] Solve. It takes only a few seconds to complete the linear solution.

[2] Select **True Scale**.

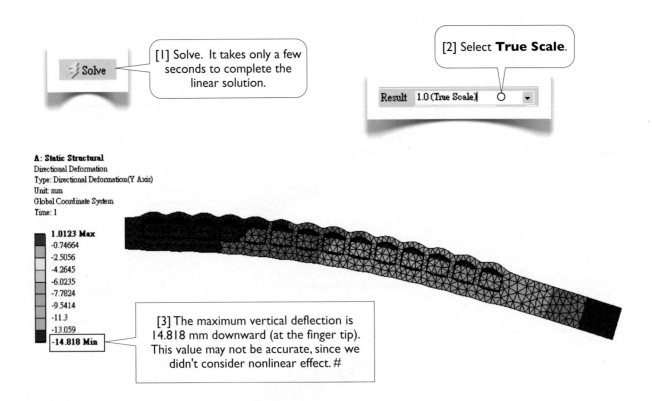

A: Static Structural
Directional Deformation
Type: Directional Deformation(Y Axis)
Unit: mm
Global Coordinate System
Time: 1

1.0123 Max
-0.74664
-2.5056
-4.2645
-6.0235
-7.7824
-9.5414
-11.3
-13.059
-14.818 Min

[3] The maximum vertical deflection is 14.818 mm downward (at the finger tip). This value may not be accurate, since we didn't consider nonlinear effect. #

Obtain a Linear Solution before Nonlinear Simulations

It is a good practice to make sure a linear solution can be obtained before a nonlinear simulation is performed. A linear simulation takes much less computational time than a nonlinear one. Nonlinearity should be considered in this case, since the deflection is large. The linear solution, which may be out of accuracy for a large amount, however, provides an effective way of model checking.

9.1-10 Obtain a Nonlinear Solution

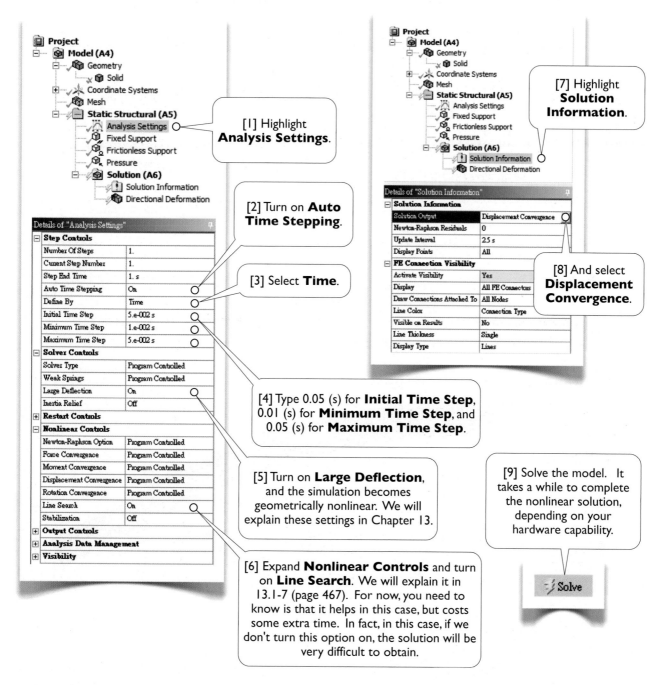

[1] Highlight **Analysis Settings**.

[2] Turn on **Auto Time Stepping**.

[3] Select **Time**.

[4] Type 0.05 (s) for **Initial Time Step**, 0.01 (s) for **Minimum Time Step**, and 0.05 (s) for **Maximum Time Step**.

[5] Turn on **Large Deflection**, and the simulation becomes geometrically nonlinear. We will explain these settings in Chapter 13.

[6] Expand **Nonlinear Controls** and turn on **Line Search**. We will explain it in 13.1-7 (page 467). For now, you need to know is that it helps in this case, but costs some extra time. In fact, in this case, if we don't turn this option on, the solution will be very difficult to obtain.

[7] Highlight **Solution Information**.

[8] And select **Displacement Convergence**.

[9] Solve the model. It takes a while to complete the nonlinear solution, depending on your hardware capability.

Displacement Convergence

Each substep of a nonlinear simulation involves an iterative process. Force and displacement values are used as convergence criteria. Most of cases, you should look at force convergence behavior during the solution, because the convergence is usually governed by force criterion. However, for this particular case, the convergence is governed by displacement criterion. That is why we look at **Displacement Convergence** instead of **Force Convergence**. These concepts will be further explained in 13.1-5 (page 465).

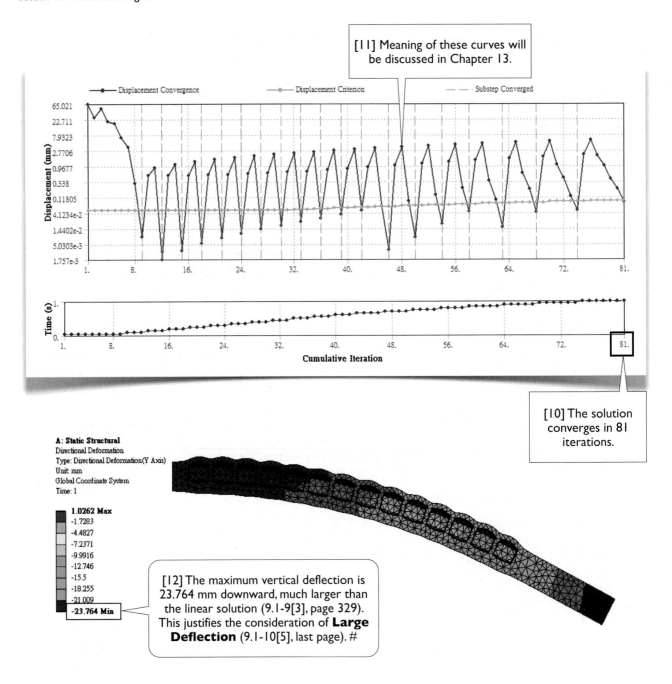

[11] Meaning of these curves will be discussed in Chapter 13.

[10] The solution converges in 81 iterations.

A: Static Structural
Directional Deformation
Type: Directional Deformation(Y Axis)
Unit: mm
Global Coordinate System
Time: 1

1.0262 Max
-1.7283
-4.4827
-7.2371
-9.9916
-12.746
-15.5
-18.255
-21.009
-23.764 Min

[12] The maximum vertical deflection is 23.764 mm downward, much larger than the linear solution (9.1-9[3], page 329). This justifies the consideration of **Large Deflection** (9.1-10[5], last page). #

Element Shapes

In many cases, nonlinear simulations can be challenging. Meshing quality plays an important role in the convergence of nonlinear solution. The mesh metric (skewness) in 9.1-6[3] (page 327) and 9.1-7[3, 6, 9] (page 328) is a measure of mesh quality. Skewness often can be improved by refining elements. It sometimes needs a large number of elements to achieve a mesh quality that is good enough to make the solution converge. In other cases, it may never achieve an acceptable mesh quality by simply refining elements.

Another factor affecting convergence is the shapes of elements. In general, hexahedra are more efficient than tetrahedra (see 9.3-13 and 9.3-14, page 353). In the following exercises, let's try to mesh the model with hexahedra as many as possible.

9.1-11 Mesh with **Hex Dominant** Method

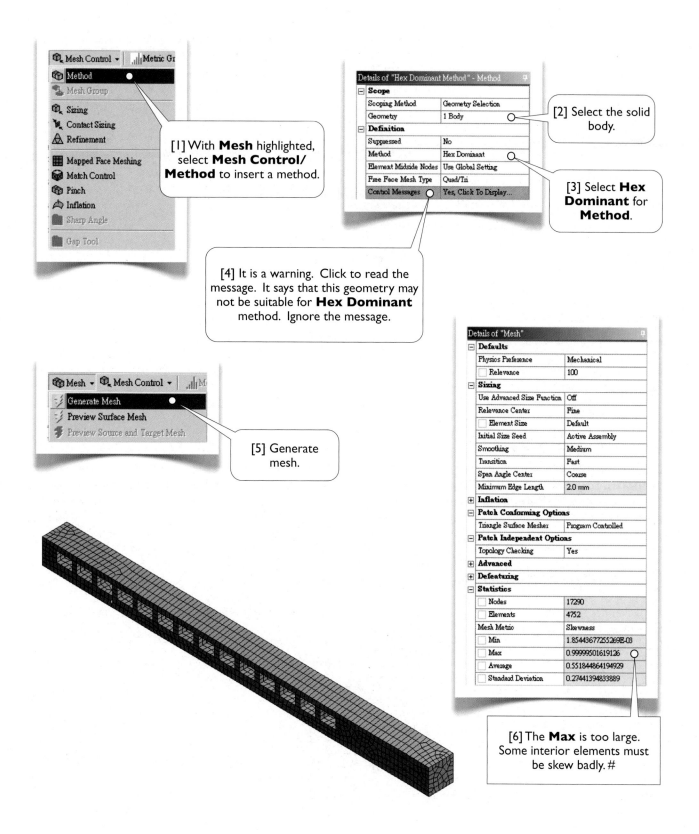

[1] With **Mesh** highlighted, select **Mesh Control/ Method** to insert a method.

[2] Select the solid body.

[3] Select **Hex Dominant** for **Method**.

[4] It is a warning. Click to read the message. It says that this geometry may not be suitable for **Hex Dominant** method. Ignore the message.

[5] Generate mesh.

[6] The **Max** is too large. Some interior elements must be skew badly. #

9.1-12 Try **Sweep** Method

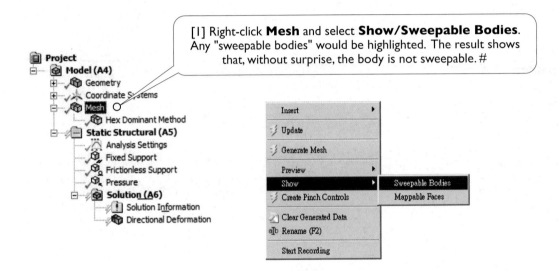

[1] Right-click **Mesh** and select **Show/Sweepable Bodies**. Any "sweepable bodies" would be highlighted. The result shows that, without surprise, the body is not sweepable. #

Sweepable Bodies

As mentioned in 5.3-2 (page 220), a simple idea of creating hexahedral elements is to mesh a face (or faces) of a body with quadrilaterals and then "sweep" along its depth direction to the other end face (or faces) of the body. The starting faces are called the *source faces* and the ending faces are called the *target faces*. The source or target faces can be either manually or automatically selected.

Not all bodies are sweepable. In our case, there is only one body, and it is not sweepable.

Mesh with **MultiZone** Method

For non-sweepable bodies, Workbench provides a sophisticated method of creating hexahedral elements, called **MultiZone** method. The idea of **MultiZone** method is to divide a non-sweepable body into several sweepable bodies, and then apply **Sweep** method on each of bodies. Actually, we already applied this method on the beam bracket model in 5.1-13[1-5] (page 209).

9.1-13 Try **MultiZone** Method

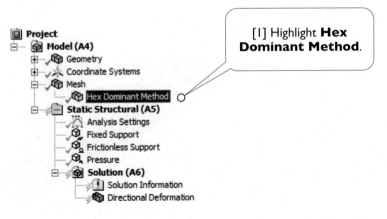

[1] Highlight **Hex Dominant Method**.

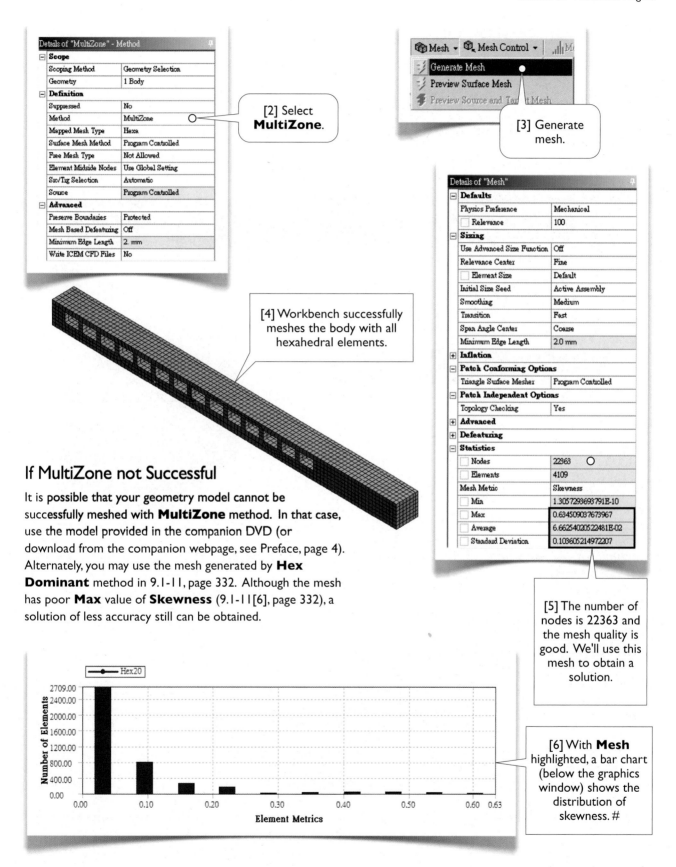

Details of "MultiZone" - Method

Scope	
Scoping Method	Geometry Selection
Geometry	1 Body
Definition	
Suppressed	No
Method	MultiZone
Mapped Mesh Type	Hexa
Surface Mesh Method	Program Controlled
Free Mesh Type	Not Allowed
Element Midside Nodes	Use Global Setting
Src/Trg Selection	Automatic
Source	Program Controlled
Advanced	
Preserve Boundaries	Protected
Mesh Based Defeaturing	Off
Minimum Edge Length	2. mm
Write ICEM CFD Files	No

[2] Select **MultiZone**.

Mesh ▾ | Mesh Control ▾ | M
- Generate Mesh
- Preview Surface Mesh
- Preview Source and Target Mesh

[3] Generate mesh.

[4] Workbench successfully meshes the body with all hexahedral elements.

Details of "Mesh"

Defaults	
Physics Preference	Mechanical
Relevance	100
Sizing	
Use Advanced Size Function	Off
Relevance Center	Fine
Element Size	Default
Initial Size Seed	Active Assembly
Smoothing	Medium
Transition	Fast
Span Angle Center	Coarse
Minimum Edge Length	2.0 mm
Inflation	
Patch Conforming Options	
Triangle Surface Mesher	Program Controlled
Patch Independent Options	
Topology Checking	Yes
Advanced	
Defeaturing	
Statistics	
Nodes	22363
Elements	4109
Mesh Metric	Skewness
Min	1.305 7293693791E-10
Max	0.634509037673967
Average	6.6625402052 2481E-02
Standard Deviation	0.103605214972207

If MultiZone not Successful

It is possible that your geometry model cannot be successfully meshed with **MultiZone** method. In that case, use the model provided in the companion DVD (or download from the companion webpage, see Preface, page 4). Alternately, you may use the mesh generated by **Hex Dominant** method in 9.1-11, page 332. Although the mesh has poor **Max** value of **Skewness** (9.1-11[6], page 332), a solution of less accuracy still can be obtained.

[5] The number of nodes is 22363 and the mesh quality is good. We'll use this mesh to obtain a solution.

[6] With **Mesh** highlighted, a bar chart (below the graphics window) shows the distribution of skewness. #

9.1-14 Examine Mesh Using Section View

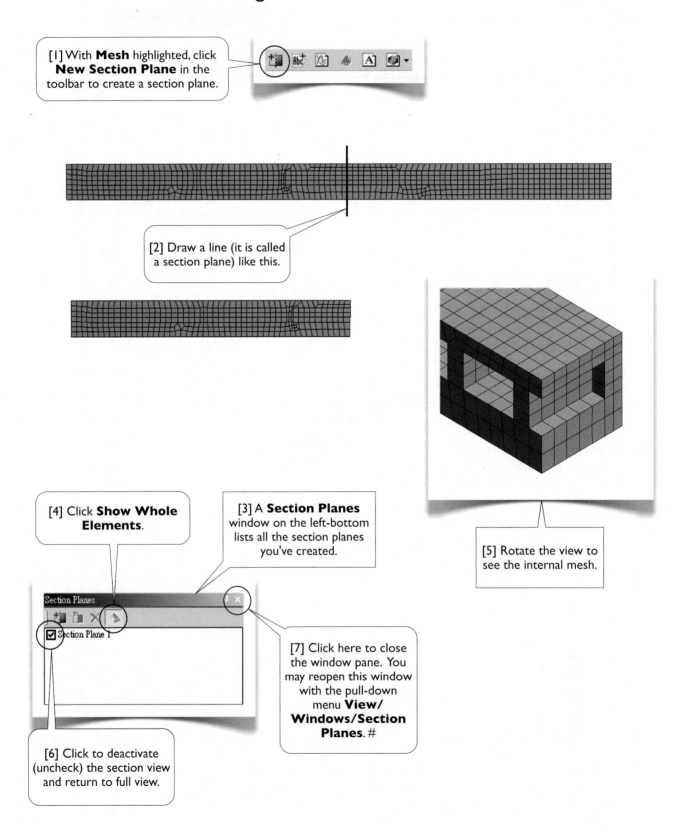

[1] With **Mesh** highlighted, click **New Section Plane** in the toolbar to create a section plane.

[2] Draw a line (it is called a section plane) like this.

[4] Click **Show Whole Elements**.

[3] A **Section Planes** window on the left-bottom lists all the section planes you've created.

[5] Rotate the view to see the internal mesh.

Section Planes

☑ Section Plane 1

[7] Click here to close the window pane. You may reopen this window with the pull-down menu **View/ Windows/Section Planes**. #

[6] Click to deactivate (uncheck) the section view and return to full view.

9.1-15 Obtain a Nonlinear Solution

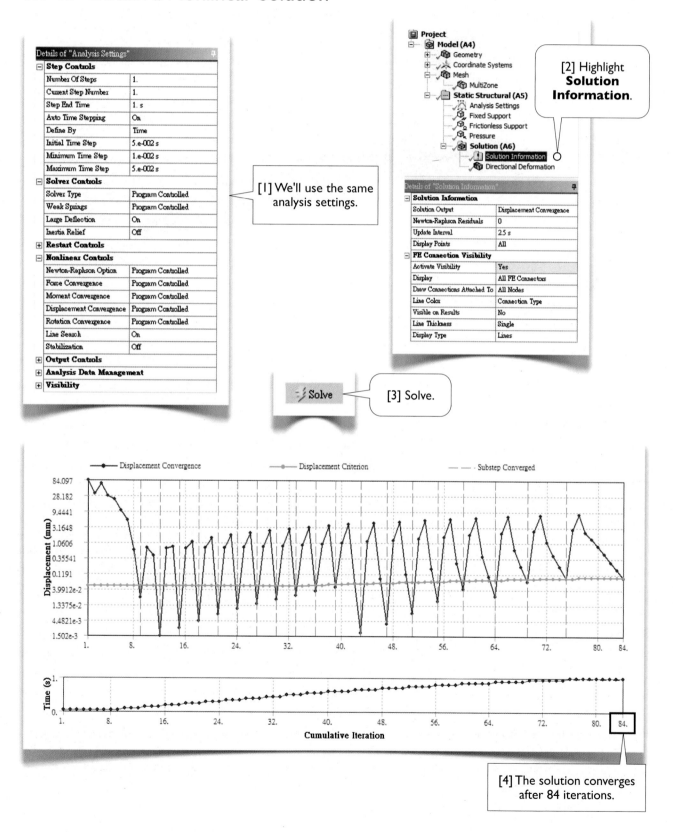

If the Solution Doesn't Converge...

Nonlinear solution is sensitive to mesh variation. Your solution convergence behavior may not be the same as shown here. For example, the number of iterations may be less or more than 84. Even that happens, the convergence curve should be similar to the one shown here. It is also possible that your solution doesn't converge. In that case, you can increase **Initial Time Step** or descrease **Minimum Time Step** (9.1-10[4], page 330), and rerun again.

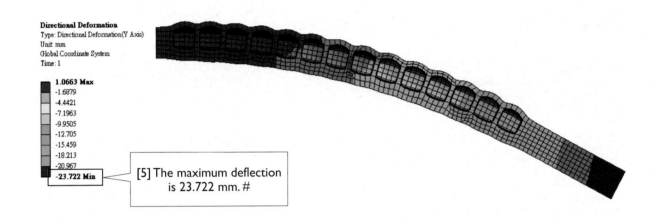

Directional Deformation
Type: Directional Deformation(Y Axis)
Unit: mm
Global Coordinate System
Time: 1

1.0663 Max
-1.6879
-4.4421
-7.1963
-9.9505
-12.705
-15.459
-18.213
-20.967
-23.722 Min

[5] The maximum deflection is 23.722 mm. #

Wrap Up

Save the project and exit Workbench.

Remark

As mentioned in 1.1-8 (page 15), when assuming a linear material, we are also assuming the compressive behavior is the same as tensile behavior, but this is usually not true for an elastomer under such a large deformation. (Note that the upper portion of the finger is subject to tension, while the lower portion is subject to compression.) Hyperelasticity, a more accurate material model for elastomer under large deformation, will be introduced in 14.1-7 to 14.1-9 (pages 523-526).

References

1. This exercise is adapted from an unpublished work led by Prof. Chao-Chieh Lan of the Department of Mechanical Engineering, NCKU.
2. ANSYS Documentation//Meshing User's Guide//Global Mesh Controls//Statistics Group//Mesh Metric//Skewness
3. ANSYS Documentation//Mechanical APDL//Theory Reference//12. Element Tools//12.1. Element Shape Testing

Section 9.2

Cover of Pressure Cylinder

9.2-1 About the Cylinder Cover

In this section, we will use the cylinder cover (Sections 4.2 and 5.2) to demonstrate some additional meshing techniques.

The geometry of the cover is relatively complicated. It seems that a free mesh with tetrahedral elements is the only meshing method. There is nothing wrong with a tetrahedral mesh as long as the mesh quality is good enough. Examining the mesh generated in 5.2-5 (page 213), we will see that the mesh quality is bad (9.2-3[3], page 339). The mesh quality needs to be improved. The simplest way is to globally adjust the relevance values. That sometimes works, although increasing the problem size, but sometimes fails. For a linear static simulation, problem size seems no big deal, but for a nonlinear or dynamic simulation, the problem size should be kept as minimum as possible, to maintain an acceptable computing time.

Note also that the purpose of this section is to demonstrate meshing techniques, rather to find the best mesh for the cylinder cover.

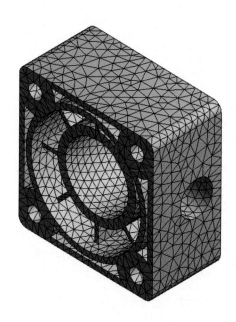

9.2-2 Open the Project **Cover**

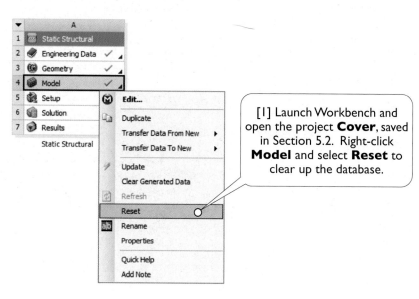

[1] Launch Workbench and open the project **Cover**, saved in Section 5.2. Right-click **Model** and select **Reset** to clear up the database.

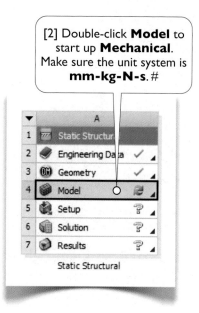

[2] Double-click **Model** to start up **Mechanical**. Make sure the unit system is **mm-kg-N-s**. #

9.2-3 Increase Mesh Density Using Global Mesh Controls

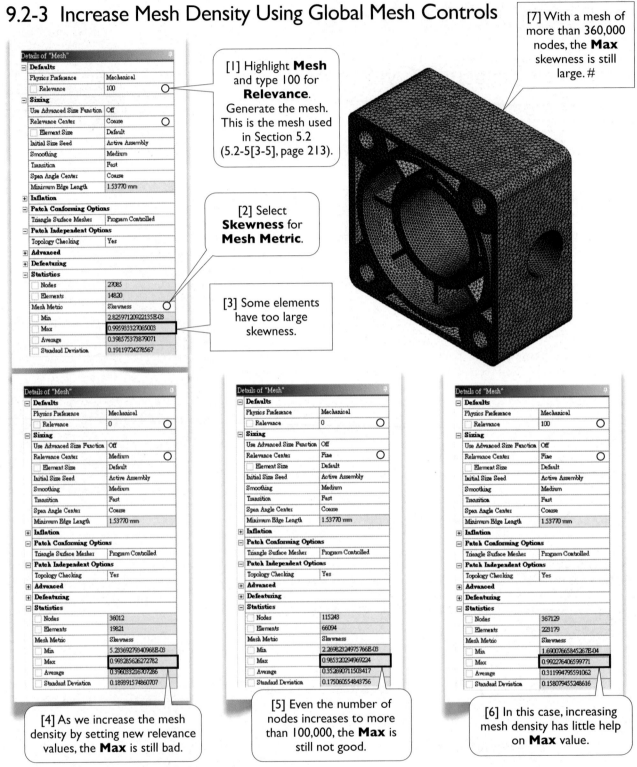

[1] Highlight **Mesh** and type 100 for **Relevance**. Generate the mesh. This is the mesh used in Section 5.2 (5.2-5[3-5], page 213).

[2] Select **Skewness** for **Mesh Metric**.

[3] Some elements have too large skewness.

[7] With a mesh of more than 360,000 nodes, the **Max** skewness is still large. #

[4] As we increase the mesh density by setting new relevance values, the **Max** is still bad.

[5] Even the number of nodes increases to more than 100,000, the **Max** is still not good.

[6] In this case, increasing mesh density has little help on **Max** value.

Increasing Mesh Density Is Not a Panacea

The lesson we learned here is that increasing mesh density, although reducing average skewness, is not a universal remedy for eliminating large skewness. We need to learn other meshing techniques.

9.2-4 Mesh with **Patch Conforming** Method

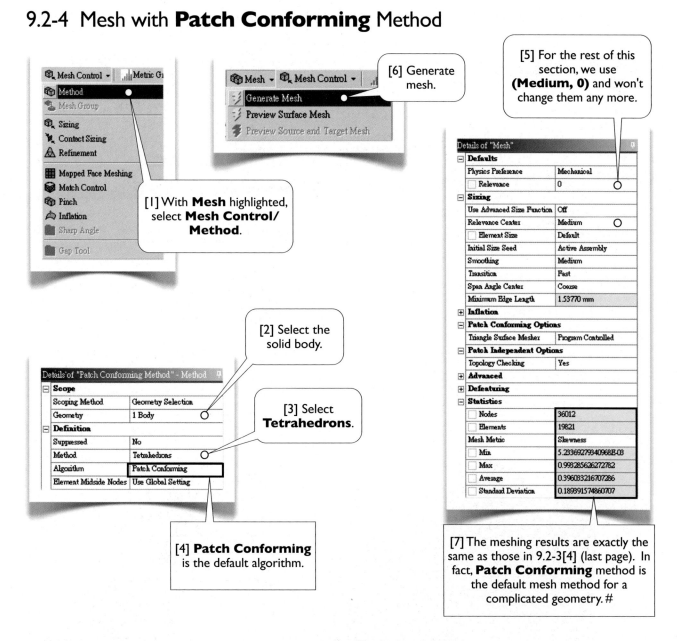

[6] Generate mesh.

[5] For the rest of this section, we use **(Medium, 0)** and won't change them any more.

[1] With **Mesh** highlighted, select **Mesh Control/Method**.

[2] Select the solid body.

[3] Select **Tetrahedrons**.

[4] **Patch Conforming** is the default algorithm.

[7] The meshing results are exactly the same as those in 9.2-3[4] (last page). In fact, **Patch Conforming** method is the default mesh method for a complicated geometry. #

Patch Conforming and **Patch Independent** Methods

In CAD jargon, faces of solid bodies are also called *patches*. The basic idea of **Patch Conforming** is to mesh all the faces of the body with triangles and then "grow" inward to create tetrahedra. In this way, the exterior shape of the body (i.e., shapes of its faces) are respected (preserved); that is the implication of the name **Patch Conforming**. For complicated geometry, this is the default method.

On the other hand, **Patch Independent** creates tetrahedra from inside out. The outermost nodes are then projected onto the boundary faces and the element edges are created. In this way, the mesh's exterior shape may be different from the original geometry; that is the implication of the name **Patch Independent**.

In some cases, when too many details exist that cause meshing difficulty, we may resort to **Patch Independent** algorithm and ignore these details. However, it is your responsibility to make sure that ignoring those details wouldn't distort the geometry too much.

9.2-5 Mesh with **Patch Independent** Method

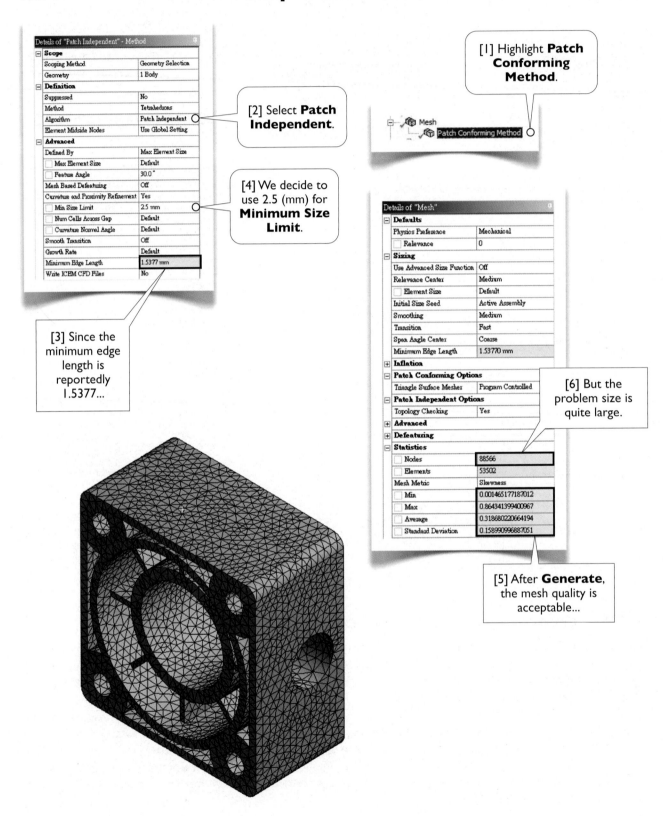

Details of "Patch Independent" - Method

Scope	
Scoping Method	Geometry Selection
Geometry	1 Body
Definition	
Suppressed	No
Method	Tetrahedrons
Algorithm	Patch Independent
Element Midside Nodes	Use Global Setting
Advanced	
Defined By	Max Element Size
Max Element Size	Default
Feature Angle	30.0 °
Mesh Based Defeaturing	Off
Curvature and Proximity Refinement	Yes
Min Size Limit	2.5 mm
Num Cells Across Gap	Default
Curvature Normal Angle	Default
Smooth Transition	Off
Growth Rate	Default
Minimum Edge Length	1.5377 mm
Write ICEM CFD Files	No

[2] Select **Patch Independent**.

[1] Highlight **Patch Conforming Method**.

Mesh
Patch Conforming Method

[4] We decide to use 2.5 (mm) for **Minimum Size Limit**.

[3] Since the minimum edge length is reportedly 1.5377...

Details of "Mesh"

Defaults	
Physics Preference	Mechanical
Relevance	0
Sizing	
Use Advanced Size Function	Off
Relevance Center	Medium
Element Size	Default
Initial Size Seed	Active Assembly
Smoothing	Medium
Transition	Fast
Span Angle Center	Coarse
Minimum Edge Length	1.53770 mm
Inflation	
Patch Conforming Options	
Triangle Surface Mesher	Program Controlled
Patch Independent Options	
Topology Checking	Yes
Advanced	
Defeaturing	
Statistics	
Nodes	88566
Elements	53502
Mesh Metric	Skewness
Min	0.001465177187012
Max	0.864341399400967
Average	0.318680220664194
Standard Deviation	0.158990996887051

[6] But the problem size is quite large.

[5] After **Generate**, the mesh quality is acceptable...

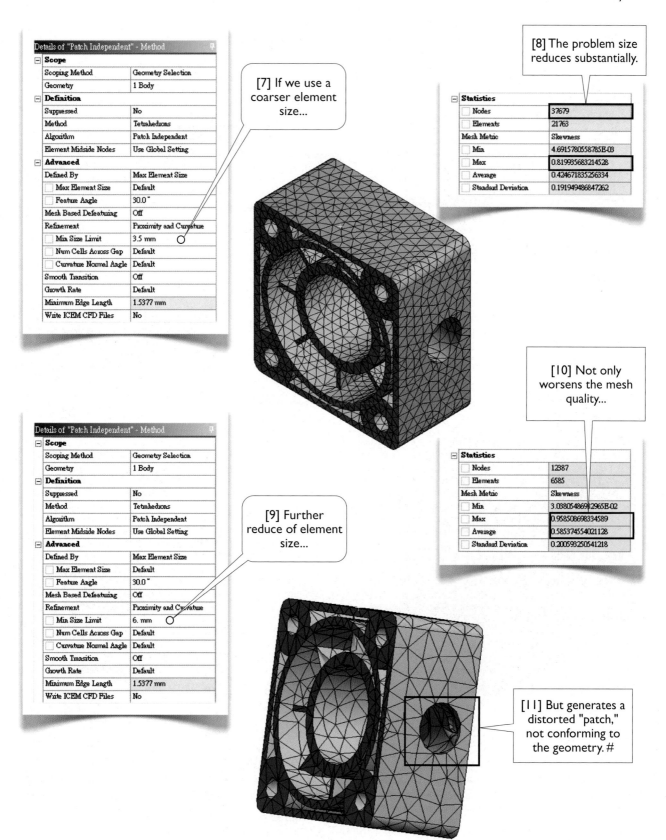

Details of "Patch Independent" - Method

Scope	
Scoping Method	Geometry Selection
Geometry	1 Body
Definition	
Suppressed	No
Method	Tetrahedrons
Algorithm	Patch Independent
Element Midside Nodes	Use Global Setting
Advanced	
Defined By	Max Element Size
Max Element Size	Default
Feature Angle	30.0 °
Mesh Based Defeaturing	Off
Refinement	Proximity and Curvature
Min Size Limit	3.5 mm
Num Cells Across Gap	Default
Curvature Normal Angle	Default
Smooth Transition	Off
Growth Rate	Default
Minimum Edge Length	1.5377 mm
Write ICEM CFD Files	No

[7] If we use a coarser element size...

[8] The problem size reduces substantially.

Statistics	
Nodes	37679
Elements	21763
Mesh Metric	Skewness
Min	4.6915780558785E-03
Max	0.819985683214528
Average	0.424671835256334
Standard Deviation	0.191949486847262

Details of "Patch Independent" - Method

Scope	
Scoping Method	Geometry Selection
Geometry	1 Body
Definition	
Suppressed	No
Method	Tetrahedrons
Algorithm	Patch Independent
Element Midside Nodes	Use Global Setting
Advanced	
Defined By	Max Element Size
Max Element Size	Default
Feature Angle	30.0 °
Mesh Based Defeaturing	Off
Refinement	Proximity and Curvature
Min Size Limit	6. mm
Num Cells Across Gap	Default
Curvature Normal Angle	Default
Smooth Transition	Off
Growth Rate	Default
Minimum Edge Length	1.5377 mm
Write ICEM CFD Files	No

[9] Further reduce of element size...

[10] Not only worsens the mesh quality...

Statistics	
Nodes	12387
Elements	6585
Mesh Metric	Skewness
Min	3.0380548690862965E-02
Max	0.958508698334589
Average	0.585374554021128
Standard Deviation	0.200593250541218

[11] But generates a distorted "patch," not conforming to the geometry. #

9.2-6 Mesh with **Hex Dominant** Method

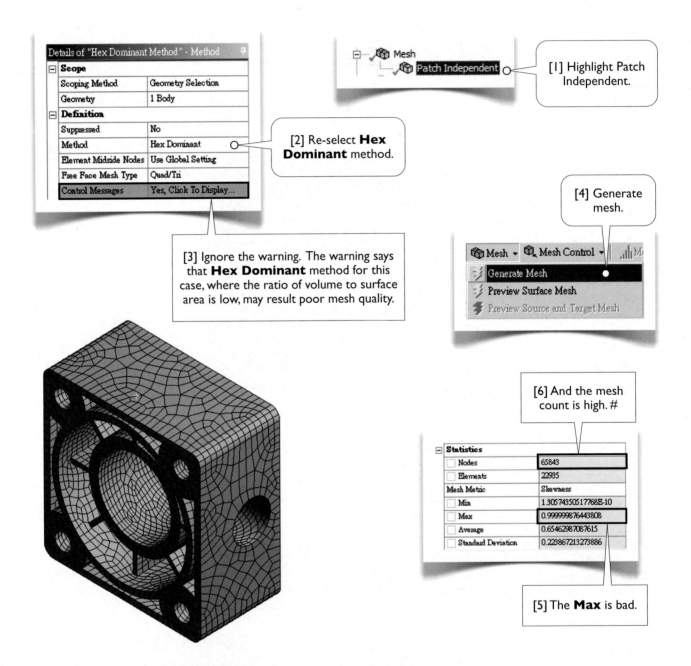

Details of "Hex Dominant Method" - Method	
Scope	
Scoping Method	Geometry Selection
Geometry	1 Body
Definition	
Suppressed	No
Method	Hex Dominant
Element Midside Nodes	Use Global Setting
Free Face Mesh Type	Quad/Tri
Control Messages	Yes, Click To Display…

[1] Highlight Patch Independent.

[2] Re-select **Hex Dominant** method.

[3] Ignore the warning. The warning says that **Hex Dominant** method for this case, where the ratio of volume to surface area is low, may result poor mesh quality.

[4] Generate mesh.

Mesh ▾ | Mesh Control ▾ | M
Generate Mesh
Preview Surface Mesh
Preview Source and Target Mesh

[6] And the mesh count is high. #

Statistics	
Nodes	65843
Elements	22935
Mesh Metric	Skewness
Min	1.30574350517768E-10
Max	0.999999876443808
Average	0.65462987087615
Standard Deviation	0.223867213273886

[5] The **Max** is bad.

Hex Dominant Method

An idea of **Hex Dominant** is to mesh the body with **Patch Conforming** first and then combine tetrahedra to form hexahedra. It usually leaves some tetrahedra that cannot be combined to form hexahedra; that is how the name **Hex Dominant** comes from. After forming hexahedra, the algorithm tries to adjust the nodes to improve the mesh quality further.

Note that, **Hex Dominant** method, by its nature, is a method of patch conforming, that is, the faces are not distorted. In fact, all methods except **Patch Independent** are patch conforming.

9.2-7 Mesh with **MultiZone** Method

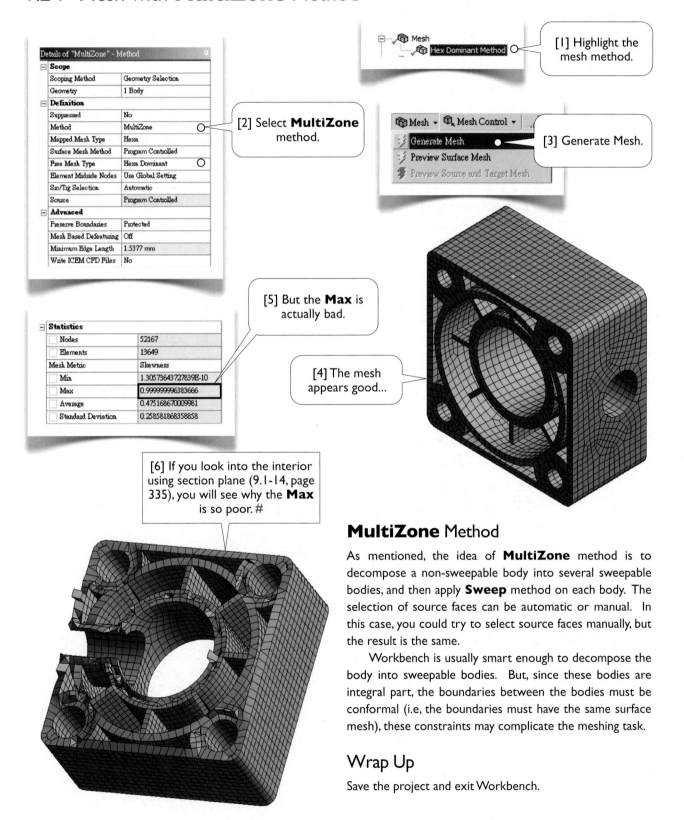

Details of "MultiZone" - Method	
Scope	
Scoping Method	Geometry Selection
Geometry	1 Body
Definition	
Suppressed	No
Method	MultiZone
Mapped Mesh Type	Hexa
Surface Mesh Method	Program Controlled
Free Mesh Type	Hexa Dominant
Element Midside Nodes	Use Global Setting
Src/Trg Selection	Automatic
Source	Program Controlled
Advanced	
Preserve Boundaries	Protected
Mesh Based Defeaturing	Off
Minimum Edge Length	1.5377 mm
Write ICEM CFD Files	No

[1] Highlight the mesh method.

[2] Select **MultiZone** method.

[3] Generate Mesh.

Statistics	
Nodes	52167
Elements	13649
Mesh Metric	Skewness
Min	1.30573643727839E-10
Max	0.999999996383666
Average	0.475168670009981
Standard Deviation	0.258581868358858

[5] But the **Max** is actually bad.

[4] The mesh appears good...

[6] If you look into the interior using section plane (9.1-14, page 335), you will see why the **Max** is so poor. #

MultiZone Method

As mentioned, the idea of **MultiZone** method is to decompose a non-sweepable body into several sweepable bodies, and then apply **Sweep** method on each body. The selection of source faces can be automatic or manual. In this case, you could try to select source faces manually, but the result is the same.

Workbench is usually smart enough to decompose the body into sweepable bodies. But, since these bodies are integral part, the boundaries between the bodies must be conformal (i.e, the boundaries must have the same surface mesh), these constraints may complicate the meshing task.

Wrap Up

Save the project and exit Workbench.

Section 9.3

Convergence Study of 3D Solid Elements

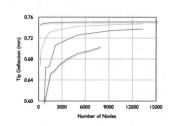

The main purpose of this section is to study 3D solid elements convergence behavior. A secondary purpose is to serve as an exercise for mesh controls techniques. A cantilever beam of rectangular cross section is used for these purposes. The conclusions drawn from the convergence study are very important for CAE engineers. Although the concepts are already introduced in Section 3.5, a CAE engineer should have more insight about elements convergence behaviors after the hands-on exercises in this section.

9.3-1 About the Cantilever Beam

The cantilever beam is made of steel and of size 100 mm x 10 mm x 10 mm [1-2]. A uniform load of 1 MPa applies on the upper face of the beam. Convergence of three solid element shapes will be compared, namely hexahedron, prism, and tetrahedron (1.3-3[1, 2. 4], page 34).

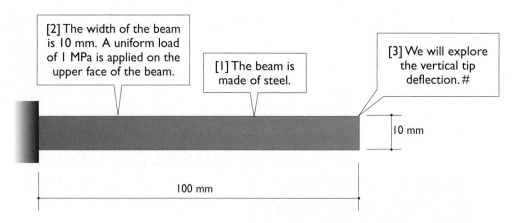

[2] The width of the beam is 10 mm. A uniform load of 1 MPa is applied on the upper face of the beam.

[1] The beam is made of steel.

[3] We will explore the vertical tip deflection. #

10 mm

100 mm

9.3-2 Start Up a New Project

Launch Workbench. Create a **Static Structural** system. Save the project as **Cantilever**. Start up DesignModeler. Select **Millimeter** as length unit and make sure **Auto Constraints** are turned on.

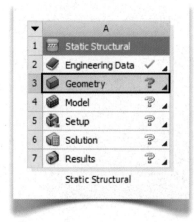

9.3-3 Create a 3D Model in DesignModeler

On **XYPlane**, create a rectangle [1]. Extrude the sketch to create a 3D model [2]. Close DesignModeler.

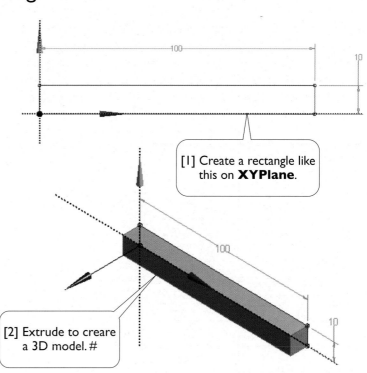

[1] Create a rectangle like this on **XYPlane**.

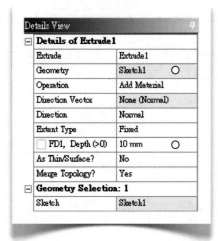

Details View	
Details of Extrude1	
Extrude	Extrude1
Geometry	Sketch1
Operation	Add Material
Direction Vector	None (Normal)
Direction	Normal
Extent Type	Fixed
FD1, Depth (>0)	10 mm
As Thin/Surface?	No
Merge Topology?	Yes
Geometry Selection: 1	
Sketch	Sketch1

[2] Extrude to creare a 3D model. #

9.3-4 Set Up Support, Load, and Solution Objects

Start up **Mechanical** and select **mm-kg-N-s** unit system. Specify a **Fixed Support** on the left face [1]. Apply a pressure of 1 MPa on the upper face [2]. Insert a **Directional Deformation** under the solution branch and set the orientation to **Y Axis** [3].

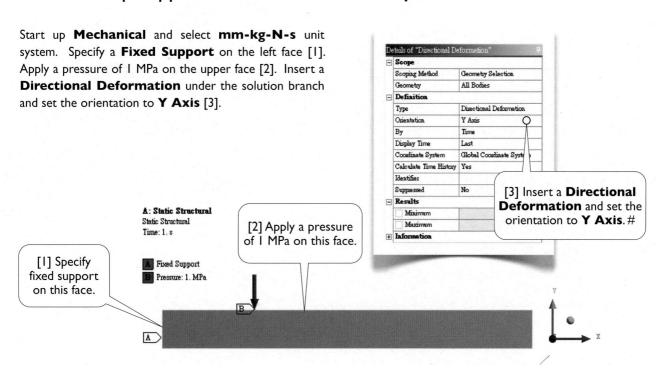

Details of "Directional Deformation"	
Scope	
Scoping Method	Geometry Selection
Geometry	All Bodies
Definition	
Type	Directional Deformation
Orientation	Y Axis
By	Time
Display Time	Last
Coordinate System	Global Coordinate Syste
Calculate Time History	Yes
Identifier	
Suppressed	No
Results	
Minimum	
Maximum	
Information	

[3] Insert a **Directional Deformation** and set the orientation to **Y Axis**. #

A: Static Structural
Static Structural
Time: 1. s

A Fixed Support
B Pressure: 1. MPa

[2] Apply a pressure of 1 MPa on this face.

[1] Specify fixed support on this face.

9.3-5 Lower-Order Hexahedra

For a model of such a regular geometry, the default mesh control settings will result in an all-hexahedra mesh [1]. To generate lower-order hexahedra, drop element midside nodes [2]. For each run, all we have to do is to change the element size [3]. Resulting tip deflections are recorded in the table below. The convergence curve is shown in [4].

Element Size (mm)	Number of Nodes	Tip Deflection (mm)
5	189	0.74571
4	416	0.74693
3	875	0.74850
2	1836	0.74980
1.5	4352	0.75048
1.3	6318	0.75072
1.2	8500	0.75086
1	12221	0.75106
0.9	19097	0.75120
0.8	24696	0.75129

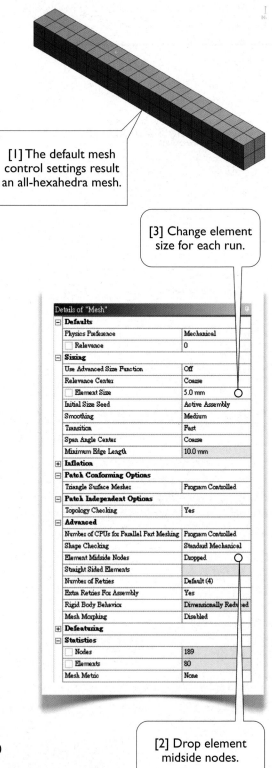

[1] The default mesh control settings result an all-hexahedra mesh.

[3] Change element size for each run.

[2] Drop element midside nodes.

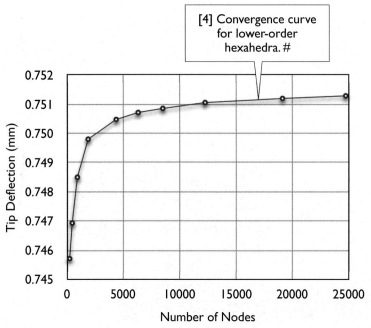

[4] Convergence curve for lower-order hexahedra. #

347

9.3-6 Lower-Order Tetrahedra

Highlight **Mesh** in the project tree and select **Mesh Control/Method** to insert a mesh control method. Select **Tetrahedrons** method [1]. Now, change element size for each run [2]. Resulting tip deflections are recorded in the table below. The convergence curve is shown in [3].

Element Size (mm)	Number of Nodes	Tip Deflection (mm)
5	102	0.30817
4	186	0.48933
2.5	339	0.53847
2	507	0.58067
1.5	736	0.59445
1.2	1543	0.65267
1	1994	0.65665
0.8	2677	0.66903
0.7	4040	0.68296
0.6	5079	0.69088
0.55	6544	0.69587
0.5	7948	0.70279

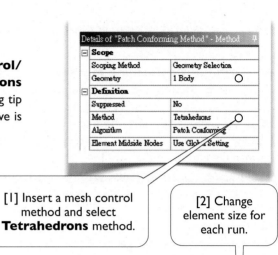

[1] Insert a mesh control method and select **Tetrahedrons** method.

[2] Change element size for each run.

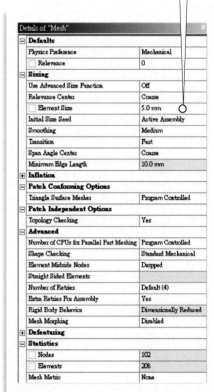

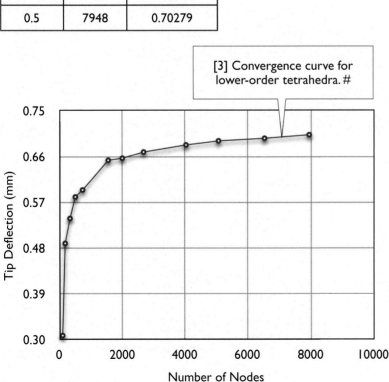

[3] Convergence curve for lower-order tetrahedra. #

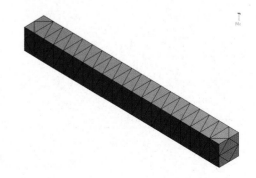

9.3-7 Lower-Order Prisms (Parallel to Loading Direction)

Highlight **Patch Conforming Method**. In the details view, change to **Sweep** method [1]. Set up the source face [2, 3]. For each run, change both sweep element sizes [4] and global element size [5]. Resulting tip deflections are recorded in the table below. The convergence curve is shown in the bottom of this page [6].

Note that the prisms are oriented such that their heights are parallel to the loading (bending) direction. We will refer to the elements oriented in this way as "parallel prisms" for the rest of this section. We will show in 9.3-8 (next page) that the convergence curve will be different if the prisms are oriented differently.

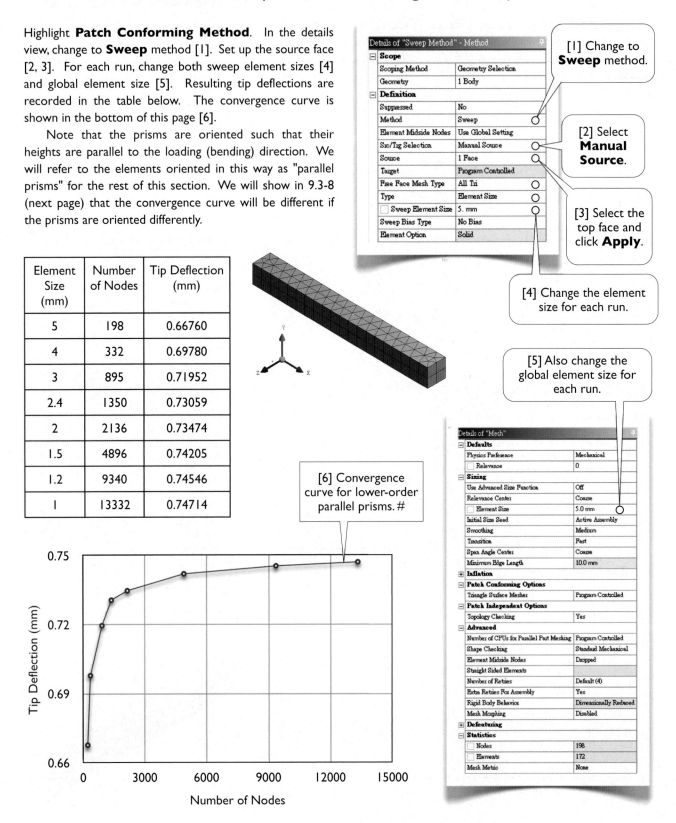

[1] Change to **Sweep** method.

[2] Select **Manual Source**.

[3] Select the top face and click **Apply**.

[4] Change the element size for each run.

[5] Also change the global element size for each run.

[6] Convergence curve for lower-order parallel prisms. #

Element Size (mm)	Number of Nodes	Tip Deflection (mm)
5	198	0.66760
4	332	0.69780
3	895	0.71952
2.4	1350	0.73059
2	2136	0.73474
1.5	4896	0.74205
1.2	9340	0.74546
1	13332	0.74714

9.3-8 Lower-Order Prisms (Perpendicular to Loading Direction)

In the details view of **Sweep Method**, re-select the source face [1]. For each run, change both sweep element size [2] and global element size [3]. Resulting tip deflections are recorded in the table below. The convergence curve is shown in the bottom of this page [4].

Note that the prisms are oriented such that their heights are perpendicular to the loading (bending) direction. We will refer to the elements oriented in this way as "perpendicular prisms" for the rest of this section.

Element Size (mm)	Number of Nodes	Tip Deflection (mm)
5	198	0.50355
4	332	0.48425
3	895	0.66307
2.4	1350	0.66495
2	2136	0.70619
1.5	4896	0.72534
1.2	9340	0.73440
1	13332	0.73719

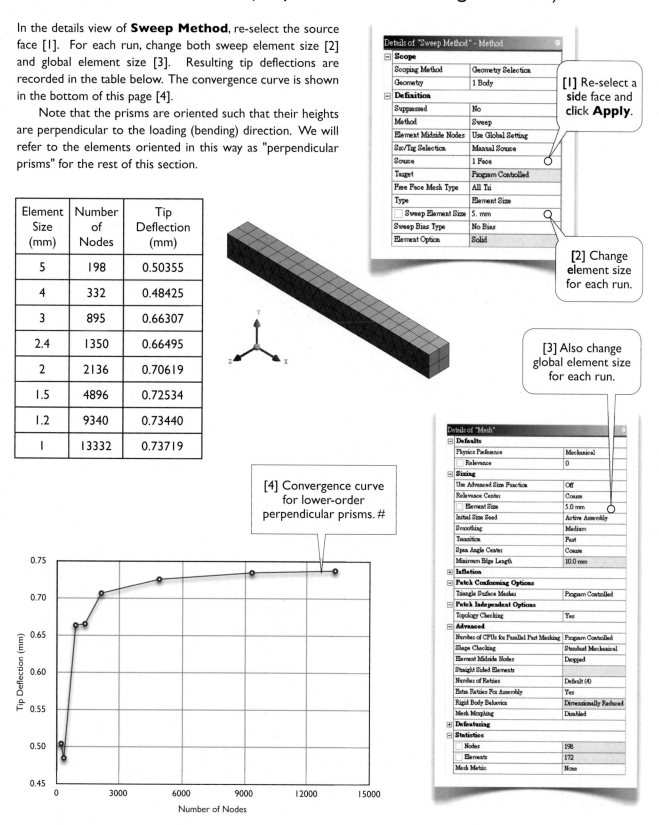

[1] Re-select a side face and click **Apply**.

[2] Change element size for each run.

[3] Also change global element size for each run.

[4] Convergence curve for lower-order perpendicular prisms. #

9.3-9 Higher-Order Hexahedra

Delete the **Sweep Method**. Repeat what you have done in 9.3-5 (page 347), except keeping the element midside nodes [1] and change the element sizes as shown below.

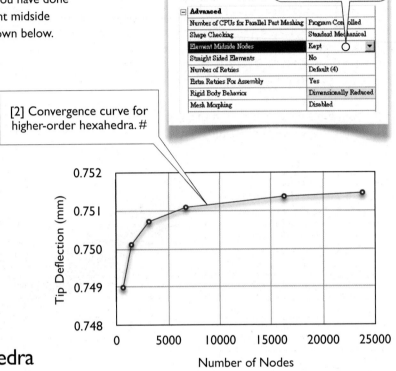

[1] Keep element midside nodes.

Element Size (mm)	Number of Nodes	Tip Deflection (mm)
5	621	0.74899
4	1440	0.75011
3	3125	0.75071
2	6696	0.75108
1.5	16256	0.75136
1.3	23787	0.75145

[2] Convergence curve for higher-order hexahedra. #

9.3-10 Higher-Order Tetrahedra

Repeat what you have done in 9.3-6 (page 348), except keeping the element midside nodes (9.3-9[1]) and change the element sizes as shown below.

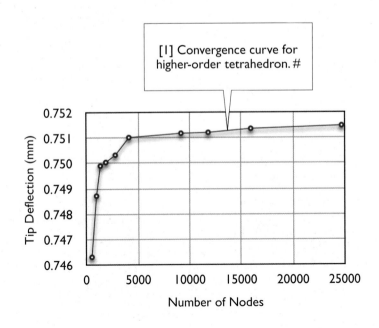

[1] Convergence curve for higher-order tetrahedron. #

Element Size (mm)	Number of Nodes	Tip Deflection (mm)
5	501	0.74631
4	954	0.74872
3	1329	0.74990
2.5	1855	0.75004
2	2768	0.75033
1.5	4077	0.75102
1.2	9151	0.75118
1	11805	0.75121
0.8	15937	0.75136
0.7	24693	0.75148

9.3-11 Higher-Order Parallel Prisms

Repeat what you have done in 9.3-7 (page 349), except keeping the element
midside nodes (9.3-9[1], last page) and change the element sizes as shown below.

Element Size (mm)	Number of Nodes	Tip Deflection (mm)
5	783	0.74964
4	1341	0.75033
3.3	2899	0.75089
3	3901	0.75090
2.4	5943	0.75115
2	9646	0.75122
1.8	12915	0.75134
1.6	19058	0.75143
1.4	27705	0.75151

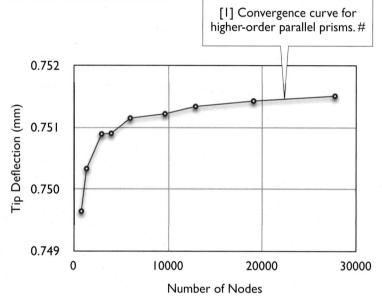

[1] Convergence curve for higher-order parallel prisms. #

9.3-12 Higher-Order Perpendicular Prisms

Repeat what you have done in 9.3-8 (page 350), except keeping the element
midside nodes (9.3-9[1], last page) and change the element sizes as shown below.

Element Size (mm)	Number of Nodes	Tip Deflection (mm)
5	783	0.74953
4	1341	0.74999
3.3	2899	0.75079
3	3901	0.75089
2.4	5943	0.75116
2	9646	0.75127
1.8	12915	0.75138
1.6	19058	0.75147
1.4	27749	0.75155

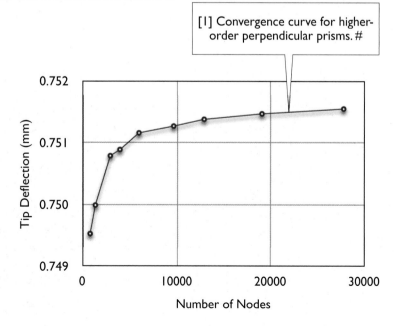

[1] Convergence curve for higher-order perpendicular prisms. #

9.3-13 Comparison: Lower-Order Elements

The chart below is made by a collection of the convergence curves in 9.3-5 to 9.3-8 (pages 347-350) to compare the convergence behaviors of the lower-order elements. The order of the convergence speed is, from fast to slow, hexahedron, parallel prism, perpendicular prism, tetrahedron. The differences between them are obvious and quite evenly spaced. The lower-order tetrahedron converges so poorly that it is not practically useful. As a guideline, NEVER use lower-order tetrahedral elements.

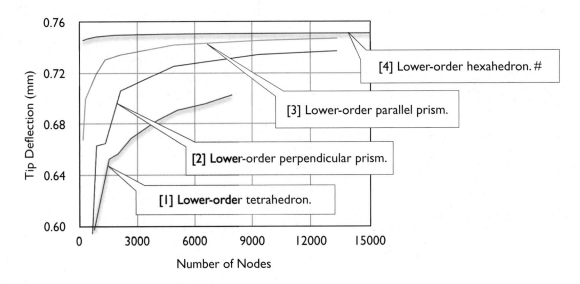

9.3-14 Comparison: Higher-Order Elements

The chart below is made by a collection of the convergence curves in 9.3-9 to 9.3-12 (pages 351-352) to compare the convergence behaviors of the higher-order elements. Except the tetrahedron, the convergence speeds are quite comparable. The tetrahedron performs poorly only when the mesh is coarse (below 4000 nodes, in this case), otherwise it may be as good as other elements. Contrasting to the lower-order tetrahedron, the higher-order tetrahedron is still practically useful as long as the mesh is fine enough.

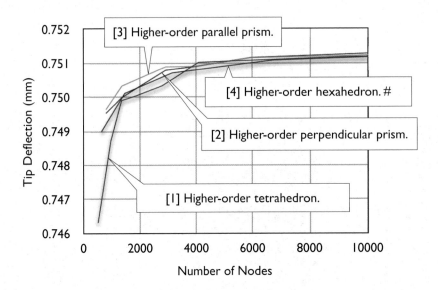

9.3-15 Comparison: Hexahedra

The chart below is made by a collection of the convergence curves in 9.3-5 (page 347) and 9.3-9 (page 351) to compare the convergence behaviors between the lower- and higher-order hexahedra. It is obvious that higher-order hexahedral is better than the lower-order hexahedral, but the difference is not so dramatic as tetrahedral (see 9.3-16).

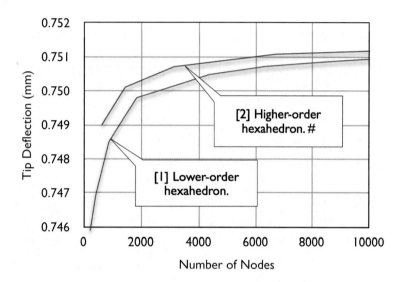

9.3-16 Comparison: Tetrahedra

The chart below is made by a collection of the convergence curves in 9.3-6 (page 348) and 9.3-10 (page 351) to compare the convergence behaviors between the lower- and higher-order tetrahedra. It is obvious that higher-order element is much better than the lower-order one. Remind you again: NEVER use lower-order tetrahedral elements.

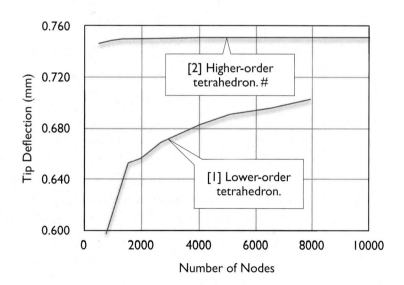

9.3-17 Comparison: Parallel Prisms

The chart below is made by a collection of the convergence curves in 9.3-7 (page 349) and 9.3-11 (page 352) to compare the convergence behaviors between the lower- and higher-order parallel prisms. It is obvious that higher-order element is much better than the lower-order one. Like lower-order tetrahedral, lower-order prismatic elements are not recommended.

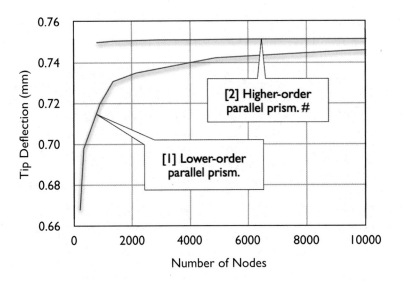

9.3-18 Comparison: Perpendicular Prisms

The chart below is made by a collection of the convergence curves in 9.3-8 (page 350) and 9.3-12 (page 352) to compare the convergence behaviors between the lower- and higher-order perpendicular prisms. It is obvious that higher-order element is much better than the lower-order one. Like lower-order tetrahedral, lower-order prismatic elements are not recommended.

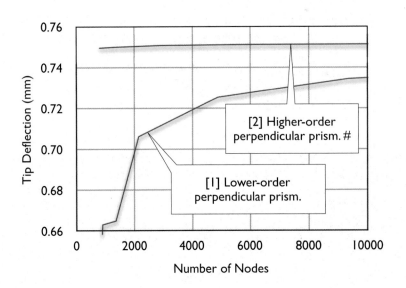

9.3-19 Summary and Guidelines

Combining the observations in Section 3.5 and this section, we may summarize the conclusions as follows: (a) Never use lower-order tetrahedra or triangles. (b) Higher-order tetrahedra or triangles can be as good as other elements as long as the mesh is fine enough. In cases of coarse mesh, however, they perform poorly and are not recommended. (c) Lower-order prisms are not recommended. (d) Lower-order hexahedra and quadrilaterals can be used, but they are not as efficient as their higher-order counterparts. (e) Higher-order hexahedra, prisms, and quadrilaterals are among the most efficient elements so far we have discussed. Mesh your models with these elements whenever possible. If that is not possible, then at least try to achieve a higher-order hexahedra-dominant or quadrilateral-dominant mesh.

Besides the above guidelines, mesh quality requirements, in terms of mesh metrics such as skewness, should also be met.

Remark

In Section 3.5 and this section, comparisons among elements are made under the same number of nodes. More reasonable comparisons should be made under the same CPU time. For a simulation task, the CPU time consists of three parts. First, the time required to establish Eq. 1.3-1(1) (page 31). It involves numerical integrations element by element. This part of CPU time depends on the total number of elements as well as the number of integration points of each element. Second, the time required to solve the equation. This part of CPU time is determined solely by the number of degrees of freedom, which is in turn determined by the number of nodes and the dimensionality (2D or 3D). Third, the others (housekeeping, overhead, etc).

For small problems, the overall CPU time is dominated by the third part. That is why we didn't use CPU time for comparison, since all cases are small when coarsely meshed. For large problems, the CPU time is essentially the sum of the first two parts.

Therefore, strictly speaking, our comparison was not perfectly reasonable. Nevertheless, the discussions and conclusions in this section pretty much reflect the reality. These guidelines should be useful.

Wrap Up

Save the project and exit Workbench.

Section 9.4

Review

9.4-1 Keywords

Choose a letter for each keyword from the list of descriptions

1. (   ) Convergence Criteria
2. (   ) Displacement Convergence Criterion
3. (   ) Force Convergence Criterion
4. (   ) Hex Dominant Method
5. (   ) Parallel Prisms

6. (   ) Patch Conforming Method
7. (   ) Patch Independent Method
8. (   ) Perpendicular Prisms
9. (   ) Skewness
10. (   ) Sweep Thin Method

Answers:

1. (B)   2. (C)   3. (D)   4. (H)   5. (I)   6. (F)   7. (G)   8. (J)   9. (A)   10. (E)

List of Descriptions

(A)   A measure of mesh quality, calculated for each element according to its geometry. Its value ranges from 0 to 1, the smaller the better. Elements of skewness of more than 0.95 are considered unacceptable.

(B)   In nonlinear simulation, the loading is divided into substeps and applied substep by substep. By default, a substep is said to be complete when both displacement convergence criterion and force convergence criterion are met during the iterations.

(C)   During the iterations of a substep of a nonlinear simulation, the displacement convergence criterion is met when the increment of displacement is less than a criterion, which is, by default, 0.5% of maximum displacement.

(D)   During the iterations of a substep of a nonlinear simulation, the force convergence criterion is met when the unbalanced force is less than a criterion, which is, by default, 0.5% of applied force.

(E)   **Sweep** mesh control method can be classified into **Sweep** and **Sweep Thin**. **Sweep** allows a more complex sweeping path while **Sweep Thin** allows only a simple sweeping path. The advantage of **Sweep Thin** is that it allows multiple faces as source or target while **Sweep** allows only one face for both source and target.

(F) A mesh control method. It meshes all the faces of the body with triangles; the triangles then "grow" inward to create tetrahedra. In this way, the shapes of the faces are respected (preserved).

(G) A mesh control method. It creates tetrahedra from inside out. The outermost nodes are then projected onto the boundary faces and the element edges are created. In this way, the mesh's outline may be different from the original geometry.

(H) A mesh control method. It meshes a body with Patch Conforming method first and then combines tetrahedra to form hexahedra. It usually leaves some tetrahedra that cannot be combined to form hexahedra.

(I) When a body is meshed with prismatic elements and the prisms are oriented such that their heights are parallel to the bending direction, the prismatic elements oriented in this way are referred as parallel prisms. (Note: this term is used only in this books.)

(J) When a body is meshed with prismatic elements and the prisms are oriented such that their heights are perpendicular to the bending direction, the prismatic elements oriented in this way are referred as perpendicular prisms. (Note: this term is used only in this books.)

9.4-2 Additional Workbench Exercises

Convergence Study for Higher-Order 2D Elements

In Section 9.3, we study the convergence of 3D elements, both higher-order and lower-order elements. In Section 3.5, we study the convergence of 2D elements, only for the lower-order elements. We haven't studied the higher-order 2D elements yet. Conduct a study of the higher-order 2D elements.

Chapter 10
Buckling and Stress Stiffening

Functionality, safety, and reliability are the main purposes of structural simulations. Resulting stresses usually relate to safety and reliability. In the 3D truss example (Section 7.2), calculated stresses are well below the material's yield strength (10.2-1, page 371). Can we conclude that the design is safe? Not yet. For any structural members (particularly slender or thin members) subject to compressive stresses, we need to check their stability before concluding their safety. This chapter mainly deals with *stability analysis*, or *buckling analysis*.

Buckling can be viewed as an ultimate case of a more general effect, called *stress stiffening*: a structure member's bending stiffness increases with increasing axial tensile stress, and, on the other hand, the member's bending stiffness decreases with the increasing compressive stress. Buckling occurs when the compressive stress reaches a level such that the bending stiffness reduces to zero; in that situation, the applying load is called a *buckling load* and the corresponding deformation is called a *buckling mode*. The purpose of buckling analyses is to find the buckling loads and the corresponding *buckling mode*.

Purpose of This Chapter

The main purpose of this chapter is to introduce *linear buckling analysis*. Since buckling can be viewed as an ultimate case of stress stiffening, the discussion will start with a thorough understanding of stress stiffening. As usual, the concepts are introduced using step-by-step exercises.

About Each Section

Section 10.1 introduces the stress stiffening effects, using a simply supported beam as an example. The results of these nonlinear analyses can be used to predict the buckling load using an extrapolation method; this procedure is called a *nonlinear buckling analysis*. A linear buckling analysis is then carried out to find the buckling loads and buckling modes. Comparing with nonlinear buckling analyses, linear buckling analysis tends to overestimate the buckling load.

Section 10.2 performs a linear buckling analysis on the 3D truss structure introduced in Section 7.2. Section 10.3 carries out a linear buckling analysis on the beam bracket which has been discussed in Sections 4.1, 5.1, and 6.2.

Section 10.1

Stress Stiffening

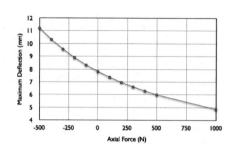

This section introduces *stress stiffening effect*, which is closely related to buckling. Stress stiffening effect is often observed in slender or thin structures or structural members, such as cables, shells, columns, walls, towers, trusses, etc. In a slender structural member, such as a column, the bending stiffness can be affected by its axial stress. In a thin structure member, such as a wall, the bending stiffness is affected by its in-plane stress. More specifically, when subject to axial (or in-plane) tension, the bending stiffness tends to increase, while subject to axial (or in-plane) compression, the bending stiffness tends to decrease.

In this section, we will use a simply supported slender beam to demonstrate the stress stiffening effects.

10.1-1 About the Simply Supported Beam

Consider a simply supported beam shown below [1-3]. The beam is made of steel and has a uniform cross section of 10 mm x 10 mm [1]. A uniformly distributed load of 0.1 N/mm is applied downward on the beam [2]. An axial force is applied at the beam's end which is free to move horizontally [3].

The vertical load, which contributes to the bending, will be fixed at 0.1 N/mm, while the horizontal force P will be varied from -500 N to 1000 N. Note that the negative P produces a compressive axial stress while the positive P produces a tensile axial stress. We will focus on the beam's maximum vertical deflection, which occurs at the middle of the span.

Let δ_0 be the beam's maximum deflection when $P = 0$. Then, when the beam is subject to a positive P, we will obtain a deflection less than δ_0. We can then conclude that the bending stiffness increases with the increasing tensile axial stress. On the other hand, when the beam is subject to a negative (compressive) P, we will obtain a deflection larger than δ_0. We can then conclude that the bending stiffness decreases with the increasing compressive axial stress. This effect is called the *stress stiffening*.

Since the bending stiffness decreases with the increasing compressive axial stress, you may raise a question: With how large the compressive force P, does the bending stiffness reduce to zero? A zero bending stiffness implies an unstable structure: a small lateral load would cause an infinitely large deflection. This phenomenon is called the *buckling*.

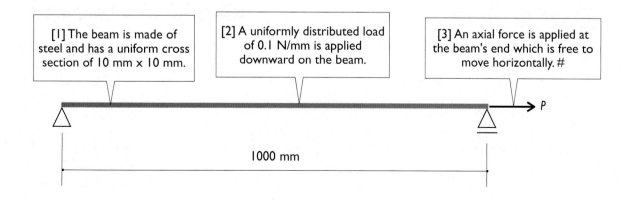

[1] The beam is made of steel and has a uniform cross section of 10 mm x 10 mm.

[2] A uniformly distributed load of 0.1 N/mm is applied downward on the beam.

[3] An axial force is applied at the beam's end which is free to move horizontally. #

1000 mm

10.1-2 Start a New Project

Launch Workbench. Create a **Static Structural** system. Save the project as **SimpleBeam**. Start up DesignModeler [1].

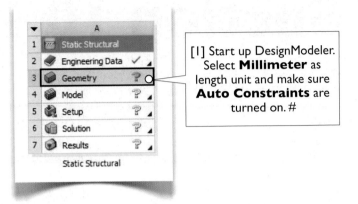

[1] Start up DesignModeler. Select **Millimeter** as length unit and make sure **Auto Constraints** are turned on. #

10.1-3 Create a Line Model in DesignModeler

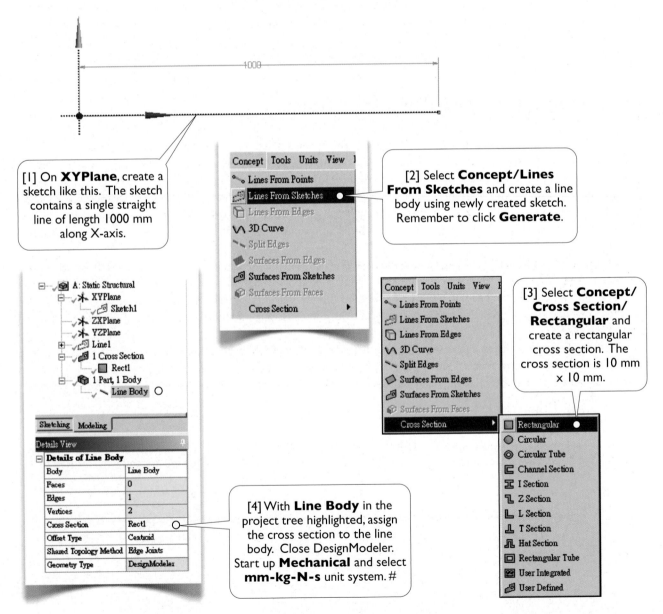

[1] On **XYPlane**, create a sketch like this. The sketch contains a single straight line of length 1000 mm along X-axis.

[2] Select **Concept/Lines From Sketches** and create a line body using newly created sketch. Remember to click **Generate**.

[3] Select **Concept/ Cross Section/ Rectangular** and create a rectangular cross section. The cross section is 10 mm x 10 mm.

[4] With **Line Body** in the project tree highlighted, assign the cross section to the line body. Close DesignModeler. Start up **Mechanical** and select **mm-kg-N-s** unit system. #

361

10.1-4 Set Up Supports

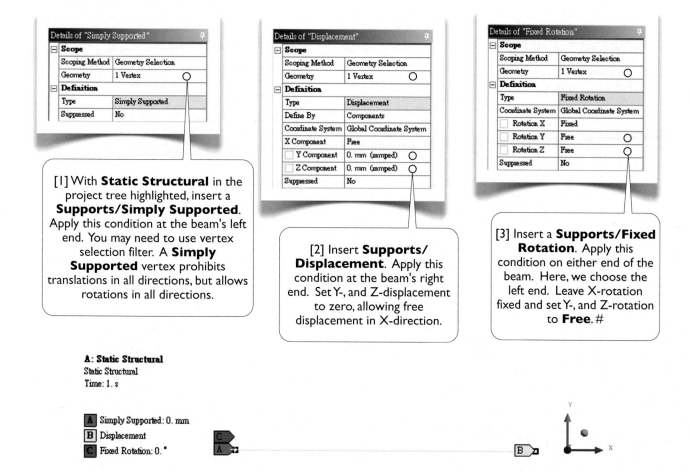

[1] With **Static Structural** in the project tree highlighted, insert a **Supports/Simply Supported**. Apply this condition at the beam's left end. You may need to use vertex selection filter. A **Simply Supported** vertex prohibits translations in all directions, but allows rotations in all directions.

[2] Insert **Supports/ Displacement**. Apply this condition at the beam's right end. Set Y-, and Z-displacement to zero, allowing free displacement in X-direction.

[3] Insert a **Supports/Fixed Rotation**. Apply this condition on either end of the beam. Here, we choose the left end. Leave X-rotation fixed and set Y-, and Z-rotation to **Free**. #

Provide Enough Supports to Avoid Rigid Body Modes

In 3.1-8 (page 111), we mentioned that it is a good practice to provide enough supports. This becomes a necessity when working on buckling or modal analyses, where rigid body modes are not automatically eliminated by using weak springs.

In this case, the newcomers often fail to fix Z-displacements in steps [1, 2] and the X-rotation in step [3]. Without these supports, a nonlinear simulation would run into convergence difficulties. In case of buckling or modal analyses, rigid body modes would present.

What's wrong with the presence of rigid body modes? The answer depends on the type of simulation. For **Static Structural** simulations, we already answer that question: Workbench will add weak springs to prevent uncontrolled large amount of rigid body motions (3.1-8, page 111), and allow a small amount of rigid body motion (3.1-10[2], page 112).

For **Linear Buckling** and **Modal** simulations, rigid body modes are trivial and appear no harm. The buckling load corresponding to a rigid body mode is zero, and the natural frequency corresponding a rigid body mode is also zero. However, presence of rigid body modes may deteriorate the numerical accuracy.

For **Transient Structural** simulations, we usually don't need to artificially eliminate rigid body modes if they exists by their natural. Let the rigid body modes be present, and the program will take care of them nicely. The exercises in Sections 12.4, 15.2, and 15.3 provide good examples for these situations.

10.1-5 Set Up Loads

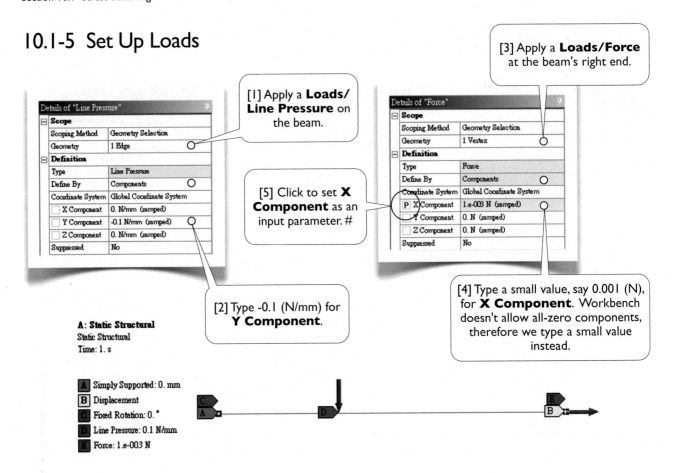

[1] Apply a **Loads/ Line Pressure** on the beam.

[3] Apply a **Loads/Force** at the beam's right end.

[5] Click to set **X Component** as an input parameter. #

[2] Type -0.1 (N/mm) for **Y Component**.

[4] Type a small value, say 0.001 (N), for **X Component**. Workbench doesn't allow all-zero components, therefore we type a small value instead.

A: Static Structural
Static Structural
Time: 1. s

A Simply Supported: 0. mm
B Displacement
C Fixed Rotation: 0. °
D Line Pressure: 0.1 N/mm
E Force: 1.e-003 N

10.1-6 Set Up Solution Objects

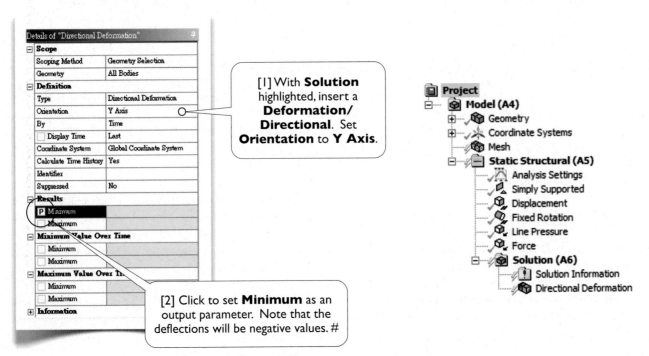

[1] With **Solution** highlighted, insert a **Deformation/ Directional**. Set **Orientation** to **Y Axis**.

[2] Click to set **Minimum** as an output parameter. Note that the deflections will be negative values. #

363

10.1-7 Set Up Mesh and Turn on Large Deflection Effect

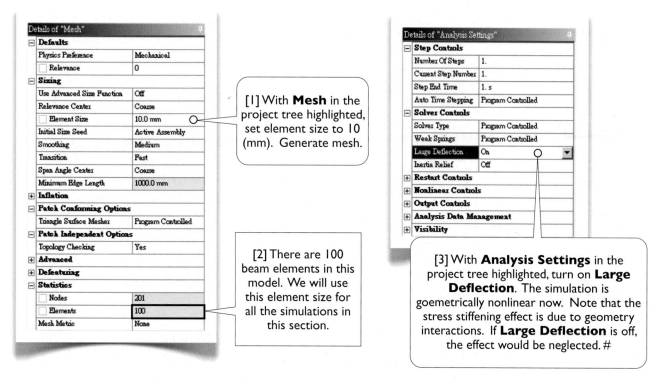

Details of "Mesh"

Defaults	
Physics Preference	Mechanical
☐ Relevance	0
Sizing	
Use Advanced Size Function	Off
Relevance Center	Coarse
☐ Element Size	10.0 mm
Initial Size Seed	Active Assembly
Smoothing	Medium
Transition	Fast
Span Angle Center	Coarse
Minimum Edge Length	1000.0 mm
Inflation	
Patch Conforming Options	
Triangle Surface Mesher	Program Controlled
Patch Independent Options	
Topology Checking	Yes
Advanced	
Defeaturing	
Statistics	
☐ Nodes	201
☐ Elements	100
Mesh Metric	None

[1] With **Mesh** in the project tree highlighted, set element size to 10 (mm). Generate mesh.

[2] There are 100 beam elements in this model. We will use this element size for all the simulations in this section.

Details of "Analysis Settings"

Step Controls	
Number Of Steps	1.
Current Step Number	1.
Step End Time	1. s
Auto Time Stepping	Program Controlled
Solver Controls	
Solver Type	Program Controlled
Weak Springs	Program Controlled
Large Deflection	On
Inertia Relief	Off
Restart Controls	
Nonlinear Controls	
Output Controls	
Analysis Data Management	
Visibility	

[3] With **Analysis Settings** in the project tree highlighted, turn on **Large Deflection**. The simulation is goemetrically nonlinear now. Note that the stress stiffening effect is due to geometry interactions. If **Large Deflection** is off, the effect would be neglected. #

10.1-8 Create Design Points

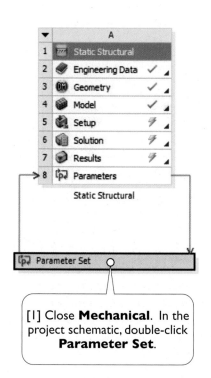

	A
1	Static Structural
2	Engineering Data ✓
3	Geometry ✓
4	Model ✓
5	Setup 🗲
6	Solution 🗲
7	Results 🗲
8	Parameters

Static Structural

Parameter Set

[1] Close **Mechanical**. In the project schematic, double-click **Parameter Set**.

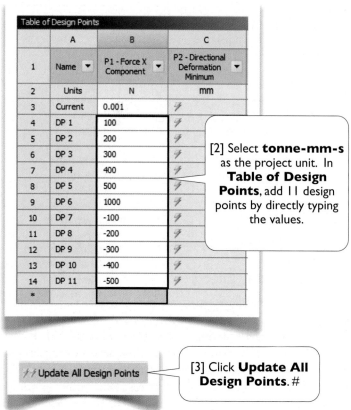

Table of Design Points

	A	B	C
1	Name	P1 - Force X Component	P2 - Directional Deformation Minimum
2	Units	N	mm
3	Current	0.001	🗲
4	DP 1	100	🗲
5	DP 2	200	🗲
6	DP 3	300	🗲
7	DP 4	400	🗲
8	DP 5	500	🗲
9	DP 6	1000	🗲
10	DP 7	-100	🗲
11	DP 8	-200	🗲
12	DP 9	-300	🗲
13	DP 10	-400	🗲
14	DP 11	-500	🗲
*			

[2] Select **tonne-mm-s** as the project unit. In **Table of Design Points**, add 11 design points by directly typing the values.

🗲🗲 Update All Design Points

[3] Click **Update All Design Points**. #

10.1-9 The Results and Discussion

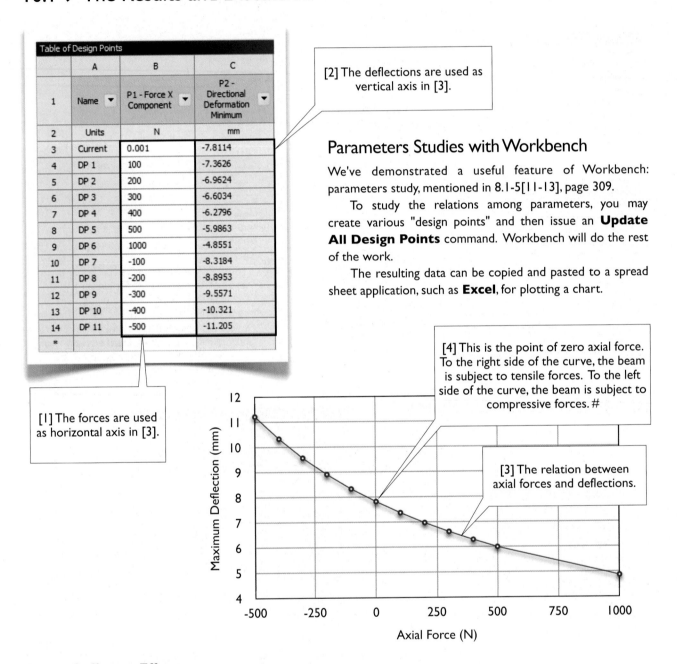

Table of Design Points

	A	B	C
1	Name	P1 - Force X Component	P2 - Directional Deformation Minimum
2	Units	N	mm
3	Current	0.001	-7.8114
4	DP 1	100	-7.3626
5	DP 2	200	-6.9624
6	DP 3	300	-6.6034
7	DP 4	400	-6.2796
8	DP 5	500	-5.9863
9	DP 6	1000	-4.8551
10	DP 7	-100	-8.3184
11	DP 8	-200	-8.8953
12	DP 9	-300	-9.5571
13	DP 10	-400	-10.321
14	DP 11	-500	-11.205
*			

[2] The deflections are used as vertical axis in [3].

Parameters Studies with Workbench

We've demonstrated a useful feature of Workbench: parameters study, mentioned in 8.1-5[11-13], page 309.

To study the relations among parameters, you may create various "design points" and then issue an **Update All Design Points** command. Workbench will do the rest of the work.

The resulting data can be copied and pasted to a spread sheet application, such as **Excel**, for plotting a chart.

[1] The forces are used as horizontal axis in [3].

[4] This is the point of zero axial force. To the right side of the curve, the beam is subject to tensile forces. To the left side of the curve, the beam is subject to compressive forces. #

[3] The relation between axial forces and deflections.

Stress Stiffening Effects

The curve [3] manifests the stress stiffening effects. As the tensile axial force increases, the deflection decreases, indicating an increase of bending stiffness. On the other hand, as the compressive axial force increases, the deflection increases, indicating a decrease of bending stiffness. When the compressive axial force reaches a certain point, the bending stiffness decreases so much that the deflection is enlarged dramatically. It naturally raises a question: what is the compressive force such that the bending stiffness completely vanishes? A zero bending stiffness implies an unstable structure: a small vertical load would cause the beam to collapse. This phenomenon is called the *buckling* and the compressive force causing the structure to buckle is called the *buckling load*, or *critical load*.

We can continue the above process and extend the curve [3] leftward until a vertical asymptote can be drawn. The force value intercepted by the asymptote would be the buckling force. And this procedure is basically a *nonlinear buckling analysis*[Ref 1].

Workbench provides a linear buckling analysis capability to assess the buckling load. A linear buckling analysis usually takes much less computing time than a nonlinear buckling analysis but usually overestimates the buckling load.

The linear buckling theory predicts the buckling load as

$$P_{buckling} = \frac{\pi^2 EI}{L^2} = \frac{\pi^2 (200,000)(833.33)}{(1000)^2} = 1645 \text{ N} \tag{1}$$

Note that the above calculation doesn't include the distributed beam load (0.1 N/mm). Although tending to overestimate the buckling load and deviate from reality, the linear buckling analysis is useful for two reasons: (a) It is computationally much cheaper than a nonlinear buckling analysis, and should be run as a first step to estimate the buckling load. (b) It can be used to determine the possible buckling mode.

We will proceed to demonstrate the linear buckling analysis for this case in the rest of the section.

10.1-10 Set Up Project Schematic

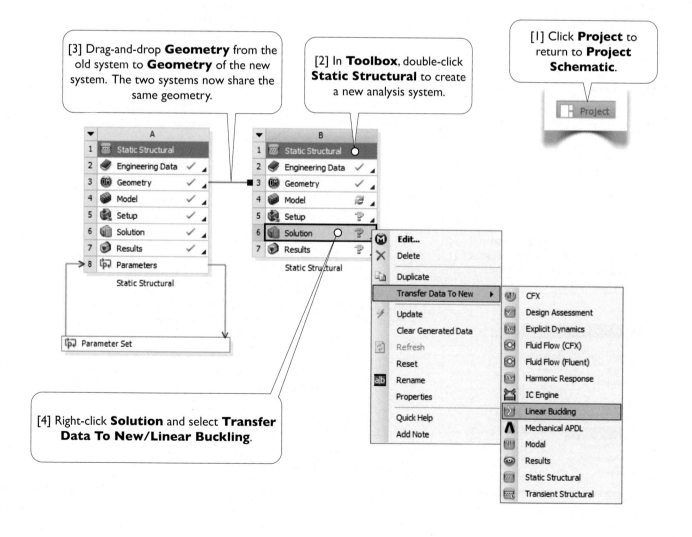

[3] Drag-and-drop **Geometry** from the old system to **Geometry** of the new system. The two systems now share the same geometry.

[2] In **Toolbox**, double-click **Static Structural** to create a new analysis system.

[1] Click **Project** to return to **Project Schematic**.

[4] Right-click **Solution** and select **Transfer Data To New/Linear Buckling**.

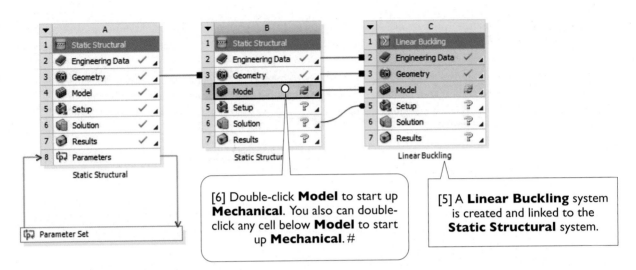

[6] Double-click **Model** to start up **Mechanical**. You also can double-click any cell below **Model** to start up **Mechanical**. #

[5] A **Linear Buckling** system is created and linked to the **Static Structural** system.

10.1-11 Set Up Supports

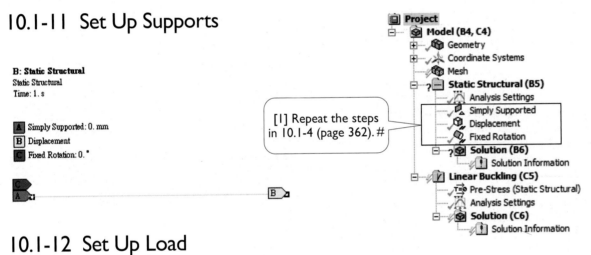

B: **Static Structural**
Static Structural
Time: 1. s

A Simply Supported: 0. mm
B Displacement
C Fixed Rotation: 0. °

[1] Repeat the steps in 10.1-4 (page 362). #

10.1-12 Set Up Load

[1] With **Static Structural** still highlighted, insert a **Loads/Force**. Select the beam's right end.

[2] Type an arbitrary compressive value, say -100 (N), for **X Component**. As a result of buckling analysis, the buckling load will be reported as a multiplier of the applied loads. #

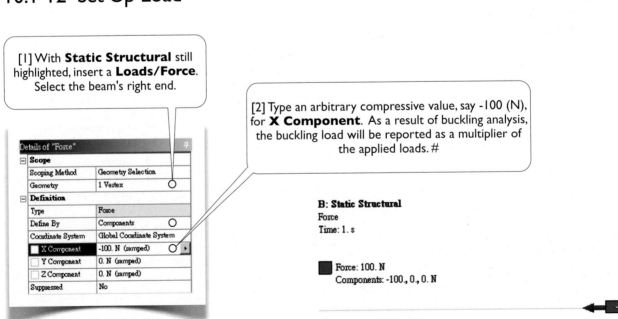

B: **Static Structural**
Force
Time: 1. s

Force: 100. N
Components: -100., 0., 0. N

10.1-13 Set Up Mesh

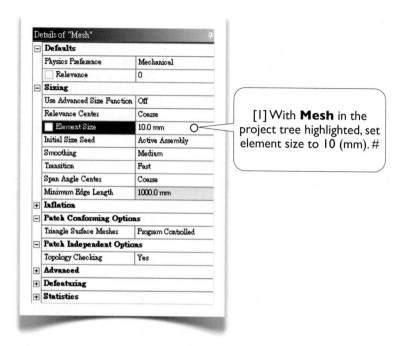

[1] With **Mesh** in the project tree highlighted, set element size to 10 (mm). #

10.1-14 Specify Number of Buckling Modes and Solve

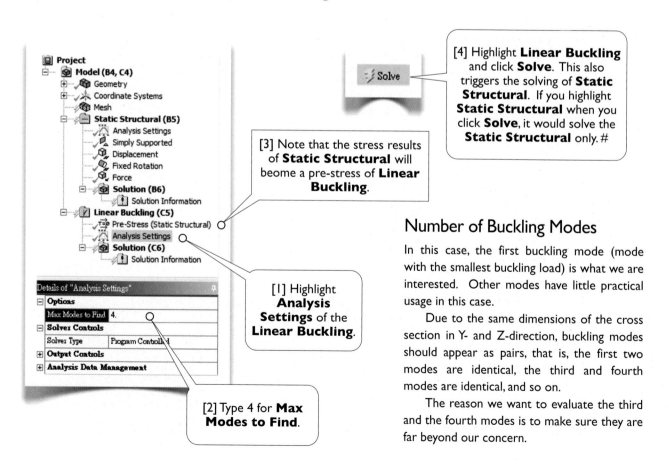

[4] Highlight **Linear Buckling** and click **Solve**. This also triggers the solving of **Static Structural**. If you highlight **Static Structural** when you click **Solve**, it would solve the **Static Structural** only. #

[3] Note that the stress results of **Static Structural** will beome a pre-stress of **Linear Buckling**.

[1] Highlight **Analysis Settings** of the **Linear Buckling**.

[2] Type 4 for **Max Modes to Find**.

Number of Buckling Modes

In this case, the first buckling mode (mode with the smallest buckling load) is what we are interested. Other modes have little practical usage in this case.

Due to the same dimensions of the cross section in Y- and Z-direction, buckling modes should appear as pairs, that is, the first two modes are identical, the third and fourth modes are identical, and so on.

The reason we want to evaluate the third and the fourth modes is to make sure they are far beyond our concern.

10.1-15 View the Results

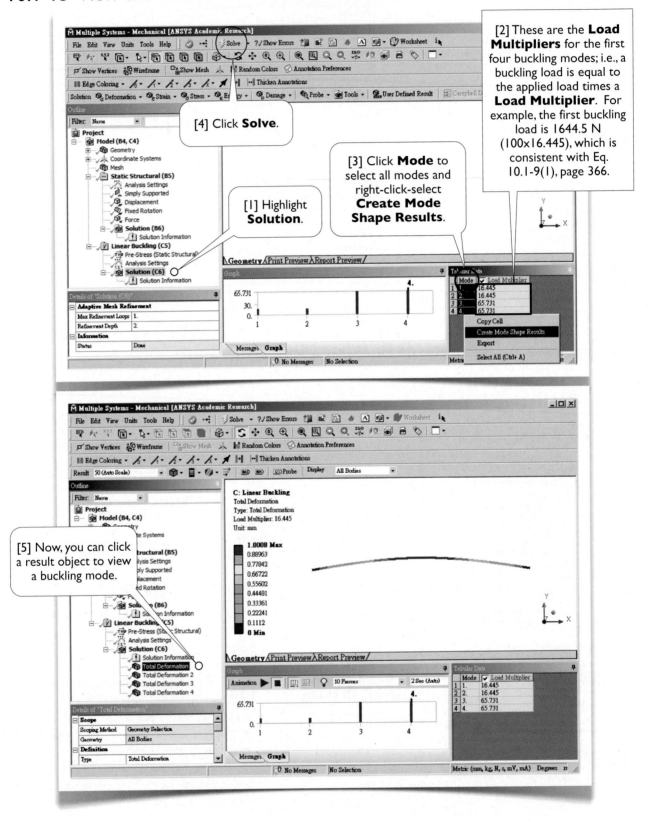

[2] These are the **Load Multipliers** for the first four buckling modes; i.e., a buckling load is equal to the applied load times a **Load Multiplier**. For example, the first buckling load is 1644.5 N (100x16.445), which is consistent with Eq. 10.1-9(1), page 366.

[4] Click **Solve**.

[3] Click **Mode** to select all modes and right-click-select **Create Mode Shape Results**.

[1] Highlight **Solution**.

[5] Now, you can click a result object to view a buckling mode.

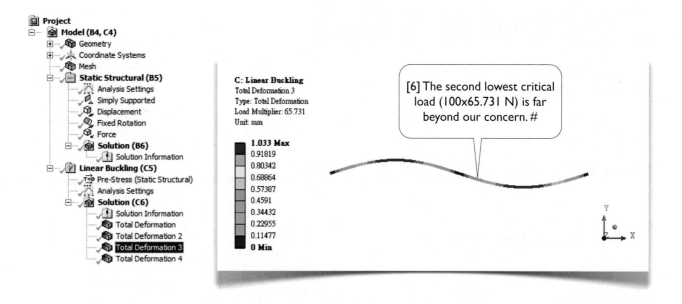

Buckling Mode Shapes

When displaying the buckling mode shapes, Workbench scales the values of deformation such that the maximum deformation is approximately 1.0. The values of deformation have no physical significance. It is the mode shapes that are useful. Similarly, the stresses or strains calculated in linear buckling analyses have no physical meaning.

Wrap Up

Save the project and exit Workbench.

Reference

1. ANSYS Documentation//Mechanical APDL//Structural Analysis Guide//7.3. Performing a Nonlinear Buckling Analysis

Section 10.2

3D Truss

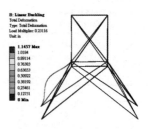

10.2-1 About the 3D Truss

In Section 7.2, we analyzed a 3D truss and obtained some preliminary results, including stresses.

The stresses range from -25,000 psi to +14,000 psi (7.2-12[4], page 284). They seem okay, since the structural steel's yield strength can be as high as 40,000 psi for both tension and compression.

Stress is only one of design considerations. Structural stability must be verified whenever compressive members are involved in the structure system.

Let's make some simple calculations to check whether structural stability should be an issue, by considering the member that has the maximum compressive force, which is $P = 15,802$ lb [1].

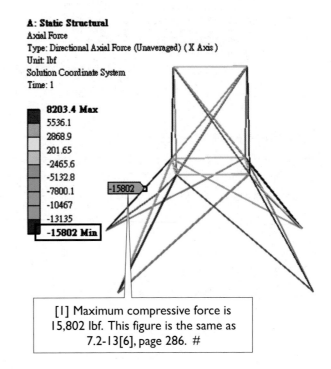

[1] Maximum compressive force is 15,802 lbf. This figure is the same as 7.2-13[6], page 286. #

The member (of maximum compressive force) has a length of 133.46 inches and a cross section (L $1\frac{1}{2} \times 1\frac{1}{2} \times \frac{1}{4}$) of an area moment of inertia of 0.13852 in^4. The structural steel's Young's modulus is 29,000,000 psi. Its buckling load, according to linear buckling theory, is estimated to be

$$P_{buckling} = \frac{\pi^2 EI}{L^2} = \frac{\pi^2(29,000,000)(0.13852)}{(133.46)^2} = 2,226 \text{ lb} = 0.14(15,802) = 0.14P$$

In other words, merely 14% of the design loads (7.2-1, page 275) would cause one of the structural members to buckle. It is now obvious that the structural stability might be a problem.

In this section, we want to perform a linear buckling analysis for the entire 3D truss structure, rather than a single member. The linear buckling analysis will result a higher multiplier than 14%. Two factors cause the deviation from the simple calculation. First, this truss is a statically indeterminate structure; buckling of one member does not necessarily cause a buckling of the entire structure. Second, the rigid joints provide additional rigidity to increase the buckling load.

The unit system used in this section is **in-lbm-lbf-s**.

10.2-2 Resume the Project **Truss**

Launch Workbench. Open the project **Truss**, saved in Section 7.2.

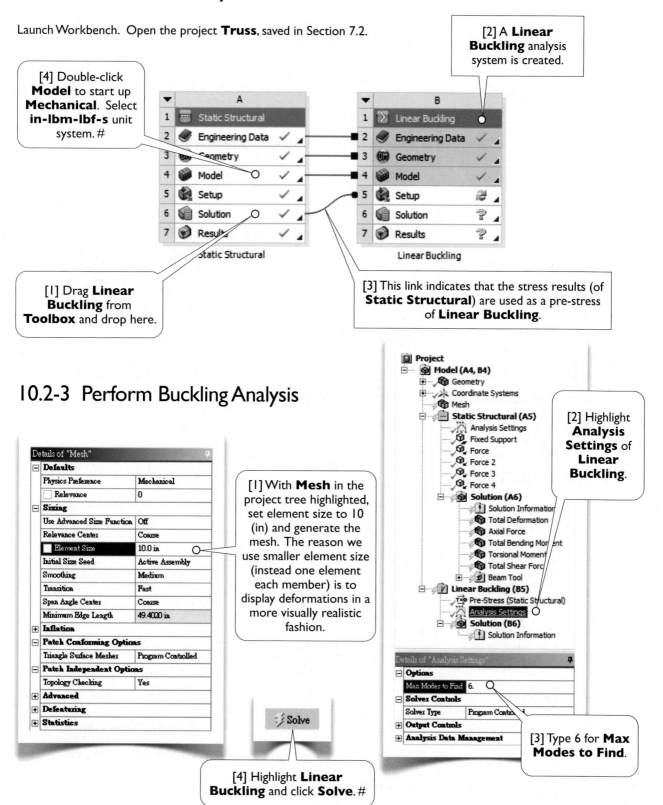

[2] A **Linear Buckling** analysis system is created.

[4] Double-click **Model** to start up **Mechanical**. Select **in-lbm-lbf-s** unit system. #

[1] Drag **Linear Buckling** from **Toolbox** and drop here.

[3] This link indicates that the stress results (of **Static Structural**) are used as a pre-stress of **Linear Buckling**.

10.2-3 Perform Buckling Analysis

[1] With **Mesh** in the project tree highlighted, set element size to 10 (in) and generate the mesh. The reason we use smaller element size (instead one element each member) is to display deformations in a more visually realistic fashion.

[2] Highlight **Analysis Settings** of **Linear Buckling**.

[3] Type 6 for **Max Modes to Find**.

[4] Highlight **Linear Buckling** and click **Solve**. #

10.2-4 View the Results

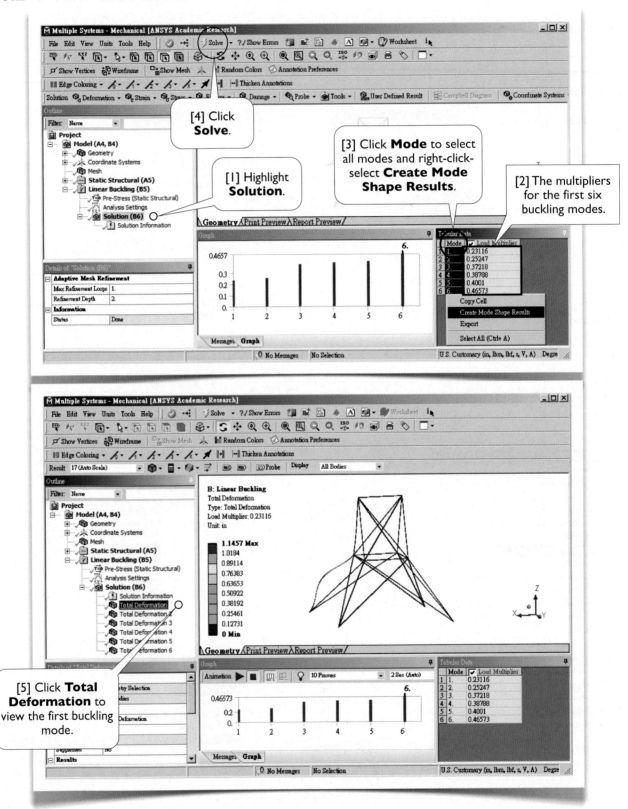

[6] Buckling will occur when 23% of design loads apply on the structure. The **Load Multiplier** can be viewed as a safety factor as far as buckling is concerned. The structure is not safe.

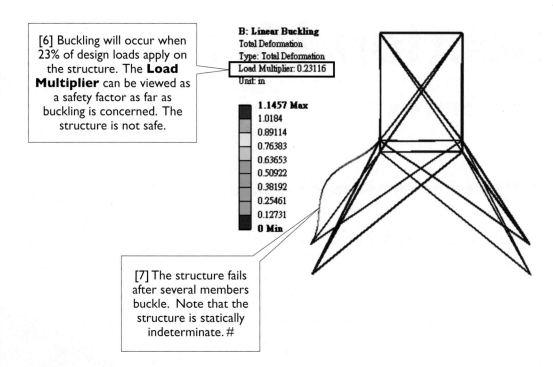

B: Linear Buckling
Total Deformation
Type: Total Deformation
Load Multiplier: 0.23116
Unit: in

1.1457 Max
1.0184
0.89114
0.76383
0.63653
0.50922
0.38192
0.25461
0.12731
0 Min

[7] The structure fails after several members buckle. Note that the structure is statically indeterminate. #

Wrap Up

Save the project and exit Workbench.

Section 10.3

Beam Bracket

10.3-1 About the Beam Bracket

In Section 5.1, we simulated a beam bracket using a 3D solid model.

The maximum von Mises stress is 83 MPa [1], well below the yield strength, 250 MPa.

By examining **Minimum Principal Stress** [2], we see that the web is subject to compressive stress; its magnitude is about 23 MPa [3]. It is a good practice that an engineer always checks the structural stability whenever compressive stresses exist, unless he has enough experience to judge that the stability checking is not necessary.

In this section, we want to make sure that, under the design load, the web does not buckle.

The unit system **mm-kg-N-s** is used in this section.

[1] The maximum von Mises stress. This figure is the same as 5.1-13[9], page 210.

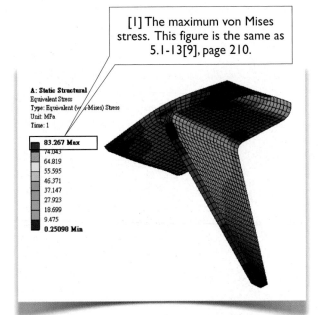

[2] **Minimum Principal Stress** can be used to examine compressive stresses.

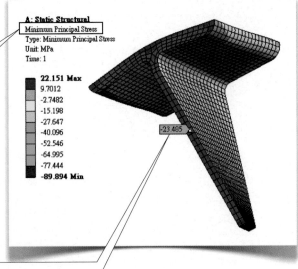

[3] Compressive stress at the web. #

10.3-2 Resume the Project **Bracket**

Launch Workbench. Open the project
Bracket, saved in Section 6.2.

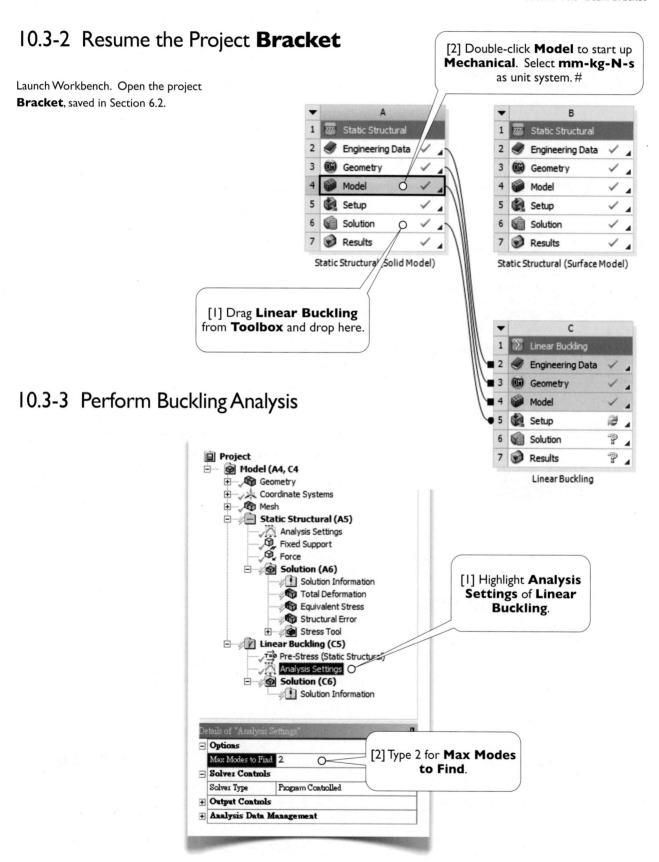

[2] Double-click **Model** to start up
Mechanical. Select **mm-kg-N-s**
as unit system. #

Static Structural (Solid Model)

Static Structural (Surface Model)

[I] Drag **Linear Buckling**
from **Toolbox** and drop here.

10.3-3 Perform Buckling Analysis

Linear Buckling

[I] Highlight **Analysis
Settings** of **Linear
Buckling**.

[2] Type 2 for **Max Modes
to Find**.

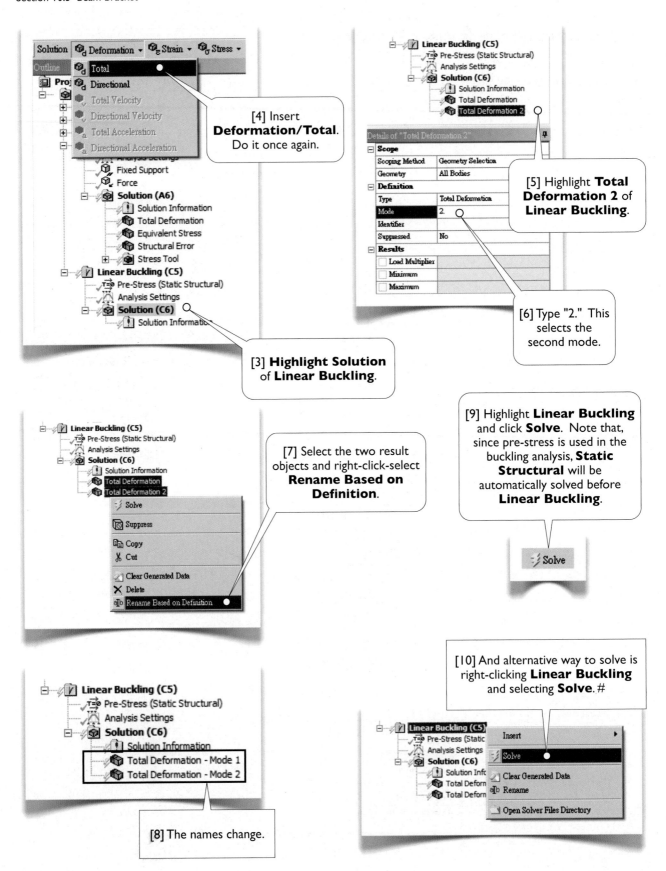

[4] Insert **Deformation/Total**. Do it once again.

[5] Highlight **Total Deformation 2** of **Linear Buckling**.

[6] Type "2." This selects the second mode.

[3] **Highlight Solution** of **Linear Buckling**.

[7] Select the two result objects and right-click-select **Rename Based on Definition**.

[9] Highlight **Linear Buckling** and click **Solve**. Note that, since pre-stress is used in the buckling analysis, **Static Structural** will be automatically solved before **Linear Buckling**.

[10] And alternative way to solve is right-clicking **Linear Buckling** and selecting **Solve**. #

[8] The names change.

10.3-4 View the Results

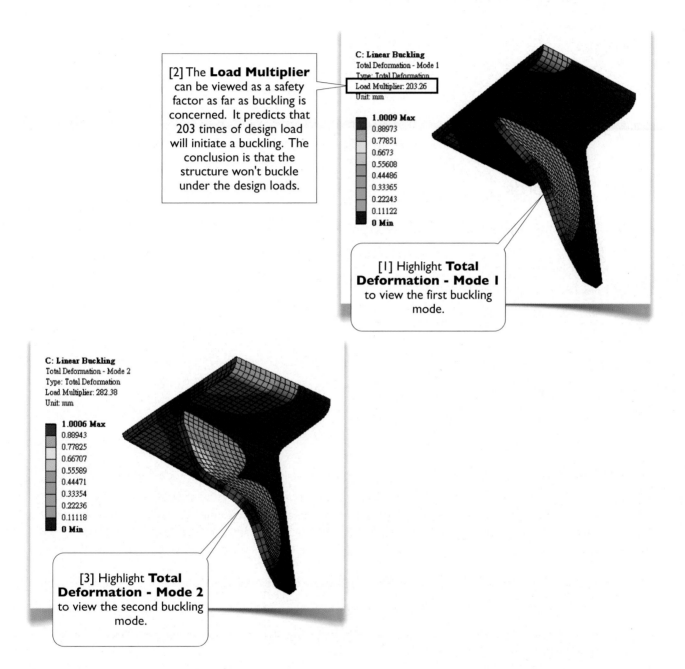

[2] The **Load Multiplier** can be viewed as a safety factor as far as buckling is concerned. It predicts that 203 times of design load will initiate a buckling. The conclusion is that the structure won't buckle under the design loads.

C: Linear Buckling
Total Deformation - Mode 1
Type: Total Deformation
Load Multiplier: 203.26
Unit: mm

1.0009 Max
0.88973
0.77851
0.6673
0.55608
0.44486
0.33365
0.22243
0.11122
0 Min

[1] Highlight **Total Deformation - Mode 1** to view the first buckling mode.

C: Linear Buckling
Total Deformation - Mode 2
Type: Total Deformation
Load Multiplier: 282.38
Unit: mm

1.0006 Max
0.88943
0.77825
0.66707
0.55589
0.44471
0.33354
0.22236
0.11118
0 Min

[3] Highlight **Total Deformation - Mode 2** to view the second buckling mode.

Wrap Up

Save the project and exit Workbench.

Section 10.4

Review

10.4-1 Keywords

Choose a letter for each keyword from the list of descriptions

1. (   ) Linear Buckling Analysis
2. (   ) Nonlinear Buckling Analysis
3. (   ) Stress Stiffening Effects

Answers:

1. (B)   2. (C)   3. (A)

List of Descriptions

(A)   A slender or thin structural member's bending stiffness increases with increasing axial tensile stress. Similarly, the member's bending stiffness decreases with increasing compressive stress. The extra (or deficient) stiffness is called the stress stiffness.

(B)   Prediction of buckling loads and buckling modes based solely on initial stress stiffening effect. It doesn't account for large deformation effect.

(C)   Prediction of buckling loads using a nonlinear analysis technique. It accounts for all nonlinear effects, including large deformation effect and nonlinear material effect.

10.4-2 Additional Workbench Exercises

Linear Buckling Analysis with Constant Loads

In the buckling analysis at the end of Section 10.1, we didn't include the lateral load (0.1 N/mm). You may wonder if the existence of the lateral load would alter the linear buckling load. Our engineering intuition tells us that a little lateral load would significantly decrease the buckling load. In linear buckling analysis, the large deformation effect is not considered. As a result, the lateral load has limited influence on the buckling loads. In our case, since the lateral load is so small, the buckling load predicted by a linear buckling analysis will be essentially the same regardless of the presence of the lateral load. This exercise requires you to verify this point.

To do this, you are performing a linear buckling analysis with constant loads. Remember that a linear buckling analysis reports a multiplier. Buckling loads are ALL the applied loads multiply by the multiplier. If the model involves a constant load, then there is no sense to multiply that constant load unless the multiplier equals a unity. Iterating on buckling analysis and trying to obtain a multiplier of unity is exactly the approach you should use when performing a linear buckling analysis with constant loads.

Buckling Pressure of Bellows Joints

The bellows joint, simulated in Section 6.1, is subject to external pressure when used in the deep ocean. If the external pressure is larger than the internal pressure, then its stability must be checked. This exercise asks you to predict the net pressure (difference between external and internal pressures) that causes the bellows joint to buckle. Also study its buckling modes.

Buckling Torque of a Beverage Can

Applying a twist on a thin beverage can will introduce tensile stress on a principal direction and compressive stress on another principal direction. Excess compressive stress may cause the skin to buckle. Model an beverage can as a cylindrical surface of length 122 mm, a diameter of 64 mm, and a thickness of 0.1 mm [1]. Assume that the can is made of AA3004, which has a Young's modulus of 68.9 MPa and a Poisson's ratio of 0.35. Predict the torque that causes the skin to buckle. Study the buckling modes [2].

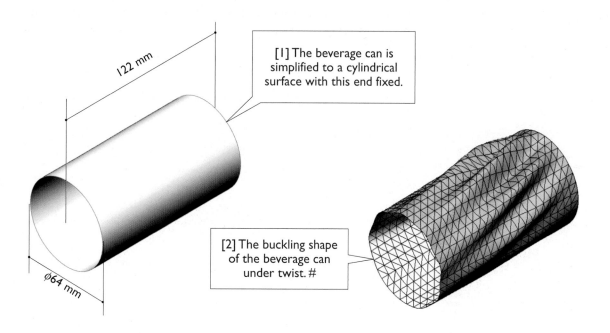

[1] The beverage can is simplified to a cylindrical surface with this end fixed.

[2] The buckling shape of the beverage can under twist. #

Chapter 11
Modal Analysis

When a structure moves or deforms very fast, dynamic effects must be included in the simulation. Modal analysis is a special type of dynamic simulation, exploring the behavior of free vibrations, vibrations without external forces. The most important characteristics of free vibrations are the natural frequencies and their corresponding vibration mode shapes.

Purpose of This Chapter

Why do we need to know the natural frequencies and the mode shapes? First, modal analysis has many usages in its own right; for example, to find the consonance frequencies of a structure in order to avoid them. This chapter provides several examples to demonstrate the usage of modal analyses. Second, these dynamic characteristics are important for further dynamic simulations; for examples, transient dynamic simulations or harmonic response analyses. Chapter 12 will have several examples to demonstrate this.

About Each Section

Section 11.1 uses the gearbox, introduced in Section 6.3, as an example to illustrate a common consideration when designing a machine involving rotary parts: avoiding resonance. Section 11.2 uses the two-story building, introduced in Section 7.3, to demonstrate the use of modal analysis to find the weakest direction of a structure and improve the stiffness of that direction. Section 11.3 discusses a case in the popular TV series *Mythbusters*, in which they succeeded in shattering CDs with a high rotational speed. Some people may have a myth that the shattering is due to the excessive centrifugal stress. This section is designed to bust that myth, by proving that the shattering is due to resonant vibrations rather than the centrifugal stress. Section 11.4 discusses the physics of music, which is closely related to modal analysis, using a guitar string as an example.

Section 11.1

Gearbox

11.1-1 About the Gearbox

In Section 6.3, we performed a static structural simulation for a gearbox. Deformation and stresses seem within safety margin under the static design loads. Dynamic behavior, however, should also be investigated.

The gearbox is designed for a speed reducer. The maximum speed at input is 630 rpm (10.5 Hz). The gearbox must be stiff enough such that, during the operation, resonance does not occur. Resonance, in this case, is harmful because it enlarges the deformation and the stresses. It also causes noises.

In this section, we want to perform a modal analysis to investigate the natural frequencies of the gearbox, to make sure these natural frequencies are much higher than the operational frequency (630 rpm).

You may raise a question: when evaluating the natural frequencies, whether the design bearing loads (6.3-7, page 253) should be applied or not. In general, the answer is yes, since the prestress will modify the stiffness (called the stress stiffening effect, see Section 10.1). In the old days, since a "prestressed modal analysis" is expensive to perform, engineers tended to neglect the prestress effect when, according to their experiences, they knew that the prestress effect is negligible.

In this section, we will perform modal analysis twice, one with prestress, another without, to show that the effect of the prestress in this case is negligible. Note that, in other cases, the effect of prestress may be significant.

11.1-2 Resume the Project **GearBox**

Launch Workbench. Open the project **Gearbox**, saved in Section 6.3.

> [2] Double-click **Model** (or any cell below) to start up **Mechanical**. Make sure the unit system is **mm-kg-N-s**. #

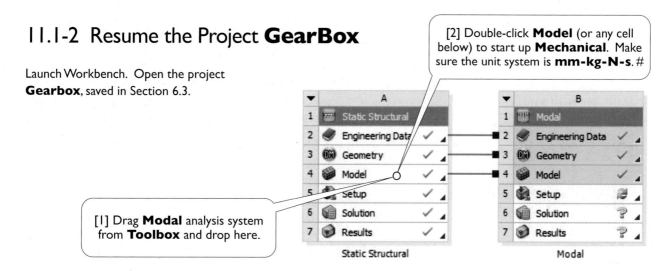

> [1] Drag **Modal** analysis system from **Toolbox** and drop here.

Static Structural simulation is not always needed for a modal analysis

We are now performing an "unprestressed modal analysis." When performing an unprestressed modal analysis, a **Static Structural** analysis is not needed (i.e., you might delete it). We keep **Static Structural** system because we want to perform a prestressed modal analysis later (11.1-5, pages 385-386).

11.1-3 Perform Modal Analysis

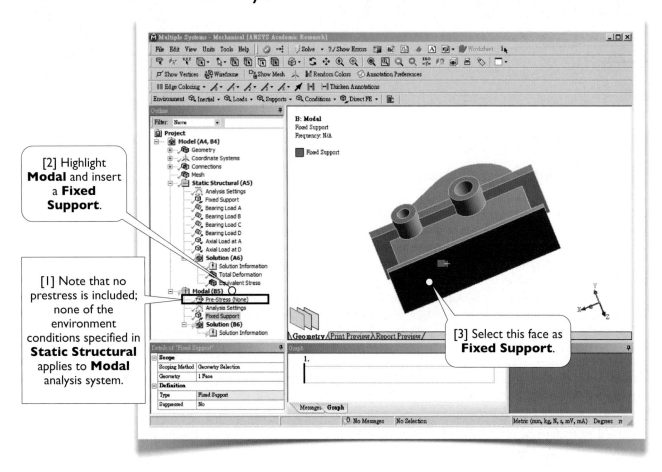

[2] Highlight **Modal** and insert a **Fixed Support**.

[1] Note that no prestress is included; none of the environment conditions specified in **Static Structural** applies to **Modal** analysis system.

[3] Select this face as **Fixed Support**.

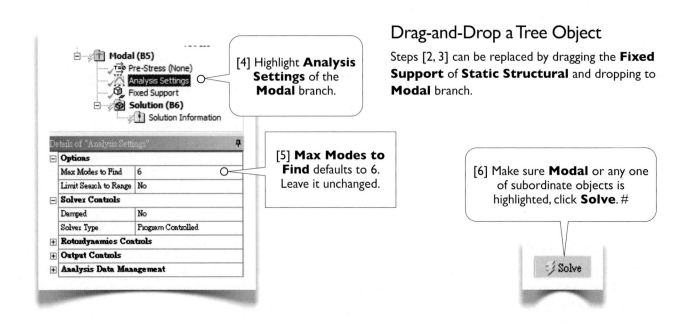

[4] Highlight **Analysis Settings** of the **Modal** branch.

[5] **Max Modes to Find** defaults to 6. Leave it unchanged.

Drag-and-Drop a Tree Object

Steps [2, 3] can be replaced by dragging the **Fixed Support** of **Static Structural** and dropping to **Modal** branch.

[6] Make sure **Modal** or any one of subordinate objects is highlighted, click **Solve**. #

11.1-4 View the Results

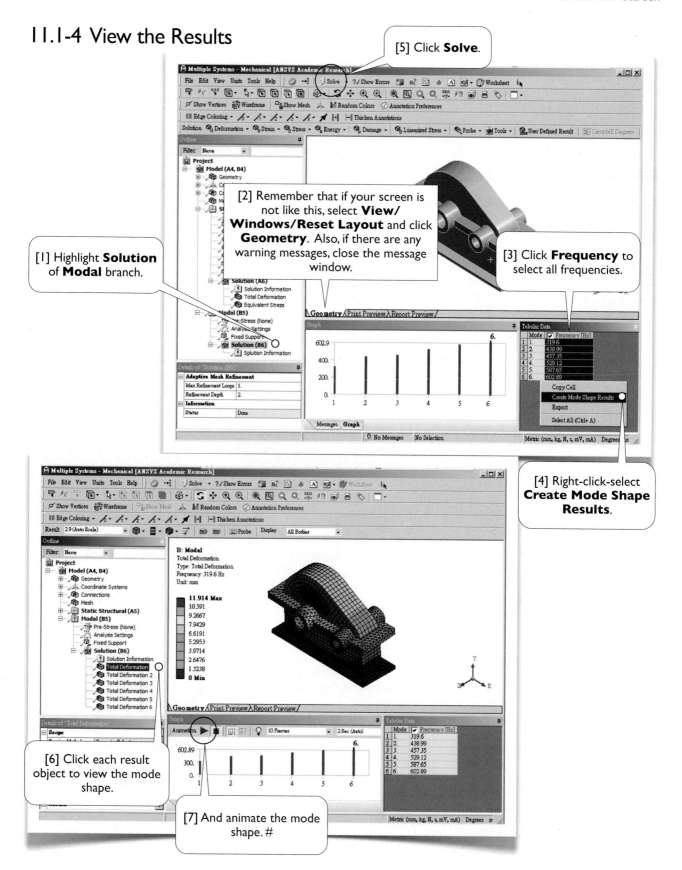

[5] Click **Solve**.

[2] Remember that if your screen is not like this, select **View/ Windows/Reset Layout** and click **Geometry**. Also, if there are any warning messages, close the message window.

[1] Highlight **Solution** of **Modal** branch.

[3] Click **Frequency** to select all frequencies.

[4] Right-click-select **Create Mode Shape Results**.

[6] Click each result object to view the mode shape.

[7] And animate the mode shape. #

Free Vibration Mode Shapes

Vibrations you've observed in [6, 7] are called *free vibrations* since no external excitations apply on the structure. Like the buckling mode shapes discussed in the last chapter, values of deformation have no physical significance. It is the "shape" of a vibration mode that is meaningful. Likewise, the stresses or strains have no physical meaning.

Natural Frequencies

The frequencies corresponding to the free vibrations are called *natural frequencies*. The lowest natural frequency is called the *fundamental natural frequency*, or simply *fundamental frequency*. In this case, the fundamental frequency is 319.6 Hz (19,176 rpm), far beyond the operational frequency (630 rpm). Before we jump to conclude that the resonance is not an issue, let's make sure that the prestress is negligible.

Rigid Body Modes

When performing a modal analysis, you should provide enough supports to avoid any rigid body motions. If you didn't, then rigid body modes would be included in the results. Rigid body modes have infinite period, or, equivalently, zero frequencies. Rigid body modes are superfluous, and should not be present (also see 10.1-4, page 362).

11.1-5 Perform Prestressed Modal Analysis

Close **Mechanical**.

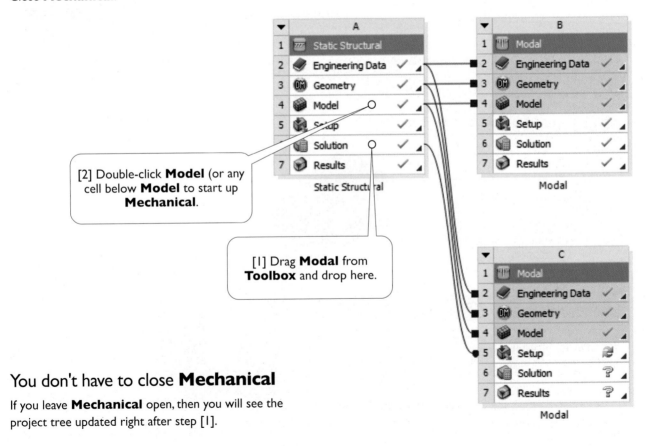

[2] Double-click **Model** (or any cell below **Model** to start up **Mechanical**.

[1] Drag **Modal** from **Toolbox** and drop here.

You don't have to close **Mechanical**

If you leave **Mechanical** open, then you will see the project tree updated right after step [1].

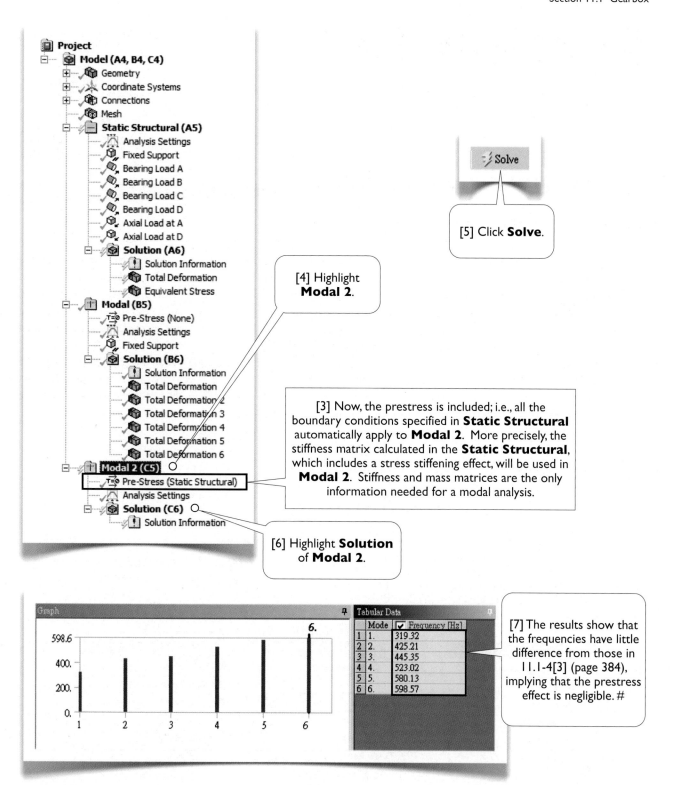

[4] Highlight **Modal 2**.

[5] Click **Solve**.

[3] Now, the prestress is included; i.e., all the boundary conditions specified in **Static Structural** automatically apply to **Modal 2**. More precisely, the stiffness matrix calculated in the **Static Structural**, which includes a stress stiffening effect, will be used in **Modal 2**. Stiffness and mass matrices are the only information needed for a modal analysis.

[6] Highlight **Solution** of **Modal 2**.

	Mode	✔ Frequency [Hz]
1	1.	319.32
2	2.	425.21
3	3.	445.35
4	4.	523.02
5	5.	580.13
6	6.	598.57

[7] The results show that the frequencies have little difference from those in 11.1-4[3] (page 384), implying that the prestress effect is negligible. #

Wrap Up

Save the project and exit Workbench.

Section 11.2

Two-Story Building

11.2-1 About the Two-Story Building

In Section 7.3, we performed a static structural simulation for a two-story building. Under the static loads, the deformation and stresses are within safety margin. Dynamic behavior, however, should be investigated for a case like this.

A structure's natural frequencies represent the structure's stiffness: the higher frequency, the stiffer. The stiffnesses of two structures can be compared using their fundamental frequencies. For a structure, stiffnesses in different directions can be compared using the lowest frequencies in the respective directions. When we want to reinforce a structure, we should first reinforce the direction which has the lowest frequency.

Local building codes usually require a minimum frequency level to avoid a building becoming too soft (i.e., not stiff enough), both for comfort and safety concerns. In this section, we will perform modal analyses using the two-story building model. Prestress is included in the simulation. We will find that the fundamental frequency of the building is too low. We then propose a simple solution to improve the stiffness of the building.

11.2-2 Resume the Project **Building**

Launch Workbench. Open the project **Building**, saved in Section 7.3.

[1] Right-click here and select **Duplicate**. We will use the duplicated system in this section and leave this original system for use in Section 12.3.

[4] Double-click **Model** (or any cell below) to start up **Mechanical**. #

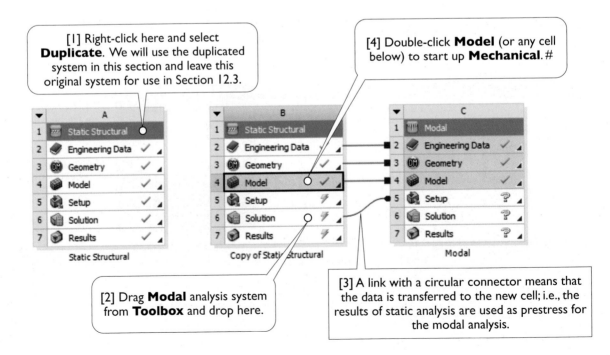

[2] Drag **Modal** analysis system from **Toolbox** and drop here.

[3] A link with a circular connector means that the data is transferred to the new cell; i.e., the results of static analysis are used as prestress for the modal analysis.

11.2-3 Perform Modal Analysis

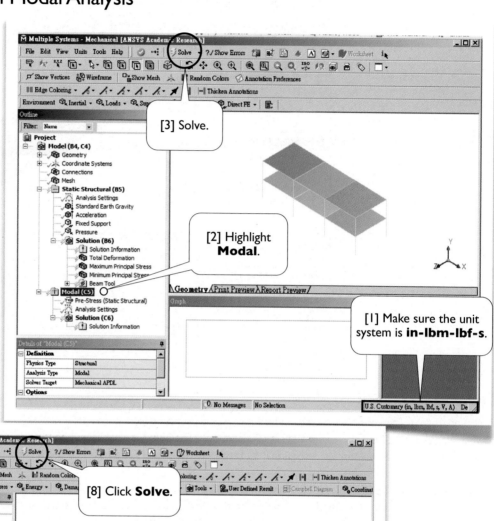

[3] Solve.

[2] Highlight **Modal**.

[1] Make sure the unit system is **in-lbm-lbf-s**.

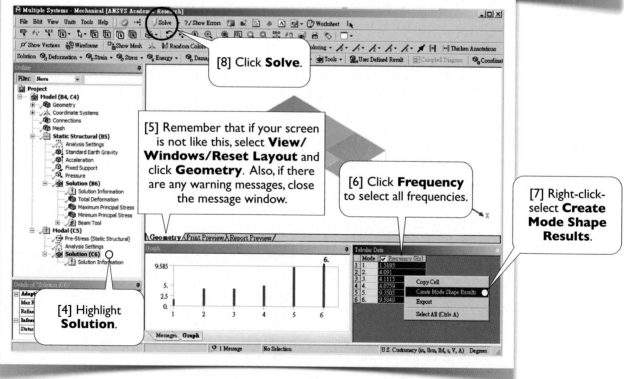

[8] Click **Solve**.

[5] Remember that if your screen is not like this, select **View/Windows/Reset Layout** and click **Geometry**. Also, if there are any warning messages, close the message window.

[6] Click **Frequency** to select all frequencies.

[7] Right-click-select **Create Mode Shape Results**.

[4] Highlight **Solution**.

388

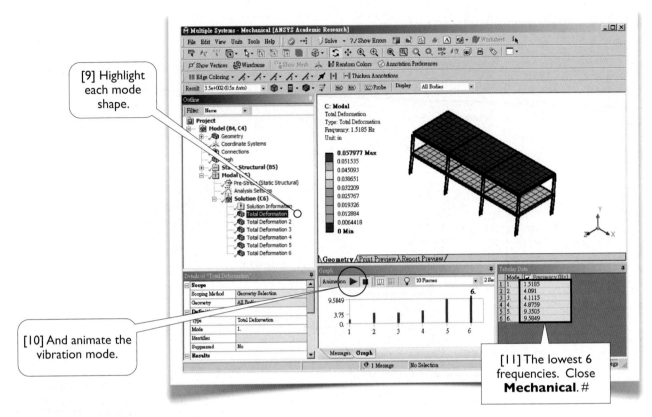

[9] Highlight each mode shape.

[10] And animate the vibration mode.

[11] The lowest 6 frequencies. Close **Mechanical**. #

Discussion of Vibration Modes

The fundamental frequency is 1.52 Hz, a vibration in X-direction (this can be observed from the animation). The second lowest frequency is 4.09 Hz, a vibration in Z-direction. The third lowest frequency mode is also in X-direction. The fourth lowest frequency mode is a torsional vibration in XZ-plane. The fifth and sixth modes are vertical (Y-direction) vibrations of the floors.

A frequency of 1.52 Hz is not only uncomfortable but sometimes unsafe. Imagine a group of young people dancing on the building's floor. The rhythmic loading of the floor may cause a safety issue, since the tempo of the music is possibly close to the building's fundamental frequency. A harmonic response analysis will be conducted to clear up this safety issue in Section 12.3. Some local building codes require a structure's natural frequency be larger than 5 Hz if the structure is to be used as a dance floor in a venue.

A simple remedy for the building is to add diagonal members to the weakest direction. Location of the diagonal members should be carefully chosen to avoid conflicting with the building's architectural functionalities.

11.2-4 Modify the Geometry

[1] Start up DesignModeler.

Close **Mechanical**.

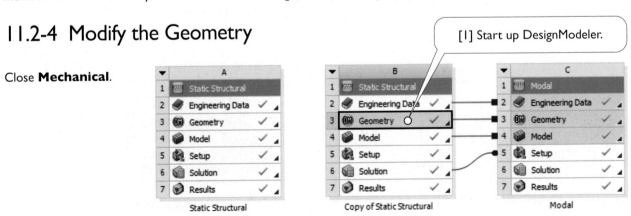

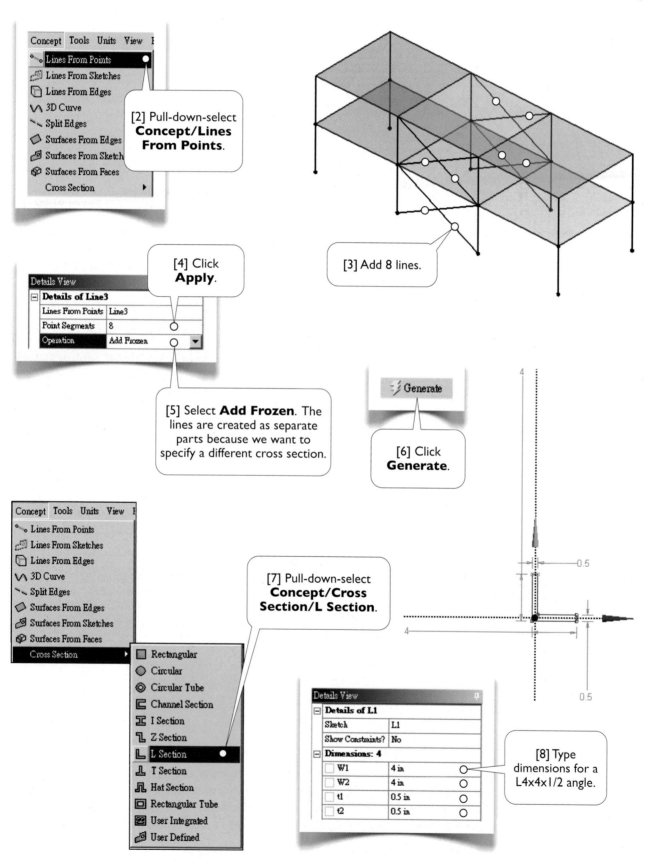

Concept Tools Units View

Lines From Points
Lines From Sketches
Lines From Edges
3D Curve
Split Edges
Surfaces From Edges
Surfaces From Sketch
Surfaces From Faces
Cross Section

[2] Pull-down-select **Concept/Lines From Points**.

[3] Add 8 lines.

Details View

Details of Line3	
Lines From Points	Line3
Point Segments	8
Operation	Add Frozen

[4] Click **Apply**.

[5] Select **Add Frozen**. The lines are created as separate parts because we want to specify a different cross section.

Generate

[6] Click **Generate**.

Concept Tools Units View

Lines From Points
Lines From Sketches
Lines From Edges
3D Curve
Split Edges
Surfaces From Edges
Surfaces From Sketches
Surfaces From Faces
Cross Section

Rectangular
Circular
Circular Tube
Channel Section
I Section
Z Section
L Section
T Section
Hat Section
Rectangular Tube
User Integrated
User Defined

[7] Pull-down-select **Concept/Cross Section/L Section**.

4

0.5

4

0.5

Details View

Details of L1	
Sketch	L1
Show Constraints?	No
Dimensions: 4	
W1	4 in
W2	4 in
t1	0.5 in
t2	0.5 in

[8] Type dimensions for a L4x4x1/2 angle.

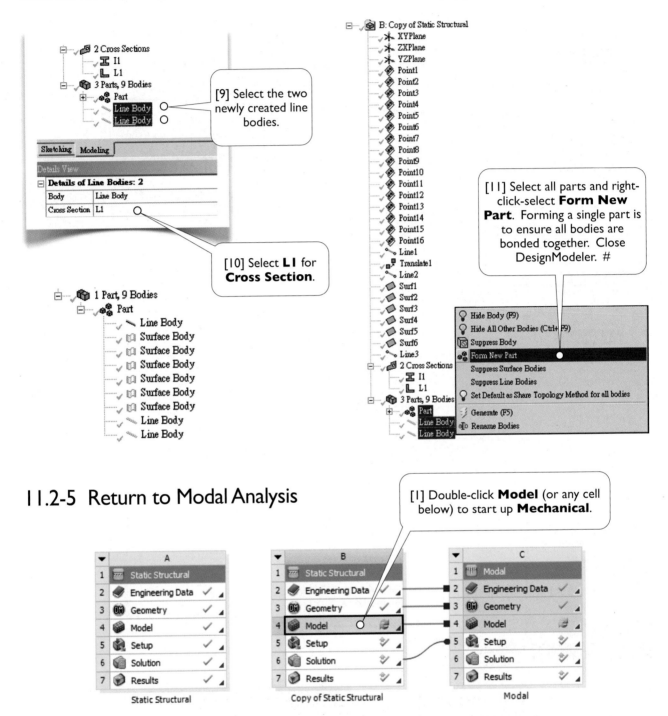

[9] Select the two newly created line bodies.

[10] Select L1 for Cross Section.

[11] Select all parts and right-click-select Form New Part. Forming a single part is to ensure all bodies are bonded together. Close DesignModeler. #

11.2-5 Return to Modal Analysis

[1] Double-click Model (or any cell below) to start up Mechanical.

Static Structural

Copy of Static Structural

Modal

Check the environment conditions every time you modify geometry

After you modify the geometry, always check environment conditions such as loads and supports. In this case, some of the fixed supports may be lost. We need to fix them [2]. Besides, delete **Acceleration** under **Static Structural** branch [3], which represents an earthquake load. We usually don't consider earthquake load as a "prestress." We do consider earth gravity as prestress. It, however, affects natural frequencies very limited. This leaves as an exercise for you (11.5-2, page 410).

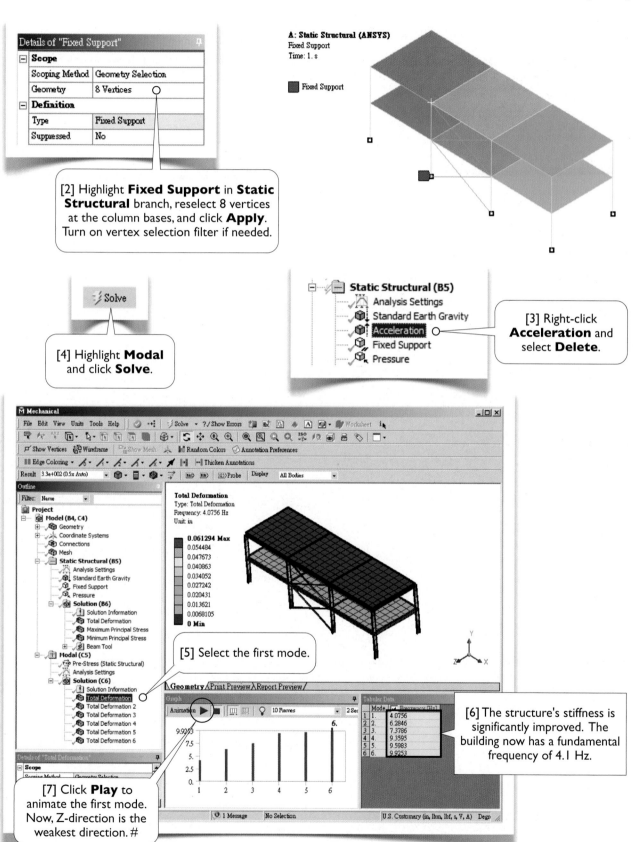

Details of "Fixed Support"

Scope	
Scoping Method	Geometry Selection
Geometry	8 Vertices
Definition	
Type	Fixed Support
Suppressed	No

A: Static Structural (ANSYS)
Fixed Support
Time: 1. s

■ Fixed Support

[2] Highlight **Fixed Support** in **Static Structural** branch, reselect 8 vertices at the column bases, and click **Apply**. Turn on vertex selection filter if needed.

Solve

[4] Highlight **Modal** and click **Solve**.

Static Structural (B5)
Analysis Settings
Standard Earth Gravity
Acceleration
Fixed Support
Pressure

[3] Right-click **Acceleration** and select **Delete**.

Total Deformation
Type: Total Deformation
Frequency: 4.0756 Hz
Unit: in

0.061294 Max
0.054484
0.047673
0.040863
0.034052
0.027242
0.020431
0.013621
0.0068105
0 Min

[5] Select the first mode.

Mode	Frequency [Hz]
1.	4.0756
2.	6.2846
3.	7.3786
4.	9.3595
5.	9.5983
6.	9.9253

[6] The structure's stiffness is significantly improved. The building now has a fundamental frequency of 4.1 Hz.

[7] Click **Play** to animate the first mode. Now, Z-direction is the weakest direction. #

Discussion

After the reinforcement of X-direction, Z-direction now becomes the weakest direction [7]. If we want to further improve the structure's stiffness, we may add bracing members in the Z-direction. This leaves for you as an exercise (11.5-2, page 410).

Wrap Up

Save the project and exit Workbench.

Section 11.3

Compact Disk

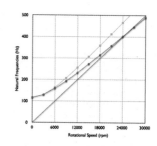

11.3-1 About the Compact Disk

A CD is made of polycarbonate (PC) [1], with density 1200 kg/m³, Young's modulus 2.2 GPa, Poisson's ratio 0.37, and tensile strength 65 MPa. The outer diameter is 120 mm, the inner (hole) diameter is 15 mm, and the thickness is 1.2 mm. For a 52x CD drive, the maximum rotational speed reaches 27,500 rpm (458 Hz) when reading the inner tracks[Ref 1].

The television series *MythBusters* conducted experiments[Refs 2, 3] in which they succeeded in shattering CDs at speeds of 23,000 rpm. When conducting the experiments, they press the CD between two nuts, which are 27 mm in diameter[Ref 4].

In this section, we first want to find out the maximum stress in the CD due to the centrifugal force when rotating in 27,500 rpm, to justify that the shattering may not be due to the centrifugal stress.

Second, we want to find the natural frequencies of the CD to investigate the possibility of resonant vibrations. We will conclude that the CD shattering may be due to sustaining vibrations rather than centrifugal stress. Also, we want to demonstrate that the natural frequencies increase with increasing rotational speed.

[1] The CD is made of polycarbonate (PC) plastic. #

φ 15 mm

φ 120 mm

11.3-2 Start Up a New Project

Launch Workbench. Save the project as **CD**. Create a **Static Structural** analysis system [1]. Make sure **SI** system is used as the project units. Double-click **Engineering Data** [2] and create a new material with the name **PC** [3]. Input the material properties: density (1200 kg/m³), Young's modulus (2.2 GPa), and Poisson's ratio (0.37) [4-6]. Click **Project** to return to **Project Schematic** [7].

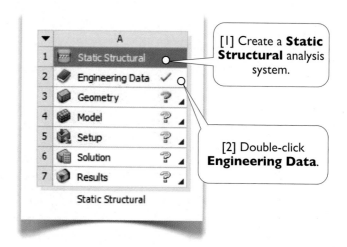

[1] Create a **Static Structural** analysis system.

[2] Double-click **Engineering Data**.

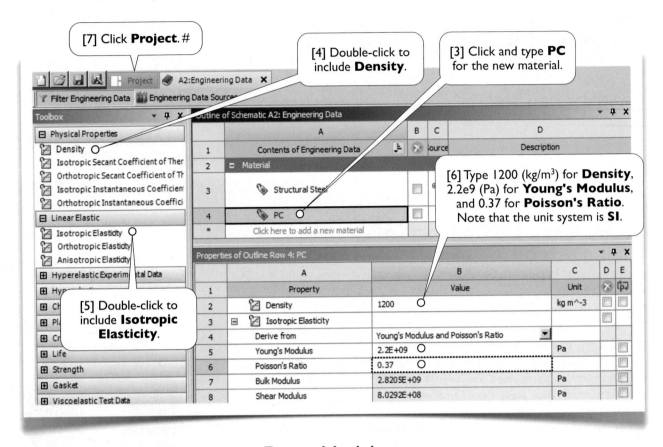

[7] Click **Project**. #

[4] Double-click to include **Density**.

[3] Click and type **PC** for the new material.

[6] Type 1200 (kg/m³) for **Density**, 2.2e9 (Pa) for **Young's Modulus**, and 0.37 for **Poisson's Ratio**. Note that the unit system is **SI**.

[5] Double-click to include **Isotropic Elasticity**.

11.3-3 Create Geometry in DesignModeler

Start up DesignModeler. Select **Millimeter** as the length unit and make sure **Auto Constraints** are turned on. Draw a sketch on **XYPlane** as shown [1]. Pull-down-select **Concept/Surfaces From Sketches** to create a surface body from the sketch [2]. Close DesignModeler.

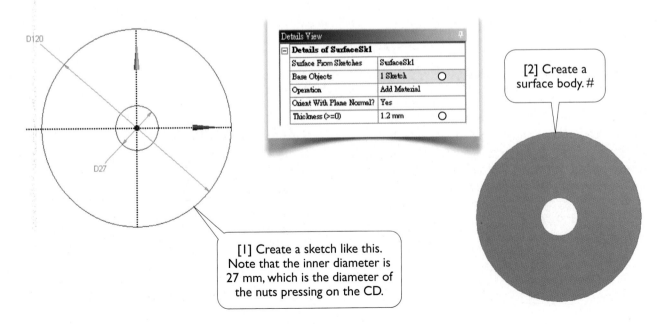

[2] Create a surface body. #

[1] Create a sketch like this. Note that the inner diameter is 27 mm, which is the diameter of the nuts pressing on the CD.

11.3-4 Assess Centrifugal Stress

Start up **Mechanical**. Make sure the unit system is **mm-kg-N-s**. Assign the material **PC** to the surface body [1]. Specify 2.0 (mm) for the global element size [2] and generate mesh.

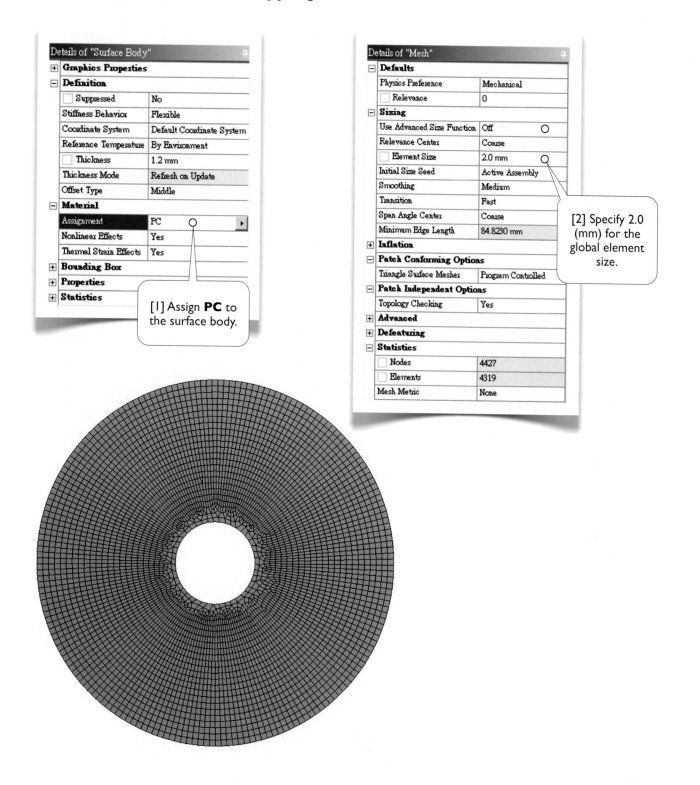

Details of "Surface Body"

Graphics Properties	
Definition	
Suppressed	No
Stiffness Behavior	Flexible
Coordinate System	Default Coordinate System
Reference Temperature	By Environment
Thickness	1.2 mm
Thickness Mode	Refresh on Update
Offset Type	Middle
Material	
Assignment	PC
Nonlinear Effects	Yes
Thermal Strain Effects	Yes
Bounding Box	
Properties	
Statistics	

[1] Assign **PC** to the surface body.

Details of "Mesh"

Defaults	
Physics Preference	Mechanical
Relevance	0
Sizing	
Use Advanced Size Function	Off
Relevance Center	Coarse
Element Size	2.0 mm
Initial Size Seed	Active Assembly
Smoothing	Medium
Transition	Fast
Span Angle Center	Coarse
Minimum Edge Length	84.8230 mm
Inflation	
Patch Conforming Options	
Triangle Surface Mesher	Program Controlled
Patch Independent Options	
Topology Checking	Yes
Advanced	
Defeaturing	
Statistics	
Nodes	4427
Elements	4319
Mesh Metric	None

[2] Specify 2.0 (mm) for the global element size.

Select the inner rim and insert a **Fixed Support** [3]. Insert a **Inertial/Rotational Velocity** load and type 27,500 (RPM) for **Z Component** [4, 5]. Insert a **Maximum Principal Stress** [6]. Solve the model [7].

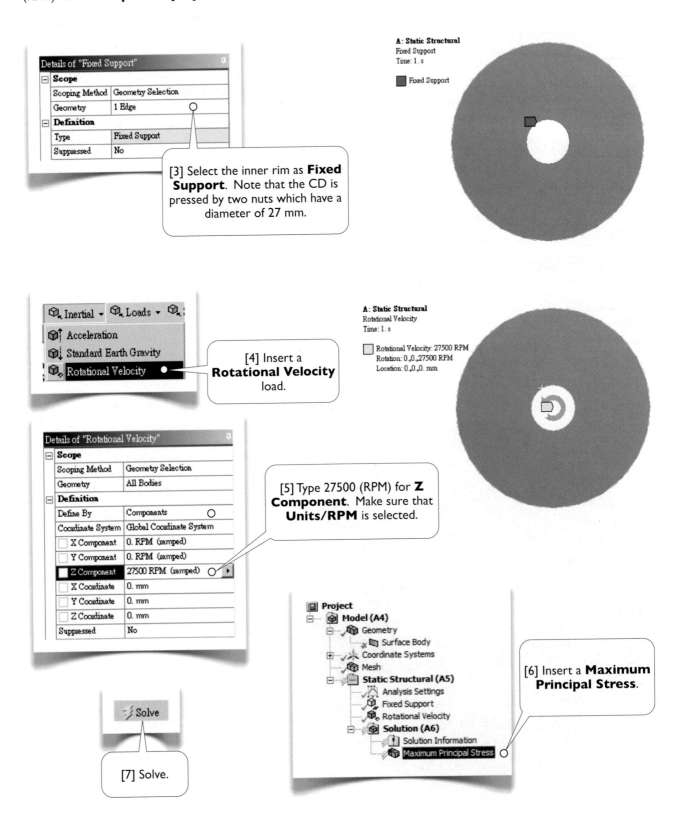

Details of "Fixed Support"

Scope	
Scoping Method	Geometry Selection
Geometry	1 Edge
Definition	
Type	Fixed Support
Suppressed	No

[3] Select the inner rim as **Fixed Support**. Note that the CD is pressed by two nuts which have a diameter of 27 mm.

A: Static Structural
Fixed Support
Time: 1. s

☐ Fixed Support

Inertial ▾ Loads ▾
Acceleration
Standard Earth Gravity
Rotational Velocity ●

[4] Insert a **Rotational Velocity** load.

A: Static Structural
Rotational Velocity
Time: 1. s

☐ Rotational Velocity: 27500 RPM
   Rotation: 0,0,27500 RPM
   Location: 0,0,0. mm

Details of "Rotational Velocity"

Scope	
Scoping Method	Geometry Selection
Geometry	All Bodies
Definition	
Define By	Components
Coordinate System	Global Coordinate System
☐ X Component	0. RPM (ramped)
☐ Y Component	0. RPM (ramped)
☐ Z Component	27500 RPM (ramped) ▸
☐ X Coordinate	0. mm
☐ Y Coordinate	0. mm
☐ Z Coordinate	0. mm
Suppressed	No

[5] Type 27500 (RPM) for **Z Component**. Make sure that **Units/RPM** is selected.

▣ **Project**
└─ ◈ **Model (A4)**
   └─ ◈ Geometry
      └─ ◻ Surface Body
   ├─ Coordinate Systems
   ├─ ◈ Mesh
   └─ ◻ **Static Structural (A5)**
      ├─ Analysis Settings
      ├─ ◈ Fixed Support
      ├─ ◈ Rotational Velocity
      └─ ◈ **Solution (A6)**
         ├─ Solution Information
         └─ ◈ Maximum Principal Stress

[6] Insert a **Maximum Principal Stress**.

⚡ Solve

[7] Solve.

The maximum tensile stress is 20.2 MPa [8], which occurs at the inner rim as expected and is far less than the material's tensile strength (65 MPa). It is unlikely that the CD shatters due to the tensile stress, unless some defects already exist in the CD before the experiments.

As next step, let's assess the CD's natural frequencies under the prestress caused by the high-speed spinning.

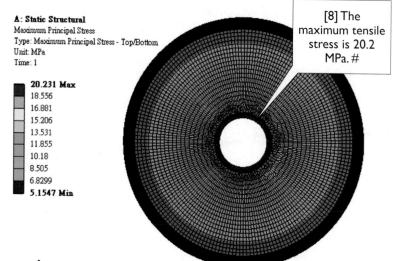

A: Static Structural
Maximum Principal Stress
Type: Maximum Principal Stress - Top/Bottom
Unit: MPa
Time: 1

20.231 Max
18.556
16.881
15.206
13.531
11.855
10.18
8.505
6.8299
5.1547 Min

[8] The maximum tensile stress is 20.2 MPa. #

11.3-5 Assess Natural Frequencies

Close **Mechanical**. Drag-and-drop a **Modal** analysis system to **Solution** cell of **Static Structural** [1]. Start up **Mechanical** by double-clicking any cells other than **Engineering Data** or **Geometry**. Highlight **Modal** and solve the model [2]. The lowest frequency [3] is close to the maximum of the operational frequencies (2,7500 rpm, or 458 Hz), which may excite the CD and may cause damage.

Let's look into this issue of resonant vibrations more thoroughly. We'll assess the prestressed natural frequencies for a range of possible operational speeds, from 0 to 30,000 rpm.

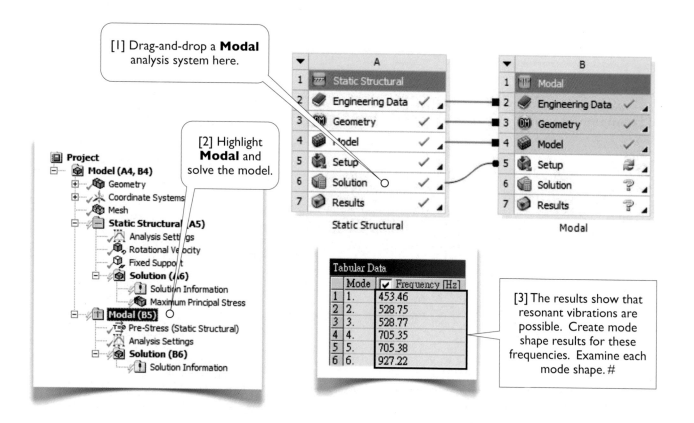

[1] Drag-and-drop a **Modal** analysis system here.

[2] Highlight **Modal** and solve the model.

Static Structural

Modal

[3] The results show that resonant vibrations are possible. Create mode shape results for these frequencies. Examine each mode shape. #

Tabular Data

	Mode	✔ Frequency [Hz]
1	1.	453.46
2	2.	528.75
3	3.	528.77
4	4.	705.35
5	5.	705.38
6	6.	927.22

11.3-6 Assess Natural Frequencies over the Range of Rotational Speeds

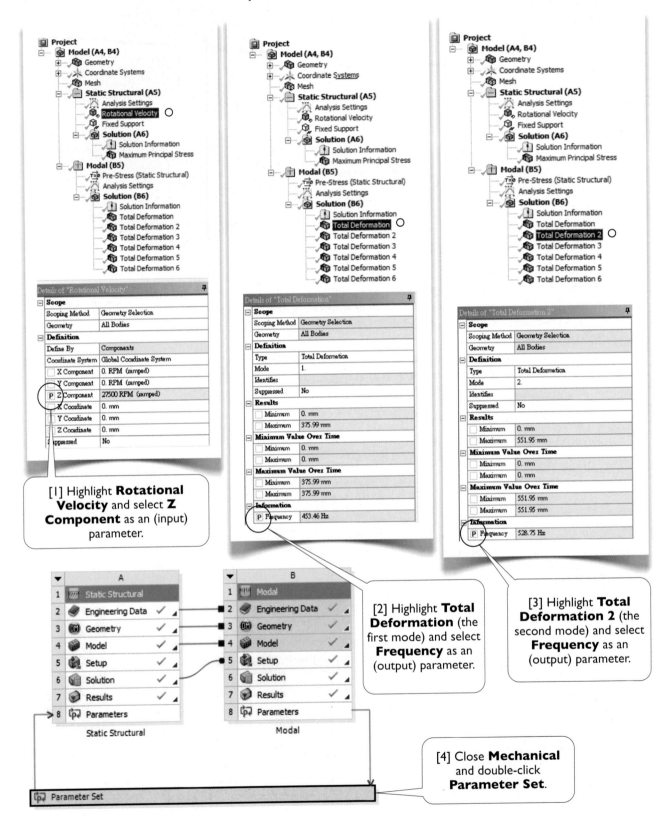

[1] Highlight **Rotational Velocity** and select **Z Component** as an (input) parameter.

[2] Highlight **Total Deformation** (the first mode) and select **Frequency** as an (output) parameter.

[3] Highlight **Total Deformation 2** (the second mode) and select **Frequency** as an (output) parameter.

[4] Close **Mechanical** and double-click **Parameter Set**.

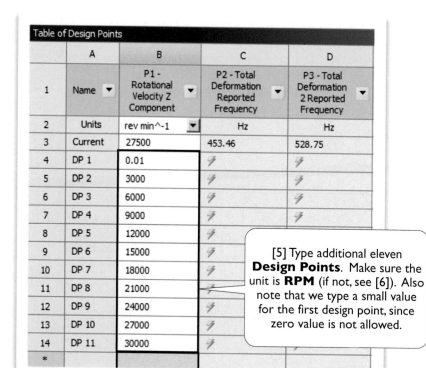

Table of Design Points

	A	B	C	D
1	Name ▼	P1 - Rotational Velocity Z Component ▼	P2 - Total Deformation Reported Frequency ▼	P3 - Total Deformation 2 Reported Frequency ▼
2	Units	rev min^-1 ▼	Hz	Hz
3	Current	27500	453.46	528.75
4	DP 1	0.01	⚡	⚡
5	DP 2	3000	⚡	⚡
6	DP 3	6000	⚡	⚡
7	DP 4	9000	⚡	⚡
8	DP 5	12000	⚡	
9	DP 6	15000	⚡	
10	DP 7	18000	⚡	
11	DP 8	21000		
12	DP 9	24000	⚡	
13	DP 10	27000	⚡	
14	DP 11	30000	⚡	
*				

[5] Type additional eleven **Design Points**. Make sure the unit is **RPM** (if not, see [6]). Also note that we type a small value for the first design point, since zero value is not allowed.

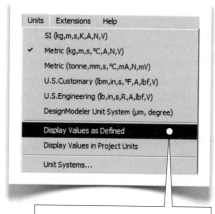

[6] If the unit of rotational velocity is not **rev min^-1** (i.e., **RPM**), turn on **Units/ Display Values as Defined**.

⚡⚡ Update All Design Points

[7] Click **Update All Design Points**.

Table of Design Points

	A	B	C	D
1	Name ▼	P1 - Rotational Velocity Z Component ▼	P2 - Total Deformation Reported Frequency ▼	P3 - Total Deformation 2 Reported Frequency ▼
2	Units	rev min^-1 ▼	Hz	Hz
3	Current	27500	453.46	528.75
4	DP 1	0.01	115.95	115.97
5	DP 2	3000	129.41	129.43
6	DP 3	6000	158.92	163.12
7	DP 4	9000	193.21	207.14
8	DP 5	12000	232.1	255.95
9	DP 6	15000	273.3	307.09
10	DP 7	18000	315.72	359.46
11	DP 8	21000	358.83	412.52
12	DP 9	24000	402.36	466.01
13	DP 10	27000	446.14	519.77
14	DP 11	30000	490.1	573.71
*				

[8] We'll use these data to plot a chart in the next page.

The data [8] are plotted in a chart as shown [9-11], which shows that the CD may be excited when the rotational speed reaches about 25000 rpm [12]. The excitation may in turn cause damage.

According to these studies, we may conclude that the CD shattering in the experiments conducted by TV series *Mythbusters* is probably due to resonant vibration effects, rather than the centrifugal stress.

A real CD drive usually provides supports for the CD, to reduce the vibrations as well as the radial deformations; therefore, the vibrations may not occur in a real CD drive.

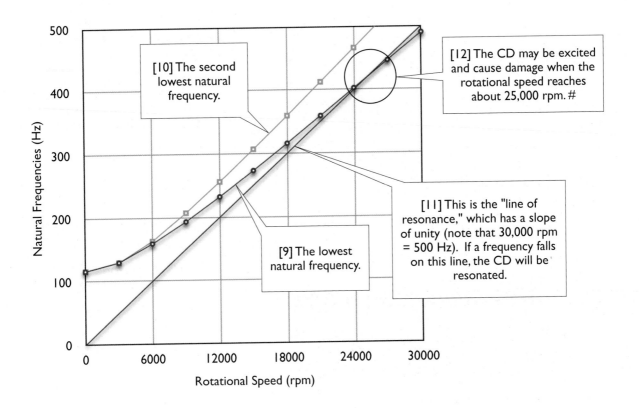

Wrap Up

Save the project and exit Workbench.

References

1. Wikipedia>Optical disc drive
2. http://www.youtube.com/watch?v=JOb6Z5Tja68
3. Wikipedia>MythBusters (2003 season)>Episode 2.3 Exploding CDs.
4. Acknowledgement: Thanks to Professor Per Blomqvist of the University of Gavle in Sweden, who pointed out this fact and suggested an more realistic simulation of this exercise. In the earlier versions of the book, I assumed that the CD was fixed in the inner rim, which has a diameter of 15 mm.

Section 11.4

Guitar String

This section tries to give engineers a lesson about the physics of music. When designing or improving a musical instrument, an engineer must know the physics of music. On the other hand, to fully appreciate the theory of music, a musician needs to know the physics behind the music.

We will use a guitar string to demonstrate some of the physics of music in this section and Section 12.5. For those students who are not interested in music theory, you may read only 11.4-1 to 11.4-3 (pages 402-405) and skip the rest of material in this section. On the other hand, if you want to introduce this article to a friend who does not have enough background in modal analyses, he may skip 11.4-1 and 11.4-2 and jump to 11.4-3 (page 405) directly.

11.4-1 About the Guitar String

The guitar string in our case is made of steel, which has a mass density of 7850 kg/m³, a Young's modulus of 200 GPa, and a Poisson's ratio of 0.3. It has a circular cross section of diameter 0.28 mm and a length of 1.0 m. The string is stretched with a tension T, and is in tune with a standard A note (*la*), which is defined as 440 Hz in modern music. In 11.4-2, we will perform a modal analysis to find the required tension T.

Before performing the simulation, let's make some simple calculation. According to the basic physics, the wave traveling on a string has a speed of

$$v = \sqrt{\frac{T}{\mu}} \qquad (1)$$

Where μ is the linear density (kg/m) of the string. The standing wave corresponding to the lowest frequency is called the *first harmonic mode*, which has a wavelength of twice the string length (2L). According to the relation between the velocity, the frequency, and the wavelength,

$$f = \frac{v}{\lambda} = \frac{v}{2L} \qquad (2)$$

According to (1) and (2), we can estimate the required tension,

$$T = \mu\left(2fL\right)^2 = 7850 \times \frac{\pi(0.00028)^2}{4}\left(2 \times 440 \times 1.0\right)^2 = 374.32 \text{ N} \qquad (3)$$

11.4-2 Perform Modal Analysis

Launch Workbench. Save the project as **String**. Create a **Static Structural** system. Drag-and-drop **Modal** analysis system to the **Solution** cell of the **Static Structural** system. In **Engineering Data**, make sure the material properties for **Structural Steel** are consistent with those of the guitar string.

Start up DesignModeler. Select **Millimeter** as length unit and make sure **Auto Constraints** are turned on. Create a sketch consisting of a line of 1000 mm on X-axis. Create a line body from the sketch. Create a circular cross section of radius 0.14 mm, and assign the cross secion to the line body. Close DesignModeler.

Start up **Mechanical**. Make sure the unit system is **mm-kg-N-s**. Under **Static Structural**, specify environment conditions [1]: a **Fixed Support** [2], a **Displacement** [3], and a **Force** [4]. Note that there should be no rigid body modes.

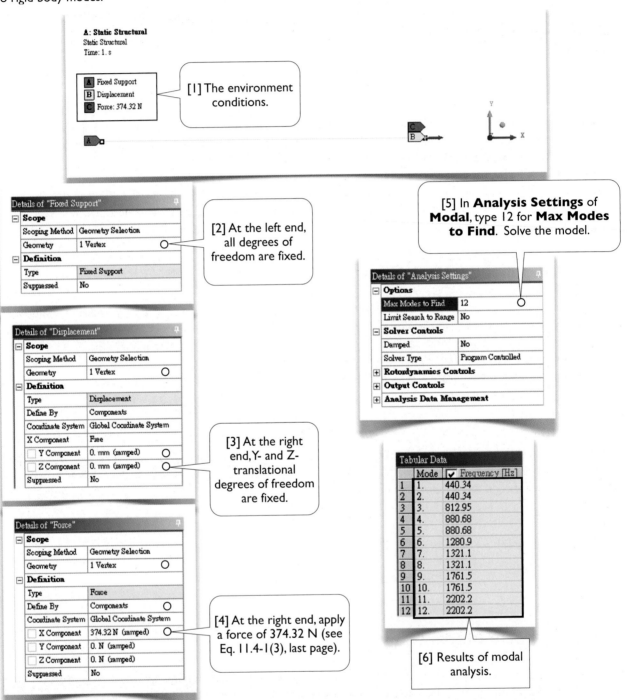

The results of the modal analysis [6] show that the lowest frequency is not exactly 440 Hz. It slightly deviates from what we've predicted (Eq. 11.4-1(3), last page). The main reason is that when applying the tension, the length of the string increases (about 30 mm). Another reason is due to the use of beam model, which is slightly different from the pure tension-only model used in our hand-calculation.

After several trial-and-errors (try this on your own), we come up with a tension of 373.74 N that exactly produces a fundamental frequency of 440 Hz [7, 8].

As expected, the spectrum of frequencies [8] includes 440 Hz, 880 Hz, 1320 Hz, 1760 Hz, 2200 Hz (which are all integral multiplications of the fundamental frequency, 440 Hz), with negligible numerical errors. Before we discuss these "harmonic modes" in 11.4-3 (next page), let's take a look at those non-harmonic modes first.

The third mode (812.9 Hz) is a rotation mode in X-direction [9, 10], visible in an animation. In the real-world, it exists, but its magnitude is usually very small, and our ears hardly sense it since the air pressure is hard to be excited in this way.

The sixth mode (1280.9 Hz) is a stretching mode [11], visible in an animation. In the real-world, it exists too, but again, we hardly sense it since the air pressure is hard to be excited in this way.

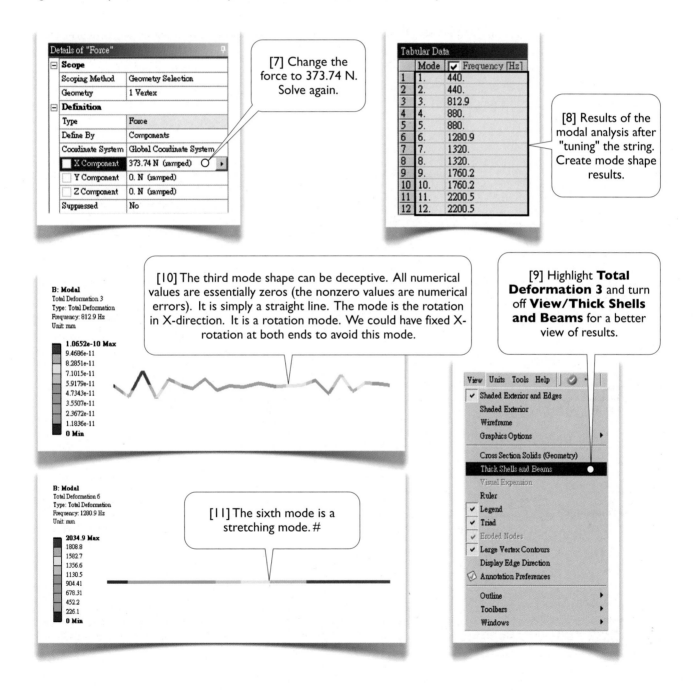

11.4-3 Harmonic Series

A *harmonic mode* has a frequency that is an integral multiplication of the fundamental frequency. The first 5 harmonic modes of the guitar string are shown below [1-5].

If you pluck a string, you will produce a tone made up of all harmonic modes. Although all the plucks produce the same note (in this case, note A), the *harmonic mixes* determine the quality of the note. If you pluck the string near the midpoint of the string, you will produce a tone dominated by the first harmonic [1]. If you pluck the string near the quarter point (in a guitar, that is near the sound hole), you will produce a tone dominated by the second harmonic [2]. And so forth. You can produce an "overtone" (a tone made up of harmonics that are all above the fundamental mode) by touching the string lightly at the midpoint and, at the same time, plucking the string; the first harmonic mode will be suppressed. You can produce other overtones in a similar way.

Different musical instruments generate tones that have different harmonic mixes. A trumpet usually produces much of higher harmonics in its frequency spectrum. This gives the trumpet a "brassy" sound. A flute usually produces a tone dominated by the first harmonic with almost no higher harmonics. This gives the flute a unique "pure" sound.

Knowing these physics, an engineer should be able to produce sounds of any musical instruments (including human voices) using a frequency-generating device.

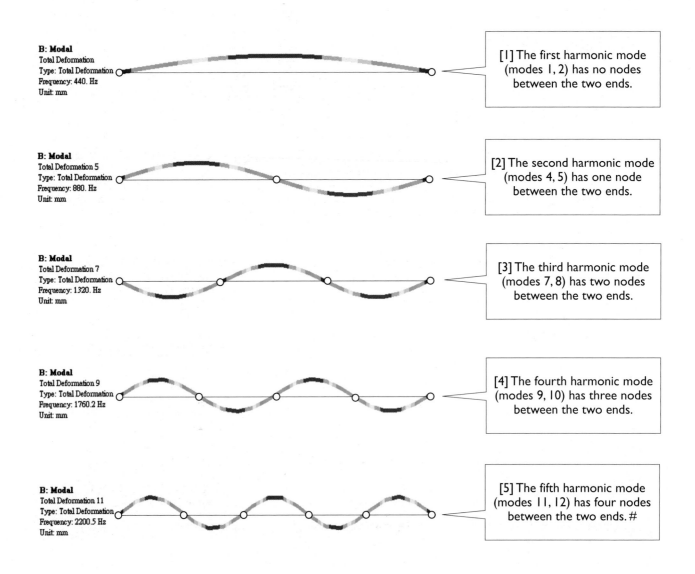

B: Modal
Total Deformation
Type: Total Deformation
Frequency: 440. Hz
Unit: mm

[1] The first harmonic mode (modes 1, 2) has no nodes between the two ends.

B: Modal
Total Deformation 5
Type: Total Deformation
Frequency: 880. Hz
Unit: mm

[2] The second harmonic mode (modes 4, 5) has one node between the two ends.

B: Modal
Total Deformation 7
Type: Total Deformation
Frequency: 1320. Hz
Unit: mm

[3] The third harmonic mode (modes 7, 8) has two nodes between the two ends.

B: Modal
Total Deformation 9
Type: Total Deformation
Frequency: 1760.2 Hz
Unit: mm

[4] The fourth harmonic mode (modes 9, 10) has three nodes between the two ends.

B: Modal
Total Deformation 11
Type: Total Deformation
Frequency: 2200.5 Hz
Unit: mm

[5] The fifth harmonic mode (modes 11, 12) has four nodes between the two ends. #

11.4-4 Just Tuning System[Ref 1]

Why do some notes sound pleasing to our ears when played together, while others do not? We know from the experience that when two notes have a simple frequency ratio, they sound harmonious with each other. The simpler the ratio, the more harmonious it sounds; we'll explain this in 11.4-6 (page 408).

In Western music, an 8-tone musical scale has traditionally been used. When learning to sing, we identify the eight tones in the scale by the syllables *do, re, mi, fa, sol, la, ti, do*. For a C-major scale in a piano, there are 8 white keys from a *C* to the higher pitch of *C* [1]. The two *C*'s have a frequency ratio of 2:1, and are said to be an *octave* apart. If we play two notes an octave apart, they sound very similar. In fact, we often have difficulty telling the difference between two notes having an octave apart. This is because that, except the fundamental harmonic of the lower note, two notes have most of the same higher harmonics.

For the following discussion, let's arbitrarily assume the frequency of the lower pitch *C* as 1. (In a modern piano, the middle *C* has a frequency of 261.63 Hz; see the table in 11.4-5.) Then the frequency of the higher pitch *C* is 2. Before being replaced by the "equal temperament" (11.4-5) in the early 20th century, the "just tuning" systems prevail in the music world. In a just tuning music system, the frequencies of the notes between the 2 *C*'s are chosen according to the "simple ratio" rule, in order to be harmonious to each other. The results are shown below [2]. Note that we didn't show the frequency ratios for the black keys (the semitones) to simplify our discussion.

Now, you can appreciate that if we play the notes *do* and *sol* together, the sound is pleasing to our ears, since they have the simplest frequency ratio between 1 and 2. You also can appreciate that the major cord C consists of the notes *do, me, sol, do*, the simplest frequency ratios (but not too "close," to avoid *beats*; see 11.4-6, page 408) between 1 and 2.

The problem of the just tuning system is that it is almost impossible to play in another key. For example, when we play in *D* key, then the frequency ratio between *D* and its *fifth* (*A*) is no longer 3/2. Instead, the frequency ratio is an awkward 40/27; the two notes are not harmonious enough any more.

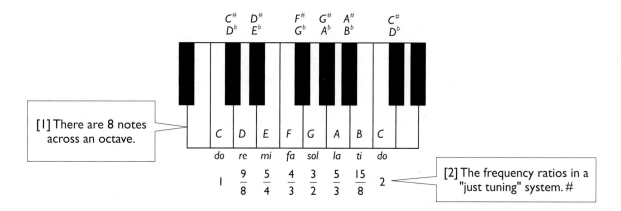

[1] There are 8 notes across an octave.

[2] The frequency ratios in a "just tuning" system. #

11.4-5 Twelve-Tone Equally Tempered Tuning System[Ref 2]

Modern Western music is dominated by a 12-*tone equally tempered tuning system*, or simply *equal temperament*. The idea is to compromise the frequency ratios between the notes, so that they can be played in different keys. In this system, an octave is equally divided into 12 tones (including semitones) in logarithmic scale. In other words, the adjacent tones have a frequency ratio of $2^{1/12}$, or 1.05946. For example, the frequency ratio between the $C^{\#}$ and the *C* is $2^{1/12}$; the frequency ratio between an *A* and the lower *C* is $2^{9/12}$. According to this idea, frequencies of the notes can be calculated and listed in the table below [1]. For comparison, we also list the frequencies of the notes in the just tuning system. The data in the table are plotted into a chart as shown [2, 3]. The compromised frequencies are close enough to the just tuning system that most of musicians are satisfied with this system for centuries.

Note	Just Tuning		Equal Temperament	
	Frequency Ratio	Frequency	Frequency Ratio	Frequency
C	1	264.00	1	261.63
C# (Db)			$2^{1/12}$	277.18
D	9/8	297.00	$2^{2/12}$	293.66
D# (Eb)			$2^{3/12}$	311.13
E	5/4	330.00	$2^{4/12}$	329.63
F	4/3	352.00	$2^{5/12}$	349.23
F# (Gb)			$2^{6/12}$	369.99
G	3/2	396.00	$2^{7/12}$	392.00
G# (Ab)			$2^{8/12}$	415.30
A	5/3	440.00	$2^{9/12}$	440.00
A# (Bb)			$2^{10/12}$	466.16
B	15/8	495.00	$2^{11/12}$	493.88
C	2	528.00	2	523.25

[1] The frequencies of the notes.

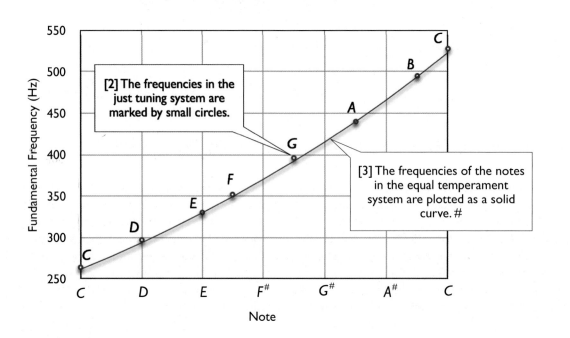

[2] The frequencies in the just tuning system are marked by small circles.

[3] The frequencies of the notes in the equal temperament system are plotted as a solid curve. #

11.4-6 Beat Frequency[Refs 3, 4]

Back to the question in the beginning of 11.4-4 (page 406): Why do some notes sound pleasing to our ears when played together, while others do not? Why does a simple frequency ratio imply a harmonious sound? To explain this, another physical phenomenon called *beats* is at work.

When two waves of different frequencies are combined, they interfere with each other. When they are in phase, the combined wave has a large amplitude. When they are out of phase, the amplitude becomes smaller. This fluctuation in amplitude of the combined wave is called *beats*, and the frequency is called the *beat frequency*.

The beat frequency is equal to the frequency difference of the two waves. If the two notes are very close in frequency, the beat frequency is slow enough to be heard as a variation in amplitude, that is, you can hear the sound getting louder and softer in a repetitive pattern. This effect can be useful in tuning a musical instrument, since the beats disappear when two frequencies are in tune.

What happens when we play *C* and *D* together? The beat frequency is 32.03 Hz (293.66 - 261.63, see 11.4-5[1], last page), which is large enough that we hear a harsh buzz, which is unpleasant for our ears. In music, however, a dissonant sound is sometimes used to produce desired effects, for example, a sad mood.

What happens when we play *C* and *G* together? The beat frequency is 130.37 Hz (392.00 - 261.63, see 11.4-5[1], last page), which is very close to one half of the middle C (261.63 Hz). In other words, the beat frequency is an octave below middle *C*. Therefore, it fits nicely within a chord containing *C* and *G*. Thus, beats can explain the harmony when we play major chords.

Wrap Up

Save the project and exit Workbench.

References

1. Wikipedia>Just Intonation.
2. Wikipedia>Equal Temperament.
3. Wikipedia>Beat.
4. Griffith, W. T., The Physics of Everyday Phenomena, Fourth Edition, McGraw-Hill, 2004.

Section 11.5

Review

11.5-1 Keywords

Choose a letter for each keyword from the list of descriptions

1. (   ) Dynamic Behavior
2. (   ) Free Vibration
3. (   ) Fundamental Natural Frequency
4. (   ) Harmonic Mode
5. (   ) Modal Analysis

6. (   ) Natural Frequencies
7. (   ) Rigid Body Modes
8. (   ) Structural Dynamic Analysis
9. (   ) Transient Structural Analysis

Answers:

1. (B)   2. (E)   3. (G)   4. (I)   5. (C)   6. (F)   7. (H)   8. (A)   9. (D)

List of Descriptions

(A) Technique used to determine the dynamic behavior of a structure.

(B) Structural dynamic behavior includes vibration characteristics, effect of harmonic loads, and effect of general time-varying loads.

(C) A dynamic analysis to investigate the vibration characteristics, specifically the frequencies and shapes, of a structure without any time-varying loads. The frequencies are called natural frequencies.

(D) A dynamic analysis to investigate the response of a structure under general time-varying loads.

(E) The vibration of a structure without any external forces.

(F) The frequencies corresponding to the free vibrations are called natural frequencies.

(G) The lowest frequency of free vibration.

(H) Rigid body modes have infinite period, or, equivalently, zero frequencies. In general, rigid body modes should not be present.

(I) A harmonic mode has a frequency that is an integer multiple of the fundamental frequency.

11.5-2 Additional Workbench Exercises

Gravity Is Negligible When Evaluating Frequencies of the Two-Story Building

In 11.2-5 (page 391), we mentioned that the earth gravity affects natural frequencies of the two-story building very limited. Verify this.

Improving the Stiffness of the Two-Story Building

In the end of Section 11.2 (page 393), we mentioned that if we want to further improve the structure's solidity, we may add bracing members in the Z-direction. Implement this idea.

Model Airplane Wing

A wing of a model airplane is detailed in the ANSYS Help System[Ref 1]. Carry out the modal analysis for the airplane wing. How do you stiffen the wing further without increasing the weight?

Reference

1. ANSYS Documentation//Mechanical APDL//Introductory Tutorials//8. Modal Tutorial

---

Chapter 12
Transient Structural Simulations

In the real world, all loads are time-varying, so are the structural responses. For example, imagine that you hang a block on a spring and slowly release it. The force on the spring increases gradually from zero until it reaches the weight of the block, and then the block moves up and down for a while, and finally steadies at a certain position, due to the damping of the system. To know the whole process, you need to perform a dynamic simulation. If you are concerned about only the final state (final position), called the *steady state*, of the response, a static simulation is adequate. The response before the steady state is called a *transient state*. A situation that a dynamic simulation can be replaced by a series of static simulations is that if the structure displaces so slowly that the dynamic effects (inertia and damping effects) are negligible. A series of time-varying static simulations is called a *quasistatic simulation*.

Other than these two categories of cases (cases of finding a steady state solution or cases that the structure displaces slowly), dynamic effects must be taken into account and dynamic simulations are needed.

Purpose of This Chapter

The purpose of this chapter is to provide background knowledge as well as practical examples for students to master techniques for dynamics simulations. It is a sequel of the last chapter, modal analysis, in which no external forces are involved. Except Section 12.3, in which **Harmonic Response** analyses are carried out , all the other exercises are performed with **Transient Structural** analysis system, which uses an implicit integration method to calculate the response. **Explicit Dynamics** analysis system, on the other hand, uses an explicit integration method. The explicit dynamics will be introduced in Chapter 15, where differences between implicit and explicit methods will be discussed.

About Each Section

Section 12.1 intends to equip students with background knowledge of structural dynamics. Again, it is the concepts, rather than the mathematics details, that we want to emphasize. We first use a single-degree-of-freedom structural system to illustrate some ideas, and then conceptually generalize these ideas for multiple-degrees-of-freedom structural systems.

Section 12.2 provides a practical example to demonstrate the application of dynamic loads and other considerations such as integration time steps and damping. Section 12.3 uses the two-story building (Sections 7.3, 11.2) as an example, demonstrating the procedure of harmonic response analysis. Section 12.4 performs an impact simulation. One of its purposes is to demonstrate how to specify a simple initial condition, namely uniform velocity. Another purpose is to show the limitation of implicit integration methods and the necessity of explicit methods for high-speed impact simulations.

Section 12.5 is a sequel of Section 11.4. As mentioned in 11.4-3 (page 405), if you pluck a string, you will produce a tone made up of all harmonic modes. We will apply plucks at different locations of the string to observe the responses. Another purpose of Section 12.5 is to demonstrate how to specify a more general initial condition. Some initial conditions need a static simulation themselves. In this section, we will use the results of a static simulation as an initial condition for a transient simulation.

Section 12.1
Basics of Structural Dynamics

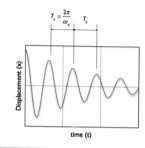

The purpose of this section is to provide basics of structural dynamics, so that students can understand the exercises in this chapter. The concepts and terminologies introduced in this section will be used throughout this chapter and the rest of the book. We first use a single-degree-of-freedom lumped mass model to explain some basic behavior of dynamic response. The results will be conceptually extended to multiple-degrees-of-freedom cases (and not limiting to lumped mass models).

12.1-1 Lumped Mass Model

In the old days of hand-calculations, many dynamics problems were simplified as lumped mass systems of a few degrees of freedom. As an example, to find the lateral displacements, the two-story building (Sections 7.3, 11.2) is modeled as a two-degrees-of-freedom system as shown [1-7]. The lumped mass models are still useful even in the modern days.

The parameters (m_1, m_2, k_1, k_2, c_1, c_2) must be reasonably evaluated to obtain an acceptable solution. Each of the masses m_1 and m_2 [2, 3] may include the floor mass, part of the columns mass, and the equivalent mass of the loads on the floor. Each of the spring constants k_1 and k_2 [4, 5] may be calculated according to the bending stiffnesses of the columns and beams at the floor. Each of the damping coefficients c_1 and c_2 [6, 7] represents all energy dissipating mechanisms of the floor.

The energy dissipating mechanisms include frictions between the building and the surrounding air (viscous damping), material's internal frictions (material damping), and the frictions in the joints connecting structural members (Coulomb damping). Evaluating damping is one of the most challenging tasks of engineering practice. Fortunately, damping in most real-world structures are usually very small. Consequently, in many cases, it may not be crucial and we may choose a reasonable value according to engineering experiences. In other cases, conducting experiments to evaluate damping values may be needed. We'll discuss about damping in 12.1-3, pages 415-417.

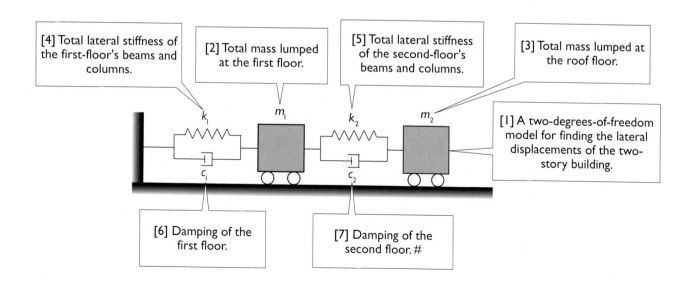

[4] Total lateral stiffness of the first-floor's beams and columns.

[2] Total mass lumped at the first floor.

[5] Total lateral stiffness of the second-floor's beams and columns.

[3] Total mass lumped at the roof floor.

[1] A two-degrees-of-freedom model for finding the lateral displacements of the two-story building.

[6] Damping of the first floor.

[7] Damping of the second floor. #

12.1-2 Single Degree of Freedom Model

Consider a single-degree-of-freedom (SDOF) model as shown [1]. Apply Newton's law of motion on the block of mass m,

$$\sum F = ma$$
$$p - kx - c\dot{x} = m\ddot{x}$$

or

$$m\ddot{x} + c\dot{x} + kx = p \qquad (1)$$

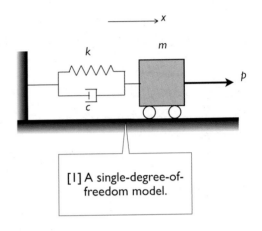

[1] A single-degree-of-freedom model.

Eq. (1) is the governing equation of the SDOF model. We'd assumed that the spring force is linearly proportional to the displacement, and the damping force is linearly proportional to the velocity. This kind of damping is called a *viscous damping*. We will discuss this assumption of damping in 12.1-3, pages 415-417.

If no external forces exist, Eq. (1) becomes

$$m\ddot{x} + c\dot{x} + kx = 0 \qquad (2)$$

Eq. (2) represents a *free vibration* system.

If the damping is negligible, then the equation becomes

$$m\ddot{x} + kx = 0 \qquad (3)$$

Eq. (3) represents a *undamped free vibration* system.

Undamped Free Vibration

The general solution of Eq. (3) is

$$x = A\sin(\omega t + B) \qquad (4)$$

where A and B are arbitrary real numbers and

$$\omega = \sqrt{\frac{k}{m}} \qquad (5)$$

Verification of this solution can be done by substituting Eqs. (4, 5) into Eq. (3). A typical plot of Eq. (4) is shown in [2], where we arbitrarily assume $B = 0$, since it is irrelevant. The vibration can be generated by holding up the mass of an undamped SDOF system [1], displacing an arbitrary amount A, releasing it, and letting it vibrate freely. The meaning of the *natural angular frequency* ω is also shown in the plot [2]. Relations between the natural angular frequency ω (rad/s), *natural frequency* f (Hz), and *natural period* T (s) are

$$f = \frac{\omega}{2\pi} \qquad (6)$$

$$T = \frac{1}{f} \qquad (7)$$

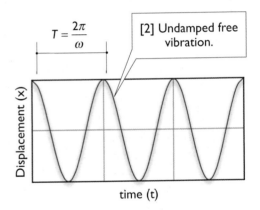

$T = \dfrac{2\pi}{\omega}$

[2] Undamped free vibration.

Displacement (x)

time (t)

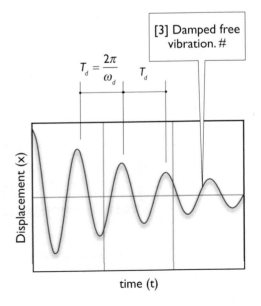

$T_d = \dfrac{2\pi}{\omega_d}$ T_d

[3] Damped free vibration. #

Displacement (x)

time (t)

413

Damped Free Vibration

Assume the damping c is smaller than the critical damping, defined in Eq. (11), then the general solution of Eq. (2) is

$$x = Ae^{-\xi\omega t}\sin\left(\omega_d t + B\right)$$

(8)

where A and B are arbitrary real numbers and

$$\omega_d = \omega\sqrt{1-\xi^2}$$

(9)

$$\xi = \frac{c}{c_c}$$

(10)

$$c_c = 2m\omega$$

(11)

Verification of this solution can be done by substituting Eqs. (8-11) into Eq. (2). A typical plot of Eq. (8) is shown in [3]. Again, this vibration can be generated by holding up the mass of a SDOF system [1], displacing an arbitrary amount A, releasing it, and letting it vibrate freely.

The quantity c_c, defined in Eq. (11), is called the *critical damping coefficient*. It can be shown that if the damping c is larger than or equal to c_c, then the Eq. (8) is no longer valid. Actually, the motion is no longer oscillatory, and the system is called *over-damped* (if $c > c_c$) or *critically damped* (if $c = c_c$). For most of structures, we may reasonably assume that the system is under-damped ($c < c_c$), and the Eq. (8) is always valid.

The quantity ξ, defined in Eq. (10), is called the *damping ratio*. Values of damping ratio for typical structures range from about 0.02 (e.g., piping systems) to about 0.07 (e.g., bolted structures, reinforced concrete)[Ref 1].

The quantity ω_d, defined in Eq. (9), is called *damped natural angular frequency*. Note that, for a small damping ratio, ω_d and ω is practically no difference. For example, with $\xi = 0.07$, $\omega_d = 0.9975\omega$.

Measuring Damping Coefficient

Given the mass m and the spring constant k of a SDOF system, and if we can obtain a damped free vibration curve [3] from an experiment, then we can calculate the damping coefficient from these information, as follows. From Eq. (8), the peak displacement is

$$x_{peak} = Ae^{-\xi\omega t}$$

Recognizing the time between two peaks is $T = 2\pi/\omega_d \approx 2\pi/\omega$, we can write down the *displacement ratio R* between two consecutive peaks

$$R = \frac{x_{peak2}}{x_{peak1}} = \frac{Ae^{-\xi\omega(t+\frac{2\pi}{\omega})}}{Ae^{-\xi\omega t}} = e^{-2\pi\xi}$$

(12)

For example, with $\xi = 0.07$, the displacement ratio between two consecutive peaks is $R = 0.64$. In practice, the displacement ratio can be calculated by averaging the displacement ratios of *several* cycles of vibrations. The damping ratio ξ then can be calculated from Eq. (12)

$$\xi = \frac{-\ln R}{2\pi}$$

(13)

The damping coefficient, if desired, can be calculated using Eqs. (10) and (11),

$$c = \frac{-\ln R}{\pi}m\omega$$

(14)

414

Viscous damping coefficient c is not an intrinsic property of material

Imaging that the SDOF system [1] represents a cantilever beam made of a material and you want to characterize the damping property for the material. You make a piece of specimen, generate a free vibration curve [3], and calculate the damping coefficient according to Eq. (14). It seems easy. The problem is that, in this way, the damping coefficient depends on the geometry of the specimen: different geometries have different damping coefficients. To characterize damping for a material, we need more knowledge of damping mechanism.

12.1-3 Damping

Damping Mechanisms

As mentioned, damping includes all energy dissipating mechanisms. In a structural system, all energy dissipating mechanisms boil down to one word: friction. In a structure, three categories of frictions can be identified: (a) friction between the structure and its surrounding fluid, called *viscous damping*; (b) internal friction in the material, called *material damping*, *solid damping*, or *elastic hysteresis*; and (c) friction in the connection between structural members, called *dry friction* or *Coulomb friction*. When the structure is surrounded by the air, the viscous damping is usually very small, and the major sources of damping are material damping and Coulomb friction.

Viscous Damping

Viscous damping is the friction between a structure and its surrounding fluid. If small, viscous damping force can reasonably assume to be proportional to the velocity of the structural displacement

$$F_D = c\dot{x} \tag{1}$$

The viscous damping coefficient c can be input directly as an element parameter, such as a spring element [1]. For each material, you can include a constant damping ratio as a material property (see [9]). In some analysis systems, the viscous damping can be specified using a global damping ratio ξ [2]. The damping coefficient c is then calculated according to Eqs. 12.1-2(10, 11) (last page),

$$c = 2m\omega\xi = 2\xi\sqrt{mk} \tag{2}$$

In general, to introduce viscous damping to a structure, you may add individual elements involving viscous damping, such as spring elements [1]. In **Harmonic Response** analysis (Section 12.3), which calculates the response under various frequencies, viscous damping can be specified using a global damping ratio [2].

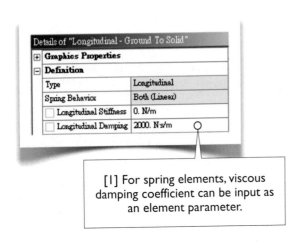

[1] For spring elements, viscous damping coefficient can be input as an element parameter.

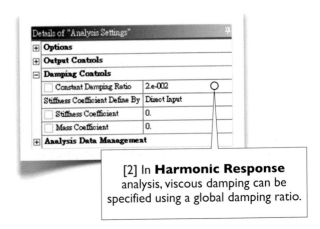

[2] In **Harmonic Response** analysis, viscous damping can be specified using a global damping ratio.

Material Damping

Material damping is the internal friction between molecules or grains of the material. Other names of material damping include *solid damping* and *elastic hysteresis*. It is often the major sources of damping in a real-world structure.

To understand the material damping, consider a typical stress-strain relation in a uniaxial material test [3-5]. Unlike the tests shown in 1.4-1[1-5] (page 36), here we repeatedly increase the stress [3] and then release the stress [4]. If the plastic deformation (permanent deformation) is not present, the strain will return to the original state when releasing the stress, and the curve of releasing stress is likely to be a straight line. A material in which the strain state returns to the original state when the stress is released is called an *elastic material*.

The area enclosed by the curves [5] represents the energy dissipation (to the environment) due to the internal friction of the material. This area is typically very small; the plot [3-5] is exaggerated for instructional purpose.

The plot [3-5] is analogous to a *B-H* curve of a magnetic material [6], where *B* is the *magnetic flux density* and *H* is the *magnetic field intensity*. A magnetic field *H* is applied on a magnetic material to create a *B* field within the material (i.e., the material is magnetized). The area enclosed by the *B-H* curve represents an energy dissipation, and is called the *magnetic hysteresis*. Because of the analogy, the energy dissipation in a stress-strain curve is called the *elastic hysteresis*.

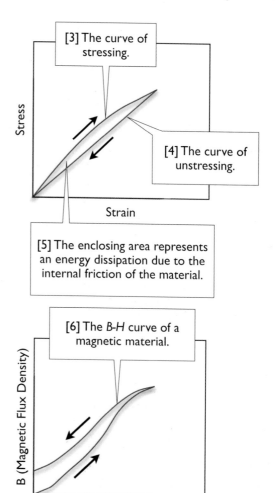

[3] The curve of stressing.

[4] The curve of unstressing.

[5] The enclosing area represents an energy dissipation due to the internal friction of the material.

[6] The *B-H* curve of a magnetic material.

Stress

Strain

B (Magnetic Flux Density)

H (Magnetic Field Intensity)

Recognizing that the damping is small for a structure and the global behavior is similar regardless of the sources of damping, Workbench assumes, as a simplification, that the material damping force is proportional to the structural velocity (this somehow deviates from reality, in which the material damping force is more likely proportional to the structural displacement rather than the structural velocity), the same as the viscous damping,

$$F_D = c\dot{x} \tag{3}$$

However, we cannot characterize a material using a damping coefficient c, since, as mentioned in the end of 12.1-2 (page 415), the damping coefficient c is not an intrinsic property of a material. To filter out factor of geometry, we need more elaboration. Eq. (2) shows how the coefficient c relates to the mass m and stiffness k for the case of single degree of freedom; for cases of multiple degrees of freedom, the relation is not so simple. In engineering practice, an efficient way to characterize a material is proposing a mathematics form with parameters and then determining the parameters using data fitting. With this idea, the coefficient c is assumed a linear combination of the mass m and the stiffness k, that is[Ref 1],

$$c = \alpha m + \beta k \tag{4}$$

Now, the parameters α and β are used to characterize the damping property of a material. Eq. (4) is based on an observation that damping is related to the mass and the stiffness of the structure.

Using $c = 2m\omega\xi$ in Eq. (2) and $k = m\omega^2$ in Eq. 12.1-2(5) (page 413), We may rewrite Eq. (4) in terms of frequency and damping ratio,

$$2\omega\xi = \alpha + \beta\omega^2 \qquad (5)$$

If we can make a single material specimen and measure the damping ratios ξ_i under different excitation frequencies ω_i, or make several material specimens of different sizes, and measure the damping ratios ξ_i under their respective fundamental frequencies ω_i, or, even better, a combination of the above ideas, then we can evaluate the material parameters α and β by a standard data fitting procedure.

In the Workbench, it allows you to input α and β value for each material as material properties [7]. A **Transient Structural** analysis system also allows you to input a global β value (**Stiffness Coefficient**) and a global α value (**Mass Coefficient**), in a details view of **Analysis Settings** [8].

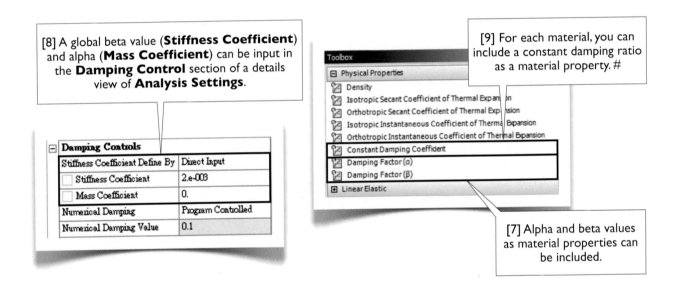

[8] A global beta value (**Stiffness Coefficient**) and alpha (**Mass Coefficient**) can be input in the **Damping Control** section of a details view of **Analysis Settings**.

[9] For each material, you can include a constant damping ratio as a material property. #

[7] Alpha and beta values as material properties can be included.

If we assume $\alpha = 0$, Eq. (5) becomes

$$\beta = \frac{2\xi}{\omega} \qquad (6)$$

Eq. (6) is a simple relation between the β value and the damping ratio ξ. It can be used to estimate one value when knowing the other one, if the frequency is also known.

Although the damping ratio is meant to be used for the viscous damping [2] rather than the material damping, in practice, the engineers often use a damping ratio to simplify the overall damping effect, when the damping is not critical to the response. As mentioned in 12.1-2 (page 414), values of damping ratio for typical structures range from 0.02 to 0.07.

Coulomb Friction

Another major source of damping is the friction in the connection between structural members. It is called the *dry friction* or *Coulomb friction*. In Workbench, it is implemented as *frictional contacts*. To include the Coulomb friction, you have to specify frictional contacts between parts, which will be discussed in Chapter 13.

12.1-4 Analysis Systems

Generally, we are dealing with multiple-degrees-of-freedom systems. The foregoing concepts may be extended for general cases. Specifically, Eq. 12.1-2(1) (page 413) can be generalized for multiple-degrees-of-freedom cases,

$$[M]\{\ddot{D}\}+[C]\{\dot{D}\}+[K]\{D\}=\{F\} \tag{1}$$

Where $\{D\}$ is the nodal displacements vector, $\{F\}$ is the nodal external forces vector, $[M]$ is the *mass matrix*, $[C]$ is the *damping matrix*, and $[K]$ is the *stiffness matrix*. Eq. (1) also can be viewed as a generalization of Eq. 1.3-1(1) (page 31).

Eq. (1) represents the governing equation of a transient structural simulation. It can be viewed as a force equilibrium relation. On the right hand side of the equation is the external force $\{F\}$. On the left hand side, the first item $[M]\{\ddot{D}\}$ is the *inertia force*, the second item $[C]\{\dot{D}\}$ is the *damping force*, and the third item $[K]\{D\}$ is the *elastic force*. Combination of the inertia force, damping force, and the elastic force balances with the external force.

Let's look at some specialized cases of Eq. (1) (also see [1]).

Modal Analysis

Imagine that you displace a structure a certain amount and then release. There is no external force involved; it is called a free vibration. Eq. (1), since there is no external forces, becomes

$$[M]\{\ddot{D}\}+[C]\{\dot{D}\}+[K]\{D\}=0 \tag{2}$$

Solution of Eq. (2) is not unique. For a problem of n degrees of freedom, it has at most n solutions, denoted by $\{D_i\}, i=1,2,...,n$. These solutions are called *mode shapes* of the structure. Each mode shape $\{D_i\}$ can be resonantly excited by an external excitation of frequency ω_i, called the *natural frequency* of the mode. The lowest frequency is called the *fundamental natural frequency*, or simply *fundamental frequency*. Finding all or some of the mode shapes and their corresponding natural frequencies is called a *modal analysis*.

In a modal analysis, since we are usually interested only in the natural frequencies and the relative shapes of the vibration modes (the absolute values of deformation depend on the energy that excites the structure), the damping effect is usually neglected (see Eq. 12.1-2(9), page 414) to simplify the calculation; Eq. (2) becomes

$$[M]\{\ddot{D}\}+[K]\{D\}=0 \tag{3}$$

It is Eq. (3) that Workbench solves in a **Modal** analysis system. Note that modal analysis is a linear analysis; all nonlinearities are ignored.

We've performed some modal analyses in Chapter 11.

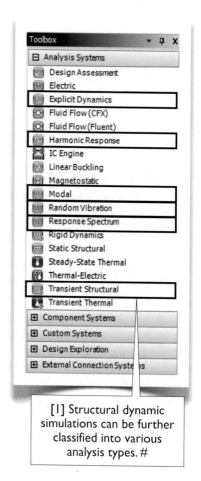

[1] Structural dynamic simulations can be further classified into various analysis types. #

Transient Structural Analysis

Transient Structural analysis solves the general form of Eq. (1). External force $\{F\}$ can be time-dependent forces. All nonlinearities can be included. It uses a *direct integration method* to calculate the dynamic response. In this chapter, all exercises, except Section 12.3, are carried out with **Transient Structural** system [1]. Through these exercises, we will demonstrate how to specify initial conditions, dynamic loads, and set up **Analysis Settings**.

Harmonic Response Analysis

Imagine that a rotatory machine is installed on the floor of the two-story building (Sections 7.3, 11.2), its operational speed 3000 rpm. Due to an inevitable eccentricity of the rotation, the machine generates an up-and-down harmonic force on the floor. After started up, the machine's speed increases from zero up to 3000 rpm. Any natural frequencies of the building ranging from zero to 3000 rpm may be excited by the harmonic force. While vibrations of the building or its structural components are unavoidable, the question is how large amplitude of the vibrations will happen. Do the vibrations cause any safety concern or psychological annoyance? A **Harmonic Response** analysis can be carried out to answer that question.

Harmonic Response analysis solves a special form of Eq. (1), in which the external force on ith degree of freedom is of the form

$$F_i = A_i \sin(\Omega t + \phi_i) \tag{4}$$

where A_i is the amplitude of the force, ϕ_i is the phase angle of the force, and Ω is the angular frequency of the external force. Due to the special form of the external forces, the calculation is much more efficient than a general transient response analysis. The steady-state solution of the equation will be of the form

$$D_i = B_i \sin(\Omega t + \varphi_i) \tag{5}$$

The goal of the harmonic response analysis is to find the magnitude B_i and the phase angle φ_i of the response for each degree of freedom, under a range of frequencies of the external forces. Note that harmonic response analysis is a linear analysis; all nonlinearities are ignored.

In Section 12.3, we will use the two-story building to demonstrate the procedure of harmonic response analysis.

Explicit Dynamics

Similar to **Transient Structural**, **Explicit Dynamics** also solves the general form of Eq. (1). External force $\{F\}$ can be time-dependent forces. All nonlinearities can be included. It also uses a direct integration method to calculate the dynamic response. The difference is described as follows.

The direct integration method used in **Transient Structural** analysis is called an *implicit integration method*. The implicit method works fine for most of applications except for high-speed impact simulations.

In high-speed impact simulations, the duration of impact time is so short that the integration time needs to be extremely small (e.g., micro to nano seconds) to catch the details behavior. If the implicit integration method is used, the total number of time steps becomes huge that the computational time is unbearable. That calls for an *explicit integration method*, implemented in **Explicit Dynamics** analysis system.

For many of transient dynamic simulations, the explicit method is not popular for one reason: it requires very small integration time steps to achieve an accurate solution. A small integration time is exactly what a high-speed impact simulation needs, therefore it is not a disadvantage any more. The advantage, on the other hand, is that the calculation is very efficient in each time step. Overall, the high-speed impact simulations are possible only with the explicit integration method.

There are cases, other than high-speed impact simulations, that are benefited by using the explicit method. Highly nonlinear simulations usually require very small time steps to overcome the convergence difficulties. In such cases, explicit method may be used.

In this chapter, we will restrict the discussion in implicit dynamics only (i.e., using **Transient Structural** analysis system). The applications of **Explicit Dynamics** will be postponed until Chapter 15. The reason is that, since explicit dynamics usually involves nonlinearities, we need more background on nonlinear simulations, which will be covered in Chapters 13 and 14.

For the beginners, the origination of the names "implicit" or "explicit" may not be important. You may regard them as code names. For those students with strong curiosity, you will learn them in Section 15.1.

Response Spectrum Analysis

We often design an engineering object such that it can withstand oscillatory or repeated loadings. For example, a building must withstand the strikes of earthquakes. When designing a building, we may use a well-recorded earthquake as a "design earthquake." The history of the earthquake (typically a history of acceleration versus time) can be input as loads and a transient structural simulation is then carried out. After the simulation, the maximum stress (or any other responses) at each location of the structure is collected. The members of the structure are designed according to these maximum values of responses. If the maximum response is the sole purpose of the simulation, a much inexpensive way of simulation is available: response spectrum analysis.

A time history of earthquake can be transformed to a response spectrum, a maximum response versus frequency (or period) plot. The response spectrum is then input to a **Response Spectrum** analysis system. The output is the maximum response (e.g., maximum stress) at each location of the structure.

Random Vibration Analysis

There exists a design methodology that consider probabilistic loads, instead of deterministic loads. The probabilistic loads are described by a spectrum representing probability distribution of excitation at varying frequencies in known directions. The technique is used to design structures withstanding probabilistic loadings, for example, a space vehicle subject to probabilistic strikes of meteorites or asteroids.

References

1. Cook, R.D., Milkus, D. S., Plesha, M. E., and Witt, R. J., *Concepts and Applications of Finite Element Analysis, Fourth Edition*, John Wiley & Sons, Inc., 2002; Section 11.5 Damping.
2. ANSYS Documentation//Mechanical APDL//Theory Reference//14.3. Damping Matrices

Section 12.2

Lifting Fork

12.2-1 About the Lifting Fork

In Section 4.3, we built a model for a lifting fork and glass, in which the fork was modeled as solid body and the glass as surface body. The lifting fork [1] is used in an LCD factory to handle a glass panel [2], which is so large and so thin that the engineers are concerned about its vertical deflections during dynamic handling.

The fork is made of steel with a density of 7850 kg/m³, Young's modulus of 200 GPa, and Poisson's ratio of 0.3. The glass has a density of 2370 kg/m³, Young's modulus of 70 GPa, and Poisson's ratio of 0.22.

In this section we will perform a static structural simulation first, to evaluate the vertical deflection of the glass panel under the gravitational force. This is critical when determining the clearance of the processing machine [3]. During the dynamic handling, the fork accelerates upward from rest to a velocity of 6 m/s in 0.3 seconds and then decelerates to a full stop in another 0.3 seconds, causing the glass panel to vibrate [4]. We want to know the time duration when the vibration is settled to a certain amount so that the glass can be moved into the processing machine [3]. We also want to know the maximum stress during the handling.

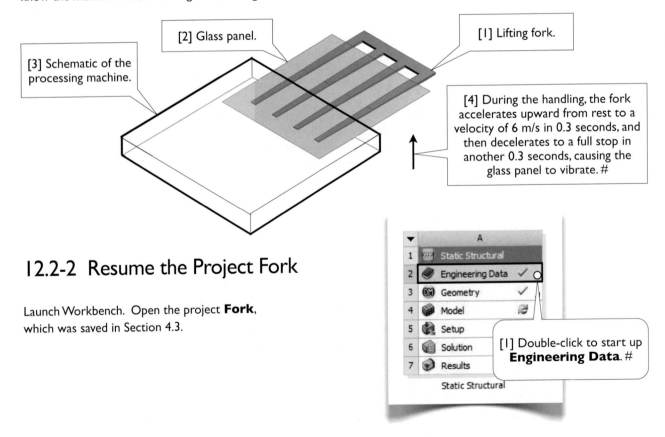

[3] Schematic of the processing machine.

[2] Glass panel.

[1] Lifting fork.

[4] During the handling, the fork accelerates upward from rest to a velocity of 6 m/s in 0.3 seconds, and then decelerates to a full stop in another 0.3 seconds, causing the glass panel to vibrate. #

12.2-2 Resume the Project Fork

Launch Workbench. Open the project **Fork**, which was saved in Section 4.3.

[1] Double-click to start up **Engineering Data**. #

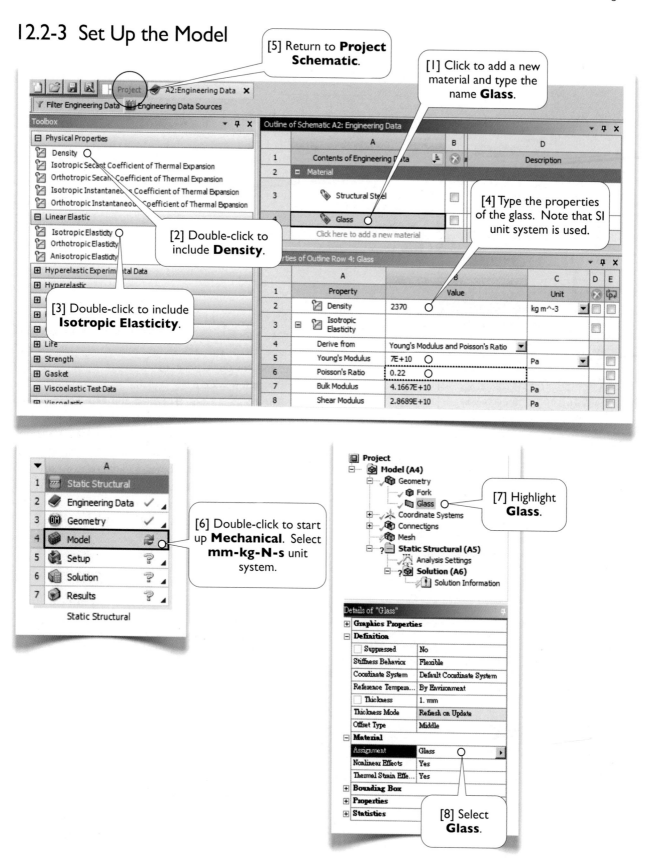

12.2-3 Set Up the Model

[5] Return to **Project Schematic**.

[1] Click to add a new material and type the name **Glass**.

[2] Double-click to include **Density**.

[3] Double-click to include **Isotropic Elasticity**.

[4] Type the properties of the glass. Note that SI unit system is used.

[6] Double-click to start up **Mechanical**. Select **mm-kg-N-s** unit system.

[7] Highlight **Glass**.

[8] Select **Glass**.

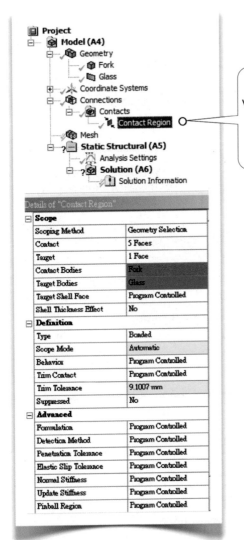

[9] Highlight **Contact Region**. Workbench automatically establish contact between the glass and fork. We don't need to change anything for this.

About the Contact Region

In this case, we use **Bonded** rather than more realistic contact types such as **Frictionless** or **Frictional**, because **Bonded** should be accurate enough for this case, and we don't want to introduce nonlinearity into the simulation system.

The contact types **Bonded** and **No Separation** are the only two contact types that do not introduce nonlinearity. **No Separation** contact condition between two surfaces prohibits separation in their normal direction, but allows sliding relative to each other. The sliding is assumed very small such that the small-deformation theory applies and the simulation remains linear.

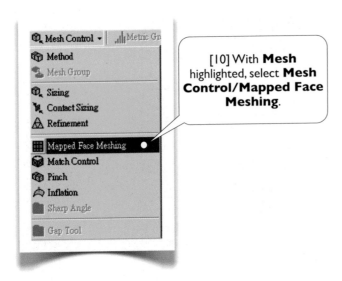

[10] With **Mesh** highlighted, select **Mesh Control/Mapped Face Meshing**.

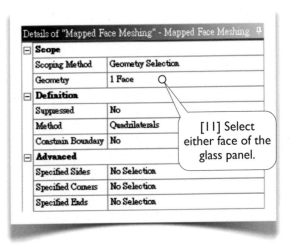

[11] Select either face of the glass panel.

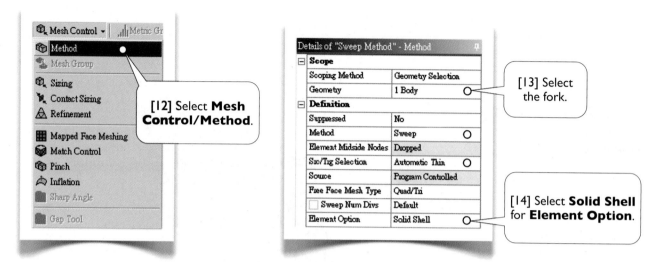

[12] Select **Mesh Control/Method**.

[13] Select the fork.

[14] Select **Solid Shell** for **Element Option**.

Thin Sweep Method

Sweep method can be further classified into ordinary **Sweep** and **Thin Sweep** (by selecting **Automatic Thin**). The ordinary **Sweep** allows a more complex sweeping path while **Thin Sweep** allows only a simple sweeping path. A feature of **Thin Sweep** is that it allows multiple faces as source or target while the ordinary **Sweep** allows only one face for both source and target.

Solid Shell

When **Solid Shell** is selected for **Element Option** [14], Workbench will mesh the body with **SOLSH190**[Ref 1] elements. The solid shell element is fully compatible with other types of solid elements and has extra degrees of freedom to account for bending modes. These extra degrees of freedom are treated as "internal" degrees of freedom and eliminated before the element is added to the global matrices. As a guideline, when a solid body is meshed with only one layer of elements in a direction and the deformation is dominated by bending, then **Solid Shell** is an appropriate choice.

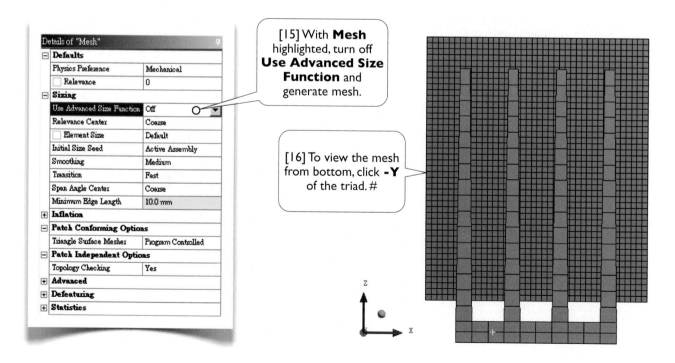

[15] With **Mesh** highlighted, turn off **Use Advanced Size Function** and generate mesh.

[16] To view the mesh from bottom, click **-Y** of the triad. #

12.2-4 Evaluate Deflection under Gravity

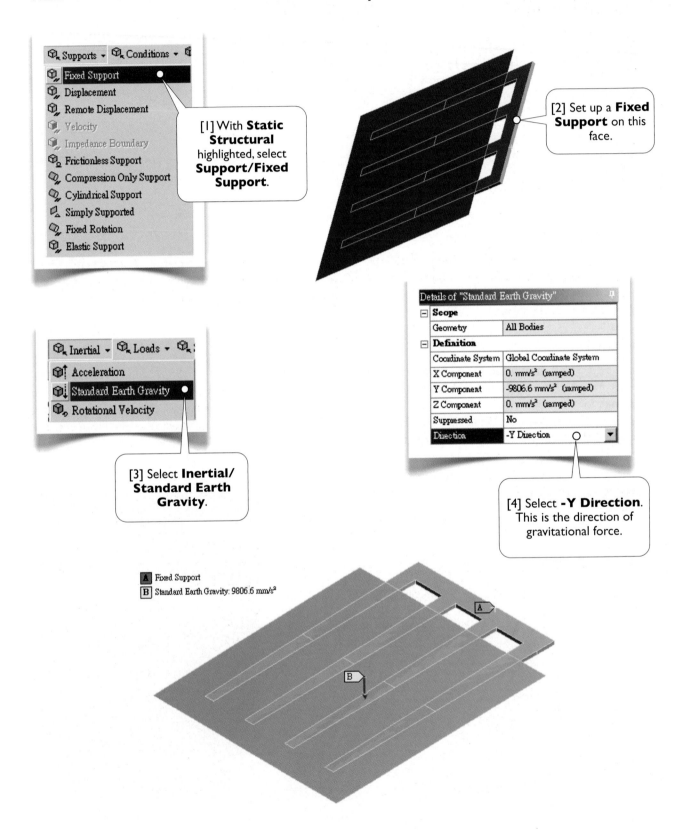

Supports ▾ **Conditions** ▾

- Fixed Support ●
- Displacement
- Remote Displacement
- Velocity
- Impedance Boundary
- Frictionless Support
- Compression Only Support
- Cylindrical Support
- Simply Supported
- Fixed Rotation
- Elastic Support

[1] With **Static Structural** highlighted, select **Support/Fixed Support**.

[2] Set up a **Fixed Support** on this face.

Inertial ▾ **Loads** ▾

- Acceleration
- Standard Earth Gravity ●
- Rotational Velocity

[3] Select **Inertial/ Standard Earth Gravity**.

Details of "Standard Earth Gravity"

Scope	
Geometry	All Bodies
Definition	
Coordinate System	Global Coordinate System
X Component	0. mm/s² (ramped)
Y Component	-9806.6 mm/s² (ramped)
Z Component	0. mm/s² (ramped)
Suppressed	No
Direction	-Y Direction

[4] Select **-Y Direction**. This is the direction of gravitational force.

A Fixed Support
B Standard Earth Gravity: 9806.6 mm/s²

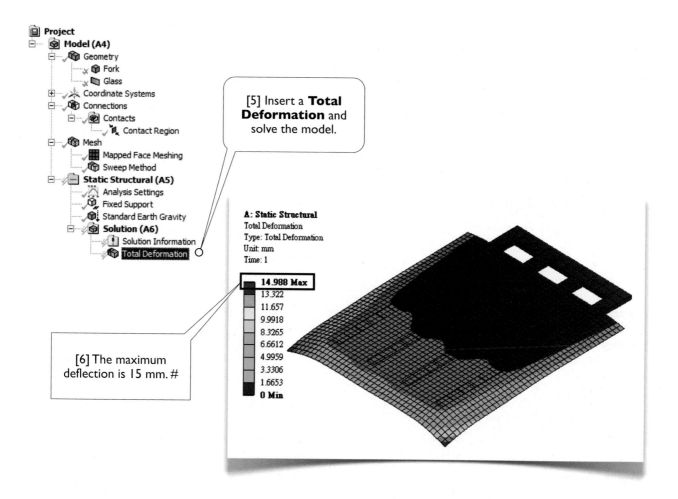

[5] Insert a **Total Deformation** and solve the model.

A: **Static Structural**
Total Deformation
Type: Total Deformation
Unit: mm
Time: 1

14.988 Max
13.322
11.657
9.9918
8.3265
6.6612
4.9959
3.3306
1.6653
0 Min

[6] The maximum deflection is 15 mm. #

12.2-5 Perform Transient Structural Simulation

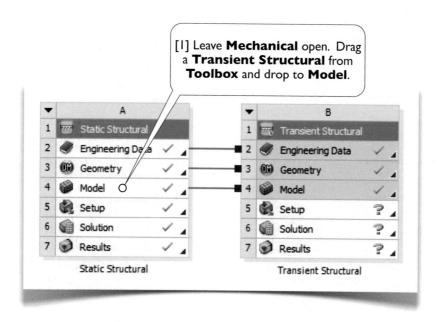

[1] Leave **Mechanical** open. Drag a **Transient Structural** from **Toolbox** and drop to **Model**.

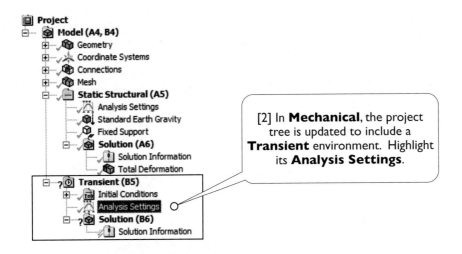

[2] In **Mechanical**, the project tree is updated to include a **Transient** environment. Highlight its **Analysis Settings**.

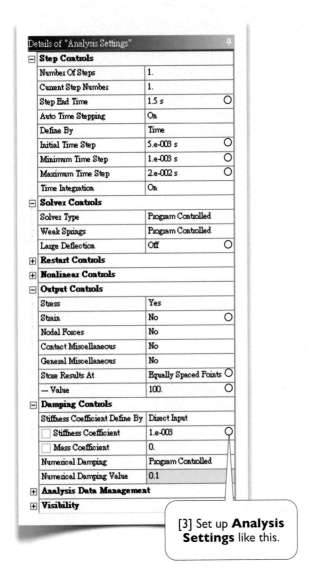

[3] Set up **Analysis Settings** like this.

Step End Time

This is the total simulation time. According to engineers' experience, the vibration should settle to a negligible amount in 1.5 sec.

Initial Time Step

As a guideline, it is suggested that the integration time step be about 1/20 of the response period, to catch the detail behavior of the structural response. The response frequency is estimated to be 10 Hz (see 12.2-5[16], page 429). According to this guideline, the integration time step is about

$$ITS = \frac{1}{20f} = \frac{1}{20(10)} = 0.005 \text{ sec}$$

Output Controls

Transient dynamic simulations usually generate huge amount of data. **Output Controls** allows users to cut down data storage space and computing time.

Beta Damping Value

The beta damping value (stiffness coefficient) is reported to be 0.001 from lab experiments (see 12.1-3, pages 415-417).

We could specify this value as a material property. Since the vibration is dominated by the glass, here, we choose to input the beta value as a global damping value, that is, neglecting the difference caused by the steel material.

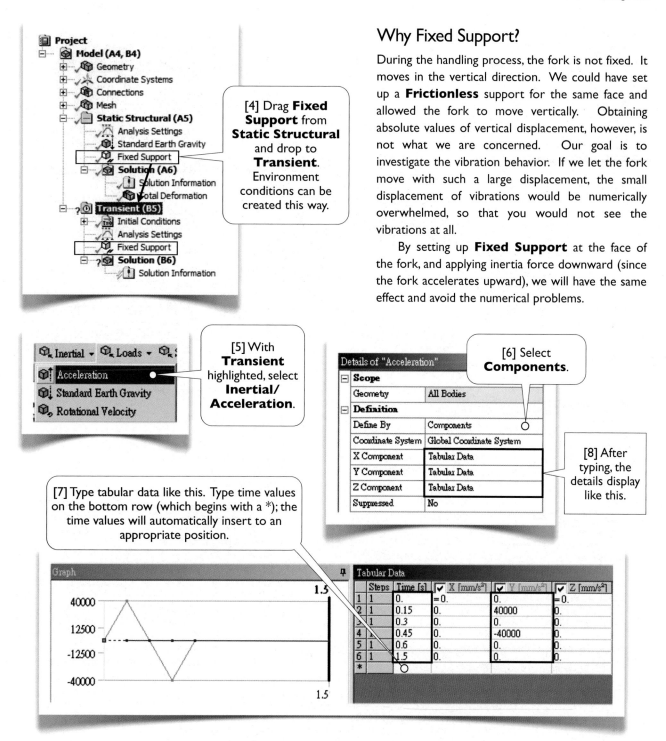

Why Fixed Support?

During the handling process, the fork is not fixed. It moves in the vertical direction. We could have set up a **Frictionless** support for the same face and allowed the fork to move vertically. Obtaining absolute values of vertical displacement, however, is not what we are concerned. Our goal is to investigate the vibration behavior. If we let the fork move with such a large displacement, the small displacement of vibrations would be numerically overwhelmed, so that you would not see the vibrations at all.

By setting up **Fixed Support** at the face of the fork, and applying inertia force downward (since the fork accelerates upward), we will have the same effect and avoid the numerical problems.

[4] Drag **Fixed Support** from **Static Structural** and drop to **Transient**. Environment conditions can be created this way.

[5] With **Transient** highlighted, select **Inertial/Acceleration**.

[6] Select **Components**.

[8] After typing, the details display like this.

[7] Type tabular data like this. Type time values on the bottom row (which begins with a *); the time values will automatically insert to an appropriate position.

How the acceleration data come from?

During the handling, the fork accelerates upward until a speed of 6 m/s in 0.3 sec and then decelerates until a full stop in another 0.3 sec. The average acceleration is thus 20 m/s², but it is not a constant acceleration. The system controls the acceleration such that it increases linearly to 40 m/s² and then decreases to zero linearly. That way, the impact will be minimized and thus the amplitude of vibrations will be minimized too. The same idea applies during the deceleration.

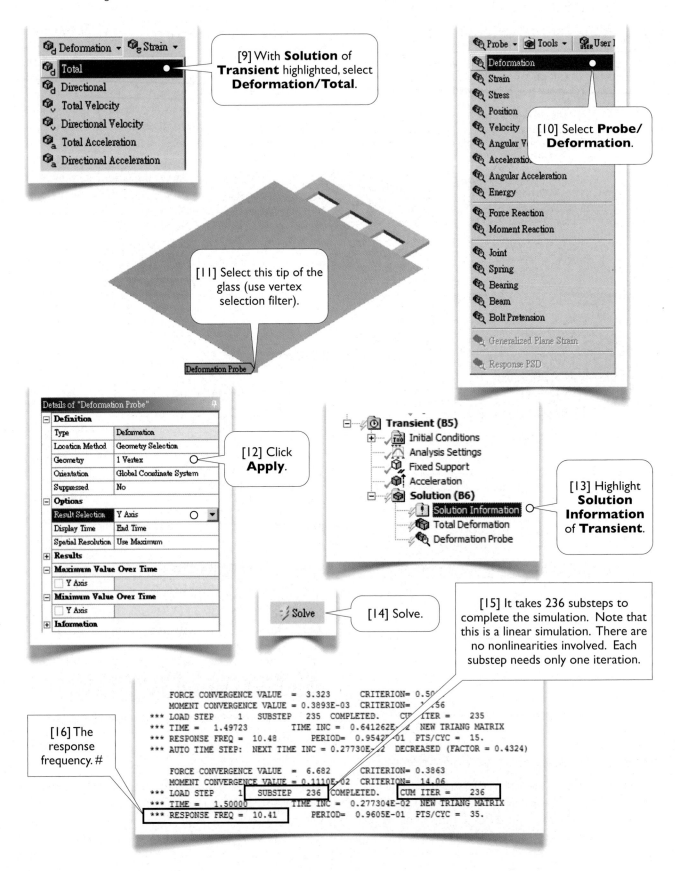

[9] With **Solution** of **Transient** highlighted, select **Deformation/Total**.

Deformation ▾   Strain ▾
  Total   ●
  Directional
  Total Velocity
  Directional Velocity
  Total Acceleration
  Directional Acceleration

Probe ▾   Tools ▾   User
  Deformation   ●
  Strain
  Stress
  Position
  Velocity
  Angular V
  Acceleration
  Angular Acceleration
  Energy
  Force Reaction
  Moment Reaction
  Joint
  Spring
  Bearing
  Beam
  Bolt Pretension
  Generalized Plane Strain
  Response PSD

[10] Select **Probe/ Deformation**.

[11] Select this tip of the glass (use vertex selection filter).

Deformation Probe

Details of "Deformation Probe"
  Definition
    Type | Deformation
    Location Method | Geometry Selection
    Geometry | 1 Vertex
    Orientation | Global Coordinate System
    Suppressed | No
  Options
    Result Selection | Y Axis
    Display Time | End Time
    Spatial Resolution | Use Maximum
  Results
  Maximum Value Over Time
    ☐ Y Axis
  Minimum Value Over Time
    ☐ Y Axis
  Information

[12] Click **Apply**.

Transient (B5)
  Initial Conditions
  Analysis Settings
  Fixed Support
  Acceleration
  Solution (B6)
    Solution Information
    Total Deformation
    Deformation Probe

[13] Highlight **Solution Information** of **Transient**.

Solve

[14] Solve.

[15] It takes 236 substeps to complete the simulation. Note that this is a linear simulation. There are no nonlinearities involved. Each substep needs only one iteration.

[16] The response frequency. #

```
         FORCE CONVERGENCE VALUE  = 3.323      CRITERION= 0.50
         MOMENT CONVERGENCE VALUE = 0.3893E-03  CRITERION=    56
***  LOAD STEP    1   SUBSTEP   235  COMPLETED.    C    ITER =     235
***  TIME =   1.49723      TIME INC = 0.641262E    NEW TRIANG MATRIX
***  RESPONSE FREQ =  10.48     PERIOD=  0.954    01  PTS/CYC =  15.
***  AUTO TIME STEP:  NEXT TIME INC = 0.27730E    DECREASED (FACTOR = 0.4324)

         FORCE CONVERGENCE VALUE  = 6.682      CRITERION= 0.3863
         MOMENT CONVERGENCE VALUE = 0.1110E-02  CRITERION=  14.06
***  LOAD STEP    1   SUBSTEP   236  COMPLETED.        CUM ITER =     236
***  TIME =   1.50000      TIME INC = 0.277304E-02  NEW TRIANG MATRIX
***  RESPONSE FREQ =  10.41     PERIOD=  0.9605E-01  PTS/CYC =  35.
```

12.2-6 View the Results

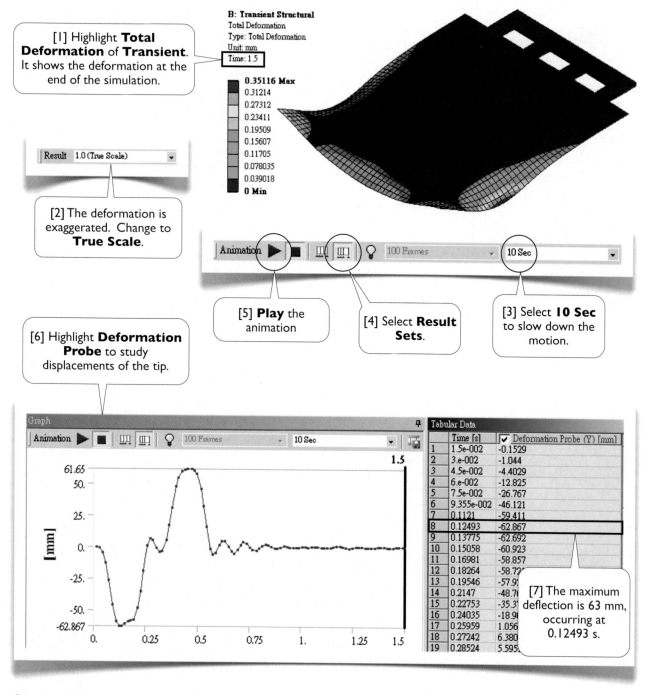

[1] Highlight **Total Deformation** of **Transient**. It shows the deformation at the end of the simulation.

B: Transient Structural
Total Deformation
Type: Total Deformation
Unit: mm
Time: 1.5

0.35116 Max
0.31214
0.27312
0.23411
0.19509
0.15607
0.11705
0.078035
0.039018
0 Min

Result 1.0 (True Scale)

[2] The deformation is exaggerated. Change to **True Scale**.

Animation ▶ ■ ⫿⫿⫿ ⫿⫿⫿ ♀ 100 Frames 10 Sec

[5] **Play** the animation

[4] Select **Result Sets**.

[3] Select **10 Sec** to slow down the motion.

[6] Highlight **Deformation Probe** to study displacements of the tip.

Graph

Animation ▶ ■ ⫿⫿⫿ ⫿⫿⫿ ♀ 100 Frames 10 Sec

	Time [s]	✓ Deformation Probe (Y) [mm]
1	1.5e-002	-0.1529
2	3.e-002	-1.044
3	4.5e-002	-4.4029
4	6.e-002	-12.825
5	7.5e-002	-26.767
6	9.355e-002	-46.121
7	0.1121	-59.411
8	0.12493	-62.867
9	0.13775	-62.692
10	0.15058	-60.923
11	0.16981	-58.857
12	0.18264	-58.72
13	0.19546	-57.9
14	0.2147	-48.7
15	0.22753	-35.3
16	0.24035	-18.9
17	0.25959	1.056
18	0.27242	6.380
19	0.28524	5.595

[7] The maximum deflection is 63 mm, occurring at 0.12493 s.

Observation

The maximum deflection is **63** mm [7], which occurs at 0.12493 sec. The vibration damps out fast and reduces to less than 1 mm in about 1 sec. At the end of 1.5 sec, the response frequency is about 10.4 Hz (12.2-5[16], last page). Workbench automatically adjusts the time step according to the response frequency.

Next, we want to investigate the maximum stress and its location.

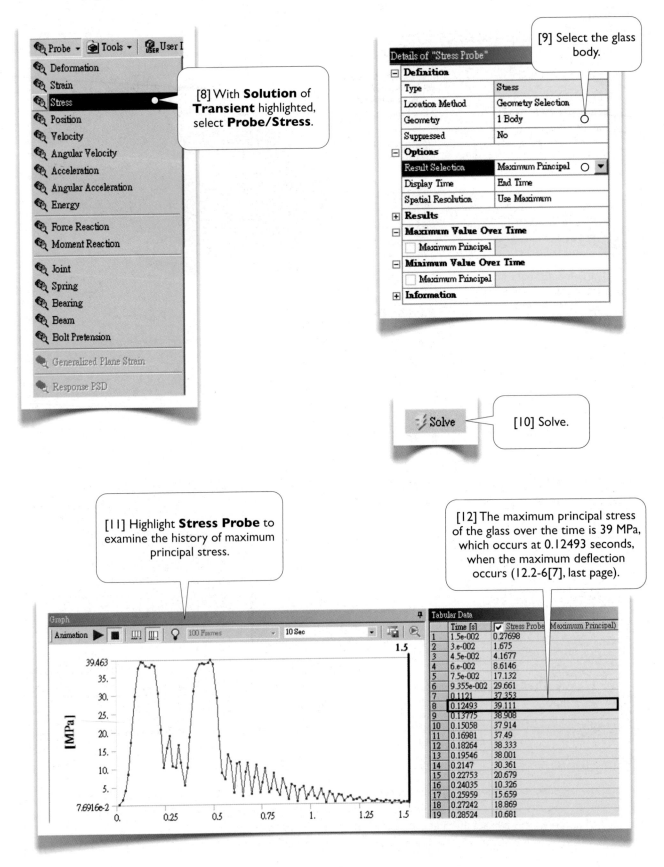

[8] With **Solution** of **Transient** highlighted, select **Probe/Stress**.

[9] Select the glass body.

[10] Solve.

[11] Highlight **Stress Probe** to examine the history of maximum principal stress.

[12] The maximum principal stress of the glass over the time is 39 MPa, which occurs at 0.12493 seconds, when the maximum deflection occurs (12.2-6[7], last page).

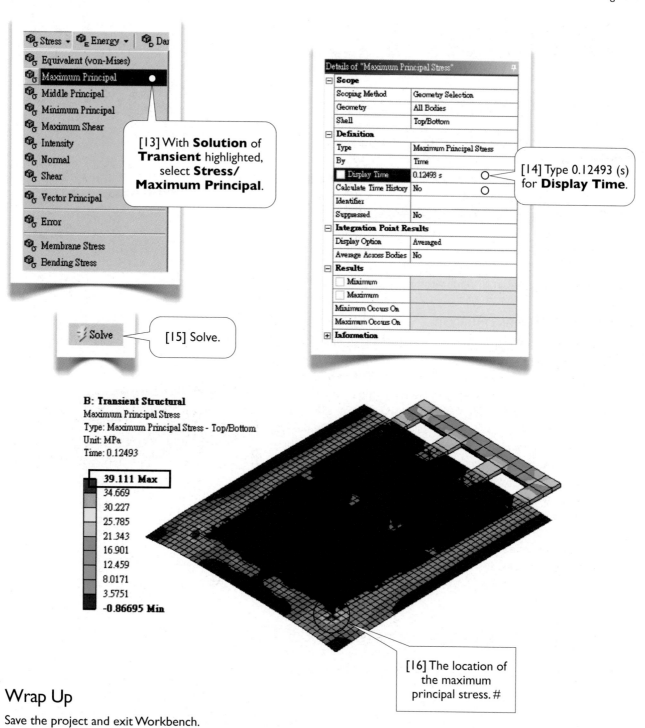

Stress ▾ Energy ▾ Dar

 Equivalent (von-Mises)
 Maximum Principal ●
 Middle Principal
 Minimum Principal
 Maximum Shear
 Intensity
 Normal
 Shear

 Vector Principal

 Error

 Membrane Stress
 Bending Stress

[13] With **Solution** of **Transient** highlighted, select **Stress/Maximum Principal**.

Details of "Maximum Principal Stress"

Scope	
Scoping Method	Geometry Selection
Geometry	All Bodies
Shell	Top/Bottom
Definition	
Type	Maximum Principal Stress
By	Time
Display Time	0.12493 s
Calculate Time History	No
Identifier	
Suppressed	No
Integration Point Results	
Display Option	Averaged
Average Across Bodies	No
Results	
Minimum	
Maximum	
Minimum Occurs On	
Maximum Occurs On	
Information	

[14] Type 0.12493 (s) for **Display Time**.

Solve

[15] Solve.

B: Transient Structural
Maximum Principal Stress
Type: Maximum Principal Stress - Top/Bottom
Unit: MPa
Time: 0.12493

39.111 Max
34.669
30.227
25.785
21.343
16.901
12.459
8.0171
3.5751
-0.86695 Min

[16] The location of the maximum principal stress. #

Wrap Up

Save the project and exit Workbench.

Reference

1. ANSYS Documentation//Mechanical APDL//Element Reference //I. Element Library//SOLSH190

Section 12.3

Harmonic Response Analysis: Two-Story Building

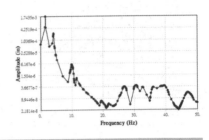

12.3-1 About the Two-Story Building

In this section, we will demonstrate the procedure of a harmonic response analysis (12.1-4, page 419). The two-story building (Sections 7.3, 11.2) will be used again to demonstrate the procedures.

Harmonic Response Analysis

At the end of 11.2-3 (page 389), we mentioned that the rhythmic loading on the floor may cause a safety issue. Is "dancing on the floor" really an issue? Since the building is designed to withstand a live load of 50 psf (lb/ft^2), we will assume that a group of young people of 50 psf is dancing on a side-span floor deck [1] to simulate an asymmetric loading that will cause the building side sway. The dancing is so hard that the young people generate a vertical periodical force of 10 psf, that is, the loading fluctuates from 40 psf to 60 psf.

Engineers usually don't consider "dancing" as a serious issue. Let's look at a more realistic engineering consideration. Imagine that an electric motor (or any rotatory machines) is installed on the floor deck [1]. The operational speed of the machine is 3000 rpm. When started up, the machine's speed increases from zero up to 3000 rpm. Is the vibration caused by the rotatory machine an issue?

In this section, we will perform a harmonic response analysis to answer these questions.

[1] Harmonic loading is applied on this floor deck. #

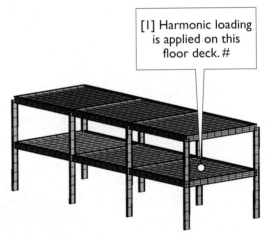

12.3-2 Perform an Unprestressed Modal Analysis

Launch Workbench. Open the project **Building**, saved in Section 11.2.

[1] In this section, we want to reuse this system. Remember the model in this system has no diagonal members.

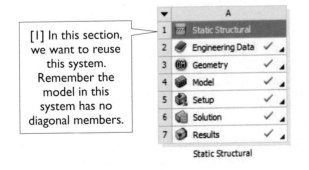

Static Structural

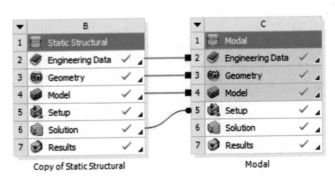

Copy of Static Structural   Modal

433

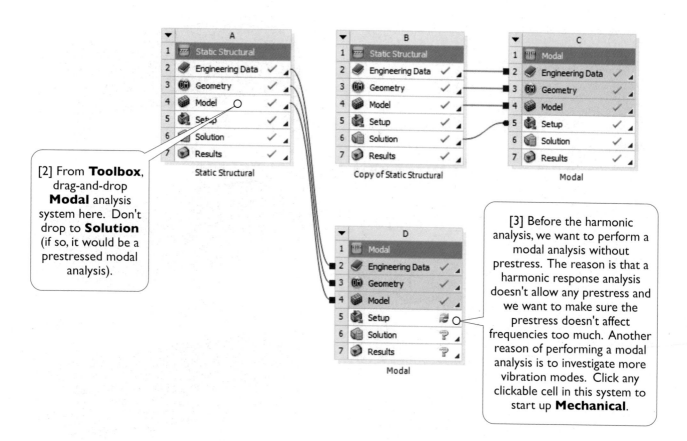

[2] From **Toolbox**, drag-and-drop **Modal** analysis system here. Don't drop to **Solution** (if so, it would be a prestressed modal analysis).

[3] Before the harmonic analysis, we want to perform a modal analysis without prestress. The reason is that a harmonic response analysis doesn't allow any prestress and we want to make sure the prestress doesn't affect frequencies too much. Another reason of performing a modal analysis is to investigate more vibration modes. Click any clickable cell in this system to start up **Mechanical**.

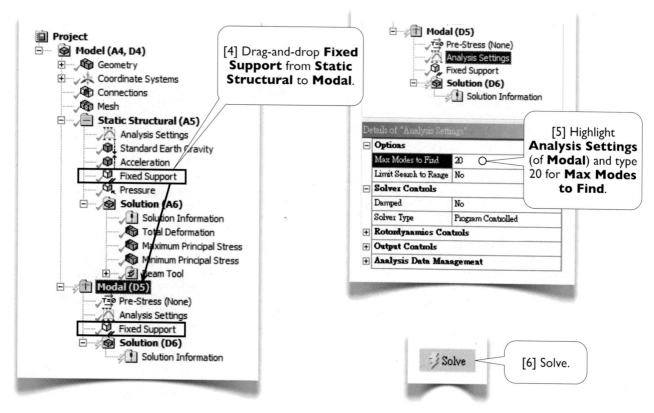

[4] Drag-and-drop **Fixed Support** from **Static Structural** to **Modal**.

[5] Highlight **Analysis Settings** (of **Modal**) and type 20 for **Max Modes to Find**.

[6] Solve.

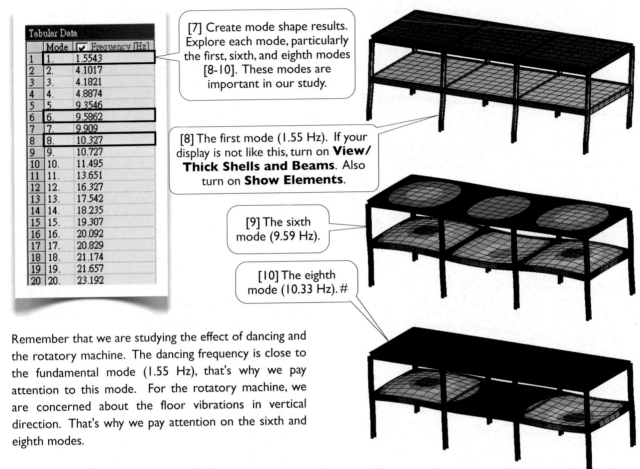

Tabular Data		
	Mode	☑ Frequency [Hz]
1	1.	1.5543
2	2.	4.1017
3	3.	4.1821
4	4.	4.8874
5	5.	9.3546
6	6.	9.5862
7	7.	9.909
8	8.	10.327
9	9.	10.727
10	10.	11.495
11	11.	13.651
12	12.	16.327
13	13.	17.542
14	14.	18.235
15	15.	19.307
16	16.	20.092
17	17.	20.829
18	18.	21.174
19	19.	21.657
20	20.	23.192

[7] Create mode shape results. Explore each mode, particularly the first, sixth, and eighth modes [8-10]. These modes are important in our study.

[8] The first mode (1.55 Hz). If your display is not like this, turn on **View/ Thick Shells and Beams**. Also turn on **Show Elements**.

[9] The sixth mode (9.59 Hz).

[10] The eighth mode (10.33 Hz). #

Remember that we are studying the effect of dancing and the rotatory machine. The dancing frequency is close to the fundamental mode (1.55 Hz), that's why we pay attention to this mode. For the rotatory machine, we are concerned about the floor vibrations in vertical direction. That's why we pay attention on the sixth and eighth modes.

12.3-3 Perform Harmonic Response Analysis

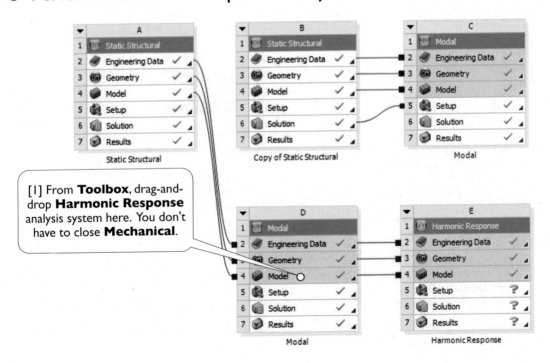

[1] From **Toolbox**, drag-and-drop **Harmonic Response** analysis system here. You don't have to close **Mechanical**.

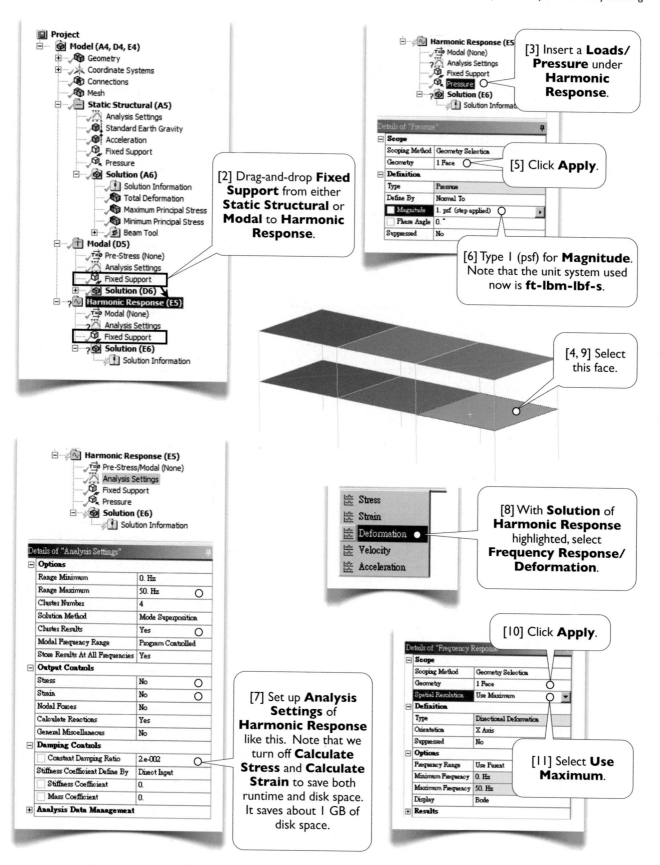

Project
- **Model (A4, D4, E4)**
  - Geometry
  - Coordinate Systems
  - Connections
  - Mesh
  - **Static Structural (A5)**
    - Analysis Settings
    - Standard Earth Gravity
    - Acceleration
    - Fixed Support
    - Pressure
    - **Solution (A6)**
      - Solution Information
      - Total Deformation
      - Maximum Principal Stress
      - Minimum Principal Stress
      - Beam Tool
  - **Modal (D5)**
    - Pre-Stress (None)
    - Analysis Settings
    - Fixed Support
    - **Solution (D6)**
  - **Harmonic Response (E5)**
    - Modal (None)
    - Analysis Settings
    - Fixed Support
    - **Solution (E6)**
      - Solution Information

[2] Drag-and-drop **Fixed Support** from either **Static Structural** or **Modal** to **Harmonic Response**.

Harmonic Response (E5
- Modal (None)
- Analysis Settings
- Fixed Support
- Pressure
- **Solution (E6)**
  - Solution Informat

[3] Insert a **Loads/ Pressure** under **Harmonic Response**.

Details of "Pressure"
Scope	
Scoping Method	Geometry Selection
Geometry	1 Face
Definition	
Type	Pressure
Define By	Normal To
Magnitude	1. psf (step applied)
Phase Angle	0. °
Suppressed	No

[5] Click **Apply**.

[6] Type 1 (psf) for **Magnitude**. Note that the unit system used now is **ft-lbm-lbf-s**.

[4, 9] Select this face.

Harmonic Response (E5)
- Pre-Stress/Modal (None)
- Analysis Settings
- Fixed Support
- Pressure
- **Solution (E6)**
  - Solution Information

Details of "Analysis Settings"
Options	
Range Minimum	0. Hz
Range Maximum	50. Hz
Cluster Number	4
Solution Method	Mode Superposition
Cluster Results	Yes
Modal Frequency Range	Program Controlled
Store Results At All Frequencies	Yes
Output Controls	
Stress	No
Strain	No
Nodal Forces	No
Calculate Reactions	Yes
General Miscellaneous	No
Damping Controls	
Constant Damping Ratio	2.e-002
Stiffness Coefficient Define By	Direct Input
Stiffness Coefficient	0.
Mass Coefficient	0.
Analysis Data Management	

[7] Set up **Analysis Settings** of **Harmonic Response** like this. Note that we turn off **Calculate Stress** and **Calculate Strain** to save both runtime and disk space. It saves about 1 GB of disk space.

- Stress
- Strain
- **Deformation**
- Velocity
- Acceleration

[8] With **Solution** of **Harmonic Response** highlighted, select **Frequency Response/ Deformation**.

[10] Click **Apply**.

Details of "Frequency Response"
Scope	
Scoping Method	Geometry Selection
Geometry	1 Face
Spatial Resolution	Use Maximum
Definition	
Type	Directional Deformation
Orientation	X Axis
Suppressed	No
Options	
Frequency Range	Use Parent
Minimum Frequency	0. Hz
Maximum Frequency	50. Hz
Display	Bode
Results	

[11] Select **Use Maximum**.

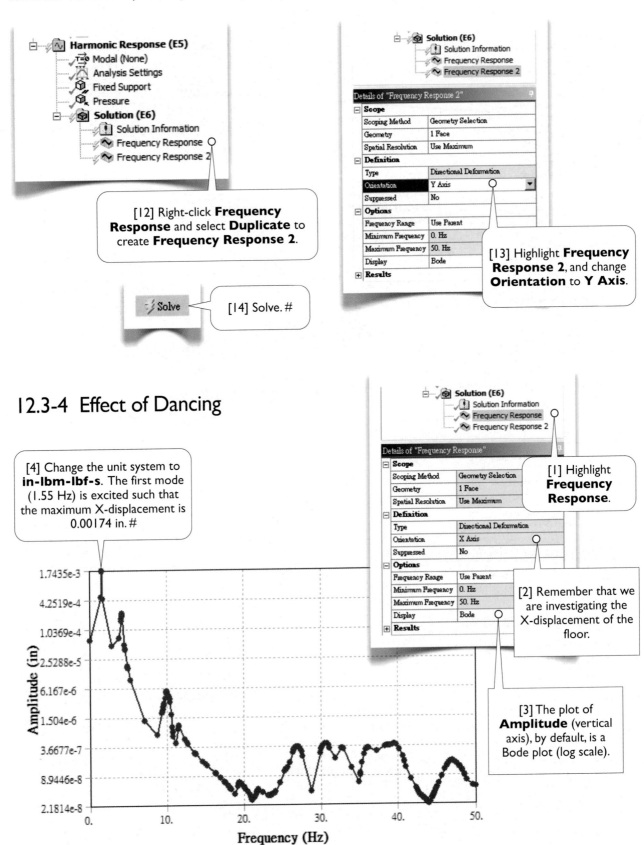

[12] Right-click **Frequency Response** and select **Duplicate** to create **Frequency Response 2**.

[13] Highlight **Frequency Response 2**, and change **Orientation** to **Y Axis**.

[14] Solve. #

12.3-4 Effect of Dancing

[4] Change the unit system to **in-lbm-lbf-s**. The first mode (1.55 Hz) is excited such that the maximum X-displacement is 0.00174 in. #

[1] Highlight **Frequency Response**.

[2] Remember that we are investigating the X-displacement of the floor.

[3] The plot of **Amplitude** (vertical axis), by default, is a Bode plot (log scale).

Interpretation of the Harmonic Response Plot

The harmonic response plot [4] is an amplitude versus frequency plot. Because we are investigating the effect of the dancing, we should look at the frequencies that are less than 3 or 4 Hz (people don't dance faster than that). At dancing frequency of 1.55 Hz, the structure is excited such that the maximum X-displacement is 0.00174 in. Remember that in 12.3-1 (page 433), we estimate the dancing loading is about 10 psf periodically. In 12.3-3[6] (page 436), we input a unit of harmonic load (1 psf), therefore the estimated response should be 10 times of 0.00174 in, that is 0.0174 in. (0.44 mm). Obviously, this value is too small to be worried about.

We conclude that dancing is not an issue for this building.

12.3-5 Effect of Rotatory Machine

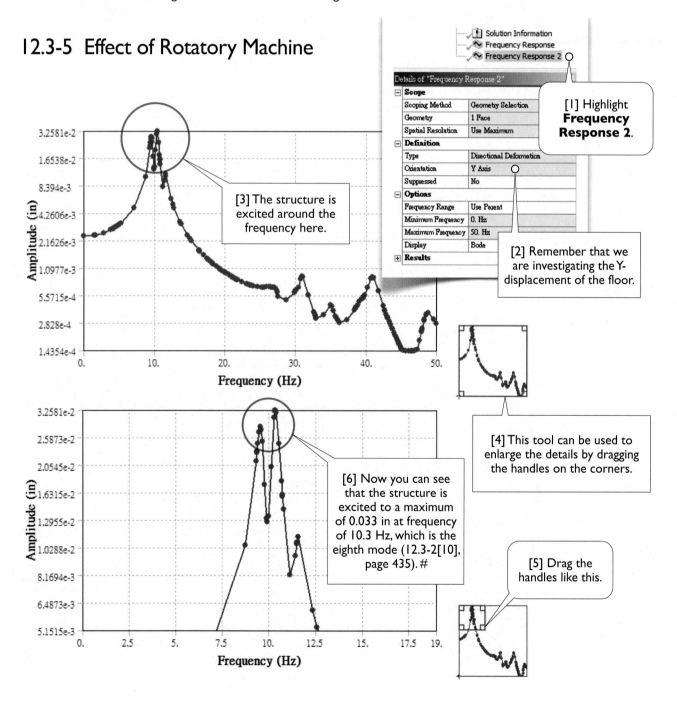

[1] Highlight **Frequency Response 2**.

[2] Remember that we are investigating the Y-displacement of the floor.

[3] The structure is excited around the frequency here.

[4] This tool can be used to enlarge the details by dragging the handles on the corners.

[5] Drag the handles like this.

[6] Now you can see that the structure is excited to a maximum of 0.033 in at frequency of 10.3 Hz, which is the eighth mode (12.3-2[10], page 435). #

Interpretation of the Harmonic Response Plot

Now we are investigating the effect of the rotatory machine from 0 to 3000 rpm (50 Hz). We estimate that the amplitude of the harmonic load (of the electric motor) should be no more than 0.1 psf distributing on the floor (that totals to 40 lb). In 12.3-3[6] (page 436), we input a unit of harmonic load (1 psf), therefore the estimated response should be 0.1 times of the response shown in [3] or [6].

Although high frequencies do excite the floor, but the values are very small. At frequency of 10.3 Hz, the excitation reaches a maximum of 0.0033 in (0.1 times of 0.033 in), or 0.084 mm, that is even smaller than the dancing effect. Therefore, the value is too small to cause an issue.

We conclude that the rotatory machine is safe for this building.

Wrap Up

Save the project and exit Workbench.

Section 12.4

Disk and Block

This exercise has two purposes: (1) to demonstrate how to apply a simple initial condition, namely uniform velocity, on a body, and (2) to show a limitation of **Transient Structural** analysis system for impact simulations, and to motivate students for learning **Explicit Dynamics** in Chapter 15.

12.4-1 About the Disk and Block

Consider a disk of radius of 40 mm and a block of 200 mm x 20 mm, both have a thickness of 10 mm, on a frictionless horizontal surface [1, 2]. Both are made of a very soft polymer of Young's modulus of 10 kPa, Poisson's ratio of 0.4, and mass density of 1000 kg/m³.

Right before the impact, the disk moves toward the block with a velocity of 0.5 m/s, and the positions of the disk and the block are as shown.

We purposely consider an very soft material (Young's modulus of 10 kPa) and a very slow-speed impact (velocity of 0.5 m/s) to relieve numerical difficulty. Increasing either of them would make the impact duration shorter, and in turn require a shorter integration time step (to find a solution). This leaves as an exercise for you at the end of this chapter.

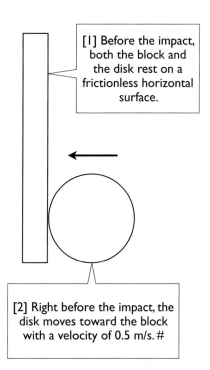

[1] Before the impact, both the block and the disk rest on a frictionless horizontal surface.

[2] Right before the impact, the disk moves toward the block with a velocity of 0.5 m/s. #

12.4-2 Start Up

Launch Workbench. Create a **Transient Structural** analysis system [1] by double-clicking it in **Toolbox**. Save the project as **Disk**.

Double-click **Engineering Data** to prepare material data [2].

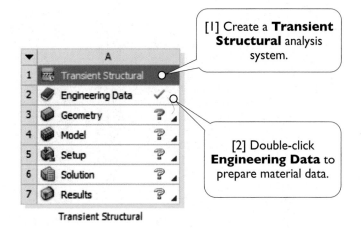

[1] Create a **Transient Structural** analysis system.

	A	
1	Transient Structural	○
2	Engineering Data	✓ ○
3	Geometry	?
4	Model	?
5	Setup	?
6	Solution	?
7	Results	?

Transient Structural

[2] Double-click **Engineering Data** to prepare material data.

Create a new material, **Polymer** [3], and input material properties as shown [4]. Return to **Project Schematic** [5] and double-click **Geometry** to start up DesignModeler. Choose **Millimeter** as length unit.

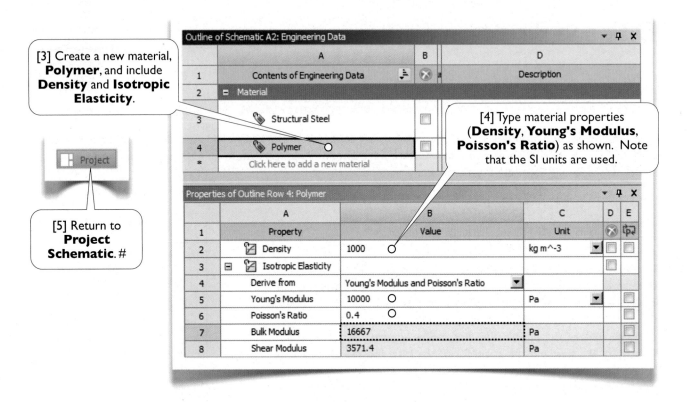

[3] Create a new material, **Polymer**, and include **Density** and **Isotropic Elasticity**.

[5] Return to **Project Schematic**. #

[4] Type material properties (**Density, Young's Modulus, Poisson's Ratio**) as shown. Note that the SI units are used.

12.4-3 Create Geometry in DesignModeler

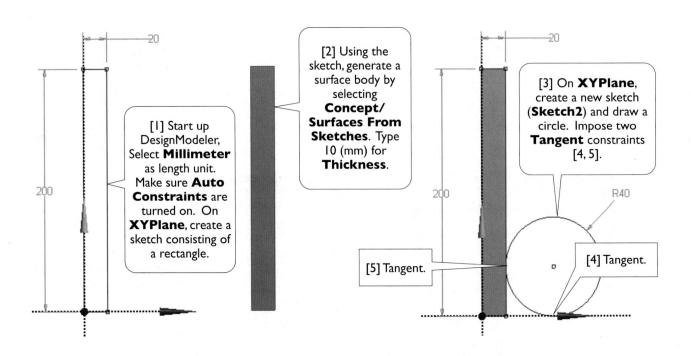

[1] Start up DesignModeler, Select **Millimeter** as length unit. Make sure **Auto Constraints** are turned on. On **XYPlane**, create a sketch consisting of a rectangle.

[2] Using the sketch, generate a surface body by selecting **Concept/ Surfaces From Sketches**. Type 10 (mm) for **Thickness**.

[3] On **XYPlane**, create a new sketch (**Sketch2**) and draw a circle. Impose two **Tangent** constraints [4, 5].

[5] Tangent.

[4] Tangent.

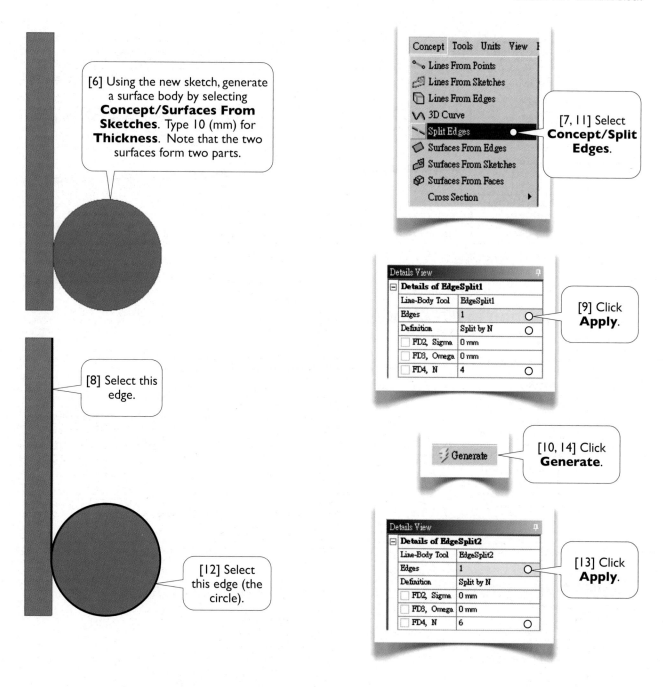

[6] Using the new sketch, generate a surface body by selecting **Concept/Surfaces From Sketches**. Type 10 (mm) for **Thickness**. Note that the two surfaces form two parts.

[7, 11] Select **Concept/Split Edges**.

[8] Select this edge.

[9] Click **Apply**.

[10, 14] Click **Generate**.

[12] Select this edge (the circle).

[13] Click **Apply**.

Why Split Edges?

The purpose of splitting the edges into segments is that, when meshed, each segment can have its own mesh density. We need finer mesh around the contact region.

How to Make Sure an Edge is Successfully Split?

Move your mouse on a split edge, which is then highlighted. Go through each split edge to make sure an edge is successfully split.

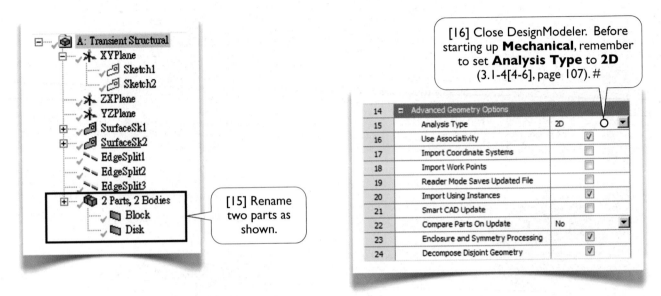

[16] Close DesignModeler. Before starting up **Mechanical**, remember to set **Analysis Type** to **2D** (3.1-4[4-6], page 107). #

[15] Rename two parts as shown.

12.4-4 Simulation in **Mechanical**

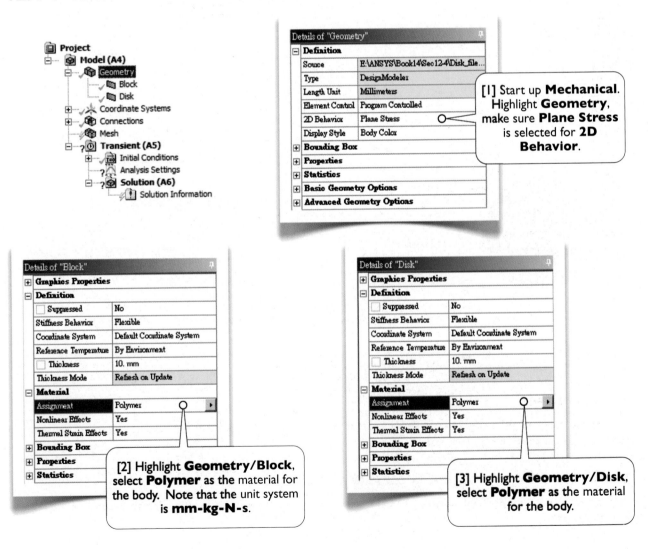

[1] Start up **Mechanical**. Highlight **Geometry**, make sure **Plane Stress** is selected for **2D Behavior**.

[2] Highlight **Geometry/Block**, select **Polymer** as the material for the body. Note that the unit system is **mm-kg-N-s**.

[3] Highlight **Geometry/Disk**, select **Polymer** as the material for the body.

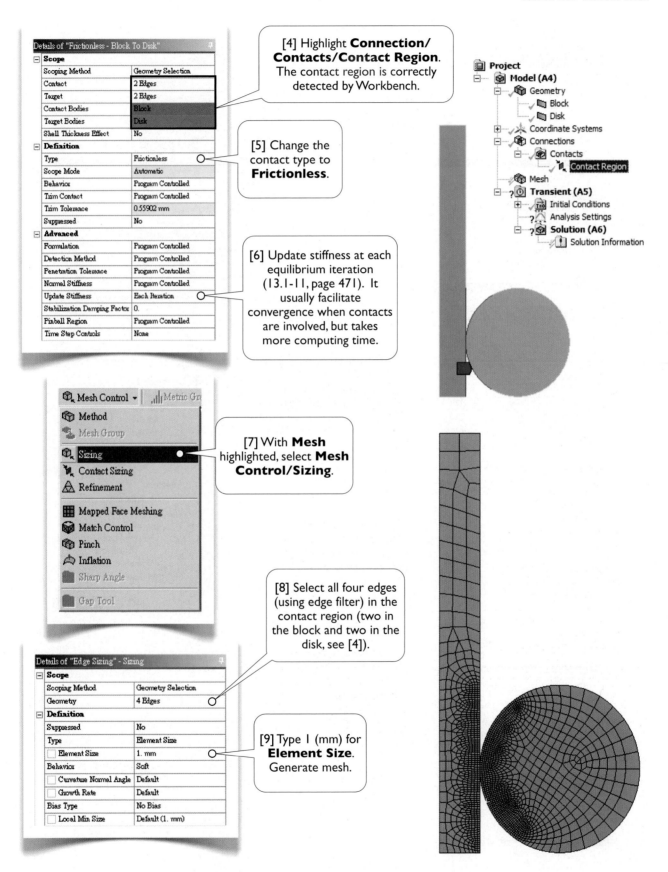

Details of "Frictionless - Block To Disk"

Scope	
Scoping Method	Geometry Selection
Contact	2 Edges
Target	2 Edges
Contact Bodies	Block
Target Bodies	Disk
Shell Thickness Effect	No
Definition	
Type	Frictionless
Scope Mode	Automatic
Behavior	Program Controlled
Trim Contact	Program Controlled
Trim Tolerance	0.55902 mm
Suppressed	No
Advanced	
Formulation	Program Controlled
Detection Method	Program Controlled
Penetration Tolerance	Program Controlled
Normal Stiffness	Program Controlled
Update Stiffness	Each Iteration
Stabilization Damping Factor	0.
Pinball Region	Program Controlled
Time Step Controls	None

[4] Highlight **Connection/ Contacts/Contact Region**. The contact region is correctly detected by Workbench.

[5] Change the contact type to **Frictionless**.

[6] Update stiffness at each equilibrium iteration (13.1-11, page 471). It usually facilitate convergence when contacts are involved, but takes more computing time.

Project
  Model (A4)
    Geometry
      Block
      Disk
    Coordinate Systems
    Connections
      Contacts
        Contact Region
    Mesh
  Transient (A5)
    Initial Conditions
    Analysis Settings
    Solution (A6)
      Solution Information

Mesh Control ▾   Metric Gr

- Method
- Mesh Group
- Sizing
- Contact Sizing
- Refinement
- Mapped Face Meshing
- Match Control
- Pinch
- Inflation
- Sharp Angle
- Gap Tool

[7] With **Mesh** highlighted, select **Mesh Control/Sizing**.

[8] Select all four edges (using edge filter) in the contact region (two in the block and two in the disk, see [4]).

Details of "Edge Sizing" - Sizing

Scope	
Scoping Method	Geometry Selection
Geometry	4 Edges
Definition	
Suppressed	No
Type	Element Size
Element Size	1. mm
Behavior	Soft
Curvature Normal Angle	Default
Growth Rate	Default
Bias Type	No Bias
Local Min Size	Default (1. mm)

[9] Type 1 (mm) for **Element Size**. Generate mesh.

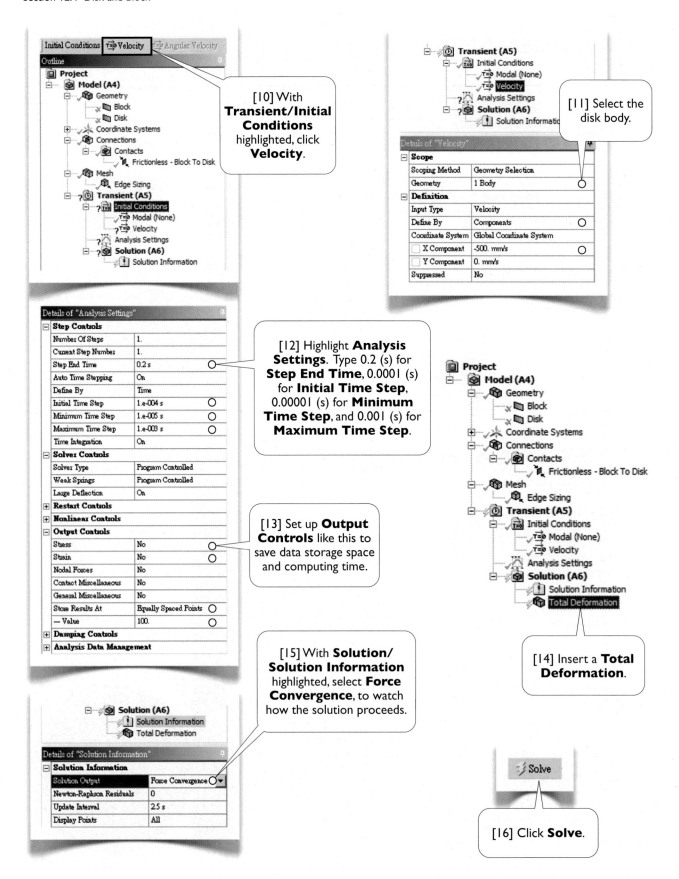

[10] With **Transient/Initial Conditions** highlighted, click **Velocity**.

[11] Select the disk body.

Details of "Velocity"

Scope	
Scoping Method	Geometry Selection
Geometry	1 Body
Definition	
Input Type	Velocity
Define By	Components
Coordinate System	Global Coordinate System
X Component	-500. mm/s
Y Component	0. mm/s
Suppressed	No

Details of "Analysis Settings"

Step Controls	
Number Of Steps	1.
Current Step Number	1.
Step End Time	0.2 s
Auto Time Stepping	On
Define By	Time
Initial Time Step	1.e-004 s
Minimum Time Step	1.e-005 s
Maximum Time Step	1.e-003 s
Time Integration	On
Solver Controls	
Solver Type	Program Controlled
Weak Springs	Program Controlled
Large Deflection	On
Restart Controls	
Nonlinear Controls	
Output Controls	
Stress	No
Strain	No
Nodal Forces	No
Contact Miscellaneous	No
General Miscellaneous	No
Store Results At	Equally Spaced Points
— Value	100.
Damping Controls	
Analysis Data Management	

[12] Highlight **Analysis Settings**. Type 0.2 (s) for **Step End Time**, 0.0001 (s) for **Initial Time Step**, 0.00001 (s) for **Minimum Time Step**, and 0.001 (s) for **Maximum Time Step**.

[13] Set up **Output Controls** like this to save data storage space and computing time.

[15] With **Solution/ Solution Information** highlighted, select **Force Convergence**, to watch how the solution proceeds.

[14] Insert a **Total Deformation**.

[16] Click **Solve**.

Details of "Solution Information"

Solution Information	
Solution Output	Force Convergence
Newton-Raphson Residuals	0
Update Interval	2.5 s
Display Points	All

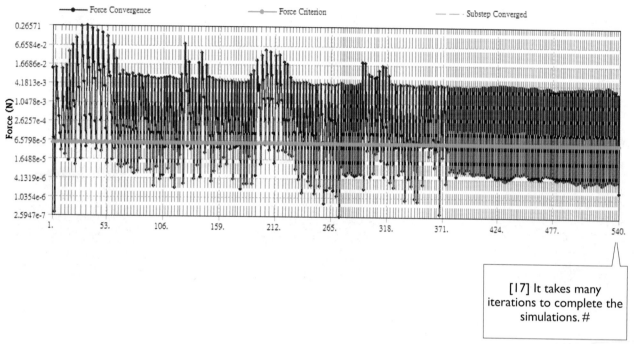

[17] It takes many iterations to complete the simulations. #

12.4-5 Animate the Impact

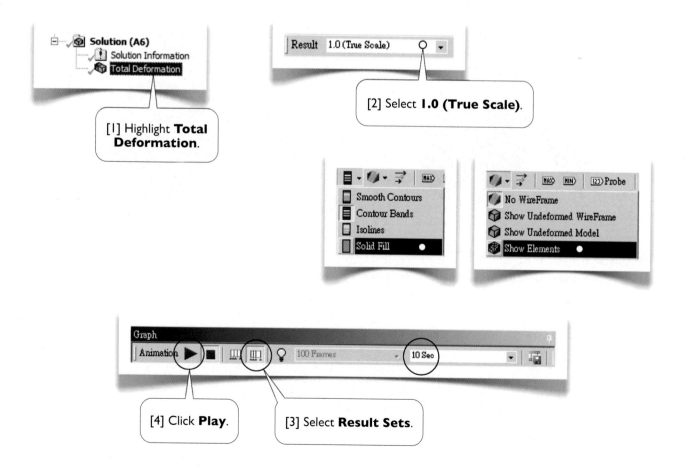

[1] Highlight **Total Deformation**.

[2] Select **1.0 (True Scale)**.

[4] Click **Play**.

[3] Select **Result Sets**.

446

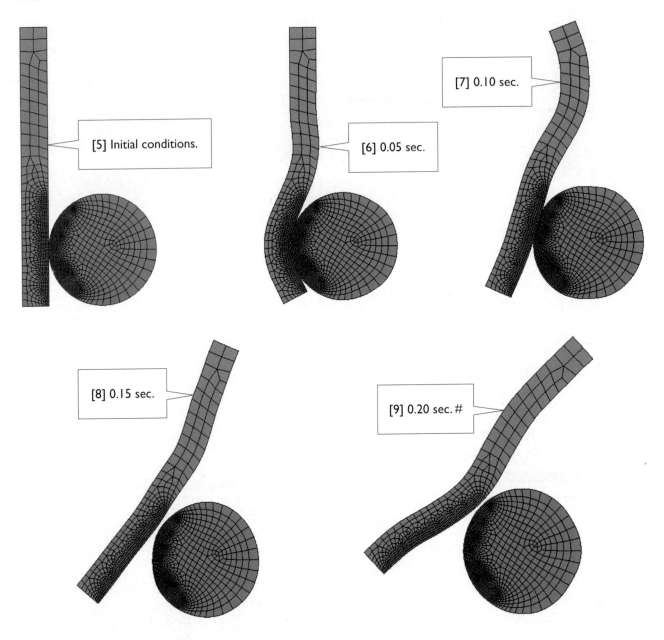

[5] Initial conditions.

[6] 0.05 sec.

[7] 0.10 sec.

[8] 0.15 sec.

[9] 0.20 sec. #

How to Obtain a Snapshot at a Specific Time?

Highlight **Total Deformation**, type the time for **Display Time**, and click **Solve**.

Wrap Up

Save the project and exit Workbench.

Section 12.5

Guitar String

This exercise has two purposes: (1) to demonstrate how to use the results of a static simulation as an initial condition of a transient dynamic simulation, and (2) to show the transient behavior of a guitar string, as a sequel of Section 11.4.

12.5-1 About the Guitar String

As mentioned, in 11.4-3 (page 405), if you pluck a string, you will produce a tone made up of all harmonic modes. It is the harmonic mixes that differentiate the sound of one music instrument from others. A good guitar player knows very well that he/she can control the quality of sound by plucking the string at different locations. We will explore this phenomenon in this section, using the guitar string introduced in Section 11.4.

The string is plucked at the middle, quarter, and eighth points respectively, and its transient responses are observed. More precisely, the string is applied a vertical displacement of 10 mm at respective locations, and then instantaneously released to produce vibrations. The vibrations will eventually reach a steady state, that is, free vibrations, after a certain duration of time. It, however, is the transient vibrations that impress our ears most.

12.5-2 Using Results of Static Analysis as Initial Condition

Applying the vertical displacement requires a **Static Structural** analysis. Releasing the string to produce vibrations requires a **Transient Structural** analysis using the results of the static analysis as an initial condition. In a **Transient Structural**, it is possible to specify the first step as a static simulation. The results become the initial condition of the next step of transient simulation. This two-step method has become a standard procedure, when the simulation requires a static simulation as an initial condition.

Before we demonstrate the two-step method, let's clear up some important concepts. In transient dynamic simulation, each time step needs an initial condition to carry it on. The results of the last time step become the initial condition of the next time step. The initial condition, more specifically, is the position and velocity of each node.

In a **Transient Structural** simulation, Workbench allows you to turn off/on **Time Integration** [1]. Turning **Time Integration** off means turning the dynamic effects (1.1-10, page 17) off, and Eq. 12.1-4(1) (page 418) reduces to Eq. 1.3-1(1) (page 31); i.e., a static simulation. You can turn on/off **Time Integration** for any load step.

Make sure you don't confuse with load steps (or simply called steps), substeps, and equilibrium iterations. Multiple load steps can be created when you need to specify different **Analysis Settings** for each step. Each step is further divided into substeps (also called time steps) for two reasons: (a) in a transient dynamic simulation, each substep is a time integration step, and (b) in a nonlinear simulation, a substep must be small enough to achieve convergence within that substep. For nonlinear simulations (either dynamic or static), each substep may need multiple equilibrium iterations to achieve convergence.

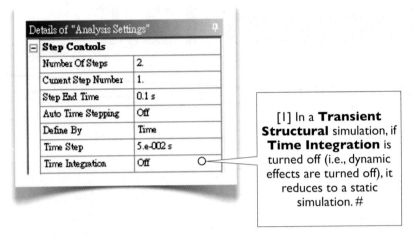

Details of "Analysis Settings"	
Step Controls	
Number Of Steps	2.
Current Step Number	1.
Step End Time	0.1 s
Auto Time Stepping	Off
Define By	Time
Time Step	5.e-002 s
Time Integration	Off

[1] In a **Transient Structural** simulation, if **Time Integration** is turned off (i.e., dynamic effects are turned off), it reduces to a static simulation. #

12.5-3 Resume the Project **String**

Launch Workbench. Open the project **String**, which was saved in Section 11.4.

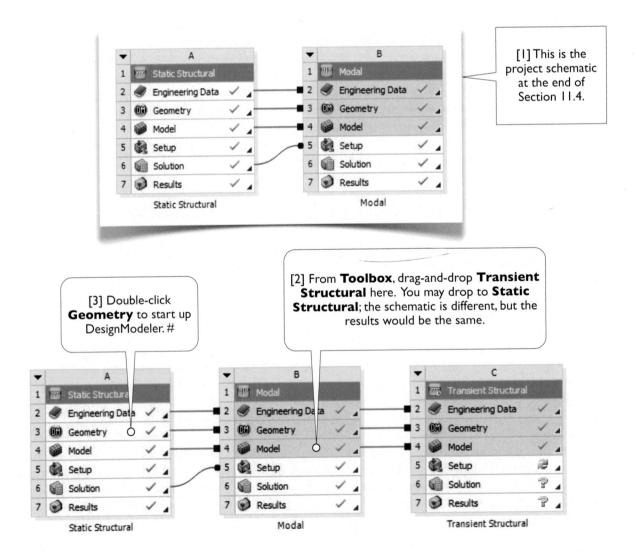

[1] This is the project schematic at the end of Section 11.4.

[2] From **Toolbox**, drag-and-drop **Transient Structural** here. You may drop to **Static Structural**; the schematic is different, but the results would be the same.

[3] Double-click **Geometry** to start up DesignModeler. #

449

12.5-4 Modify Geometry in DesignModeler

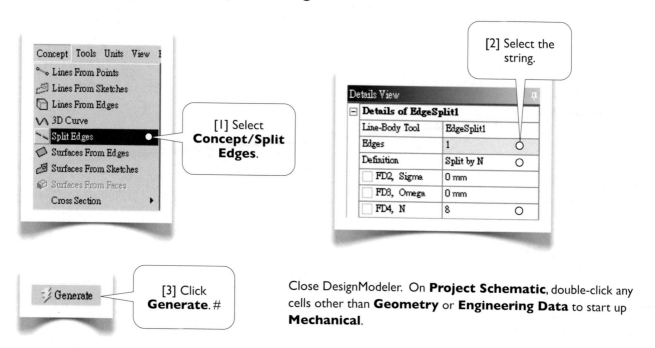

[1] Select **Concept/Split Edges**.

[2] Select the string.

[3] Click **Generate**. #

Close DesignModeler. On **Project Schematic**, double-click any cells other than **Geometry** or **Engineering Data** to start up **Mechanical**.

12.5-5 Set Up Environment Conditions

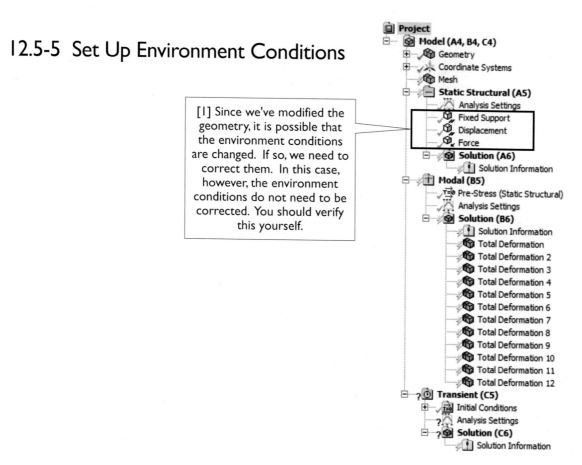

[1] Since we've modified the geometry, it is possible that the environment conditions are changed. If so, we need to correct them. In this case, however, the environment conditions do not need to be corrected. You should verify this yourself.

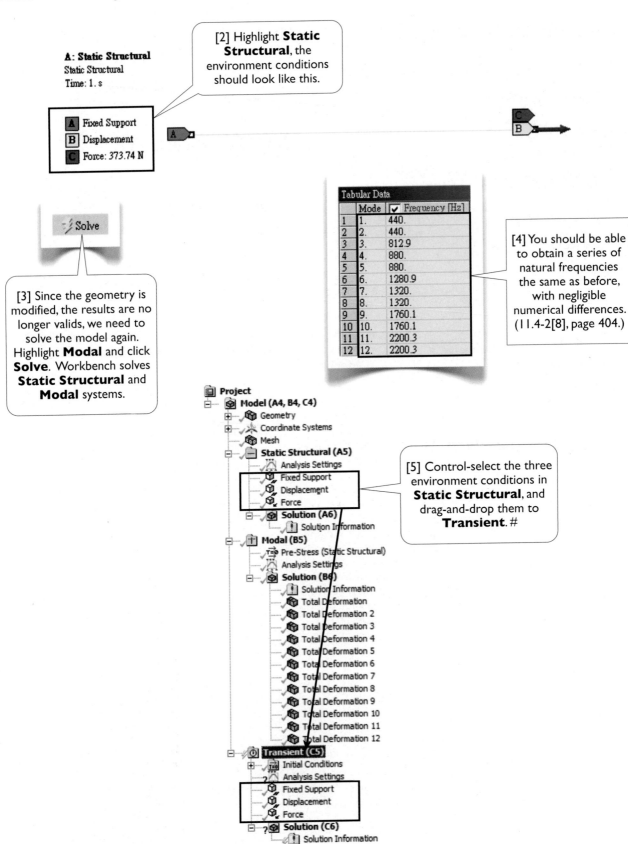

A: Static Structural
Static Structural
Time: 1. s

[A] Fixed Support
[B] Displacement
[C] Force: 373.74 N

[2] Highlight **Static Structural**, the environment conditions should look like this.

Solve

[3] Since the geometry is modified, the results are no longer valids, we need to solve the model again. Highlight **Modal** and click **Solve**. Workbench solves **Static Structural** and **Modal** systems.

Tabular Data

	Mode	✔ Frequency [Hz]
1	1.	440.
2	2.	440.
3	3.	812.9
4	4.	880.
5	5.	880.
6	6.	1280.9
7	7.	1320.
8	8.	1320.
9	9.	1760.1
10	10.	1760.1
11	11.	2200.3
12	12.	2200.3

[4] You should be able to obtain a series of natural frequencies the same as before, with negligible numerical differences. (11.4-2[8], page 404.)

Project
Model (A4, B4, C4)
  Geometry
  Coordinate Systems
  Mesh
  Static Structural (A5)
    Analysis Settings
    Fixed Support
    Displacement
    Force
    Solution (A6)
      Solution Information
  Modal (B5)
    Pre-Stress (Static Structural)
    Analysis Settings
    Solution (B6)
      Solution Information
      Total Deformation
      Total Deformation 2
      Total Deformation 3
      Total Deformation 4
      Total Deformation 5
      Total Deformation 6
      Total Deformation 7
      Total Deformation 8
      Total Deformation 9
      Total Deformation 10
      Total Deformation 11
      Total Deformation 12
  Transient (C5)
    Initial Conditions
    Analysis Settings
    Fixed Support
    Displacement
    Force
    Solution (C6)
      Solution Information

[5] Control-select the three environment conditions in **Static Structural**, and drag-and-drop them to **Transient**. #

12.5-6 Set Up **Analysis Settings**

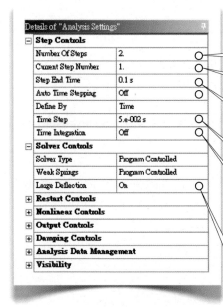

[1] Highlight **Analysis Settings** of **Transient** and type 2 for total number of steps.

[2] We are now working on the first step.

[3] The ending time of the first step is arbitrary.

[4] The step is further divided into two substeps. We will explain this later on.

[5] Turn off **Time Integration** to perform a static simulation for this step.

[6] Make sure **Large Deflection** is on. Prestress (373.74 N, 11.4-2[7], page 404) can be included only in a nonlinear simulation.

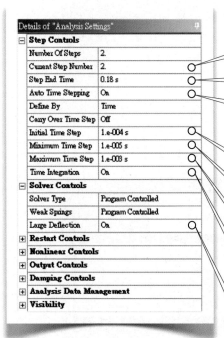

[7] Type 2 (or click Step 2 in **Tabular Data**) to switch to the second step.

[8] The total time for transient simulation is 0.08 seconds (0.18 - 0.10 = 0.08).

[9] Make sure Auto **Time Stepping** is on to let Workbench decide integration time steps according to response frequencies.

[10] Set the initial time step to 0.0001 seconds.

[11] Set the minimum time step to one tenth of the initial time step.

[12] Set the maximum time step to ten times of the initial time step.

[13] Make sure **Time Integration** is on for the second step.

[14] Make sure **Large Deflection** is on so that the prestress can be included. #

12.5-7 Set Up Initial Condition

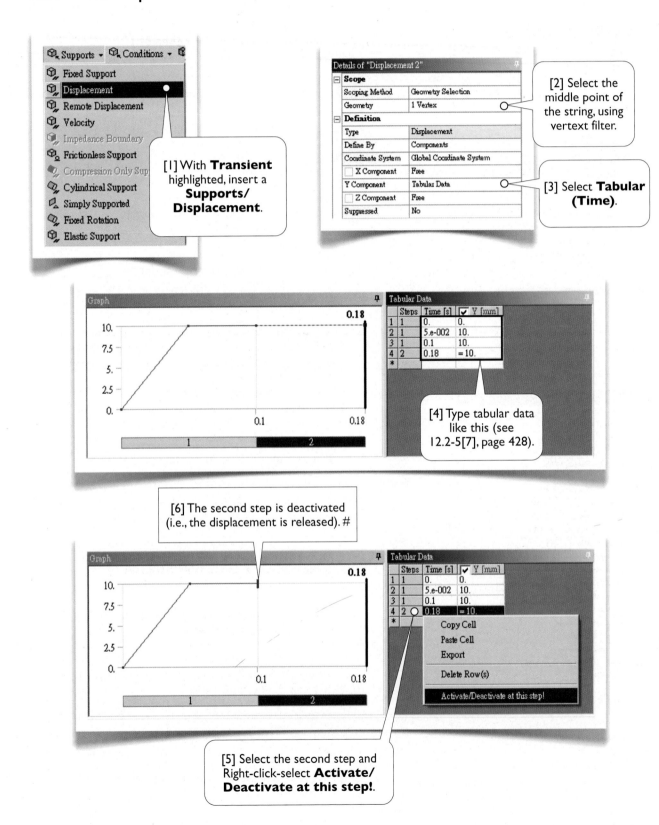

12.5-8 Insert Result Objects and Solve the Model

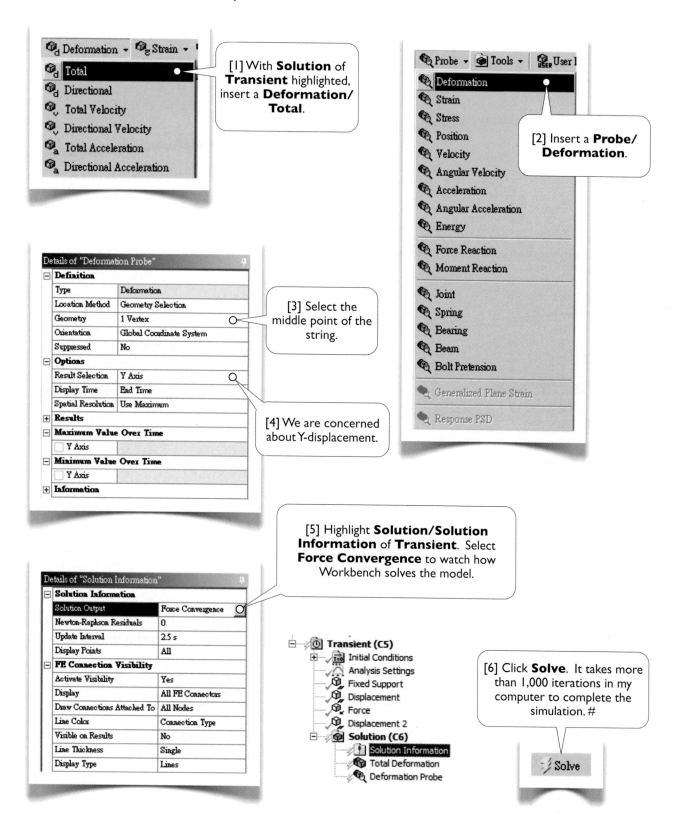

[1] With **Solution** of **Transient** highlighted, insert a **Deformation/ Total**.

[2] Insert a **Probe/ Deformation**.

[3] Select the middle point of the string.

[4] We are concerned about Y-displacement.

[5] Highlight **Solution/Solution Information** of **Transient**. Select **Force Convergence** to watch how Workbench solves the model.

[6] Click **Solve**. It takes more than 1,000 iterations in my computer to complete the simulation. #

454

12.5-9 View the Results

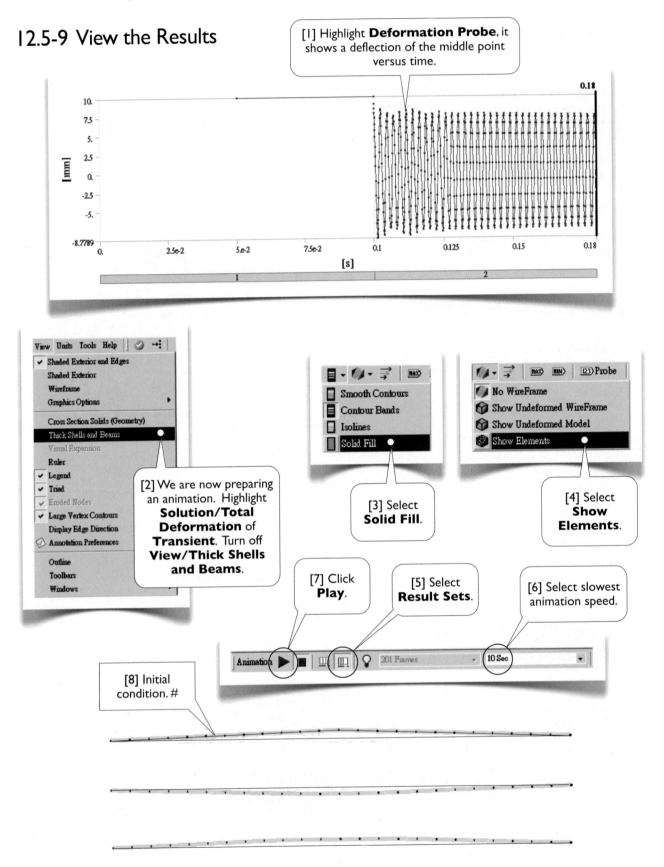

[1] Highlight **Deformation Probe**, it shows a deflection of the middle point versus time.

[2] We are now preparing an animation. Highlight **Solution/Total Deformation** of **Transient**. Turn off **View/Thick Shells and Beams**.

[3] Select **Solid Fill**.

[4] Select **Show Elements**.

[7] Click **Play**.

[5] Select **Result Sets**.

[6] Select slowest animation speed.

[8] Initial condition. #

12.5-10 Pluck at Quarter Point

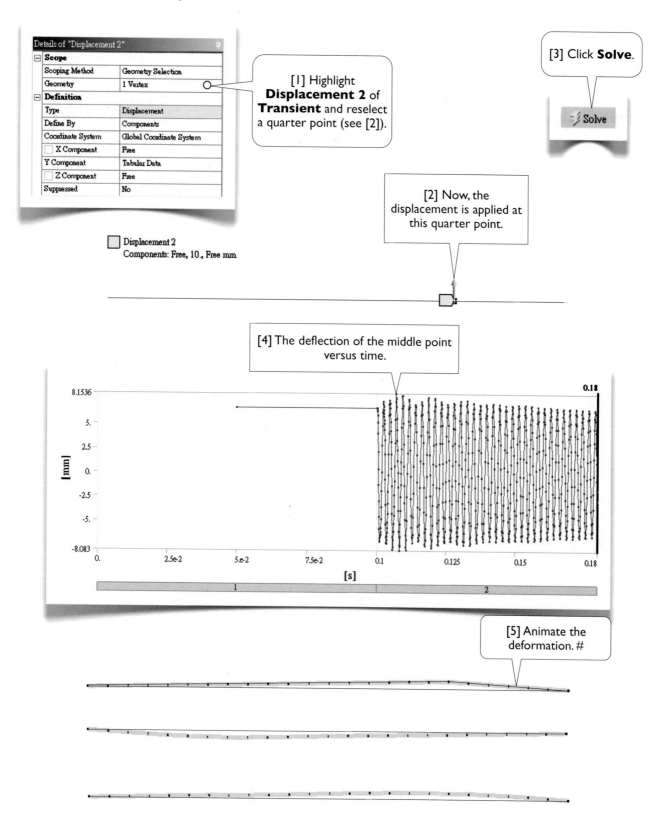

Details of "Displacement 2"

Scope	
Scoping Method	Geometry Selection
Geometry	1 Vertex
Definition	
Type	Displacement
Define By	Components
Coordinate System	Global Coordinate System
☐ X Component	Free
Y Component	Tabular Data
☐ Z Component	Free
Suppressed	No

[1] Highlight **Displacement 2** of **Transient** and reselect a quarter point (see [2]).

[3] Click **Solve**.

⚡ Solve

Displacement 2
Components: Free, 10., Free mm

[2] Now, the displacement is applied at this quarter point.

[4] The deflection of the middle point versus time.

[5] Animate the deformation. #

456

12.5-11 Pluck at Eighth Point

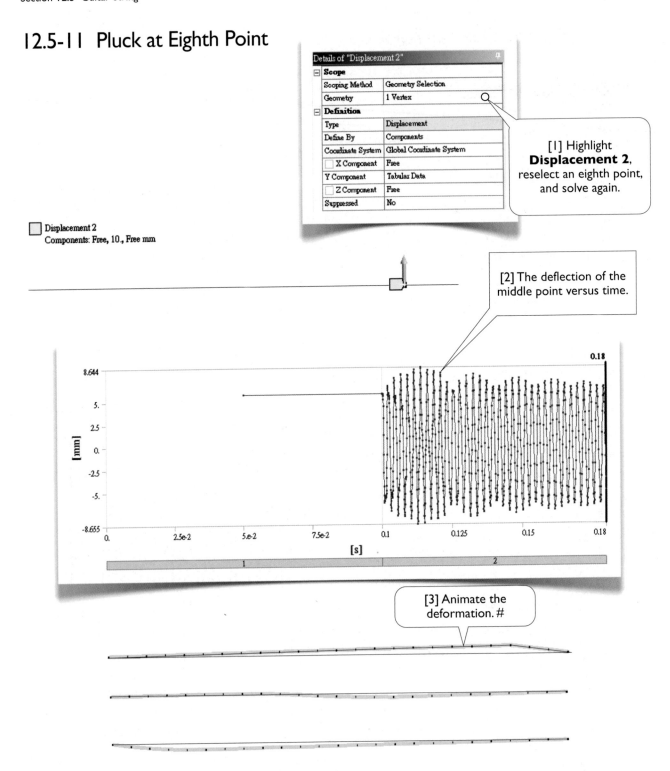

Details of "Displacement 2"

Scope	
Scoping Method	Geometry Selection
Geometry	1 Vertex
Definition	
Type	Displacement
Define By	Components
Coordinate System	Global Coordinate System
☐ X Component	Free
Y Component	Tabular Data
☐ Z Component	Free
Suppressed	No

[1] Highlight **Displacement 2**, reselect an eighth point, and solve again.

☐ Displacement 2
Components: Free, 10., Free mm

[2] The deflection of the middle point versus time.

[3] Animate the deformation. #

Wrap Up

Save the project and exit Workbench.

Section 12.6

Review

12.6-1 Keywords

Choose a letter for each keyword from the list of descriptions

1. (   ) Coulomb Damping
2. (   ) Critical Damping
3. (   ) Explicit Dynamics
4. (   ) Harmonic Response Analysis
5. (   ) Lumped Mass Model
6. (   ) Material Damping

7. (   ) No Separation Contact
8. (   ) Random Vibration Analysis
9. (   ) Response Spectrum Analysis
10. (   ) Solid Shell Element
11. (   ) Steps, Substeps, and Equilibrium Iterations
12. (   ) Viscous Damping

Answers:

1. (E)   2. (B)   3. (I)   4. (F)   5. (A)   6. (D)   7. (K)   8. (H)   9. (G)   10. (L)
11. (J)   12. (C)

List of Descriptions

(A) Analysis models that assume the mass concentrated at certain locations. The discrete masses are typically connected by springs and dampers. The models are often used to study dynamic systems.

(B) When the damping of a dynamic system is smaller than a critical value, its free vibration is oscillatory. On the other hand, if the damping is larger than the critical value, the motion is not oscillatory.

(C) The damping due to the friction between the structure and its surrounding fluid. The damping force typically assumes to be proportional to the velocity of the structural displacement.

(D) Also called solid damping or elastic hysteresis. The damping due to the internal friction in the material. The behavior is still an open research topic. In the Workbench, assuming small material damping, we may express it by an equivalent viscous damping. Further, the viscous damping coefficient (of this equivalent viscous damping) is assumed to be a linear combination of the stiffness matrix and the mass matrix. The form of linear combination (alpha and beta) is then determined by lab experiments and data fittings.

(E) Also called Coulomb friction or dry friction. The damping due to the friction in the connection between structural members. In the Workbench, it can be modeled using frictional contact.

(F) A dynamic analysis to investigate the maximum response of a structure under steady harmonic (sinusoidal) loads.

(G) A dynamic analysis to evaluate the maximum response of a structure under loading conditions described by a spectrum representing the maximum response at varying frequencies in known directions to a specific time history. The technique is often used to design structures withstanding multiple short-duration loadings, such as earthquakes.

(H) A dynamic analysis to evaluate the probabilistic response of a structure under probabilistic loads described by a spectrum representing probability distribution of excitation at varying frequencies in known directions. The technique is used to design structures withstanding random loadings.

(I) A technique for transient dynamic analysis. Explicit integration method is used, which requires a very small integration time step to achieve solution accuracy. For a single time step, it is much more efficient than an implicit method, which is used in a **Transient Structural** analysis system. These features make it very efficient for a high-speed impact simulation, or highly nonlinear simulations.

(J) We may divide the whole loading history into steps, to specify different **Analysis Settings** for each step. Each step can be further divided into substeps. For transient dynamic simulation, each substep is an integration time step. For nonlinear simulation, dividing into substeps is to expedite convergence. In nonlinear simulations, each substep may need several equilibrium iterations to find its solution.

(K) Two surfaces with **No Separation** contact condition prohibit separation in their normal direction, but allow small sliding relative to each other. Both **Bonded** and **No Separation** contact types do not introduce contact nonlinearity.

(L) A solid element that includes additional deformation modes to improve the bending accuracy; useful when meshing thin solid bodies.

12.6-2 Additional Workbench Exercises

High-Speed Impact Simulation

The impact simulation in the Section 12.4 is actually not useful, since the material (of Young's Modulus 10 kPa) is not realistic and the impact speed (0.5 m/s) is slow. Try more realistic simulations by yourself using the same geometric model. Gradually increase the Young's modulus and the impact speed. Each time you may need to decrease the integration time step. This exercise is to experience the limitation of the implicit method, which is used in a **Transient Structural** analysis system. High-speed impact simulations are more suitable by using **Explicit Dynamics** analysis system (Chapter 15).

Chapter 13
Nonlinear Simulations

When the relationship between the response and the load of a structure is linear, the structure is a *linear structure*, and the simulation is a *linear simulation*. Otherwise, the structure is a *nonlinear structure* and the simulation is a *nonlinear simulation*. In the real world, all structures are more or less nonlinear. In many cases, however, when nonlinearities are negligible, we may predict their behavior using linear simulations. For other cases, when nonlinearities are not negligible, nonlinear simulations are needed.

Structural nonlinearities come from three sources: large deformation, change of connectivity, and nonlinear stress-strain relations. Nonlinearity due to large deformation is called *geometry nonlinearity*. Nonlinearity due to the change of connectivity is called *topology nonlinearity*, which includes failure of structural components and change of contact status. In this chapter, we will discuss the change of contact status only, which can be termed *contact nonlinearity*. Nonlinearity due to nonlinear stress-strain relations is called *material nonlinearity* and will be covered in Chapter 14.

In general, nonlinear simulations are much more challenging than linear simulations. They not only take much of computing time, but sometimes fail to find a solution. Solution behaviors of nonlinear simulations highly depend on settings of solution parameters. A thorough comprehension of these solution parameters becomes critical when you want to adjust these parameters to reduce the computing time, or try to successfully find a solution.

Purpose of This Chapter

This chapter discusses nonlinear solution algorithms and focuses on geometry nonlinearity and contact nonlinearity. Material nonlinearity will be covered in Chapter 14. This chapter provides some basics of nonlinear simulations process, so that you can understand and use various solution parameters.

About Each Section

Section 13.1 provides some basics of nonlinear simulations, including Newton-Raphson method, the solution method used in Workbench. Concepts of convergence follow the introduction. Section 13.2 provides a step-by-step example of geometric nonlinearity. Section 13.3 provides a step-by-step example of contact nonlinearity. Section 13.4 provides an additional exercise of contact nonlinearity.

Section 13.1

Basics of Nonlinear Simulations

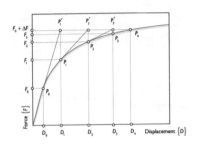

PART A. NONLINEAR SOLUTION METHODS

13.1-1 What Are Nonlinear Simulations

In a linear simulation, the response is linearly proportional to the load. More specifically, let's use Eq. 1.3-1(1) for the upcoming discussion

$$[K]\{D\} = \{F\}$$

Copy of 1.3-1(1), page 31

The nodal force $\{F\}$ may represent the load; the nodal displacement $\{D\}$ may represent the response (stress and strain can be computed from the displacement); the stiffness matrix $[K]$ contains the proportionality coefficients between the force and the displacement. Eq. 1.3-1(1) can be conceptually plotted as shown below [1]. Note that both the horizontal axis and the vertical axis are actually multi-dimensional (i.e., vectors), and $[K]$ is actually the gradient of $\{F\}$ with respect to $\{D\}$. $[K]$ is a matrix of dimension $n \times n$, where n is the degrees of freedom of the system. In cases of single degree of freedom, $[K]$ is a scalar and is the "spring constant" of the structure.

Performing a linear static simulation means to solve Eq. 1.3-1(1), once and for all. For linear transient dynamic simulations, Eq. 12.1-4(1) (page 418) is solved instead; it involves integrations over time domain. In each integration time step, the equation is solved exactly once.

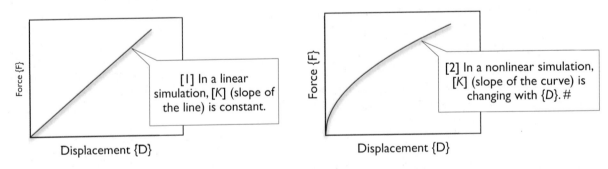

In a nonlinear simulation, the relation between $\{F\}$ and $\{D\}$ is nonlinear, as shown in the figure above [2]. Note that, in the [2], we've shown a "concave down" curve, but it may also be a "concave up" curve, or even a curve with inflection points. In nonlinear cases, $[K]$ matrix in Eq. 1.3-1(1) is no longer a constant matrix, it changes with $\{D\}$; that is, $[K]$ is a function of $\{D\}$. To emphasize this, we may rewrite

$$[K(D)]\{D\} = \{F\}$$

(1)

Challenges of nonlinear simulations come from the difficulties of solving Eq. (1).

13.1-2 Causes of Structural Nonlinearities

As mentioned in the opening of this chapter, sources of structural nonlinearities can be classified into three categories: geometry nonlinearity, topology nonlinearity, and material nonlinearity. A problem may include more than one category of nonlinearities.

A linear simulation implies that no nonlinearities are present; it in turn implies: (a) the deformation is very small, (b) there is no topological changes, and (c) the stress-strain relation is linear; i.e., it can be described by Hooke's law. It also implies that the principle of superposition is applicable and the solution is independent of loading history. On the other hand, in nonlinear problems, the principle of superposition is not applicable and the solution may depend on loading history.

Geometry Nonlinearity

Geometry nonlinearity is due to large deformation of structures. The stiffness matrix $[K]$ is composed by element stiffness matrices, and each element stiffness is a function of the element's material properties as well as geometry. When the deformation of structure is so large that the stiffness matrix $[K]$ is changed substantially, geometry nonlinearity must be considered. To include geometry nonlinearity, simply turn on **Large Deflection** in **Analysis Settings** [1]. Among the three sources of nonlinearities, geometry nonlinearity is usually the easiest to tackle: reducing time steps is usually enough to improve the convergence. Some exceptions (e.g., the pneumatic finger of Section 9.1) need special treatments. In this book, so far, we've experienced many simulations involving geometry nonlinearity. Section 13.2 provides one more exercise for geometry nonlinearity.

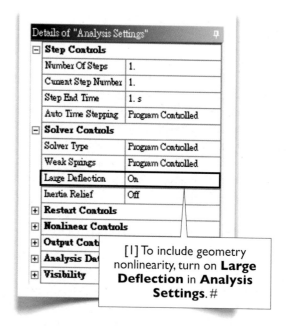

[1] To include geometry nonlinearity, turn on **Large Deflection** in **Analysis Settings**. #

Topology Nonlinearity

When the topology (connectivity) of a structure changes, its stiffness matrix also changes. Possible topology changes include failure of structural members or materials, and the changes of contact status. In this chapter, we will cover only contact nonlinearity; i.e., change of contact status.

Contact nonlinearity itself is challenging and, moreover, it is usually accompanied by large deformation. In this book, so far, we've experienced many simulations involving contact nonlinearity. Sections 13.3 and 13.4 are two more exercises for the contact nonlinearity.

Material Nonlinearity

When its material's stress-strain relation is not linear (i.e., it cannot be described by Hooke's law), the problem involves material nonlinearity. In these cases, we need other ways of describing the relation between stress and strain (or stress and strain rate, when dealing with viscous materials). A mathematical model used to describe a stress-strain relationship is called a material model. Hooke's law, Eqs. 1.2-8(1, 4) (pages 27, 28), is an example of material models. A material model is usually a mathematic form with some parameters; these parameters are usually determined by data fitting using material test data.

Besides Hooke's law, Workbench provides many other material models. Understanding and using these material models are one of the most challenging tasks in the finite element simulations. We will discuss a few of material models in Chapter 14.

13.1-3 Load Steps, Substeps, and Equilibrium Iterations

Steps (Load Steps)

You can divide the entire loading history into one or more *load steps*, or simply called *steps*. The number of steps can be specified in the details view of **Analysis Settings** [1]. To switch between steps, you can type a step number in the details view [2], click a step number in **Graph** [3], or click a step number in **Tabular Data** [4]. Each step can have its own analysis settings [5].

Time Steps (Substeps)

Each load step is further divided into substeps, or time steps. In dynamic simulations, time step is used for integration over time domain (Chapter 12); main consideration of the time step size is to capture the response characteristics. In static simulation, a load step can be divided into substeps to achieve or enhance convergence. Smaller time step size usually converges easier, but, of course, needs more number of time steps to complete a load step.

Iterations (Equilibrium Iterations)

For nonlinear problems, each time step itself needs several iterations to solve Eq. 13.1-1(1) (page 461). Each iteration involves solving a subproblem, Eq. 1.3-1(1) (page 31), the linearized equilibrium equation. Solving Eq. 1.3-1(1) is called an equilibrium iteration, or simply iteration. We will introduce the process of equilibrium iterations, known as the Newton-Raphson method, in the next subsection.

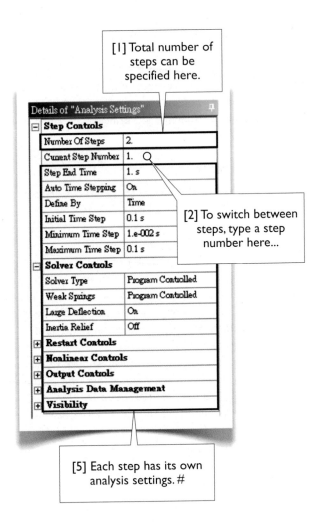

[1] Total number of steps can be specified here.

[2] To switch between steps, type a step number here...

[5] Each step has its own analysis settings. #

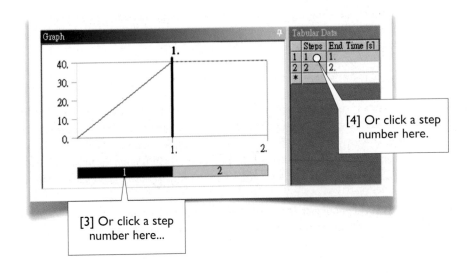

[4] Or click a step number here.

[3] Or click a step number here...

13.1-4 Newton-Raphson Method[Ref 1]

Suppose that the simulation proceeds at a certain time step [1-3], where the displacement is D_0, the external force is F_0, and P_0 represents the point at the response curve described by Eq. 13.1-1(1) (page 461). Now, the time is increased one substep further so that the external force is increased to $F_0 + \Delta F$ [4], and we want to find the displacement at next time step [5].

Starting from point P_0, Workbench calculates a *tangent stiffness* $\left[K(D_0)\right]$, the linearized stiffness, and solves the following equation

$$\left[K(D_0)\right]\{\Delta D\} = \{\Delta F\} \tag{1}$$

The displacement D_0 is increased by ΔD to become D_1. Now, in the D-F space, we are at $(D_1, F_0 + \Delta F)$ (i.e., the point P_1'), far from our goal P_4. To proceed, we need to "drive" the point P_1' back to the actual curve (i.e., P_1).

Substituting the displacement D_1 into the left-hand side of the governing equation, Eq. 13.1-1(1), we can calculate the actual force F_1 which would balance the displacement,

$$\left[K(D_1)\right]\{D_1\} = \{F_1\}$$

Now we can locate the point (D_1, F_1), which is on the actual force-displacement curve. The difference between the external force (here, $F_0 + \Delta F$) and the balanced force (here, F_1) is called the *residual force* of that equilibrium iteration,

$$F_1^R = (F_0 + \Delta F) - F_1$$

If the residual force is smaller than a criterion, then the substep is said to be converged, otherwise, another equilibrium iteration takes place. The iterations repeat until the *convergence criterion* satisfies.

The procedure described above is called the *Newton-Raphson Method*.

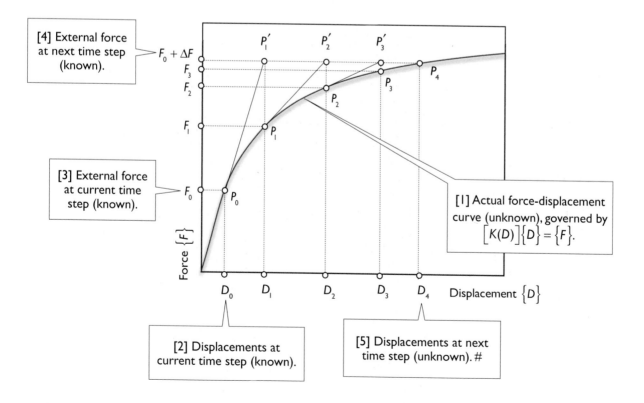

[4] External force at next time step (known).

[3] External force at current time step (known).

[1] Actual force-displacement curve (unknown), governed by $\left[K(D)\right]\{D\} = \{F\}$.

[2] Displacements at current time step (known).

[5] Displacements at next time step (unknown). #

13.1-5 Convergence Criteria[Ref 2]

In the last subsection, we stated that when the residual force F^R is smaller than a criterion, then the substep is converged. This statement is not strictly correct. There are at most four convergence criteria that can be activated under your control, namely, **force convergence** [1], **displacement convergence** [2], **moment convergence** [3], and **rotation convergence** [4]. The moment convergence and rotation convergence can be activated only when shell elements or beam elements are used. These convergence monitoring methods are all defaulted to **Program Controlled**, that is, Workbench automatically turns on any of them when it is appropriate. You may manually turn off or turn on any of them.

When you turn on any of them, you may specify a **Value**, a **Tolerance**, and a **Minimum Reference**. The criterion is then

$$\text{Criterion} = \text{Tolerance} \times \text{maximum}(\text{Value},\ \text{Minimum Reference})$$

The force (or moment) convergence satisfies when

$$\left\| F^R \right\| < \text{Criterion} \tag{1}$$

The displacement (or rotation) convergence satisfies when

$$\left\| \Delta D \right\| < \text{Criterion} \tag{2}$$

Where $\left\| \cdot \right\|$ denotes the norm of the underlying vector, and is called a **Convergence Value**. Value defaults to ANSYS Calculated, which usually means the current maximum value. For example, in the example of 13.1-4 (last page), the current maximum force value is $\left\| F_0 \right\|$, and the current maximum displacement value is $\left\| D_0 \right\|$. Tolerances default to 0.5%. Note that setting up a Minimum Reference is to avoid a never-convergent situation when Value is near zero.

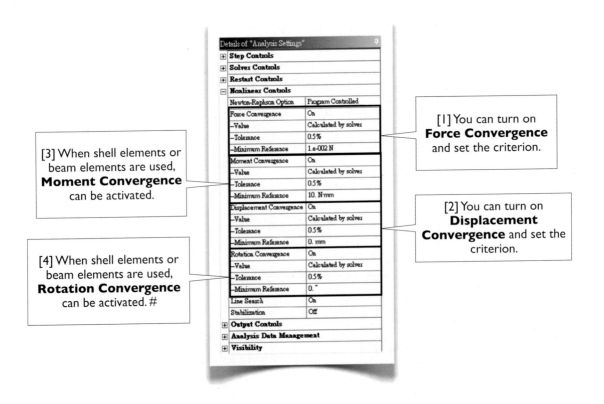

[1] You can turn on **Force Convergence** and set the criterion.

[3] When shell elements or beam elements are used, **Moment Convergence** can be activated.

[4] When shell elements or beam elements are used, **Rotation Convergence** can be activated. #

[2] You can turn on **Displacement Convergence** and set the criterion.

13.1-6 Solution Information

A text form of convergence values and criteria for each iteration is available in **Solution Information** [1]. A graphics form of this information is also available [2]. A key to tackle nonlinear problems is the ability to interpret and draw conclusions from these information.

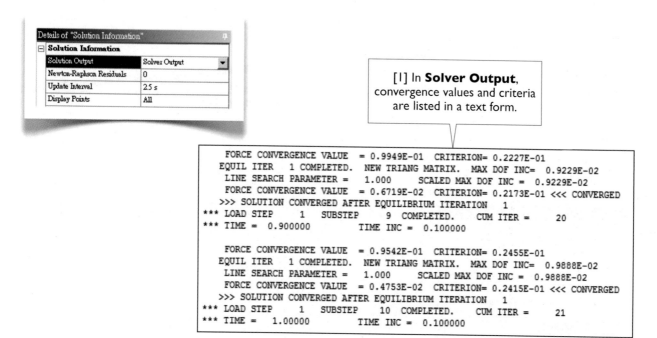

[1] In **Solver Output**, convergence values and criteria are listed in a text form.

```
FORCE CONVERGENCE VALUE    = 0.9949E-01  CRITERION= 0.2227E-01
EQUIL ITER   1 COMPLETED.  NEW TRIANG MATRIX.  MAX DOF INC=  0.9229E-02
LINE SEARCH PARAMETER =    1.000      SCALED MAX DOF INC =  0.9229E-02
FORCE CONVERGENCE VALUE    = 0.6719E-02  CRITERION= 0.2173E-01 <<< CONVERGED
>>> SOLUTION CONVERGED AFTER EQUILIBRIUM ITERATION   1
*** LOAD STEP    1    SUBSTEP     9 COMPLETED.    CUM ITER =     20
*** TIME =  0.900000          TIME INC =  0.100000

FORCE CONVERGENCE VALUE    = 0.9542E-01  CRITERION= 0.2455E-01
EQUIL ITER   1 COMPLETED.  NEW TRIANG MATRIX.  MAX DOF INC=  0.9888E-02
LINE SEARCH PARAMETER =    1.000      SCALED MAX DOF INC =  0.9888E-02
FORCE CONVERGENCE VALUE    = 0.4753E-02  CRITERION= 0.2415E-01 <<< CONVERGED
>>> SOLUTION CONVERGED AFTER EQUILIBRIUM ITERATION   1
*** LOAD STEP    1    SUBSTEP    10 COMPLETED.    CUM ITER =     21
*** TIME =  1.00000           TIME INC =  0.100000
```

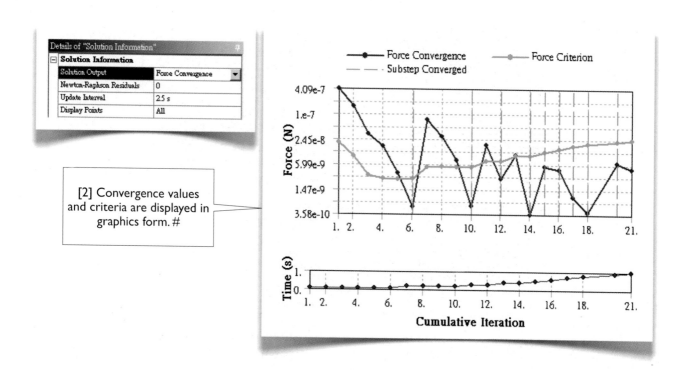

[2] Convergence values and criteria are displayed in graphics form. #

13.1-7 Line Search[Ref 3]

In the example of 13.1-4[1-5] (page 464), each equilibrium iteration (Eq. 13.1-4(1)) calculates a displacement ΔD, that is smaller than the goal, therefore several iterations are needed to reach the goal. That is true in cases when the F-D curves are monotonically "concave down," such as 13.1-4[1].

In cases when F-D curves are highly nonlinear or "concave up," [1] the calculated displacement ΔD in a single iteration may overshoot the goal. In such cases, a numerical technique called **Line Search** can be activated to "scale down" the incremental displacement [2]. In these cases (e.g., the pneumatic finger; see 1.1-8[5] (page 15) and 9.1-10[6] (page 330)), **Line Search** is helpful, but takes extra computing time.

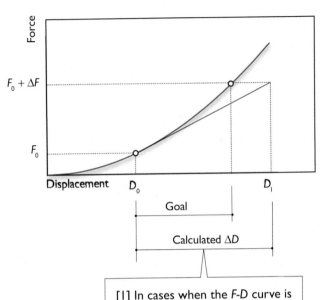

[1] In cases when the F-D curve is highly nonlinear or concave up, the calculated ΔD in a single iteration may overshoot the goal.

[2] **Line Search** can be turned on to scale down the incremental displacement. By default, it is **Program Controlled**.

[3] **Line Search Parameter** scales down the incremental displacement. #

```
FORCE CONVERGENCE VALUE   =  327.2      CRITERION= 0.2962
EQUIL ITER   1 COMPLETED.  NEW TRIANG MATRIX.  MAX DOF INC= -0.1956
LINE SEARCH PARAMETER =   0.5779      SCALED MAX DOF INC = -0.1131
FORCE CONVERGENCE VALUE   =  179.4      CRITERION= 0.2133
EQUIL ITER   2 COMPLETED.  NEW TRIANG MATRIX.  MAX DOF INC=  0.1267E-01
LINE SEARCH PARAMETER =   1.000      SCALED MAX DOF INC =  0.1267E-01
FORCE CONVERGENCE VALUE   =  34.92      CRITERION= 0.2253
EQUIL ITER   3 COMPLETED.  NEW TRIANG MATRIX.  MAX DOF INC=  0.1004
LINE SEARCH PARAMETER =   0.6958      SCALED MAX DOF INC =  0.6985E-01
FORCE CONVERGENCE VALUE   =  30.58      CRITERION= 0.2916
EQUIL ITER   4 COMPLETED.  NEW TRIANG MATRIX.  MAX DOF INC=  0.1047E-01
LINE SEARCH PARAMETER =   1.000      SCALED MAX DOF INC =  0.1047E-01
FORCE CONVERGENCE VALUE   =  5.867      CRITERION= 0.3076
EQUIL ITER   5 COMPLETED.  NEW TRIANG MATRIX.  MAX DOF INC=  0.1734E-01
LINE SEARCH PARAMETER =   0.8577      SCALED MAX DOF INC =  0.1487E-01
FORCE CONVERGENCE VALUE   =  2.724      CRITERION= 0.3313
EQUIL ITER   6 COMPLETED.  NEW TRIANG MATRIX.  MAX DOF INC=  0.3472E-03
LINE SEARCH PARAMETER =   1.000      SCALED MAX DOF INC =  0.3472E-03
FORCE CONVERGENCE VALUE   = 0.2257      CRITERION= 0.3382   <<< CONVERGED
 >>> SOLUTION CONVERGED AFTER EQUILIBRIUM ITERATION   6
*** LOAD STEP    2   SUBSTEP   114 COMPLETED.    CUM ITER =   1889
*** TIME =   1.89395        TIME INC = 0.288325E-01
*** AUTO TIME STEP: NEXT TIME INC = 0.43249E-01  INCREASED (FACTOR = 1.5000)
```

PART B. CONTACT NONLINEARITY

13.1-8 Contact Types[Ref 4]

Several contact types are available in Workbench [1]: **Bonded**, **No Separation**, **Frictionless**, **Rough**, **Frictional**, and **Forced Frictional Sliding**, described as follows according to increasing degree of nonlinearity.

Bonded

Two faces (or edges) in **Bonded** contact are coupled together both in their tangential direction and normal direction. No contact nonlinearities are introduced.

No Separation

Two faces (or edges) in **No Separation** contact are coupled in their normal direction only. The tangential direction allows a small sliding on each other. No contact nonlinearities are introduced since small displacement theory is assumed for the sliding.

Frictionless

Two faces (or edges) in **Frictionless** contact are free to separate. And, when in contact, they may slide in tangential direction without any friction force. This contact type does introduce contact nonlinearities. No small displacement theory is assumed.

Rough

Two faces (or edges) in **Rough** contact are free to separate. But, when in contact, they cannot slide in tangential direction, due to large friction between them. This contact type also introduces contact nonlinearities. No small displacement theory is assumed.

Frictional

Two faces (or edges) in **Frictional** contact are free to separate. And, when in contact, they may slide only when the shear stress between them exceeds a critical value, calculated by multiplying the normal stress by a friction coefficient, which is input as a contact property. This contact type also introduces contact nonlinearities. No small displacement theory is assumed.

Forced Frictional Sliding

Similar to **Frictional** except that there is no "sticking" state; i.e., two faces (or edges) slide even when the shear stress is below the critical value.

[1] Several types of contact are available. #

Linear versus Nonlinear Contacts

The contact types **Bonded** and **No Separation** are called linear contacts because they assume small deformation and involve no nonlinearities. The other contact types will introduce contact nonlinearities.

468

13.1-9 Contact versus Target

To specify a contact region, you need to select a set of **Contact** faces (or edges), and select a set of **Target** faces (or edges) [1]. During the solution, Workbench checks the contact status for each node (or integration point) on the **Contact** faces against the **Target** faces.

If **Behavior** is set to **Symmetric**, the roles of **Contact** and **Target** will be symmetric [2]; i.e., Workbench checks each point on the **Contact** against the **Target**, as well as each point on the **Target** against the **Contact**. If **Behavior** is set to **Asymmetric**, the checking is only one-sided.

Consider a point, on a **Contact** face, which is approaching a **Target** face. The point is called a contacting point. Workbench keeps tracing the contacting point so that it won't penetrate into the target face. When the point is in contact with the surface, Workbench starts to enforce contact compatibility (i.e., preventing penetration).

In some cases, when **Symmetric** behavior is not available, or when you want to set to **Asymmetric**, to save the run time, selection of **Contact** and **Target** becomes important. You may select **Target** faces using following guidelines: the faces belong to a body with fixed supports, the faces belong to a body with more rigid material (higher Young's modulus), the faces with less curvature, etc.

For **Symmetric**, results are reported for both contact and target sides. When **Asymmetric** is used, all result data is on the contact side.

13.1-10 Contact Formulations[Ref 5]

Workbench offers several **Formulation** [1] to enforce contact compatibility at the contact interface; i.e., preventing penetration of the contacting point into the target faces.

MPC (Multi-Point Constraint)

For linear contact types (**Bonded** and **No Separation**), a multi-point constraint (**MPC**) formulation is available. **MPC** internally adds constraint equations to couple the displacements between contacting faces. Although you can use other formulations for **Bonded** and **No Separation** contact types, **MPC** is recommended for these linear contact types, since it is a direct, efficient formulation.

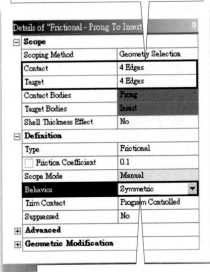

[1] To specify a contact region, you need select a set of **Contact** faces (or edges), and select a set of **Target** faces (or edges).

[2] If **Behavior** is set to **Symmetric**, the roles of **Contact** and **Target** will be symmetric. #

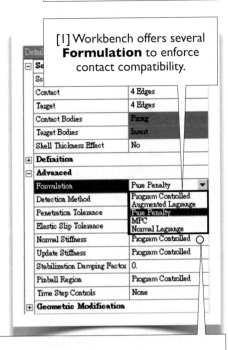

[1] Workbench offers several **Formulation** to enforce contact compatibility.

[2] **Normal Stiffness** can be input here. The input value is used to multiply a stiffness value provided by the program. #

Pure Penalty

Whenever a contacting point penetrates normally by an amount x_n into a target face, it will be pushed back by a normal force F_n,

$$F_n = k_n x_n \tag{1}$$

where k_n is called the **Normal Stiffness** of the contact region [2]. A **Normal Stiffness** has no real physical meaning; it is a numerical parameter of the penalty algorithm. Solution convergence behavior is usually sensitive to this parameter. A larger k_n usually gives a more accurate solution (less penetration), but may raise convergence issues. Reducing k_n usually helps convergence, but results an increasing penetration. As a simple rule, whenever bumping into a convergence problem, try reducing k_n first. The **Normal Stiffness** can also be automatically adjusted during the solution (see **Update Stiffness** in 13.1-11, next page).

If sliding in tangential direction is also prohibited (e.g., **Rough** contact), a similar treatment can be implemented. Whenever a contacting point slides tangentially by an amount x_t, it will be pushed back by a tangential force F_t,

$$F_t = k_t x_t \tag{2}$$

where k_t is called the *tangential stiffness* of the contact region. Unlike **Normal Stiffness**, the tangential stiffness is always controlled by the program.

Note that, in cases when **Augmented Lagrange** or **Normal Lagrange** (to be discussed) formulation is used, the formulations apply on the normal direction only. The tangential direction still uses **Pure Penalty** formulation, Eq. (2). When **MPC** is used, both normal and tangential directions use **MPC** formulation.

Normal Lagrange

Normal Lagrange formulation adds an extra degree of freedom, namely, contact pressure, to satisfy compatibility. Whenever a contacting point is in touch with the target face, the contact pressure is explicitly calculated. It is the contact pressure that prevents further penetration,

$$F_n = \lambda \tag{3}$$

where λ is the contact pressure, and is traditionally called a *Lagrange multiplier* Note that this formulation is used in the normal direction only. In the tangential direction, **Pure Penalty** formulation, Eq. (2), is used. This is where the name **Normal Lagrange** comes from.

Normal Lagrange formulation does not require a normal stiffness, and, theoretically, it can enforce zero penetration. However, since no penetration is allowed, the contact status is either open or closed (a step function). This can sometimes make convergence difficult because contact points may oscillate between open and closed status. This behavior is called *chattering*.

Augmented Lagrange

The idea is to hybrid **Pure Penalty** with **Normal Lagrange**; i.e., the push-back normal force is

$$F_n = k_n x_n + \lambda \tag{4}$$

Because of the contact pressure λ, **Augmented Lagrange** formulation is less sensitive to **Normal Stiffness** k_n. Note that again, in the tangential direction, **Pure Penalty** formulation, Eq. (2), is used.

Although **Pure Penalty** is the default setting, **Augmented Lagrange** is recommended for general frictional or frictionless contact in large deformation problems. Since Lagrange method adds extra degree of freedom, it also takes extra computing time.

13.1-11 Advanced Contact Settings[Refs 6, 7]

Pinball Region

The pinball is a sphere region, its radius can be defined in **Pinball Region**. Consider again that a contacting point approaches a target face. If a target node is within the pinball region centered at a contacting node, the contacting node is considered to be in "near" contact with the target node and will be monitored. Target nodes outside of the pinball region will not be monitored.

If **Bonded** type is specified, surfaces that have gap smaller than the pinball radius are treated as bonded.

Interface Treatment

For **Bonded** contact type, a large enough pinball radius may allow any gap between contacting faces to be ignored. For **Frictional** or **Frictionless** contact types, an initial gap is not automatically ignored, no matter how large the pinball is, since the gap may represent the real geometry.

If an initial gap is present [1] and a force is applied, one part may "fly away" relative to another part [2] if the initial contact is not established right at the end of the time step.

To alleviate situations where a gap (clearance) is modeled but needs to be ignored to establish initial contact for **Frictional** or **Frictionless** contact types, **Interface Treatment** can internally offset the contact surfaces by a specified amount. Note that this treatment is intended for small gaps. Don't apply it in a large gap.

Time Step Controls

Time Step Controls tries to enhance convergence by allowing adjustments of time step size based on contact behavior.

By default, contact behavior does not affect auto time stepping, since adjustment of time step based on contact behavior may increase computing time too much.

With **Time Step Controls** turned on, Workbench adjust the time step size based on contact behavior.

Update Stiffness

The **Normal Stiffness** can be automatically adjusted during the solution. Whenever convergence difficulties arise, **Normal Stiffness** will be reduced automatically.

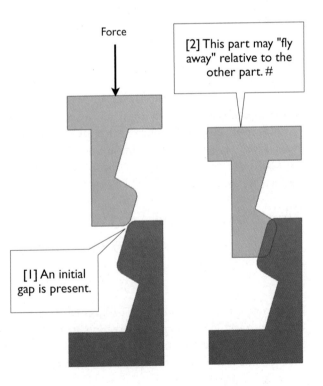

Details of "Frictional - Prong To Insert"		
Scope		
Scoping Method	Geometry Selection	
Contact	4 Edges	
Target	4 Edges	
Contact Bodies	Prong	
Target Bodies	Insert	
Shell Thickness Effect	No	
Definition		
Advanced		
Formulation	Pure Penalty	
Detection Method	Program Controlled	
Penetration Tolerance	Program Controlled	
Elastic Slip Tolerance	Program Controlled	
Normal Stiffness	Program Controlled	
Update Stiffness	Each Iteration	○
Stabilization Damping Factor	0.	
Pinball Region	Program Controlled	○
Time Step Controls	None	○
Geometric Modification		
Interface Treatment	Add Offset, No Ramping	○
Offset	0. mm	○
Contact Geometry Correction	None	

Force

[2] This part may "fly away" relative to the other part. #

[1] An initial gap is present.

References

1. ANSYS Documentation//Mechanical APDL//Theory Reference//14.12. Newton-Raphson Procedure
2. ANSYS Documentation//Mechanical APDL//Theory Reference//14.12.2. Convergence
3. ANSYS Documentation//Mechanical APDL//Theory Reference//14.12.5. Line Search
4. ANSYS Documentation//Mechanical Applications//Mechanical Users' Guide//Setting Connections//Contact//Contact Settings//Definition Settings
5. ANSYS Documentation//Mechanical Applications//Mechanical Users' Guide//Setting Connections//Contact//Contact Formulation Theory
6. ANSYS Documentation//Mechanical Applications//Mechanical Users' Guide//Setting Connections//Contact//Contact Settings//Advanced Settings
7. ANSYS Documentation//Mechanical Applications//Mechanical Users' Guide//Setting Connections//Contact//Contact Settings//Geometric Modification

Section 13.2

Translational Joint[Ref 1]

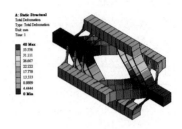

13.2-1 About the Translational Joint

A translational joint is used to connect two machine components, so that the relative motion of the two components is restricted to translation in a specific direction. Conventionally, translational joints are designed as mechanisms, composed by parts, between which the clearance or interference is inevitable; they either decrease precision or increase friction. The translational joint in this section is not a mechanism. Rather, it is a unitary flexible structure, in which no clearance or interference exist.

The translational joint [1-4] is made of POM (polyoxymethylene, a plastic), which has a Young's modulus of 2 GPa and a Poisson's ratio of 0.35. The most important design consideration is that the rigidity of translational direction should be much less than all other directions, so that the motion can be restricted in that direction only.

Here, we want to explore the geometric nonlinearity of the structure: how the applied force increases nonlinearly with the translational displacement. For this purpose, we will model the structure using line bodies entirely. The goal of the simulation is to plot a force-versus-displacement chart. The unit system used in the simulation is **mm-kg-N-s**.

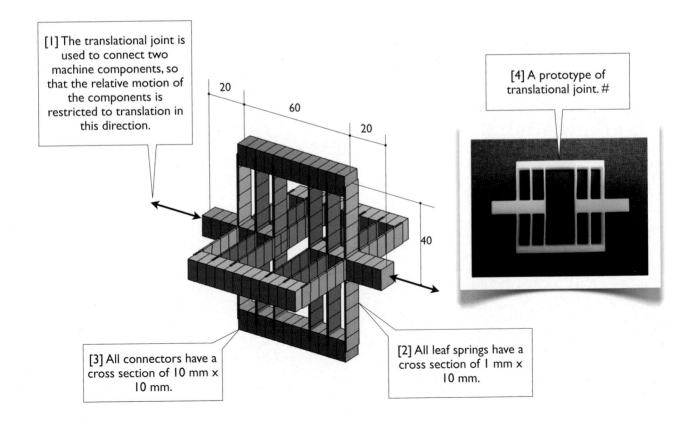

[1] The translational joint is used to connect two machine components, so that the relative motion of the components is restricted to translation in this direction.

[4] A prototype of translational joint. #

[3] All connectors have a cross section of 10 mm x 10 mm.

[2] All leaf springs have a cross section of 1 mm x 10 mm.

20

60

20

40

473

13.2-2 Start Up

Launch Workbench. Create a **Static Structural** analysis system. Save the project as **Transjoint**.

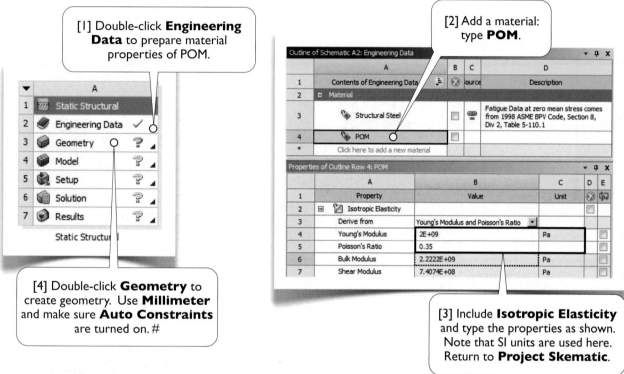

[1] Double-click **Engineering Data** to prepare material properties of POM.

[2] Add a material: type **POM**.

[4] Double-click **Geometry** to create geometry. Use **Millimeter** and make sure **Auto Constraints** are turned on. #

[3] Include **Isotropic Elasticity** and type the properties as shown. Note that SI units are used here. Return to **Project Skematic**.

13.2-3 Create Geometry in DesignModeler

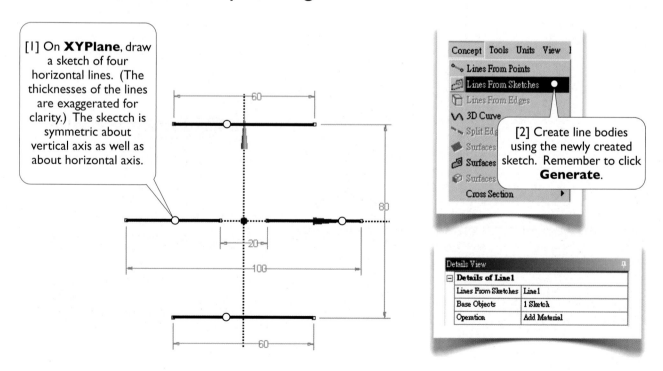

[1] On **XYPlane**, draw a sketch of four horizontal lines. (The thicknesses of the lines are exaggerated for clarity.) The sketch is symmetric about vertical axis as well as about horizontal axis.

[2] Create line bodies using the newly created sketch. Remember to click **Generate**.

[3] On **ZXPlane**, draw a sketch of two vertical lines. (The thicknesses of the lines are exaggerated for clarity.) The sketch is symmetric about vertical axis as well as about horizontal axis.

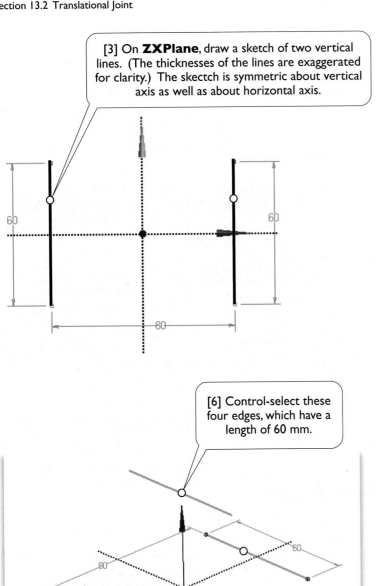

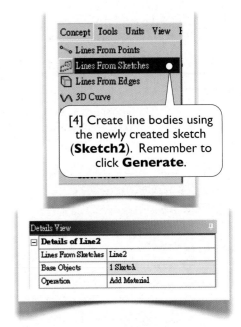

[4] Create line bodies using the newly created sketch (**Sketch2**). Remember to click **Generate**.

[6] Control-select these four edges, which have a length of 60 mm.

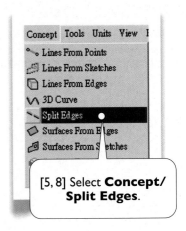

[5, 8] Select **Concept/ Split Edges**.

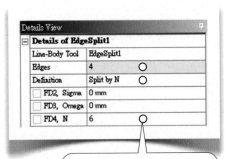

[7] Split each edge into 6 segments. The purpose of splitting is to create construction points on the edges. Remember to click **Generate**.

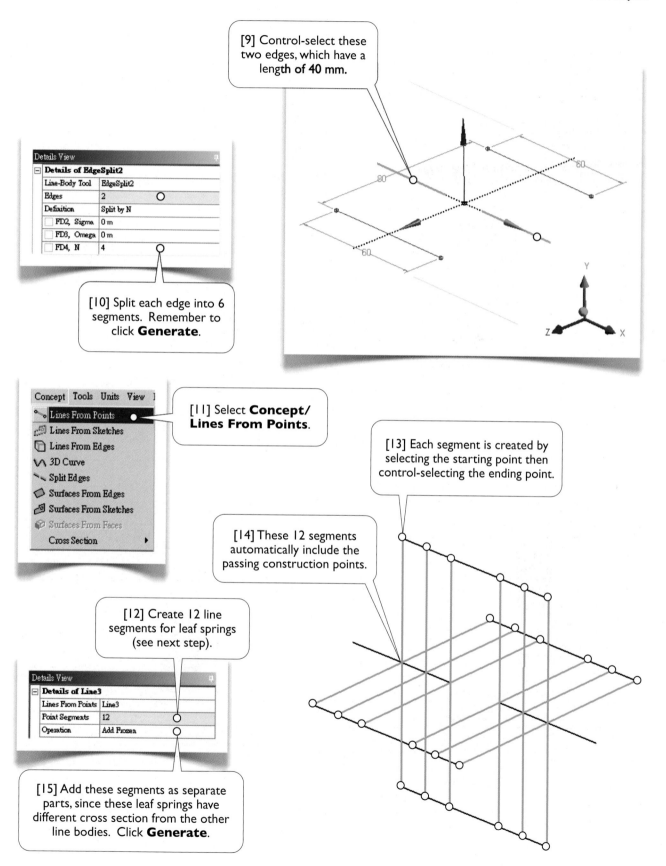

[9] Control-select these two edges, which have a length of 40 mm.

Details View

Details of EdgeSplit2

Line-Body Tool	EdgeSplit2
Edges	2
Definition	Split by N
FD2, Sigma	0 m
FD3, Omega	0 m
FD4, N	4

[10] Split each edge into 6 segments. Remember to click **Generate**.

Concept   Tools   Units   View

- Lines From Points
- Lines From Sketches
- Lines From Edges
- 3D Curve
- Split Edges
- Surfaces From Edges
- Surfaces From Sketches
- Surfaces From Faces
- Cross Section

[11] Select **Concept/ Lines From Points**.

[13] Each segment is created by selecting the starting point then control-selecting the ending point.

[14] These 12 segments automatically include the passing construction points.

[12] Create 12 line segments for leaf springs (see next step).

Details View

Details of Line3

Lines From Points	Line3
Point Segments	12
Operation	Add Frozen

[15] Add these segments as separate parts, since these leaf springs have different cross section from the other line bodies. Click **Generate**.

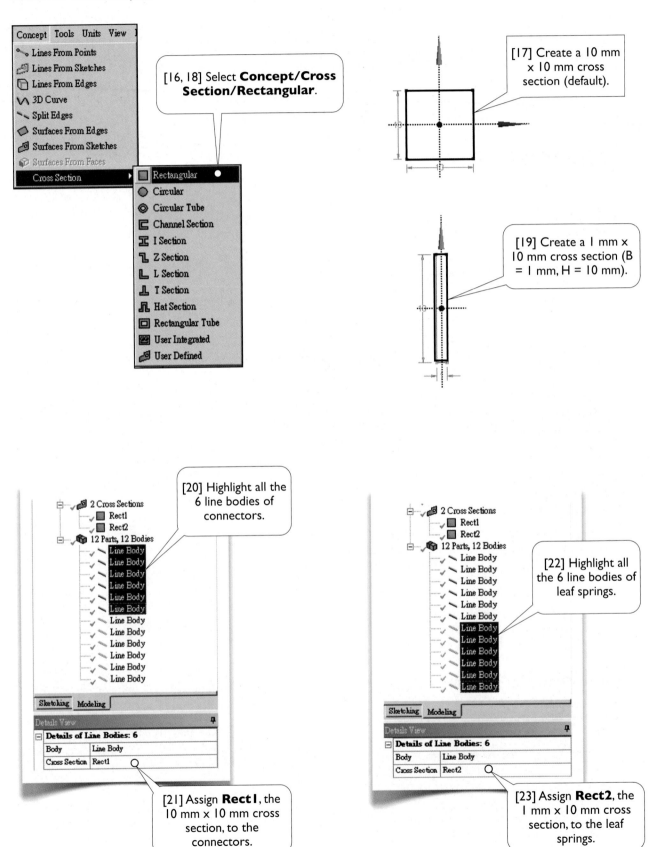

Concept Tools Units View

Lines From Points
Lines From Sketches
Lines From Edges
3D Curve
Split Edges
Surfaces From Edges
Surfaces From Sketches
Surfaces From Faces
Cross Section

Rectangular
Circular
Circular Tube
Channel Section
I Section
Z Section
L Section
T Section
Hat Section
Rectangular Tube
User Integrated
User Defined

[16, 18] Select **Concept/Cross Section/Rectangular**.

[17] Create a 10 mm x 10 mm cross section (default).

[19] Create a 1 mm x 10 mm cross section (B = 1 mm, H = 10 mm).

[20] Highlight all the 6 line bodies of connectors.

2 Cross Sections
Rect1
Rect2
12 Parts, 12 Bodies
Line Body
Line Body
Line Body
Line Body
Line Body
Line Body
Line Body
Line Body
Line Body
Line Body
Line Body
Line Body

Sketching Modeling

Details View

Details of Line Bodies: 6
| Body | Line Body |
| Cross Section | Rect1 |

[21] Assign **Rect1**, the 10 mm x 10 mm cross section, to the connectors.

[22] Highlight all the 6 line bodies of leaf springs.

2 Cross Sections
Rect1
Rect2
12 Parts, 12 Bodies
Line Body
Line Body
Line Body
Line Body
Line Body
Line Body
Line Body
Line Body
Line Body
Line Body
Line Body
Line Body

Sketching Modeling

Details View

Details of Line Bodies: 6
| Body | Line Body |
| Cross Section | Rect2 |

[23] Assign **Rect2**, the 1 mm x 10 mm cross section, to the leaf springs.

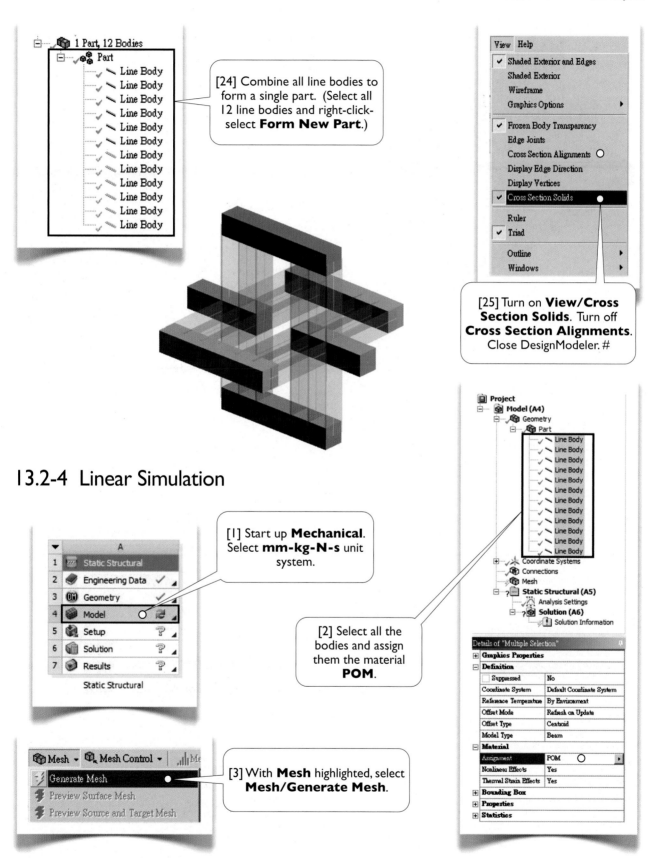

[24] Combine all line bodies to form a single part. (Select all 12 line bodies and right-click-select **Form New Part**.)

[25] Turn on **View/Cross Section Solids**. Turn off **Cross Section Alignments**. Close DesignModeler. #

13.2-4 Linear Simulation

[1] Start up **Mechanical**. Select **mm-kg-N-s** unit system.

[2] Select all the bodies and assign them the material **POM**.

[3] With **Mesh** highlighted, select **Mesh/Generate Mesh**.

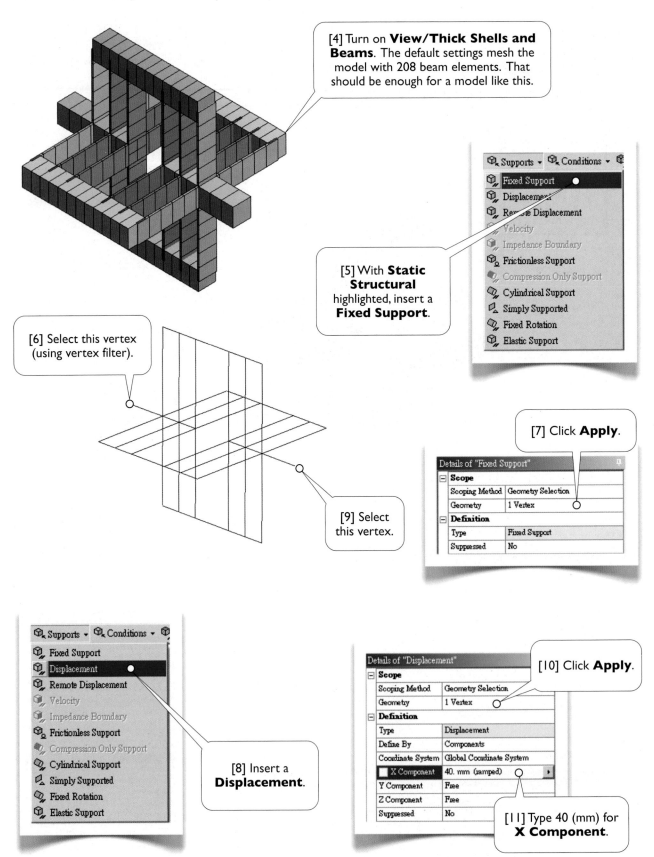

[4] Turn on **View/Thick Shells and Beams**. The default settings mesh the model with 208 beam elements. That should be enough for a model like this.

[5] With **Static Structural** highlighted, insert a **Fixed Support**.

Supports ▾   Conditions ▾

Fixed Support ●
Displacement
Remote Displacement
Velocity
Impedance Boundary
Frictionless Support
Compression Only Support
Cylindrical Support
Simply Supported
Fixed Rotation
Elastic Support

[6] Select this vertex (using vertex filter).

[9] Select this vertex.

[7] Click **Apply**.

Details of "Fixed Support"

Scope	
Scoping Method	Geometry Selection
Geometry	1 Vertex
Definition	
Type	Fixed Support
Suppressed	No

Supports ▾   Conditions ▾

Fixed Support
Displacement ●
Remote Displacement
Velocity
Impedance Boundary
Frictionless Support
Compression Only Support
Cylindrical Support
Simply Supported
Fixed Rotation
Elastic Support

[8] Insert a **Displacement**.

[10] Click **Apply**.

Details of "Displacement"

Scope	
Scoping Method	Geometry Selection
Geometry	1 Vertex
Definition	
Type	Displacement
Define By	Components
Coordinate System	Global Coordinate System
☐ X Component	40. mm (ramped) ▸
Y Component	Free
Z Component	Free
Suppressed	No

[11] Type 40 (mm) for **X Component**.

479

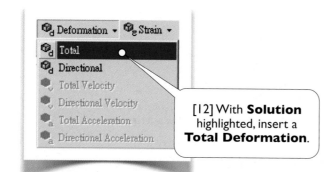

[12] With **Solution** highlighted, insert a **Total Deformation**.

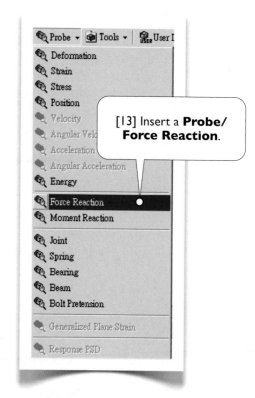

[13] Insert a **Probe/ Force Reaction**.

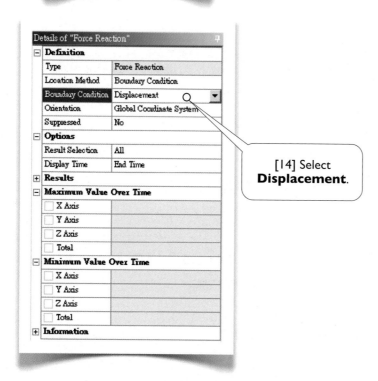

[14] Select **Displacement**.

Force Reaction

Forces are needed to maintain a displacement condition. For example, if a point of a structure is to be held still (i.e., a fixed support), forces are needed to maintain the zero-displacement condition. The forces are called *reaction forces*.

This concept can be generalized to a nonzero displacement condition. In our case, we want to maintain a displacement of 40 mm in X-direction, the force required is called the reaction force of that displacement.

The purpose of **Force Reaction** object is to evaluate the force required for the 40 mm displacement.

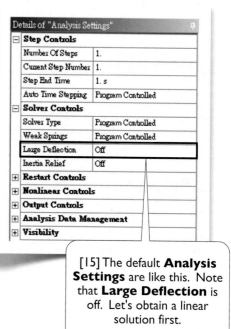

[15] The default **Analysis Settings** are like this. Note that **Large Deflection** is off. Let's obtain a linear solution first.

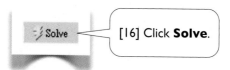

[16] Click **Solve**.

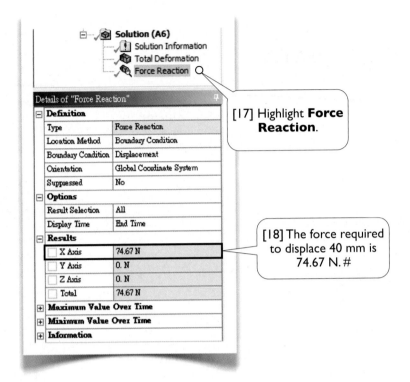

Solution (A6)
Solution Information
Total Deformation
Force Reaction

[17] Highlight **Force Reaction**.

Details of "Force Reaction"

Definition	
Type	Force Reaction
Location Method	Boundary Condition
Boundary Condition	Displacement
Orientation	Global Coordinate System
Suppressed	No
Options	
Result Selection	All
Display Time	End Time
Results	
X Axis	74.67 N
Y Axis	0. N
Z Axis	0. N
Total	74.67 N
Maximum Value Over Time	
Minimum Value Over Time	
Information	

[18] The force required to displace 40 mm is 74.67 N. #

13.2-5 Nonlinear Simulation

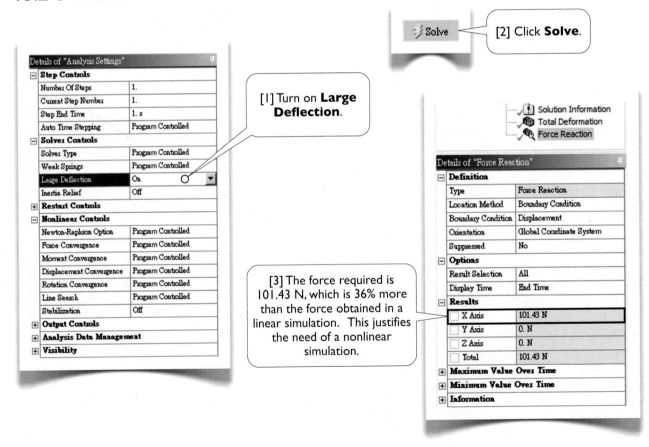

Details of "Analysis Settings"

Step Controls	
Number Of Steps	1.
Current Step Number	1.
Step End Time	1. s
Auto Time Stepping	Program Controlled
Solver Controls	
Solver Type	Program Controlled
Weak Springs	Program Controlled
Large Deflection	On
Inertia Relief	Off
Restart Controls	
Nonlinear Controls	
Newton-Raphson Option	Program Controlled
Force Convergence	Program Controlled
Moment Convergence	Program Controlled
Displacement Convergence	Program Controlled
Rotation Convergence	Program Controlled
Line Search	Program Controlled
Stabilization	Off
Output Controls	
Analysis Data Management	
Visibility	

[1] Turn on **Large Deflection**.

Solve

[2] Click **Solve**.

Solution Information
Total Deformation
Force Reaction

Details of "Force Reaction"

Definition	
Type	Force Reaction
Location Method	Boundary Condition
Boundary Condition	Displacement
Orientation	Global Coordinate System
Suppressed	No
Options	
Result Selection	All
Display Time	End Time
Results	
X Axis	101.43 N
Y Axis	0. N
Z Axis	0. N
Total	101.43 N
Maximum Value Over Time	
Minimum Value Over Time	
Information	

[3] The force required is 101.43 N, which is 36% more than the force obtained in a linear simulation. This justifies the need of a nonlinear simulation.

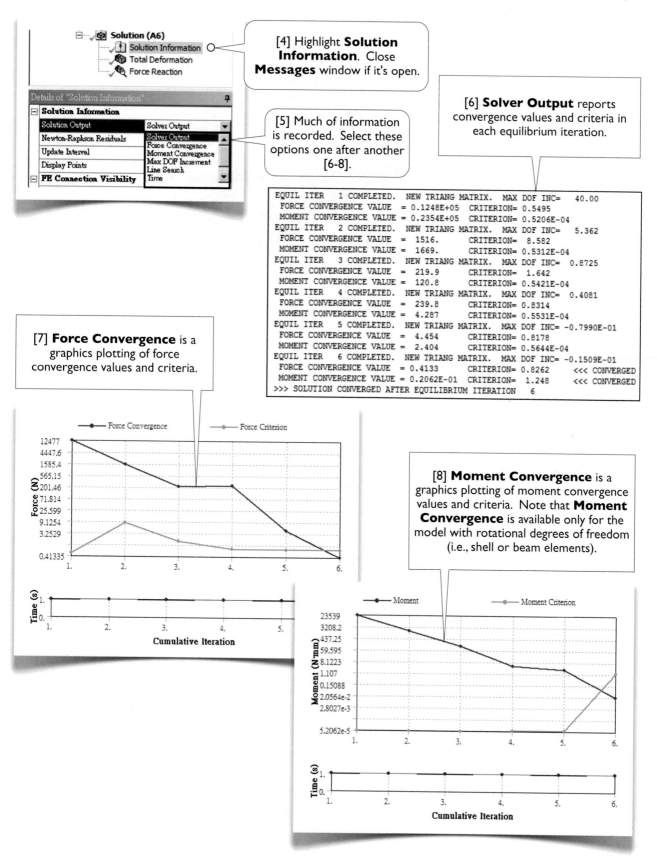

[4] Highlight **Solution Information**. Close **Messages** window if it's open.

[5] Much of information is recorded. Select these options one after another [6-8].

[6] **Solver Output** reports convergence values and criteria in each equilibrium iteration.

Details of "Solution Information"

Solution Information	
Solution Output	Solver Output
Newton-Raphson Residuals	
Update Interval	
Display Points	
FE Connection Visibility	

Solver Output
Force Convergence
Moment Convergence
Max DOF Increment
Line Search
Time

```
EQUIL ITER   1 COMPLETED.  NEW TRIANG MATRIX.  MAX DOF INC=   40.00
  FORCE CONVERGENCE VALUE  = 0.1248E+05  CRITERION= 0.5495
  MOMENT CONVERGENCE VALUE = 0.2354E+05  CRITERION= 0.5206E-04
EQUIL ITER   2 COMPLETED.  NEW TRIANG MATRIX.  MAX DOF INC=   5.362
  FORCE CONVERGENCE VALUE  = 1516.       CRITERION= 8.582
  MOMENT CONVERGENCE VALUE = 1669.       CRITERION= 0.5312E-04
EQUIL ITER   3 COMPLETED.  NEW TRIANG MATRIX.  MAX DOF INC=   0.8725
  FORCE CONVERGENCE VALUE  = 219.9       CRITERION= 1.642
  MOMENT CONVERGENCE VALUE = 120.8       CRITERION= 0.5421E-04
EQUIL ITER   4 COMPLETED.  NEW TRIANG MATRIX.  MAX DOF INC=   0.4081
  FORCE CONVERGENCE VALUE  = 239.8       CRITERION= 0.8314
  MOMENT CONVERGENCE VALUE = 4.287       CRITERION= 0.5531E-04
EQUIL ITER   5 COMPLETED.  NEW TRIANG MATRIX.  MAX DOF INC= -0.7990E-01
  FORCE CONVERGENCE VALUE  = 4.454       CRITERION= 0.8178
  MOMENT CONVERGENCE VALUE = 2.404       CRITERION= 0.5644E-04
EQUIL ITER   6 COMPLETED.  NEW TRIANG MATRIX.  MAX DOF INC= -0.1509E-01
  FORCE CONVERGENCE VALUE  = 0.4133      CRITERION= 0.8262   <<< CONVERGED
  MOMENT CONVERGENCE VALUE = 0.2062E-01  CRITERION= 1.248    <<< CONVERGED
>>> SOLUTION CONVERGED AFTER EQUILIBRIUM ITERATION   6
```

[7] **Force Convergence** is a graphics plotting of force convergence values and criteria.

[8] **Moment Convergence** is a graphics plotting of moment convergence values and criteria. Note that **Moment Convergence** is available only for the model with rotational degrees of freedom (i.e., shell or beam elements).

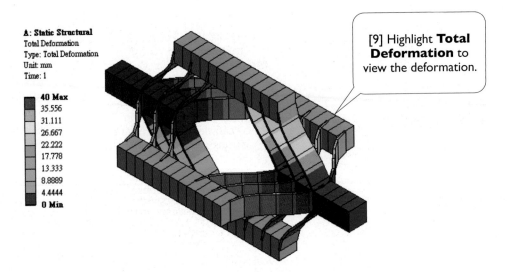

A: Static Structural
Total Deformation
Type: Total Deformation
Unit: mm
Time: 1

40 Max
35.556
31.111
26.667
22.222
17.778
13.333
8.8889
4.4444
0 Min

[9] Highlight **Total Deformation** to view the deformation.

Workbench solves this case easily in only one substep, and the substep takes 6 equilibrium iterations to converge. To plot a force-versus-displacement chart, let's specify a small initial substep value and let Workbench automatically adjust the substep values for the subsequent substeps. This feature is called **Auto Time Stepping**.

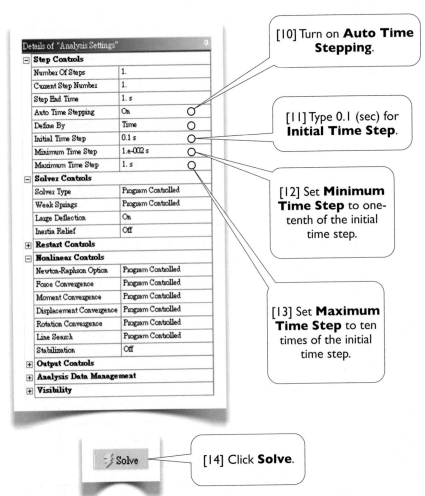

[10] Turn on **Auto Time Stepping**.

[11] Type 0.1 (sec) for **Initial Time Step**.

[12] Set **Minimum Time Step** to one-tenth of the initial time step.

[13] Set **Maximum Time Step** to ten times of the initial time step.

[14] Click **Solve**.

Auto Time Stepping

Workbench uses **Initial Time Step** for the first substep. If the convergence behavior is merely okay, it continues to use that time step for the next substep. If the convergence behavior is very good, it may increase the time step (typically 50%) for the next substep. If the convergence behavior is pretty bad, it may decrease the time step (typically 50%; called *bisection*) for the next substep.

In cases that the convergence behavior is so bad that, after several times of bisections, the time step becomes less than **Minimum Time Step**, Workbench will stop the solution procedure and report an error message for you.

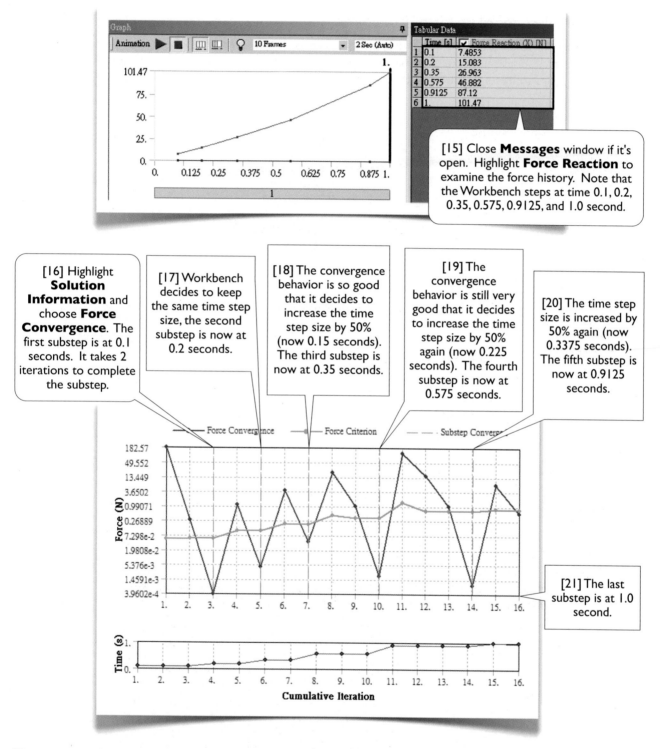

[15] Close **Messages** window if it's open. Highlight **Force Reaction** to examine the force history. Note that the Workbench steps at time 0.1, 0.2, 0.35, 0.575, 0.9125, and 1.0 second.

[16] Highlight **Solution Information** and choose **Force Convergence**. The first substep is at 0.1 seconds. It takes 2 iterations to complete the substep.

[17] Workbench decides to keep the same time step size, the second substep is now at 0.2 seconds.

[18] The convergence behavior is so good that it decides to increase the time step size by 50% (now 0.15 seconds). The third substep is now at 0.35 seconds.

[19] The convergence behavior is still very good that it decides to increase the time step size by 50% again (now 0.225 seconds). The fourth substep is now at 0.575 seconds.

[20] The time step size is increased by 50% again (now 0.3375 seconds). The fifth substep is now at 0.9125 seconds.

[21] The last substep is at 1.0 second.

Time is used as a counter in static simulations

In a static simulation, in which the time has no real-world meaning, Workbench uses **Time** as a counter. By default, it set 1.0 second for each step. You, of course, can change that value as you wish. In our case, since the total displacement is 40 mm, we may set **Step End Time** to 40 seconds, so that each second corresponds to a millimeter of displacement.

More data points; more uniform data points

In the foregoing simulation, there are only 6 data points calculated [15], and they are distributed unevenly. To plot an accurate force-versus-displacement relation, we need more number of time steps and they had better be distributed evenly. Let's modify **Analysis Settings** further.

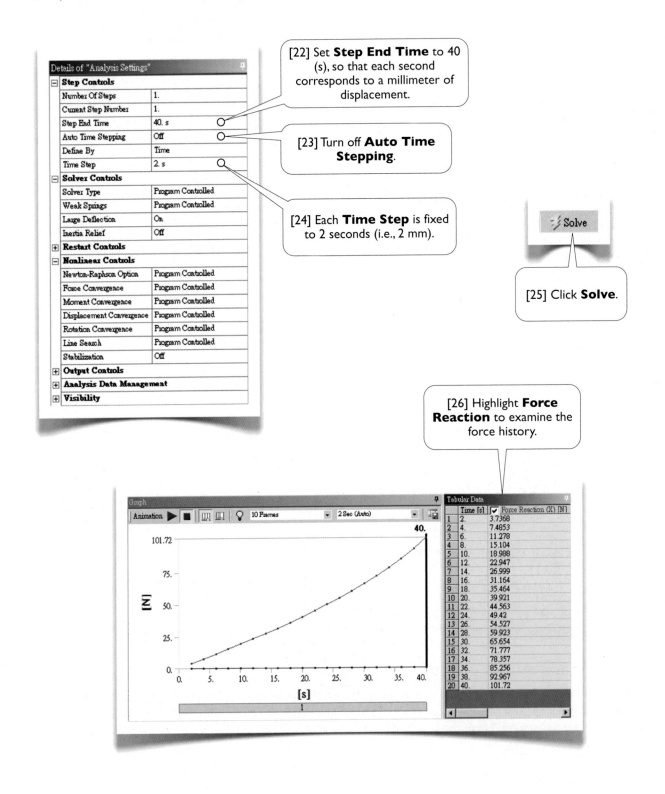

[22] Set **Step End Time** to 40 (s), so that each second corresponds to a millimeter of displacement.

[23] Turn off **Auto Time Stepping**.

[24] Each **Time Step** is fixed to 2 seconds (i.e., 2 mm).

[25] Click **Solve**.

[26] Highlight **Force Reaction** to examine the force history.

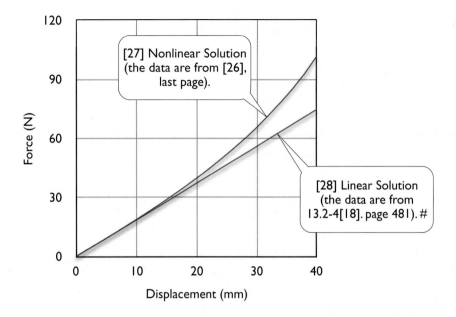

Wrap Up

Save the project and exit Workbench.

Reference

1.   Brian P. Trease, Yong-Mo Moon, and Sridhar Kota, 2005, "Design of Large-Displacement Compliant Joints," *ASME Journal of Mechanical Design*, Vol. 127, pp. 788-798.

Section 13.3

Microgripper[Refs 1, 2]

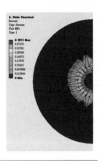

13.3-1 About the Microgripper

In Section 2.6, we introduced the microgripper and created a solid model for it. As mentioned in that section, the microgripper is made of PDMS and actuated by an SMA (shape memory alloy) actuator; it is tested by gripping a glass bead in a lab. In this section, we want to assess the gripping pressure on the glass bead under an actuation force of 40 μN exerted by the SMA. Since the glass bead is much more rigid than the PDMS material, we will model the glass bead as a rigid body. This will ease some of convergence difficulties. By taking advantage of symmetries, we will model only one quarter of the microgripper.

13.3-2 Resume the Project **Microgripper**

Launch Workbench. Open the project **Microgripper**, which was saved in Section 2.6.

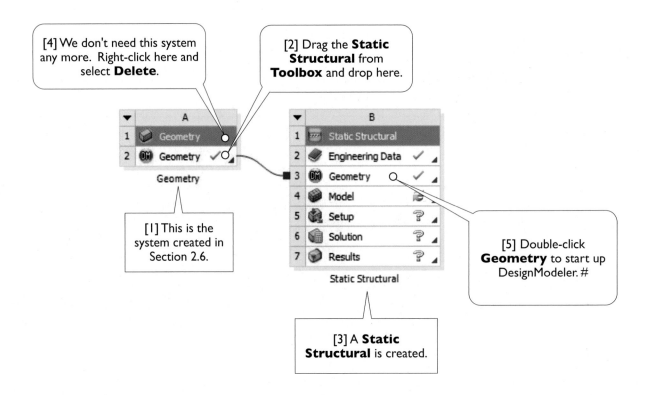

[4] We don't need this system any more. Right-click here and select **Delete**.

[2] Drag the **Static Structural** from **Toolbox** and drop here.

	A	
1	Geometry	
2	Geometry	✓

Geometry

	B		
1	Static Structural		
2	Engineering Data	✓	
3	Geometry	✓	
4	Model		
5	Setup	?	
6	Solution	?	
7	Results	?	

Static Structural

[1] This is the system created in Section 2.6.

[5] Double-click **Geometry** to start up DesignModeler. #

[3] A **Static Structural** is created.

487

13.3-3 Prepare Geometry in DesignModeler

XYPlane

[1] Click **New Plane**.

Details View

Details of Plane4	
Plane	Plane4
Type	From Plane
Base Plane	XYPlane
Transform 1 (RMB)	None
Reverse Normal/Z-Axis?	Yes
Flip XY-Axes?	No
Export Coordinate System?	No

[2] The new plane is simply a copy of **XYPlane** with its Z-Axis reversed. Click **Generate**.

Details View

Details of Symmetry1	
Symmetry	Symmetry1
Number of Planes	2
Symmetry Plane 1	Plane4
Symmetry Plane 2	YZPlane
Model Type	Full Model
Target Bodies	All Bodies

[4] There are two planes of symmetry (**Plane4** and **YZPlane**).

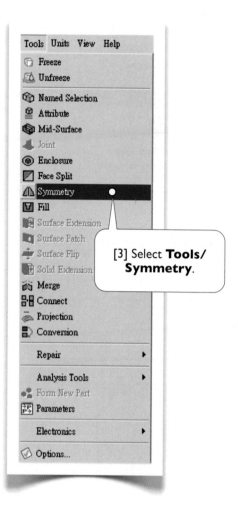

Tools Units View Help

- Freeze
- Unfreeze
- Named Selection
- Attribute
- Mid-Surface
- Joint
- Enclosure
- Face Split
- Symmetry
- Fill
- Surface Extension
- Surface Patch
- Surface Flip
- Solid Extension
- Merge
- Connect
- Projection
- Conversion
- Repair ▶
- Analysis Tools ▶
- Form New Part
- Parameters
- Electronics ▶
- Options...

[3] Select **Tools/ Symmetry**.

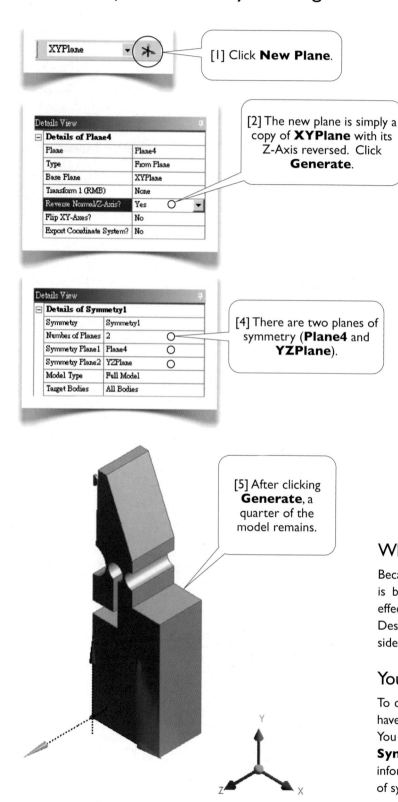

[5] After clicking **Generate**, a quarter of the model remains.

Why use Plane4 instead of XYPlane?

Because we want to keep the half of the model that is behind **XYPlane**, so that it has better visual effect. When you specify a plane of symmetry, DesignModeler keeps the portion that is in the +Z side of the plane coordinate system.

You don't have to build a full model

To create a **Tools/Symmetry** feature, you don't have to build a full model and then slice it to a half. You may just build a half model and use **Tools/ Symmetry** to specify a plane of symmetry. The information that Workbench needs are the planes of symmetry, not the other half of the model.

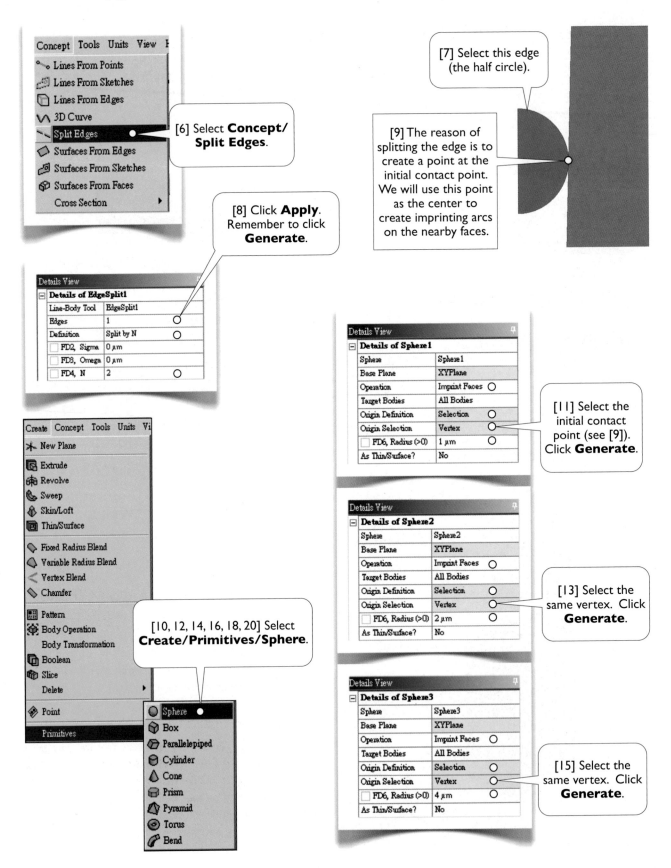

Concept Tools Units View H

- Lines From Points
- Lines From Sketches
- Lines From Edges
- 3D Curve
- Split Edges ●
- Surfaces From Edges
- Surfaces From Sketches
- Surfaces From Faces
- Cross Section

[6] Select **Concept/ Split Edges**.

[7] Select this edge (the half circle).

[9] The reason of splitting the edge is to create a point at the initial contact point. We will use this point as the center to create imprinting arcs on the nearby faces.

[8] Click **Apply**. Remember to click **Generate**.

Details View

Details of EdgeSplit1

Line-Body Tool	EdgeSplit1	
Edges	1	○
Definition	Split by N	○
☐ FD2, Sigma	0 μm	
☐ FD3, Omega	0 μm	
☐ FD4, N	2	○

Create Concept Tools Units Vi

- New Plane
- Extrude
- Revolve
- Sweep
- Skin/Loft
- Thin/Surface
- Fixed Radius Blend
- Variable Radius Blend
- Vertex Blend
- Chamfer
- Pattern
- Body Operation
- Body Transformation
- Boolean
- Slice
- Delete
- Point
- Primitives

[10, 12, 14, 16, 18, 20] Select **Create/Primitives/Sphere**.

- Sphere ●
- Box
- Parallelepiped
- Cylinder
- Cone
- Prism
- Pyramid
- Torus
- Bend

Details View

Details of Sphere1

Sphere	Sphere1	
Base Plane	XYPlane	
Operation	Imprint Faces	○
Target Bodies	All Bodies	
Origin Definition	Selection	○
Origin Selection	Vertex	○
☐ FD6, Radius (>0)	1 μm	○
As Thin/Surface?	No	

[11] Select the initial contact point (see [9]). Click **Generate**.

Details View

Details of Sphere2

Sphere	Sphere2	
Base Plane	XYPlane	
Operation	Imprint Faces	○
Target Bodies	All Bodies	
Origin Definition	Selection	○
Origin Selection	Vertex	○
☐ FD6, Radius (>0)	2 μm	○
As Thin/Surface?	No	

[13] Select the same vertex. Click **Generate**.

Details View

Details of Sphere3

Sphere	Sphere3	
Base Plane	XYPlane	
Operation	Imprint Faces	○
Target Bodies	All Bodies	
Origin Definition	Selection	○
Origin Selection	Vertex	○
☐ FD6, Radius (>0)	4 μm	○
As Thin/Surface?	No	

[15] Select the same vertex. Click **Generate**.

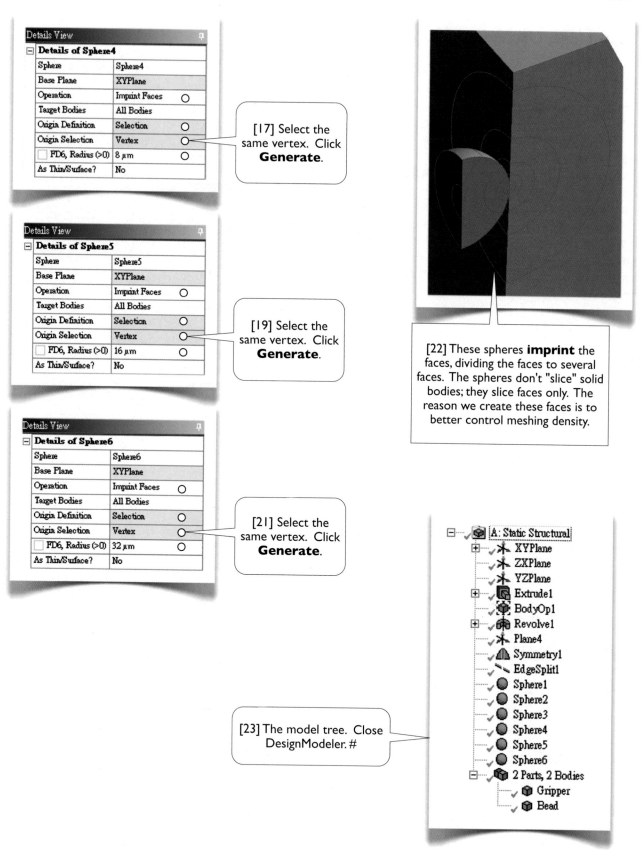

Details View

☐ **Details of Sphere4**

Sphere	Sphere4	
Base Plane	XYPlane	
Operation	Imprint Faces	○
Target Bodies	All Bodies	
Origin Definition	Selection	○
Origin Selection	Vertex	○
☐ FD6, Radius (>0)	8 μm	○
As Thin/Surface?	No	

[17] Select the same vertex. Click **Generate**.

Details View

☐ **Details of Sphere5**

Sphere	Sphere5	
Base Plane	XYPlane	
Operation	Imprint Faces	○
Target Bodies	All Bodies	
Origin Definition	Selection	○
Origin Selection	Vertex	○
☐ FD6, Radius (>0)	16 μm	○
As Thin/Surface?	No	

[19] Select the same vertex. Click **Generate**.

Details View

☐ **Details of Sphere6**

Sphere	Sphere6	
Base Plane	XYPlane	
Operation	Imprint Faces	○
Target Bodies	All Bodies	
Origin Definition	Selection	○
Origin Selection	Vertex	○
☐ FD6, Radius (>0)	32 μm	○
As Thin/Surface?	No	

[21] Select the same vertex. Click **Generate**.

[22] These spheres **imprint** the faces, dividing the faces to several faces. The spheres don't "slice" solid bodies; they slice faces only. The reason we create these faces is to better control meshing density.

[23] The model tree. Close DesignModeler. #

A: Static Structural
- XYPlane
- ZXPlane
- YZPlane
- Extrude1
- BodyOp1
- Revolve1
- Plane4
- Symmetry1
- EdgeSplit1
- Sphere1
- Sphere2
- Sphere3
- Sphere4
- Sphere5
- Sphere6
- 2 Parts, 2 Bodies
  - Gripper
  - Bead

490

13.3-4 Set Up Material in **Engineering Data**

Model the Bead as Rigid Body

The Young's modulus of glass (70 GPa) is 35,000 times of that of the PDMS (2 MPa). Whenever two bodies are so different in rigidity, you should model the more rigid one as rigid body. This not only reduces the problem size (rigid bodies won't deform, they don't need to be meshed), but also eases numerical difficulties during finite element simulations.

Rigid bodies won't deform, therefore its Young's modulus and Poisson's ratio are not relevant. For a static simulation, we can assign any material to a rigid body. For dynamic simulation, the mass density should be specified.

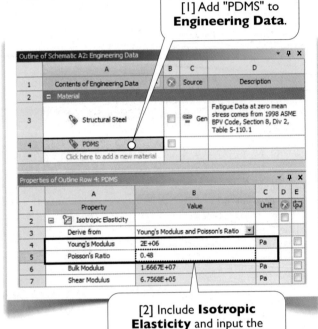

[1] Add "PDMS" to **Engineering Data**.

[2] Include **Isotropic Elasticity** and input the material properties. Return to **Project Schematic**. #

13.3-5 Set Up Model

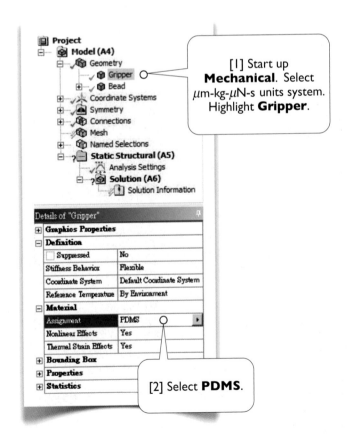

[1] Start up **Mechanical**. Select μm-kg-μN-s units system. Highlight **Gripper**.

[2] Select **PDMS**.

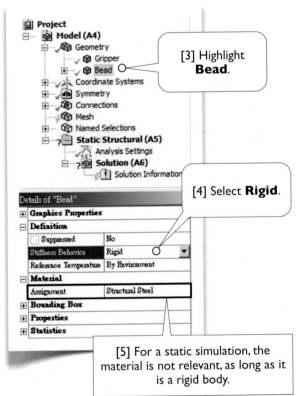

[3] Highlight **Bead**.

[4] Select **Rigid**.

[5] For a static simulation, the material is not relevant, as long as it is a rigid body.

491

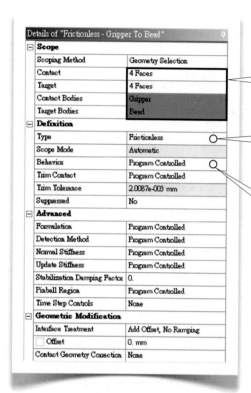

Details of "Frictionless - Gripper To Bead"

Scope	
Scoping Method	Geometry Selection
Contact	4 Faces
Target	4 Faces
Contact Bodies	Gripper
Target Bodies	Bead
Definition	
Type	Frictionless
Scope Mode	Automatic
Behavior	Program Controlled
Trim Contact	Program Controlled
Trim Tolerance	2.0087e-003 mm
Suppressed	No
Advanced	
Formulation	Program Controlled
Detection Method	Program Controlled
Normal Stiffness	Program Controlled
Update Stiffness	Program Controlled
Stabilization Damping Factor	0.
Pinball Region	Program Controlled
Time Step Controls	None
Geometric Modification	
Interface Treatment	Add Offset, No Ramping
☐ Offset	0. mm
Contact Geometry Correction	None

[6] Highlight **Connection/Contact/Contact Region**. Workbench correctly detects and establishes a contact region. You should examine these contact/target faces yourself. Note that Workbench always chooses rigid bodies as **Target Bodies**.

[7] Select **Frictionless**. Although there does exist friction between the PDMS and the glass, for our purpose of assessing gripping force, we may neglect the friction force to ease computational difficulties.

[8] Leave it **Program Controlled**. When rigid bodies are involved in the contact, Workbench always choose **Asymmetric** internally (13.1-9, page 469).

[10] Right-click **Bead** in the project tree and select **Hide Body**. Enlarge the contact region in **Gripper**.

[9] Click **-X** to rotate the model.

[11, 14, 16, 18, 20, 22] With **Mesh** highlighted, insert a **Sizing**.

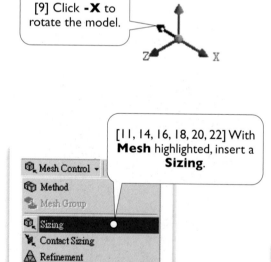

- Mesh Control ▾
- Method
- Mesh Group
- Sizing
- Contact Sizing
- Refinement
- Mapped Face Meshing
- Match Control
- Pinch
- Inflation
- Sharp Angle
- Gap Tool

[12] Select the innermost semi-circular face, which has a radius of 1 μm.

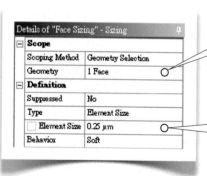

Details of "Face Sizing" - Sizing

Scope	
Scoping Method	Geometry Selection
Geometry	1 Face
Definition	
Suppressed	No
Type	Element Size
☐ Element Size	0.25 μm
Behavior	Soft

[13] Type 0.25 (μm) for **Element Size**.

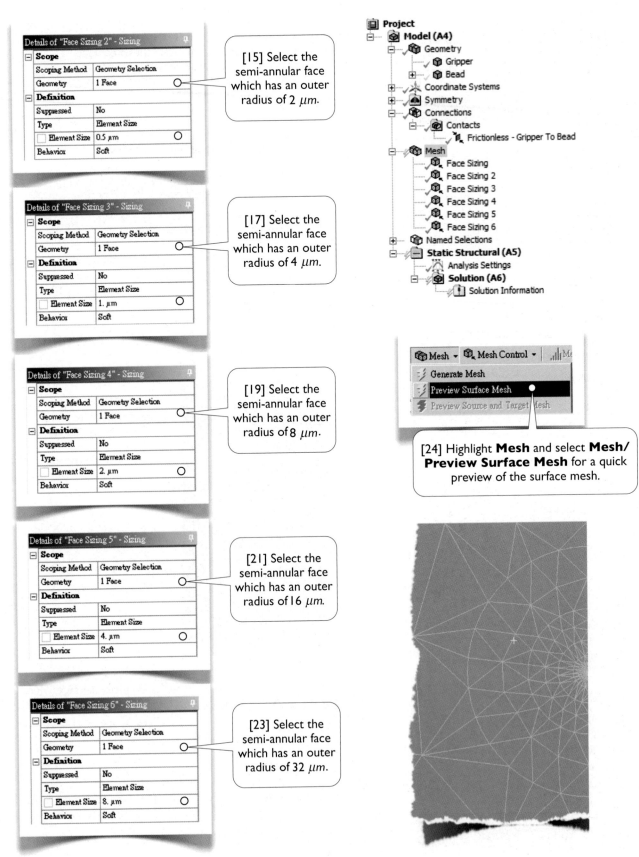

Details of "Face Sizing 2" - Sizing

Scope	
Scoping Method	Geometry Selection
Geometry	1 Face
Definition	
Suppressed	No
Type	Element Size
Element Size	0.5 μm
Behavior	Soft

[15] Select the semi-annular face which has an outer radius of 2 μm.

Details of "Face Sizing 3" - Sizing

Scope	
Scoping Method	Geometry Selection
Geometry	1 Face
Definition	
Suppressed	No
Type	Element Size
Element Size	1. μm
Behavior	Soft

[17] Select the semi-annular face which has an outer radius of 4 μm.

Details of "Face Sizing 4" - Sizing

Scope	
Scoping Method	Geometry Selection
Geometry	1 Face
Definition	
Suppressed	No
Type	Element Size
Element Size	2. μm
Behavior	Soft

[19] Select the semi-annular face which has an outer radius of 8 μm.

Details of "Face Sizing 5" - Sizing

Scope	
Scoping Method	Geometry Selection
Geometry	1 Face
Definition	
Suppressed	No
Type	Element Size
Element Size	4. μm
Behavior	Soft

[21] Select the semi-annular face which has an outer radius of 16 μm.

Details of "Face Sizing 6" - Sizing

Scope	
Scoping Method	Geometry Selection
Geometry	1 Face
Definition	
Suppressed	No
Type	Element Size
Element Size	8. μm
Behavior	Soft

[23] Select the semi-annular face which has an outer radius of 32 μm.

Project
Model (A4)
  Geometry
    Gripper
    Bead
  Coordinate Systems
  Symmetry
  Connections
    Contacts
      Frictionless - Gripper To Bead
  Mesh
    Face Sizing
    Face Sizing 2
    Face Sizing 3
    Face Sizing 4
    Face Sizing 5
    Face Sizing 6
  Named Selections
  Static Structural (A5)
    Analysis Settings
    Solution (A6)
      Solution Information

Mesh ▼   Mesh Control ▼   Me

Generate Mesh
Preview Surface Mesh
Preview Source and Target Mesh

[24] Highlight **Mesh** and select **Mesh/ Preview Surface Mesh** for a quick preview of the surface mesh.

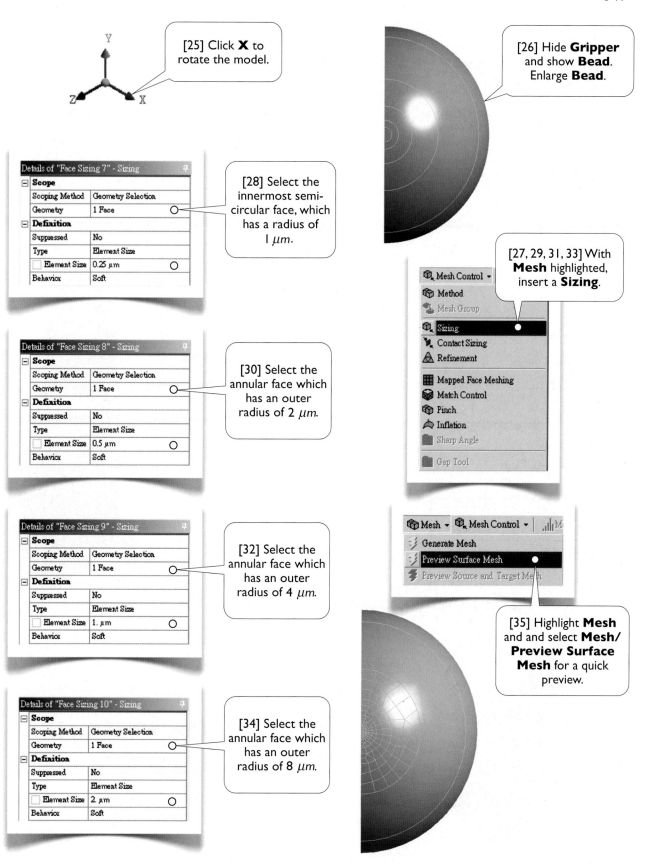

[25] Click **X** to rotate the model.

[26] Hide **Gripper** and show **Bead**. Enlarge **Bead**.

Details of "Face Sizing 7" - Sizing

Scope	
Scoping Method	Geometry Selection
Geometry	1 Face
Definition	
Suppressed	No
Type	Element Size
Element Size	0.25 μm
Behavior	Soft

[28] Select the innermost semi-circular face, which has a radius of 1 μm.

[27, 29, 31, 33] With **Mesh** highlighted, insert a **Sizing**.

Mesh Control ▾
- Method
- Mesh Group
- Sizing
- Contact Sizing
- Refinement
- Mapped Face Meshing
- Match Control
- Pinch
- Inflation
- Sharp Angle
- Gap Tool

Details of "Face Sizing 8" - Sizing

Scope	
Scoping Method	Geometry Selection
Geometry	1 Face
Definition	
Suppressed	No
Type	Element Size
Element Size	0.5 μm
Behavior	Soft

[30] Select the annular face which has an outer radius of 2 μm.

Details of "Face Sizing 9" - Sizing

Scope	
Scoping Method	Geometry Selection
Geometry	1 Face
Definition	
Suppressed	No
Type	Element Size
Element Size	1. μm
Behavior	Soft

[32] Select the annular face which has an outer radius of 4 μm.

Mesh ▾   Mesh Control ▾   M
- Generate Mesh
- Preview Surface Mesh
- Preview Source and Target Mesh

[35] Highlight **Mesh** and and select **Mesh/Preview Surface Mesh** for a quick preview.

Details of "Face Sizing 10" - Sizing

Scope	
Scoping Method	Geometry Selection
Geometry	1 Face
Definition	
Suppressed	No
Type	Element Size
Element Size	2. μm
Behavior	Soft

[34] Select the annular face which has an outer radius of 8 μm.

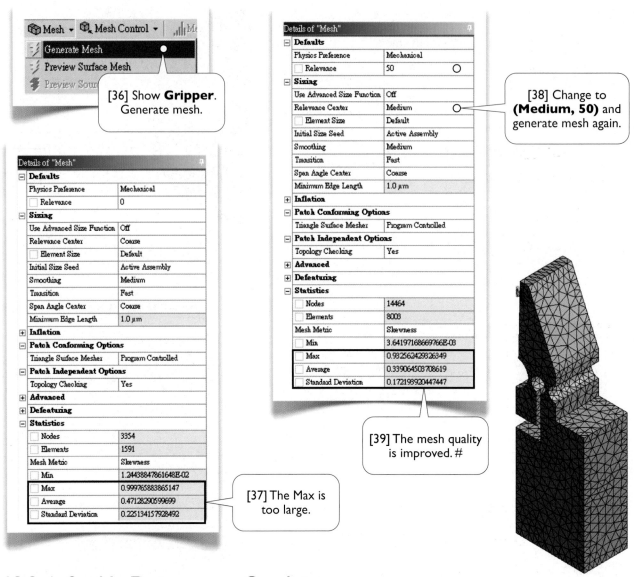

[36] Show **Gripper**. Generate mesh.

[38] Change to **(Medium, 50)** and generate mesh again.

[39] The mesh quality is improved. #

[37] The Max is too large.

13.3-6 Set Up Environment Conditions

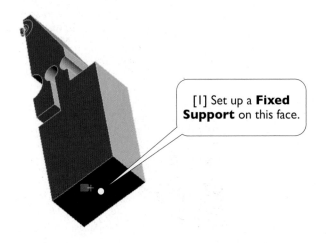

[1] Set up a **Fixed Support** on this face.

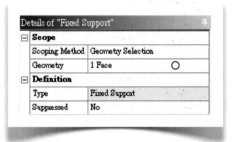

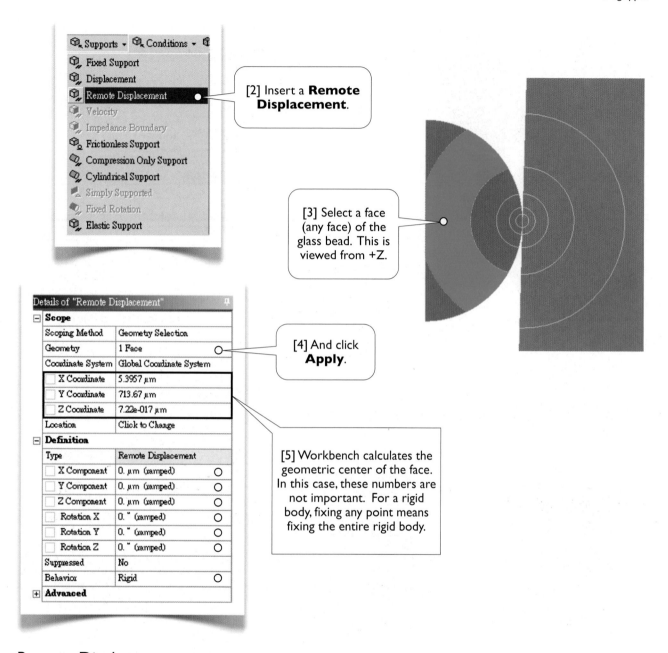

[2] Insert a **Remote Displacement**.

[3] Select a face (any face) of the glass bead. This is viewed from +Z.

[4] And click **Apply**.

[5] Workbench calculates the geometric center of the face. In this case, these numbers are not important. For a rigid body, fixing any point means fixing the entire rigid body.

Remote Displacement

A rigid body is internally represented as a node of 6 degrees of freedom, three translational and three rotational. The node is located at the geometric center of the rigid body. All environment conditions are applied at the geometric center of the rigid body. Using **Remote Displacement**, we can let Workbench transform the environment conditions to the node.

When we select a face [4] and apply a **Remote Displacement**, Workbench calculates the geometry center for that face, and the specified displacement is applied at that location. That displacement is then transformed to the geometry center of the rigid body.

In our case, we arbitrarily select a face [3]. The geometric center of that face will have zero displacements and zero rotations. These conditions will be transformed to the geometry center of the rigid body. The result is that the entire rigid body is fixed, both translationally and rotationally. That's what we intend for this rigid body.

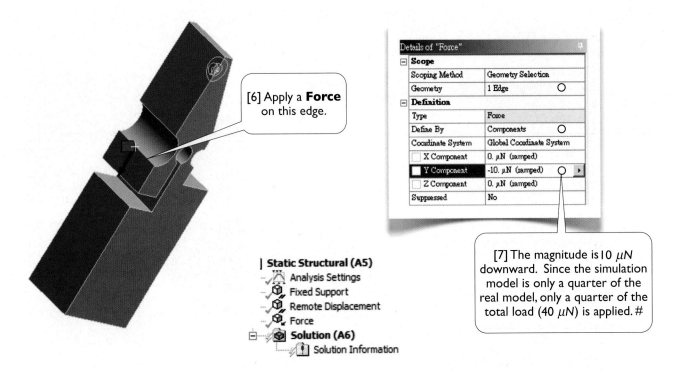

[6] Apply a **Force** on this edge.

[7] The magnitude is 10 μN downward. Since the simulation model is only a quarter of the real model, only a quarter of the total load (40 μN) is applied. #

| Static Structural (A5)
  .... Analysis Settings
  .... Fixed Support
  .... Remote Displacement
  .... Force
  └─ Solution (A6)
       .... Solution Information

13.3-7 Set Up Result Objects

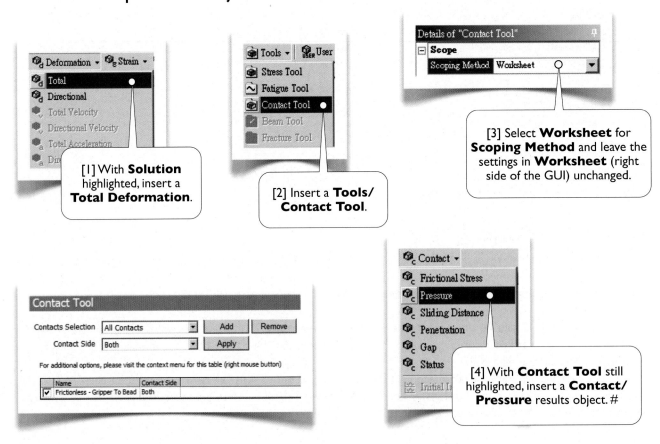

[1] With **Solution** highlighted, insert a **Total Deformation**.

[2] Insert a **Tools/ Contact Tool**.

[3] Select **Worksheet** for **Scoping Method** and leave the settings in **Worksheet** (right side of the GUI) unchanged.

[4] With **Contact Tool** still highlighted, insert a **Contact/ Pressure** results object. #

13.3-8 Solve the Model and View the Results

[2] Highlight **Solution Information**.

Details of "Analysis Settings"

Step Controls	
Number Of Steps	1.
Current Step Number	1.
Step End Time	1. s
Auto Time Stepping	Program Controlled
Solver Controls	
Solver Type	Program Controlled
Weak Springs	Program Controlled
Large Deflection	On
Inertia Relief	Off
Restart Controls	
Nonlinear Controls	
Output Controls	
Analysis Data Management	
Visibility	

- Solution (A6)
  - Solution Information
  - Total Deformation
  - Contact Tool
    - Status
    - Pressure

Details of "Solution Information"

Solution Information	
Solution Output	Force Convergence
Newton-Raphson Residuals	0
Update Interval	2.5 s
Display Points	All
FE Connection Visibility	
Activate Visibility	Yes
Display	All FE Connectors
Draw Connections Attached To	All Nodes
Line Color	Connection Type
Visible on Results	No
Line Thickness	Single
Display Type	Lines

[1] In the details view of **Analysis Settings**, turn on **Large Deflection** and leave all other settings their defaults.

[3] Select **Force Convergence**.

Solve

[4] Click **Solve**.

[5] Failing to solve with a single time step, Workbench reduces the time step to 0.35 s.

[6] This substep fails again; Workbench further reduces the time step to 0.1 s.

[7] It takes 5 iterations to complete the first substep. It advances to 0.2 s.

[8] It takes 3 iterations to complete the second substep. It advances to 0.35 s.

[9] It advances to 0.575 s, and so on (see [12]).

Legend: Force Convergence — Force Criterion — Bisection Occurred — Substep Converged

Force (μN) values: 282.05, 74.53, 19.694, 5.204, 1.3751, 0.36337, 9.6019e-2, 2.5372e-2, 6.7045e-3, 1.7716e-3, 4.6814e-4

Cumulative Iteration: 1. 2. 4. 6. 8. 10. 12. 14. 16. 18. 20. 22. 24. 25.

Time (s): 1., 0.

498

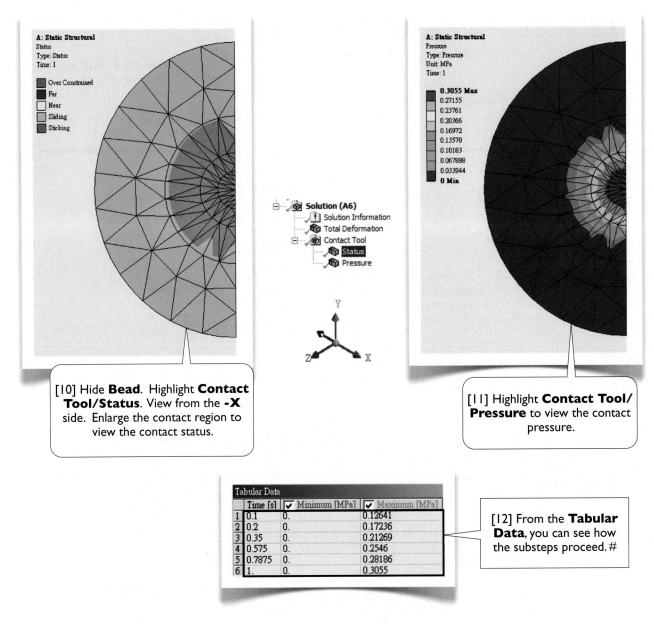

[10] Hide **Bead**. Highlight **Contact Tool/Status**. View from the **-X** side. Enlarge the contact region to view the contact status.

[11] Highlight **Contact Tool/Pressure** to view the contact pressure.

[12] From the **Tabular Data**, you can see how the substeps proceed. #

	Time [s]	✔ Minimum [MPa]	✔ Maximum [MPa]
1	0.1	0.	0.12641
2	0.2	0.	0.17236
3	0.35	0.	0.21269
4	0.575	0.	0.2546
5	0.7875	0.	0.28186
6	1.	0.	0.3055

Wrap Up

Animate the deformation. Save the project and exit Workbench.

References

1. Chang, R. J., Lin , Y. C., Shiu, C. C., and Hsieh, Y. T., "Development of SMA-Actuated Microgripper in Micro Assembly Applications," IECON, IEEE, Taiwan, 2007.
2. Shih, P. W., *Applications of SMA on Driving Micro-gripper*, MS Thesis, NCKU, ME, Taiwan, 2005.

Section 13.4

Snap Lock

13.4-1 About the Snap Lock

The snap lock consists of two parts: the insert [1] and the prongs [2]. It is fastened when pushed into position [3]. The snap lock has a thickness of 5 mm and is made of a plastic material with a Young's modulus of 2.8 GPa and a Poisson's ratio of 0.35. The coefficient of friction between the parts is 0.1. The purpose of this simulation is to find out the force required to push the insert into the position and the force required to pull it out.

We will model the problem as a plane stress problem. Due to the symmetry, only one half of the snap lock is modeled for the simulation.

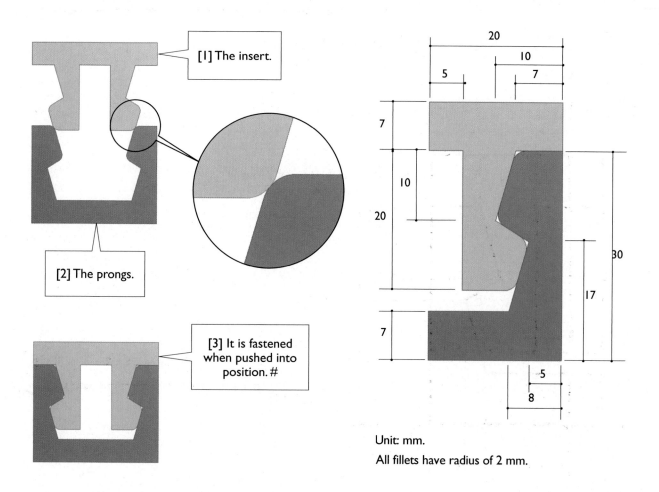

[1] The insert.

[2] The prongs.

[3] It is fastened when pushed into position. #

Unit: mm.
All fillets have radius of 2 mm.

13.4-2 Start Up

Launch Workbench. Create a **Static Structural** analysis system. Save the project as **Snap**.

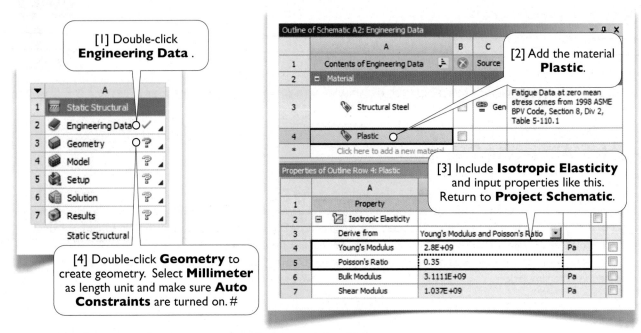

[1] Double-click **Engineering Data** .

[2] Add the material **Plastic**.

[3] Include **Isotropic Elasticity** and input properties like this. Return to **Project Schematic**.

[4] Double-click **Geometry** to create geometry. Select **Millimeter** as length unit and make sure **Auto Constraints** are turned on. #

13.4-3 Create Geometry in DesignModeler

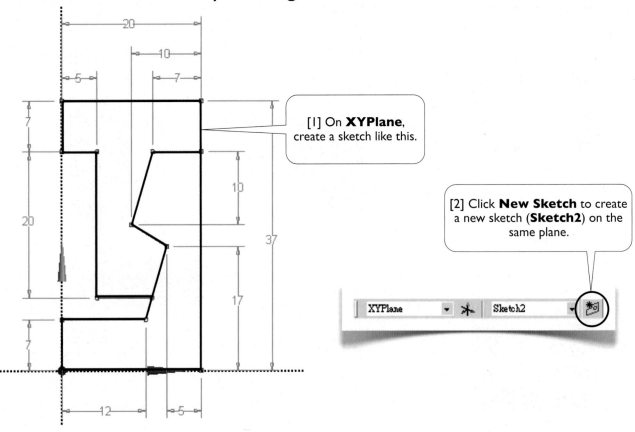

[1] On **XYPlane**, create a sketch like this.

[2] Click **New Sketch** to create a new sketch (**Sketch2**) on the same plane.

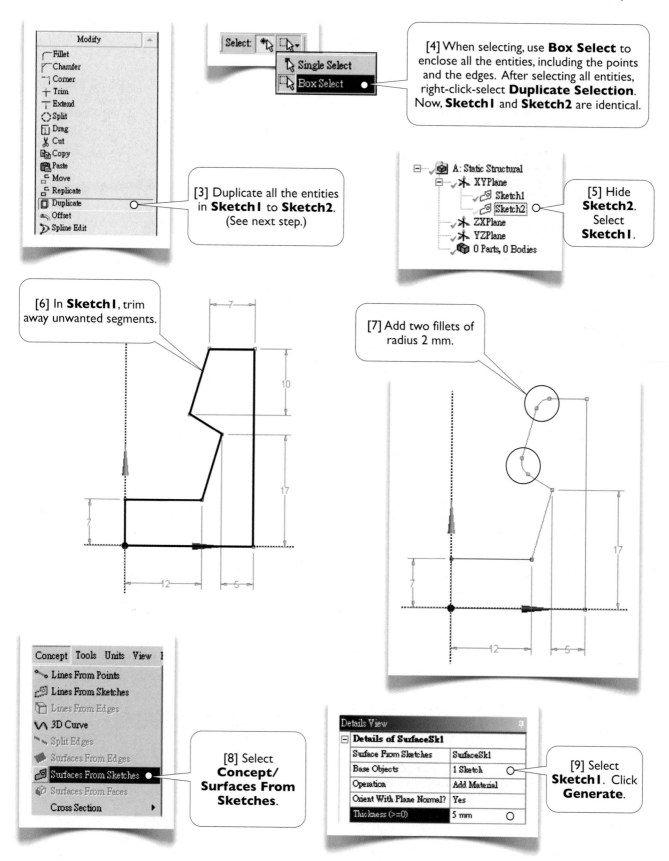

[4] When selecting, use **Box Select** to enclose all the entities, including the points and the edges. After selecting all entities, right-click-select **Duplicate Selection**. Now, **Sketch1** and **Sketch2** are identical.

[3] Duplicate all the entities in **Sketch1** to **Sketch2**. (See next step.)

[5] Hide **Sketch2**. Select **Sketch1**.

[6] In **Sketch1**, trim away unwanted segments.

[7] Add two fillets of radius 2 mm.

[8] Select **Concept/ Surfaces From Sketches**.

[9] Select **Sketch1**. Click **Generate**.

Details of SurfaceSk1	
Surface From Sketches	SurfaceSk1
Base Objects	1 Sketch
Operation	Add Material
Orient With Plane Normal?	Yes
Thickness (>=0)	5 mm

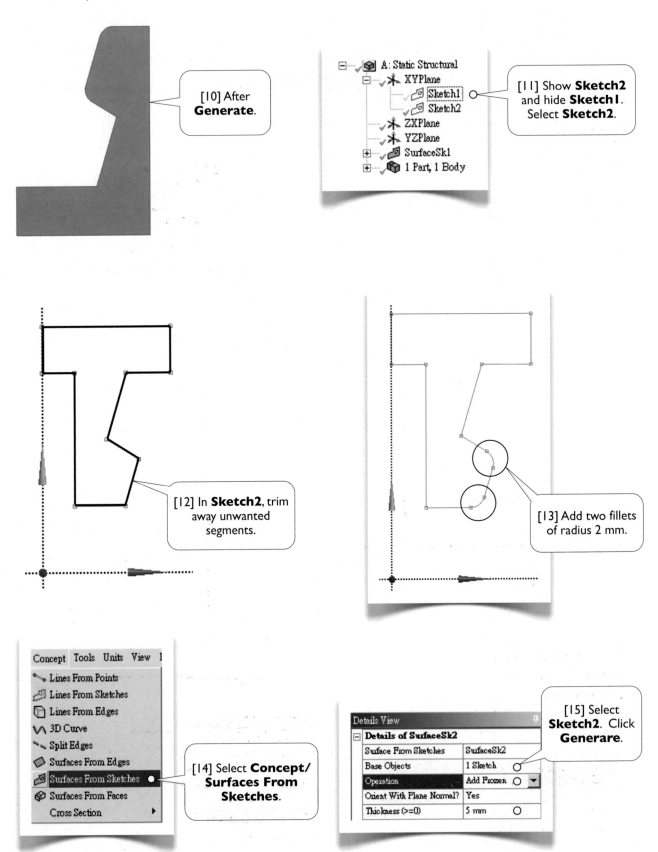

[10] After **Generate**.

[11] Show **Sketch2** and hide **Sketch1**. Select **Sketch2**.

[12] In **Sketch2**, trim away unwanted segments.

[13] Add two fillets of radius 2 mm.

[14] Select **Concept/ Surfaces From Sketches**.

[15] Select **Sketch2**. Click **Generare**.

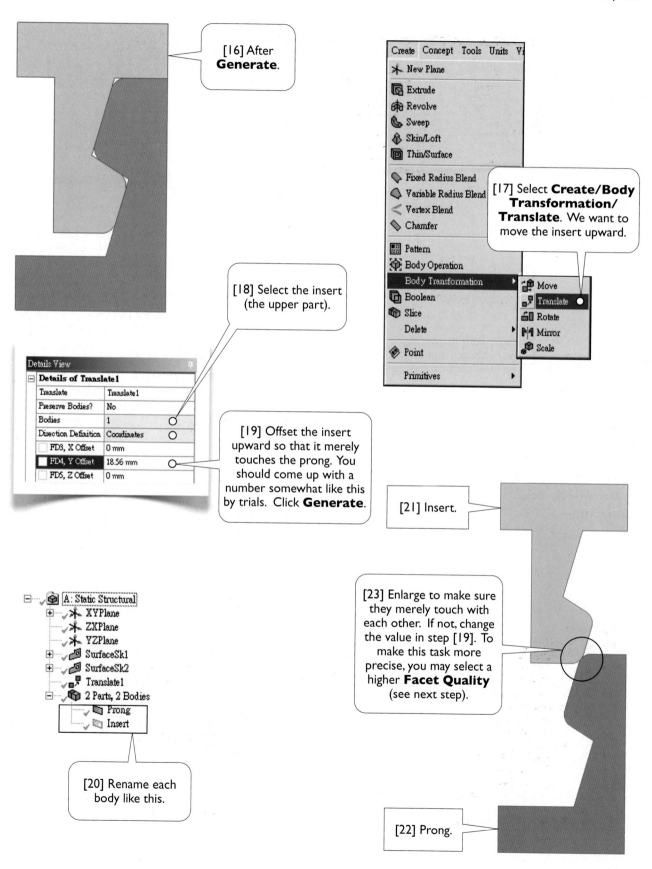

[16] After **Generate**.

[17] Select **Create/Body Transformation/ Translate**. We want to move the insert upward.

Create Concept Tools Units V:

- New Plane
- Extrude
- Revolve
- Sweep
- Skin/Loft
- Thin/Surface
- Fixed Radius Blend
- Variable Radius Blend
- Vertex Blend
- Chamfer
- Pattern
- Body Operation
- Body Transformation
  - Move
  - Translate
  - Rotate
  - Mirror
  - Scale
- Boolean
- Slice
- Delete
- Point
- Primitives

[18] Select the insert (the upper part).

Details View

Details of Translate1	
Translate	Translate1
Preserve Bodies?	No
Bodies	1
Direction Definition	Coordinates
FD8, X Offset	0 mm
FD4, Y Offset	18.56 mm
FD5, Z Offset	0 mm

[19] Offset the insert upward so that it merely touches the prong. You should come up with a number somewhat like this by trials. Click **Generate**.

[21] Insert.

[23] Enlarge to make sure they merely touch with each other. If not, change the value in step [19]. To make this task more precise, you may select a higher **Facet Quality** (see next step).

- A: Static Structural
  - XYPlane
  - ZXPlane
  - YZPlane
  - SurfaceSk1
  - SurfaceSk2
  - Translate1
  - 2 Parts, 2 Bodies
    - Prong
    - Insert

[20] Rename each body like this.

[22] Prong.

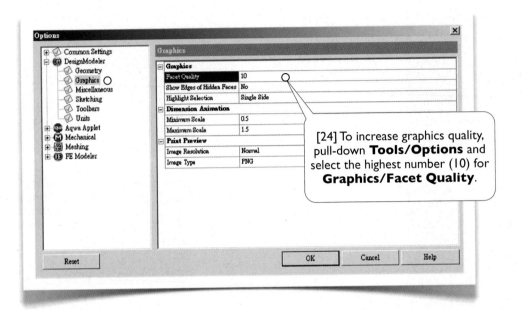

[24] To increase graphics quality, pull-down **Tools/Options** and select the highest number (10) for **Graphics/Facet Quality**.

[25] Close DesignModeler. Before attaching the geometry to **Mechanical**, remember to select **2D** option (3.1-4[4-6], page 107). #

13.4-4 Set Up Model

Start up **Mechanical**. Select the unit system **mm-kg-N-s**.

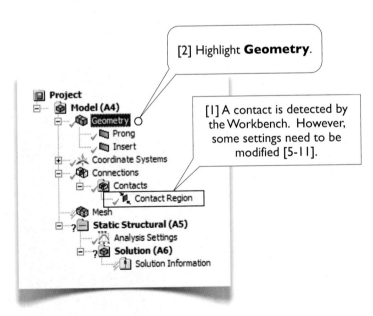

[2] Highlight **Geometry**.

[1] A contact is detected by the Workbench. However, some settings need to be modified [5-11].

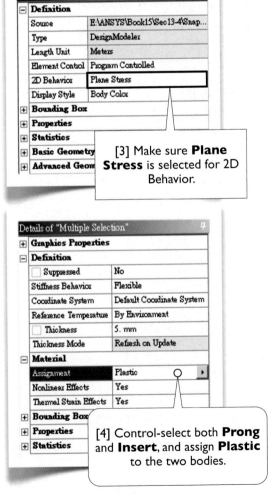

[3] Make sure **Plane Stress** is selected for 2D Behavior.

[4] Control-select both **Prong** and **Insert**, and assign **Plastic** to the two bodies.

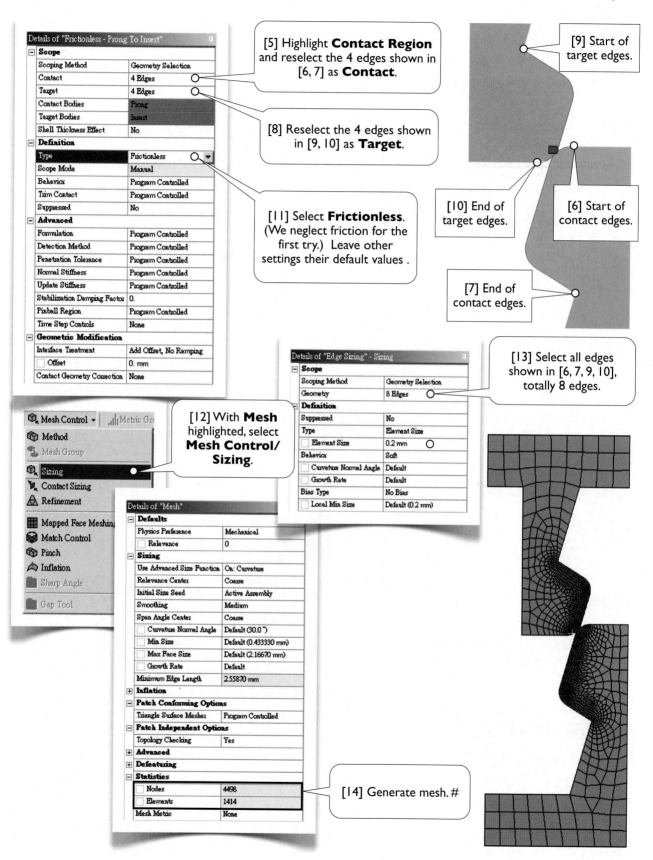

Details of "Frictionless - Prong To Insert"

Scope	
Scoping Method	Geometry Selection
Contact	4 Edges
Target	4 Edges
Contact Bodies	Prong
Target Bodies	Insert
Shell Thickness Effect	No

Definition

Type	Frictionless
Scope Mode	Manual
Behavior	Program Controlled
Trim Contact	Program Controlled
Suppressed	No

Advanced

Formulation	Program Controlled
Detection Method	Program Controlled
Penetration Tolerance	Program Controlled
Normal Stiffness	Program Controlled
Update Stiffness	Program Controlled
Stabilization Damping Factor	0.
Pinball Region	Program Controlled
Time Step Controls	None

Geometric Modification

Interface Treatment	Add Offset, No Ramping
Offset	0. mm
Contact Geometry Correction	None

[5] Highlight **Contact Region** and reselect the 4 edges shown in [6, 7] as **Contact**.

[8] Reselect the 4 edges shown in [9, 10] as **Target**.

[11] Select **Frictionless**. (We neglect friction for the first try.) Leave other settings their default values.

[9] Start of target edges.

[10] End of target edges.

[6] Start of contact edges.

[7] End of contact edges.

Mesh Control ▾ | **Metric Gr**

- Method
- Mesh Group
- Sizing
- Contact Sizing
- Refinement
- Mapped Face Meshing
- Match Control
- Pinch
- Inflation
- Sharp Angle
- Gap Tool

[12] With **Mesh** highlighted, select **Mesh Control/ Sizing**.

Details of "Edge Sizing" - Sizing

Scope	
Scoping Method	Geometry Selection
Geometry	8 Edges

Definition

Suppressed	No
Type	Element Size
Element Size	0.2 mm
Behavior	Soft
Curvature Normal Angle	Default
Growth Rate	Default
Bias Type	No Bias
Local Min Size	Default (0.2 mm)

[13] Select all edges shown in [6, 7, 9, 10], totally 8 edges.

Details of "Mesh"

Defaults

Physics Preference	Mechanical
Relevance	0

Sizing

Use Advanced Size Function	On: Curvature
Relevance Center	Coarse
Initial Size Seed	Active Assembly
Smoothing	Medium
Span Angle Center	Coarse
Curvature Normal Angle	Default (30.0 °)
Min Size	Default (0.433330 mm)
Max Face Size	Default (2.16670 mm)
Growth Rate	Default
Minimum Edge Length	2.55870 mm

Inflation

Patch Conforming Options

Triangle Surface Mesher	Program Controlled

Patch Independent Options

Topology Checking	Yes

Advanced

Defeaturing

Statistics

Nodes	4498
Elements	1414
Mesh Metric	None

[14] Generate mesh. #

13.4-5 Set Up Analysis Settings

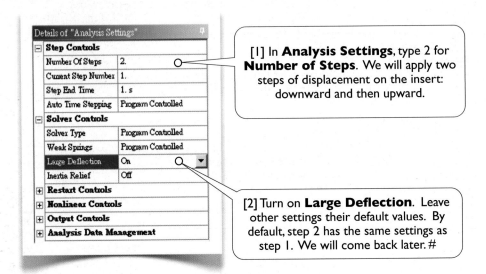

[1] In **Analysis Settings**, type 2 for **Number of Steps**. We will apply two steps of displacement on the insert: downward and then upward.

[2] Turn on **Large Deflection**. Leave other settings their default values. By default, step 2 has the same settings as step 1. We will come back later. #

13.4-6 Set Up Environment Conditions

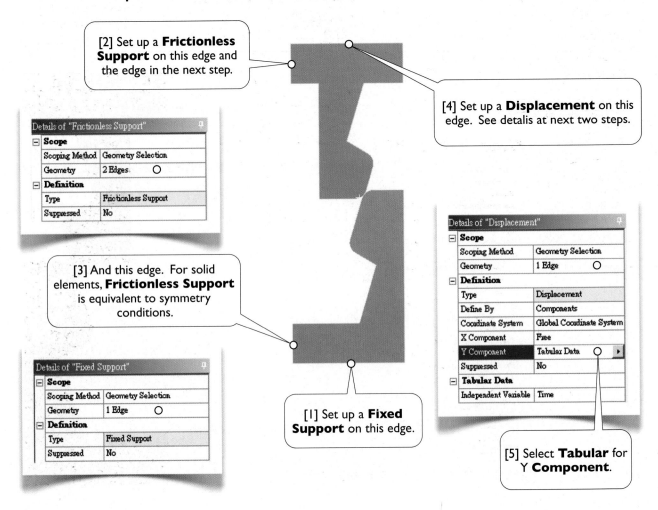

[2] Set up a **Frictionless Support** on this edge and the edge in the next step.

[4] Set up a **Displacement** on this edge. See detalis at next two steps.

[3] And this edge. For solid elements, **Frictionless Support** is equivalent to symmetry conditions.

[1] Set up a **Fixed Support** on this edge.

[5] Select **Tabular** for Y **Component**.

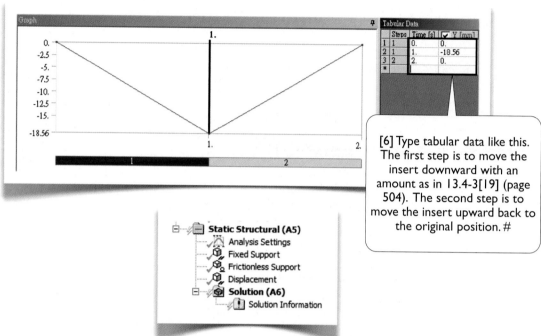

	Steps	Time [s]	✓ Y [mm]
1	1	0.	0.
2	1	1.	-18.56
3	2	2.	0.
*			

[6] Type tabular data like this. The first step is to move the insert downward with an amount as in 13.4-3[19] (page 504). The second step is to move the insert upward back to the original position. #

Static Structural (A5)
- Analysis Settings
- Fixed Support
- Frictionless Support
- Displacement
- Solution (A6)
  - Solution Information

13.4-7 Set Up Result Objects

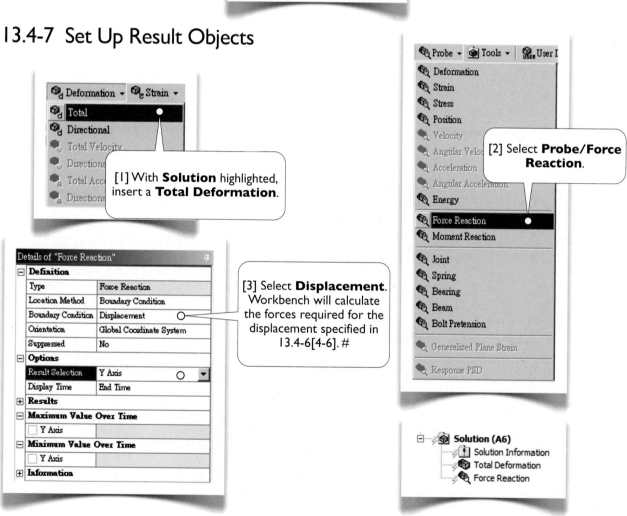

[1] With **Solution** highlighted, insert a **Total Deformation**.

[2] Select **Probe/Force Reaction**.

Details of "Force Reaction"

Definition

Type	Force Reaction
Location Method	Boundary Condition
Boundary Condition	Displacement
Orientation	Global Coordinate System
Suppressed	No

Options

Result Selection	Y Axis
Display Time	End Time

Results

Maximum Value Over Time

| ☐ Y Axis | |

Minimum Value Over Time

| ☐ Y Axis | |

Information

[3] Select **Displacement**. Workbench will calculate the forces required for the displacement specified in 13.4-6[4-6]. #

Solution (A6)
- Solution Information
- Total Deformation
- Force Reaction

13.4-8 Solve the Model and View the Results

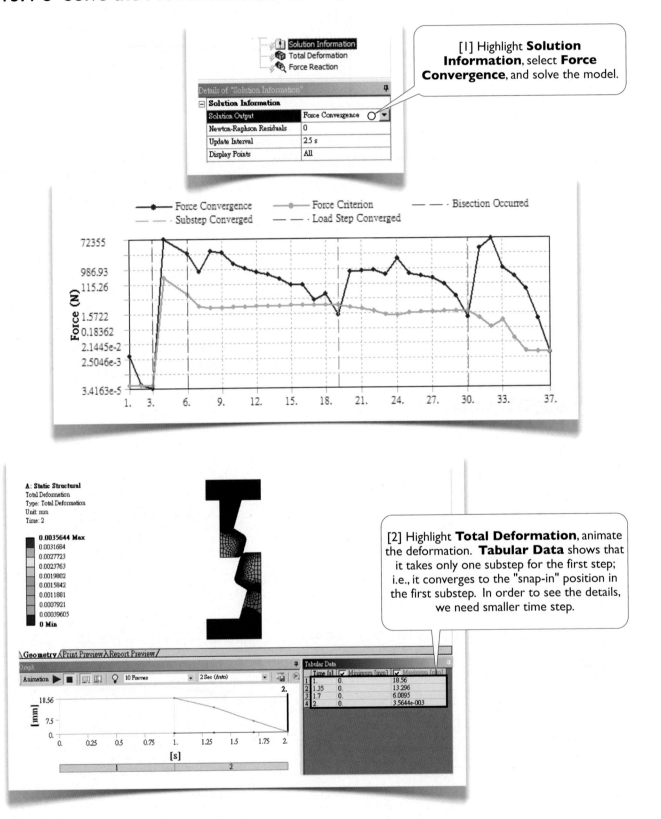

[1] Highlight **Solution Information**, select **Force Convergence**, and solve the model.

[2] Highlight **Total Deformation**, animate the deformation. **Tabular Data** shows that it takes only one substep for the first step; i.e., it converges to the "snap-in" position in the first substep. In order to see the details, we need smaller time step.

[3] Set up **Analysis Settings** for the first step.

[4] Set up **Analysis Settings** for the second step.

[5] Turning on **Carry Over Time Step** is to use the time step value of the last substep of the last step as the initial time step value for this step.

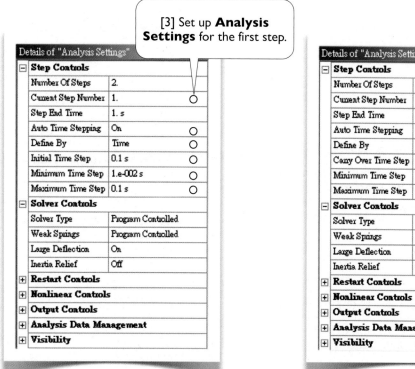

[6] Highlight **Solution Information** and solve again.

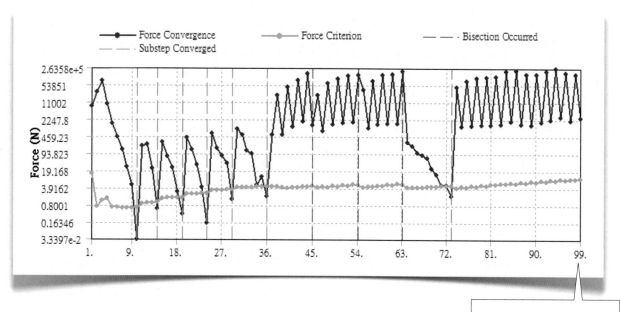

[7] Workbench fails to obtain a solution.

510

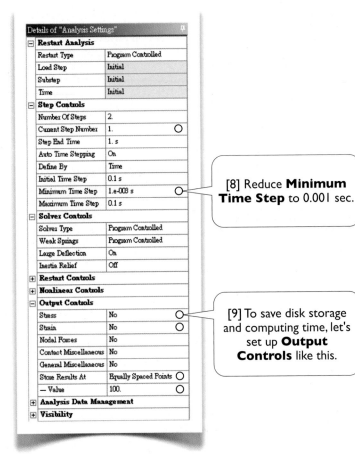

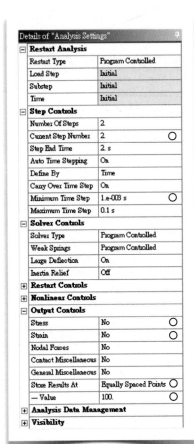

[8] Reduce **Minimum Time Step** to 0.001 sec.

[9] To save disk storage and computing time, let's set up **Output Controls** like this.

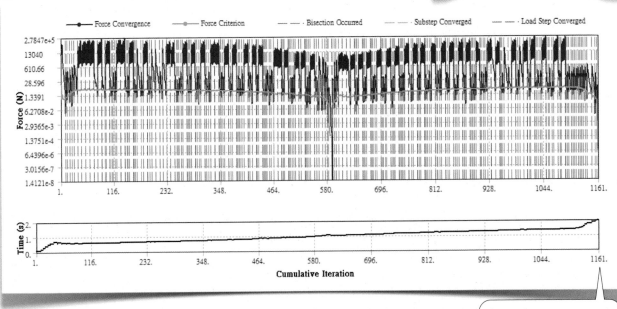

[10] Solve the model. It takes many iterations to complete the solution.

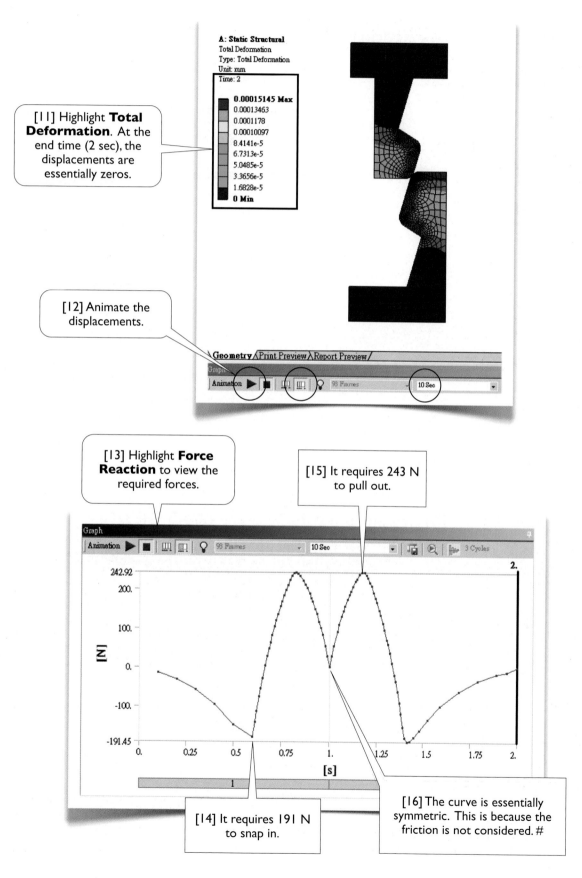

[11] Highlight **Total Deformation**. At the end time (2 sec), the displacements are essentially zeros.

[12] Animate the displacements.

[13] Highlight **Force Reaction** to view the required forces.

[15] It requires 243 N to pull out.

[14] It requires 191 N to snap in.

[16] The curve is essentially symmetric. This is because the friction is not considered. #

13.4-9 Simulation with Frictional Model

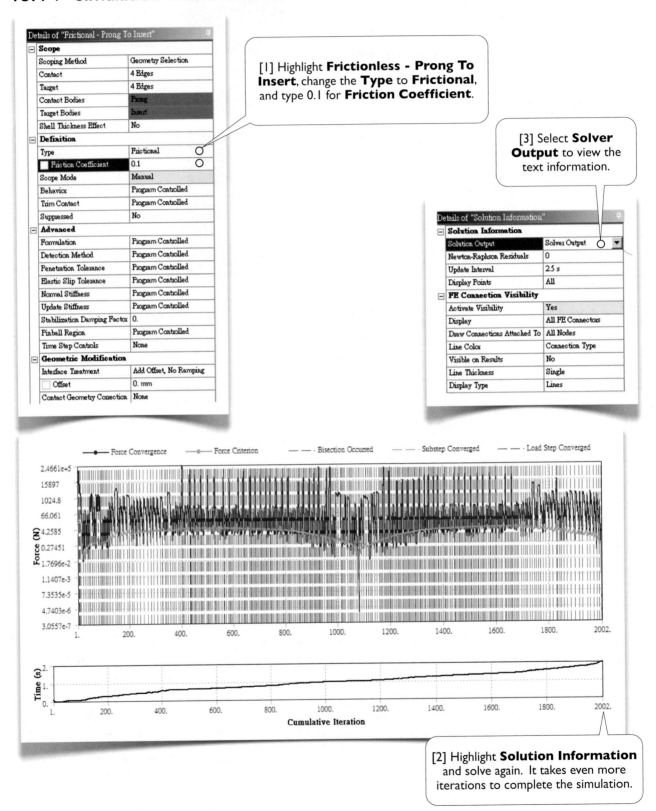

[1] Highlight **Frictionless - Prong To Insert**, change the **Type** to **Frictional**, and type 0.1 for **Friction Coefficient**.

[3] Select **Solver Output** to view the text information.

[2] Highlight **Solution Information** and solve again. It takes even more iterations to complete the simulation.

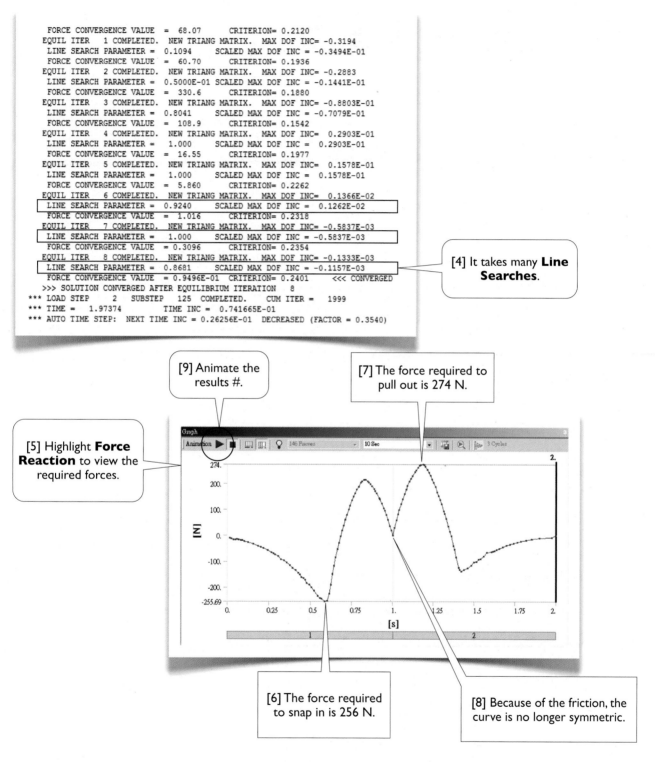

```
FORCE CONVERGENCE VALUE  =  68.07      CRITERION= 0.2120
EQUIL ITER   1 COMPLETED.  NEW TRIANG MATRIX.  MAX DOF INC= -0.3194
 LINE SEARCH PARAMETER =  0.1094     SCALED MAX DOF INC = -0.3494E-01
 FORCE CONVERGENCE VALUE  =  60.70      CRITERION= 0.1936
EQUIL ITER   2 COMPLETED.  NEW TRIANG MATRIX.  MAX DOF INC= -0.2883
 LINE SEARCH PARAMETER =  0.5000E-01 SCALED MAX DOF INC = -0.1441E-01
 FORCE CONVERGENCE VALUE  =  330.6      CRITERION= 0.1880
EQUIL ITER   3 COMPLETED.  NEW TRIANG MATRIX.  MAX DOF INC= -0.8803E-01
 LINE SEARCH PARAMETER =  0.8041     SCALED MAX DOF INC = -0.7079E-01
 FORCE CONVERGENCE VALUE  =  108.9      CRITERION= 0.1542
EQUIL ITER   4 COMPLETED.  NEW TRIANG MATRIX.  MAX DOF INC=  0.2903E-01
 LINE SEARCH PARAMETER =  1.000      SCALED MAX DOF INC =  0.2903E-01
 FORCE CONVERGENCE VALUE  =  16.55      CRITERION= 0.1977
EQUIL ITER   5 COMPLETED.  NEW TRIANG MATRIX.  MAX DOF INC=  0.1578E-01
 LINE SEARCH PARAMETER =  1.000      SCALED MAX DOF INC =  0.1578E-01
 FORCE CONVERGENCE VALUE  =  5.860      CRITERION= 0.2262
EQUIL ITER   6 COMPLETED.  NEW TRIANG MATRIX.  MAX DOF INC=  0.1366E-02
 LINE SEARCH PARAMETER =  0.9240     SCALED MAX DOF INC =  0.1262E-02
 FORCE CONVERGENCE VALUE  =  1.016      CRITERION= 0.2318
EQUIL ITER   7 COMPLETED.  NEW TRIANG MATRIX.  MAX DOF INC= -0.5837E-03
 LINE SEARCH PARAMETER =  1.000      SCALED MAX DOF INC = -0.5837E-03
 FORCE CONVERGENCE VALUE  = 0.3096      CRITERION= 0.2354
EQUIL ITER   8 COMPLETED.  NEW TRIANG MATRIX.  MAX DOF INC= -0.1333E-03
 LINE SEARCH PARAMETER =  0.8681     SCALED MAX DOF INC = -0.1157E-03
 FORCE CONVERGENCE VALUE  = 0.9496E-01  CRITERION= 0.2401    <<< CONVERGED
>>> SOLUTION CONVERGED AFTER EQUILIBRIUM ITERATION   8
*** LOAD STEP   2   SUBSTEP  125 COMPLETED.    CUM ITER =   1999
*** TIME =   1.97374      TIME INC =  0.741665E-01
*** AUTO TIME STEP:  NEXT TIME INC = 0.26256E-01 DECREASED (FACTOR = 0.3540)
```

[4] It takes many **Line Searches**.

[9] Animate the results #.

[7] The force required to pull out is 274 N.

[5] Highlight **Force Reaction** to view the required forces.

[6] The force required to snap in is 256 N.

[8] Because of the friction, the curve is no longer symmetric.

Wrap Up

Save the project and exit Workbench.

Section 13.5

Review

13.5-1 Keywords

Choose a letter for each keyword from the list of descriptions

1. (   ) Contact Nonlinearity
2. (   ) Displacement Convergence
3. (   ) Equilibrium Iterations
4. (   ) Force Convergence
5. (   ) Geometry Nonlinearity
6. (   ) Linear Structures
7. (   ) Line Search
8. (   ) Load Steps

9. (   ) Material Nonlinearity
10. (   ) Moment Convergence
11. (   ) Newton-Raphson Method
12. (   ) Nonlinear Structures
13. (   ) Residual Force
14. (   ) Rotation Convergence
15. (   ) Substeps

Answers:

1. (D)　2. (L)　3. (H)　4. (K)　5. (C)　6. (A)　7. (O)　8. (F)　9. (E)　10. (M)
11. (I)　12. (B)　13. (J)　14. (N)　15. (G)

List of Descriptions

(A) The structures in which the relation between the responses and the loads is linear.

(B) The structures in which the relation between the responses and the loads is not linear.

(C) Nonlinearity due to large deformation.

(D) Nonlinearity due to the change of contact status.

(E) Nonlinearity due to the presence of nonlinear stress-strain relation.

(F) Also called steps. In Workbench simulations, the entire loading history can be divided into one or more load steps, so that different analysis settings can be specified for each load step.

(G) Also called time steps. In dynamic simulations, time step size is used for integration over the time domain; the time step size must be small enough to capture the response characteristics. In static simulation, a load step can be divided into substeps small enough to achieve or enhance convergence.

(H) Also called iterations. For nonlinear problems, each time step needs several iterations to complete. For linear problems, each time step requires exactly one equilibrium iteration.

(I) The method used by Workbench to solve a substep in a nonlinear simulation. External force of that substep is applied. Equilibrium equation is solved for the displacement using the tangent stiffness. Internal force is calculated using updated displacement and stiffness. This completes an iteration. The process iterates until all the active convergence criteria are satisfied.

(J) During the Newton-Raphson equilibrium iterations, the difference between external force and the calculated internal force is called a residual force.

(K) A substep is said to satisfy the force convergence when the norm of the residual force is less than a criterion.

(L) A substep is said to satisfy the displacement convergence when the norm of the difference of displacements between two iterations is less than a criterion.

(M) When shell or beam elements are used in a model, moment convergence can be activated. A substep is said to satisfy the moment convergence when the norm of the residual moment is less than a criterion.

(N) When shell or beam elements are used in a model, rotation convergence can be activated. A substep is said to satisfy the rotation convergence when the norm of the difference of rotations between two iterations is less than a criterion.

(O) In some cases when a force-displacement relation is highly nonlinear or "concave up," during the Newton-Raphson iterations of a substep, the calculated displacement in a single iteration may overshoot the goal. In such cases, a numerical technique called line search can be activated to "scale down" the incremental displacement. Line search often helps convergence, but takes extra computing time.

13.5-2 Additional Workbench Exercises

Contact Stiffness Study

In 13.1-10 (page 470), when introducing **Pure Penalty** formulation, we mentioned that, in many cases, solution convergence behavior may be sensitive to **Normal Stiffness**. VM63[Ref 1] of **Verification Manual for the Mechanical APDL Application** is a good exercise to study this behavior. First, complete a simulation using all the parameters given in the verification manual, and verify the solution with the verification manual. Next, increase the Young's modulus from the original value (1 GPa) to 200 GPa (which is the Young's modulus of steel). This should create a convergence difficulty. When you run into a convergence difficulty, try to ease the difficulty by decreasing **Normal Stiffness**. This will introduce more penetration into the solution. Tabulate data (or draw curve) to show how **Normal Stiffness** affects penetration and other results.

Reference
1. ANSYS Documentation//Verification Manuals//Mechanical APDL Verification Manual//I. Verification Test Case Descriptions//VM63: Static Hertz Contact Problem

Chapter 14
Nonlinear Materials

When the stress-strain relation of a material is linear, it is called a *linear material*. On the other hand, if the stress-strain relation is not linear, it is called a *nonlinear material* Stress-strain relation of an isotropic linear material can be expressed by Hooke's law, Eq. 1.2-8(1) (page 27), in which two independent material parameters are needed to define the stress-strain relation. So far, in foregoing simulations, we assumed the materials are linear.

In reality, most of materials exhibit more or less nonlinearities. In many cases, the nonlinearities are negligible, and we use Hooke's law to describe the stress-strain relation. In some other cases, when the material nonlinearities are not negligible, *nonlinear material models* must be used to define stress-strain relations. A material model is usually a mathematics form with some parameters, called *material parameters*. To assign a material to a body, you select a material model from Workbench and provide its material parameters. The material parameters are usually obtained by data-fitting using the results of a set of material testings.

Purpose of This Chapter

Workbench provides a variety of nonlinear material models. In this chapter, we will introduce two categories of nonlinear material models: *plasticity* and *hyperelasticity*. Background knowledge will be introduced first, and two step-by-step exercises will follow to demonstrate their applications.

About Each Section

Section 14.1 gives basics of plasticity and hyperelasticity. Section 14.2 provides a step-by-step example to demonstrate the use of a plastic material model. Section 14.3 uses another step-by-step example to demonstrate the application of a hyperelastic material model.

Section 14.1

Basics of Nonlinear Materials

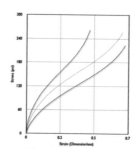

PART A. INTRODUCTION

14.1-1 Linear versus Nonlinear Materials

When the stress-strain relation of a material is linear, it is called a *linear material* [1], otherwise the material is called a *nonlinear material*. For a linear isotropic material, the stress-strain relation can be expressed by Hooke's law, Eq. 1.2-8(1) (page 27), in which two independent material parameters are needed to define the material. Note that Eq. 1.2-8(1) assumes an *isotropic material*; i.e., the Young's modulus and the Poisson's ratio are independent of directions. Orthotropic (Eq. 1.2-8(4), page 28) and anisotropic linear elasticity are also available in Workbench [2].

Besides a linear relation, the Hooke's law also assumes that the stress-strain relation is (a) elastic, (b) time-independent, and (c) rate-independent. Materials that violate any of these behaviors cannot be described by the Hooke's law, and are categorized as nonlinear materials, even though the stress-strain relation is linear for both stressing and unstressing [3].

Workbench provides non-elastic material model, called plastic material models.

Time-Dependent Stress-Strain Relations

When you apply a stress on a steel, the strain occurs instantaneously and remains the same forever. This statement is not strictly true. A more rigorous statement should be like this, "The strain occurs ALMOST instantaneously and ALMOST remains the same forever."

Consider a big water tank made of plastic material. Starting from the moment right after you fill up the water, the plastic tank begins to deform: the diameter expands slowly. It may take months until the deformation stops. This gives us a lesson that the deformation may take time. This time-dependent stress-strain behavior is called *creeping*, or *viscosity*.

For the steel under room temperature, the creeping behavior is practically negligible. In engineering practice, we often neglect creeping for solid materials under a fairly lower temperature, for example, below half of the melting point.

We will not discuss time-dependent behavior (although creep models available in Workbench) for the rest of this chapter.

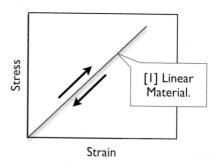

[1] Linear Material.

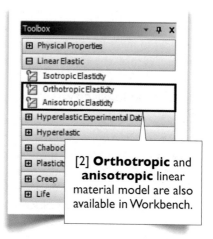

[2] **Orthotropic** and **anisotropic** linear material model are also available in Workbench.

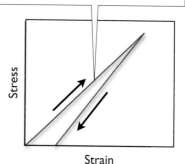

[3] This is not a linear material, even though the stress-strain relation is linear for both stressing and unstressing. #

14.1-2 Elastic versus Plastic Materials

In 14.1-1[3] (last page), the strain is not totally recovered after release of the stress. This behavior is called plasticity, and the residual strain is called the *plastic strain*. If the strain is totally recoverable, that is, if there is no residual strain after release of the stress, the behavior is called elasticity, and the material is said to be elastic.

Following this definition of elasticity, we may classify the elastic materials into three categories: (1) linear elastic, (2) nonlinear hysteresis elastic, and (3) nonlinear non-hysteresis elastic, or simply called nonlinear elastic.

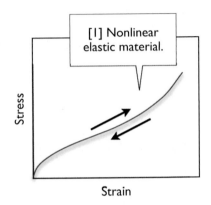

[1] Nonlinear elastic material.

Linear Elasticity

The linear elastic material is defined by Hooke's law and depicted in 14.1-1[1], last page.

Hysteresis Elasticity

The term *elastic hysteresis* has been introduced in 12.1-3[3-5] (page 416). The current version of ANSYS (including APDL) doesn't directly provide a material model to include the hysteresis behavior. You have to include the hysteresis behavior in terms of material damping (12.1-3, page 416).

Most of materials have more-or-less hysteresis behavior. However, as long as it is small enough, we may neglect the hysteresis behavior.

Nonlinear Elasticity

Nonlinear non-hysteresis elastic materials are characterized by the fact that the stressing curve and the unstressing curve are coincident [1]: the energy is conserved in stressing-unstressing cycles.

Challenge of implementing nonlinear elastic material models comes from that the strain may be as large as 100% or even 200%, such as rubber under stretching or compression. Additional consideration is that, under such large strains, the stretching and compression behaviors may not be described by the same parameters. This kind of super-large deformation elasticity is given a special name: *hyperelasticity*.

Workbench provides many hyperelastic material models and we will discuss them in Part C of this section (starting from 14.1-7, page 523). A step-by-step example is presented in Section 14.3.

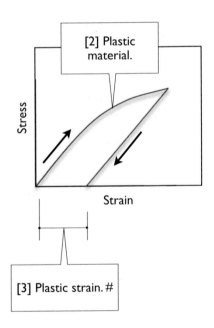

[2] Plastic material.

[3] Plastic strain. #

Plasticity

Plastic materials are characterized by the presence of the residual strain, or plastic strain [2-3]. Note that the hysteresis is always present in plastic materials: there is always energy loss in stressing-unstressing cycles.

Workbench provides many plastic material models and we will discuss them in Part B of this section (starting from 14.1-3, next page). A step-by-step example is presented in Section 14.2.

PART B. PLASTICITY

14.1-3 Idealized Stress-Strain Curve for Plasticity

Plasticity behavior typically occurs in ductile metals subject to large deformation. Plastic deformation is a result of slips between planes of atoms due to shear stresses. It is essentially a rearrangement of atoms in the crystal structure.

In Workbench, a typical stress-strain relation, such as 14.1-2[2] (last page), is idealized as shown in [1-4]. The stress-strain curve is composed of several straight segments. The slope of the first segment is the Young's modulus [3]. When the stress is released, the strain decreases with a slope equal to the Young's modulus [4]. This implies that if the stress/strain state is on the first segment, the behavior is elastic and no plastic strain remains after releasing the stress. The point at the end of the first segment is called *elastic limit*, or *initial yield point* [2]. All points higher than the initial yield point are called *subsequent yield points*, since they all represent *yield state*.

A uniaxial stress-strain relation such as [1-4] is not sufficient to fully define a plasticity behavior. There are other characteristics that must be described for general multiaxial cases: (a) What is the yield criterion? (b) What is the hardening rule?

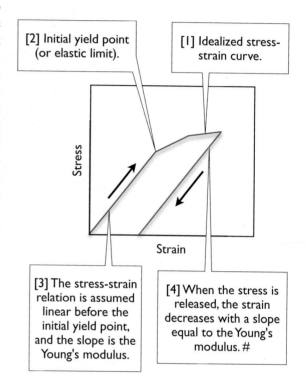

[2] Initial yield point (or elastic limit).

[1] Idealized stress-strain curve.

[3] The stress-strain relation is assumed linear before the initial yield point, and the slope is the Young's modulus.

[4] When the stress is released, the strain decreases with a slope equal to the Young's modulus. #

14.1-4 Yield Criteria

A stress-strain curve such as 14.1-3[1-4] is usually obtained by a uniaxial tensile test. It provides an initial yield strength σ_y of the uniaxial tensile test. In three-dimensional cases, the stress state is multiaxial. According to what criteria can we say that a stress state reaches a yield state? Workbench uses von Mises criterion (1.4-5, pages 39-42) as the yield criterion; i.e., a stress state reaches yield state when the von Mises stress σ_e is equal to the CURRENT *uniaxial yield strength* σ_y', or

$$\sqrt{\frac{1}{2}\left[\left(\sigma_1 - \sigma_2\right)^2 + \left(\sigma_2 - \sigma_3\right)^2 + \left(\sigma_3 - \sigma_1\right)^2\right]} = \sigma_y' \tag{1}$$

The yielding initially occurs when $\sigma_y' = \sigma_y$, and the "current" uniaxial yield strength σ_y' may change subsequently. As mentioned at the end of 1.4-5 (page 42), when plotted in the σ_1-σ_2-σ_3 space, Eq. (1) is a cylindrical surface aligned with the axis $\sigma_1 = \sigma_2 = \sigma_3$ and with a radius of $\sqrt{2}\sigma_y'$. It is called a *von Mises yield surface* [1] (next page). If the stress state is inside the cylinder, no yielding occurs. If the stress state is on the surface, yielding occurs. No stress state can be outside the yield surface. If the stress state is on the surface and the loads continue to "push" the yield surface outward, the size (radius) or the location of the yield surface will change. The rule that describes how the yield surface changes its size or location is called a *hardening rule* (14.1-5, next page).

Note that, in a uniaxial test, we are talking about "yield points" in the stress axis. In a biaxial case, the yielding states form a "yield line" in a stress plane, while in a 3D cases, the yielding states become a "yield surface" in a stress space [1].

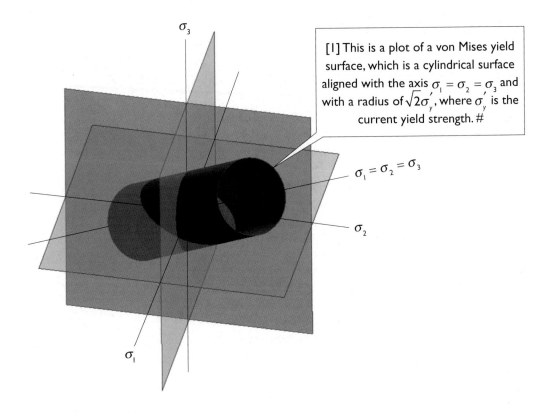

[1] This is a plot of a von Mises yield surface, which is a cylindrical surface aligned with the axis $\sigma_1 = \sigma_2 = \sigma_3$ and with a radius of $\sqrt{2}\sigma_y'$, where σ_y' is the current yield strength. #

14.1-5 Hardening Rules

Workbench implements two hardening rules: (a) kinematic hardening, and (b) isotropic hardening. In metal plasticity, hardening behavior is often a mix-up of kinematic and isotropic.

Kinematic Hardening

Kinematic hardening assumes that, If a stress state is on the yield surface and the loads continue to "push" a yield surface outward, the yield surface will change its location, according to the "pushing direction," but preserve the size of the yield surface. In a uniaxial test, it is equivalent to say that the difference between the tensile yield strength and the compressive yield strength remains a constant of $2\sigma_y$ [1].

Kinematic hardening is generally used for small strain, cyclic loading applications, especially metals.

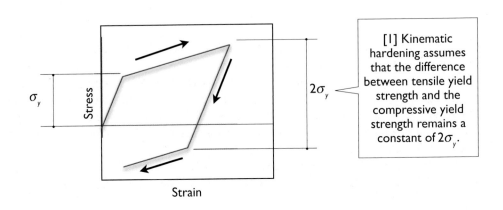

[1] Kinematic hardening assumes that the difference between tensile yield strength and the compressive yield strength remains a constant of $2\sigma_y$.

521

Isotropic Hardening

Isotropic hardening assumes that, when the loads continue to "push" a yield surface, the yield surface will expand its size, but preserve the axis of the yield surface. In a uniaxial test, it is equivalent to say that the "current" tensile yield strength and the compressive yield strength remain equal in magnitude [2].

Isotropic hardening is often used for large strain simulations. It is usually not applicable for cyclic loading applications.

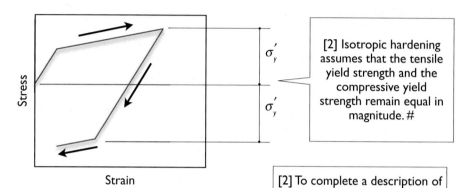

[2] Isotropic hardening assumes that the tensile yield strength and the compressive yield strength remain equal in magnitude. #

14.1-6 Workbench Plasticity Models

Workbench provides many plasticity models [1]. Besides choosing from either of hardening rules, you can choose from either type of stress-strain curves: bilinear or multilinear.

Linear Elastic Properties Must be Included

To describe a material model for plasticity, you must include a set of linear elastic properties (e.g., Young's modulus and Poisson's ratio in cases of isotropy). The Young's modulus is used as the initial slope of the stress-strain curve, in which the material initially behaves as linear elasticity.

Bilinear Stress-Strain Curve

Examples of bilinear models are shown in 14.1-5[1-2] (last page and this page); the stress-strain curve is composed by two straight segments. Besides the Young's modulus and Poisson's ratio, you need to supply an *initial yield stress* and a *tangent modulus* (the slope of the second segment).

Multilinear Stress-Strain Curve

An example of multilinear models is shown in 14.1-3[1] (page 520); the stress-strain curve is composed by several straight segments. Besides the Young's modulus and Poisson's ratio, you need to supply a tabular form of data describing the *subsequent yield stresses* and the corresponding *plastic strains*. An example is given in 14.2-6[6-10], pages 539-540.

[2] To complete a description of plasticity model, you must include its linear elastic properties (e.g., Young's modulus and Poisson's ratio in cases of isotropy). #

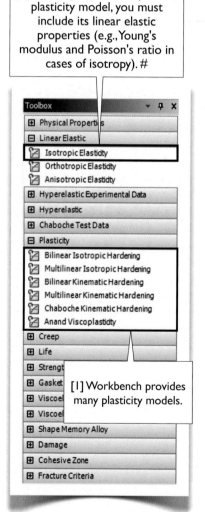

[1] Workbench provides many plasticity models.

PART C. HYPERELASTICITY

14.1-7 Test Data Needed for Hyperelasticity

As mentioned in 14.1-2 (page 519), challenge of implementing nonlinear elastic models comes from that the strain may be as large as 100% (or even 200%), such as rubber under stretching.

In plasticity or linear elasticity, we use a stress-strain curve to describe its behavior, and the stress-strain curve is usually obtained by a uniaxial tensile test. Since only tension behavior is investigated, other behaviors (e.g., compressive, shearing) must be drawn from the tensile test data. In plasticity or linear elasticity, we implicitly made following assumptions: (a) The compressive behavior is symmetric to the tension behavior; i.e., they have the same Young's modulus, and the same Poisson's ratio. The symmetry may not be true when the strain is large. We may need to conduct a compressive test to assess the compressive behavior. (b) The shear modulus G is related to the Young's modulus and the Poisson's ratio by Eq. 1.2-8(2) (page 27). Again, this assumption may not be true when the strain is large. We may need to conduct a shear test to assess the shearing behavior. (c) We also assume that the bulk modulus B is related to the Young's modulus and the Poisson's ratio by

$$B = \frac{E}{3(1 - 2v)} \tag{1}$$

Again, this assumption may not be true when the strain is large. We may need to conduct a volumetric test to assess the volumetric behavior. However, in many cases, when the bulk modulus is almost infinitely large (i.e., the material is incompressible), we usually assume incompressibility without conducting a volumetric test.

Further, when the strain is large, all the moduli (tensile, compressive, shear, and bulk) are no longer constant; they change along stress-strain curves. Nonlinear elasticity with large strain is also called *hyperelasticity*.

In summary, to describe hyperelasticity behavior, we may need following test data: (a) a set of uniaxial tensile test data, (b) a set of uniaxial compressive test data, (c) a set of shear test data, and (d) a set of volumetric test data if the material is compressible.

Often, a set of test data can be obtained by superposing two sets of other test data. For example, the set of uniaxial compressive test data can be obtained by adding a set of hydrostatic compressive test data to a set of equibiaxial tensile test data [1-3]. Reasons of doing this are: (a) Biaxial tensile test may be easier to conduct than compressive test in some testing devices; (b) For incompressible materials, hydrostatic compressive test data are trivial: all strains have zero values.

An example of test data for hyperelasticity is shown below [4-6] (next page), that will be used in Section 14.3.

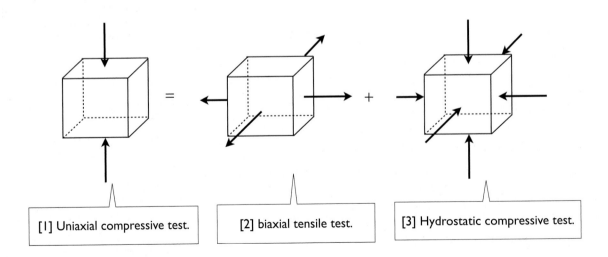

[1] Uniaxial compressive test.　　[2] biaxial tensile test.　　[3] Hydrostatic compressive test.

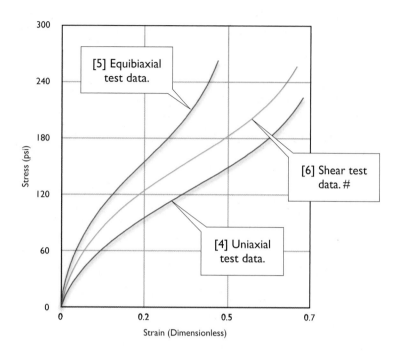

14.1-8 Strain Energy Function

Workbench provides a material model, called **Response Function** (14.1-9[1], page 526), to use experimental data (such as 14.1-7[4-6]) directly. A major disadvange of using **Response Function** model is that it may not be efficient enough; i.e., it may cost too many iterations, even if convergence is evantually achieved.

A better idea is described as follows.

As mentioned in 14.1-2 (page 519), hyperelasticity is characterized by the fact that the stressing curve and the unstressing curve are coincident (14.1-2[1], page 519). In other words, during the stressing and unstressing, the energy is conserved, or, equivalently to say, the stressing and unstressing are path independent. The stress state depends only on the strain state, and vice versa. They are independent of the stressing/unstressing history. This implies that there exists a *potential energy function* that depends on the state of the stress or strain. It reminds us the strain energy density function, which does depend only on the state of stress or strain. With these in mind, we propose a mathematical form for the strain energy

$$W = W(\varepsilon_{ij}) \tag{1}$$

And the stress can be calculated from the strain energy using

$$\sigma_{ij} = \frac{\partial W}{\partial \varepsilon_{ij}} \tag{2}$$

The strain state ε_{ij} consists of 6 strain components (Eq. 1.2-4(4), page 24). To further simply the strain energy function and develop a coordinate-independent expression, we may replace the 6 strain components (which are coordinate-dependent) with 3 strain invariants (which are coordinate-independent). Before going further, we need more background. Let's introduce some terms in solid mechanics.

Principal Stretch Ratios

The *stretch ratio* λ is defined as the ratio between fiber lengths after and before deformation,

$$\lambda = \frac{L}{L_0} = 1 + \varepsilon \tag{3}$$

When the direction is along a principal direction, the stretch ratio is called a *principal stretch ratio*. There are 3 principal stretch ratios, denoted by $\lambda_1, \lambda_2,$ and λ_3, which provide a measure of the deformation.

The *volumetric ratio* J can be defined as the volume after and before deformation,

$$J = \frac{V}{V_0} = \lambda_1 \lambda_2 \lambda_3 \tag{4}$$

Note that, if the material is incompressible, $J = 1$.

The *deviatoric principal stretch ratios* are defined as

$$\begin{aligned}
\bar{\lambda}_1 &= \lambda_1 / \sqrt[3]{J} \\
\bar{\lambda}_2 &= \lambda_2 / \sqrt[3]{J} \\
\bar{\lambda}_3 &= \lambda_3 / \sqrt[3]{J}
\end{aligned} \tag{5}$$

Strain Invariants

Let $I_1, I_2,$ and I_3 be the characteristic values (eigenvalues) of the strain state; they are also called *strain invarants*. It can be proved that

$$\begin{aligned}
I_1 &= \lambda_1^2 + \lambda_2^2 + \lambda_3^2 \\
I_2 &= \lambda_1^2 \lambda_2^2 + \lambda_2^2 \lambda_3^2 + \lambda_3^2 \lambda_1^2 \\
I_3 &= \lambda_1^2 \lambda_2^2 \lambda_3^2
\end{aligned} \tag{6}$$

The *deviatoric strain invariants* are defined as

$$\begin{aligned}
\bar{I}_1 &= \lambda_1 / \sqrt[3]{J^2} \\
\bar{I}_2 &= \lambda_2 / \sqrt[3]{J^2} \\
\bar{I}_3 &= \lambda_3 / \sqrt[3]{J^2}
\end{aligned} \tag{7}$$

Strain Energy Functions

We can replace the 6 strain components in Eq. (1) with either strain invariants or principal stretch ratios; i.e.,

$$W = W(I_1, I_2, I_3)$$

or

$$W = W(\lambda_1, \lambda_2, \lambda_3)$$

Or, we can split the strain energy into deviatoric part and volumetric part, and write

$$W = W_d(\bar{I}_1, \bar{I}_2) + W_b(J) \tag{8}$$

or

$$W = W_d(\bar{\lambda}_1, \bar{\lambda}_2, \bar{\lambda}_3) + W_b(J) \tag{9}$$

Note that $I_3 = J^2$, so $\bar{I}_3$ is not used in the definition of W.

14.1-9 Workbench Hyperelasticity Models

Workbench provides many hyperelasticity models [1]; they are based on either Eq. 14.1-8(8) or Eq. 14.1-8(9), last page.

Polynomial Form

The polynomial form is based on Eq. 14.1-8(8), last page,

$$W = \sum_{i+j=1}^{N} c_{ij}(\bar{I}_1 - 3)^i(\bar{I}_2 - 3)^j + \sum_{k=1}^{N} \frac{1}{d_k}(J-1)^{2k} \tag{1}$$

For example, **Polynomial 1st Order** ($N = 1$) is

$$W = c_{10}(\bar{I}_1 - 3) + c_{01}(\bar{I}_2 - 3)^1 + \frac{1}{d_1}(J-1)^2 \tag{2}$$

It has three parameters, c_{10}, c_{01}, and d_1. Note that, for incompressible materials, $J = 1$, $d_1 = 0$, and the last term is dropped.

Ogden Form

The Ogden form is based on Eq. 14.1-8(9), last page,

$$W = \sum_{i=1}^{N} \frac{\mu_i}{\alpha_i}(\bar{\lambda}_1^{\alpha_i} + \bar{\lambda}_2^{\alpha_i} + \bar{\lambda}_3^{\alpha_i} - 3) + \sum_{i=1}^{N} \frac{1}{d_i}(J-1)^{2i} \tag{3}$$

For example, **Ogden 1st Order** ($N = 1$) is

$$W = \frac{\mu_1}{\alpha_1}(\bar{\lambda}_1^{\alpha_1} + \bar{\lambda}_2^{\alpha_1} + \bar{\lambda}_3^{\alpha_1} - 3) + \frac{1}{d_1}(J-1)^2 \tag{4}$$

It has three parameters, μ_1, α_1, and d_1.

Mooney-Rivlin, Yeoh, and Neo-Hookean

These are reduced forms of the generalized polynomial.

Which Form to Use?

The choice depends on type of material, maximum strain, and the test data available. In general, the best form of strain energy density function is the one that produces the closest curve fit of the test data.

Section 14.3 provides an example to demonstrate some details.

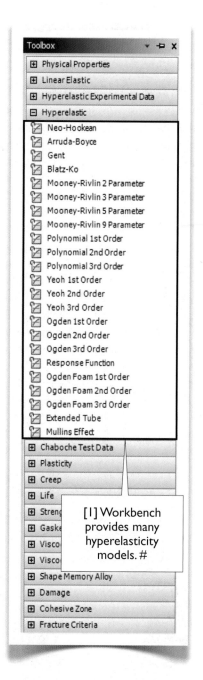

[1] Workbench provides many hyperelasticity models. #

Section 14.2

Belleville Washer

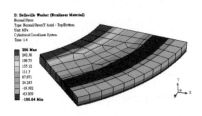

14.2-1 About the Belleville Washer

A Belleville washer [1] is also called a *Belleville spring*. The dimensions of the Belleville washer discussed in this section is shown below [2]. The washer is made of steel with a Young's modulus of 200 GPa, a Poisson's ratio of 0.3, and an initial yield stress of 250 MPa. Beyond the initial yield stress, it displays plasticity behavior as shown [3].

In this section, we will compress the Belleville spring by 1.0 mm and then release the displacement completely. As a preliminary study, before performing a simulation with plasticity, we will assume a linear material to see if the maximum stress exceeds the yield stress. If so, then we will explore the plasticity behavior of the Belleville spring. Specifically, we will examine the residual stress after the spring is completely released. A force-displacement curve will be plotted.

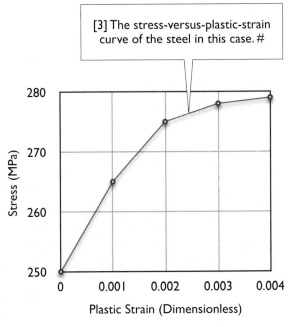

[1] A Belleville washer[Ref 1].

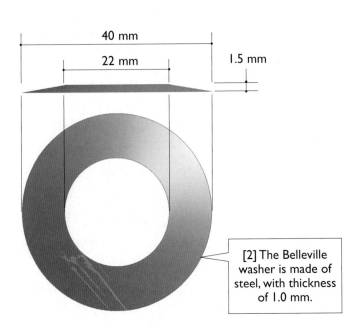

[2] The Belleville washer is made of steel, with thickness of 1.0 mm.

40 mm

22 mm

1.5 mm

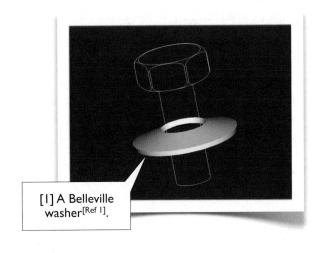

[3] The stress-versus-plastic-strain curve of the steel in this case. #

Stress (MPa)

Plastic Strain (Dimensionless)

14.2-2 Start Up

Launch Workbench. Create a **Static Structural** analysis system. Save the project as **Belleville**.

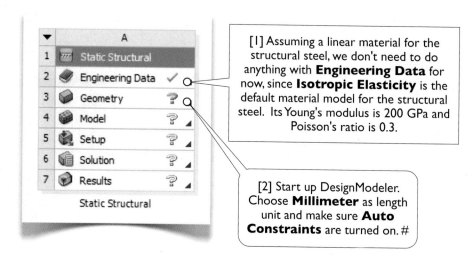

[1] Assuming a linear material for the structural steel, we don't need to do anything with **Engineering Data** for now, since **Isotropic Elasticity** is the default material model for the structural steel. Its Young's modulus is 200 GPa and Poisson's ratio is 0.3.

[2] Start up DesignModeler. Choose **Millimeter** as length unit and make sure **Auto Constraints** are turned on. #

14.2-3 Create Geometry

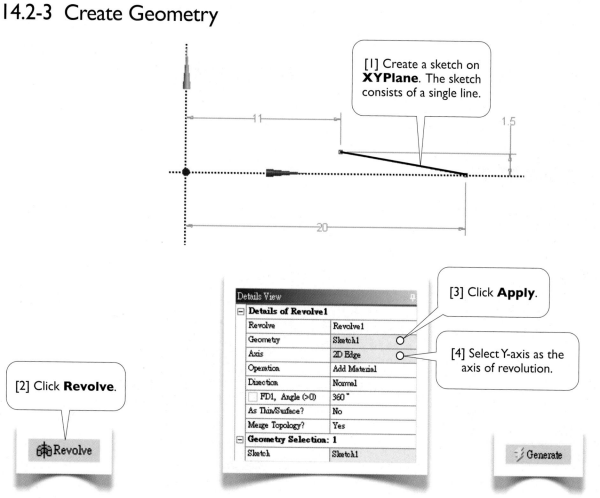

[1] Create a sketch on **XYPlane**. The sketch consists of a single line.

[3] Click **Apply**.

[4] Select Y-axis as the axis of revolution.

[2] Click **Revolve**.

528

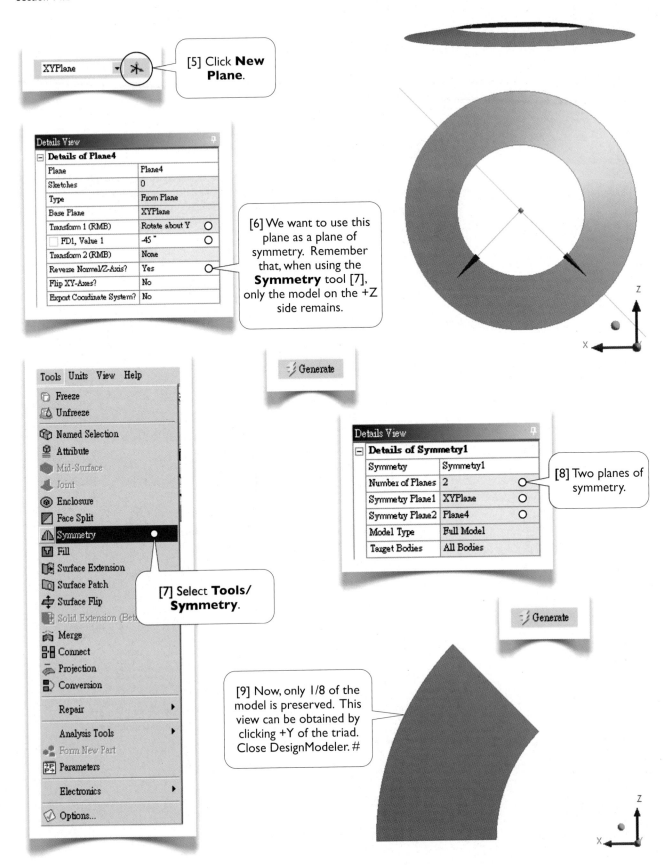

[5] Click **New Plane**.

Details View

Details of Plane4

Plane	Plane4
Sketches	0
Type	From Plane
Base Plane	XYPlane
Transform 1 (RMB)	Rotate about Y
FD1, Value 1	-45 °
Transform 2 (RMB)	None
Reverse Normal/Z-Axis?	Yes
Flip XY-Axes?	No
Export Coordinate System?	No

[6] We want to use this plane as a plane of symmetry. Remember that, when using the **Symmetry** tool [7], only the model on the +Z side remains.

Generate

Tools Units View Help

Freeze
Unfreeze
Named Selection
Attribute
Mid-Surface
Joint
Enclosure
Face Split
Symmetry
Fill
Surface Extension
Surface Patch
Surface Flip
Solid Extension (Bet...
Merge
Connect
Projection
Conversion

Repair ▶

Analysis Tools ▶

Form New Part
Parameters

Electronics ▶

Options...

[7] Select **Tools/ Symmetry**.

Details View

Details of Symmetry1

Symmetry	Symmetry1
Number of Planes	2
Symmetry Plane1	XYPlane
Symmetry Plane2	Plane4
Model Type	Full Model
Target Bodies	All Bodies

[8] Two planes of symmetry.

Generate

[9] Now, only 1/8 of the model is preserved. This view can be obtained by clicking +Y of the triad. Close DesignModeler. #

In this case, any plane derived from rotating **XYPlane** about Y-axis will be a plane of symmetry. There are infinitely many planes of symmetry for this case. In fact, an axisymmetric geometry has infinitely number of planes of symmetry. We can choose any two of them. In this case, we decide to use 1/8 of the model. In practice, we often choose a 45° sector of the model. The other possibility is to model it as an axisymmetric problem using 2D option. This leaves an exercise for you (14.4-2, page 559).

14.2-4 Set Up for Simulation with Linear Material

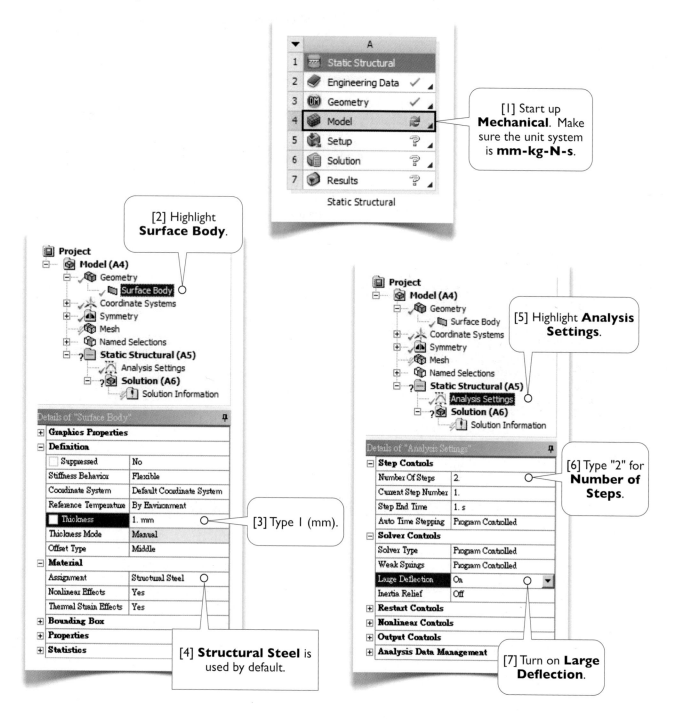

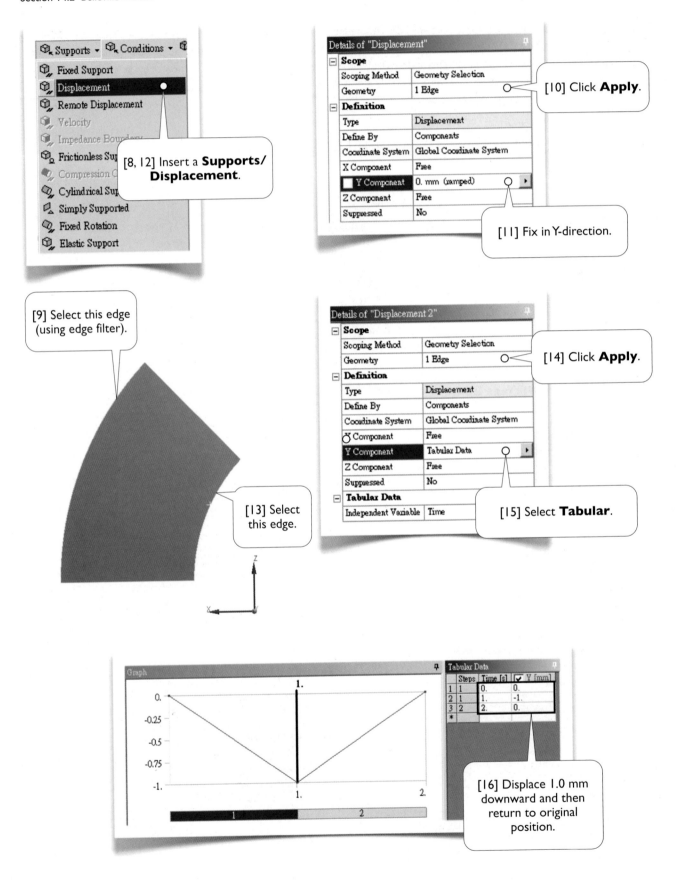

[8, 12] Insert a **Supports/Displacement**.

[10] Click **Apply**.

[11] Fix in Y-direction.

[9] Select this edge (using edge filter).

[14] Click **Apply**.

[13] Select this edge.

[15] Select **Tabular**.

[16] Displace 1.0 mm downward and then return to original position.

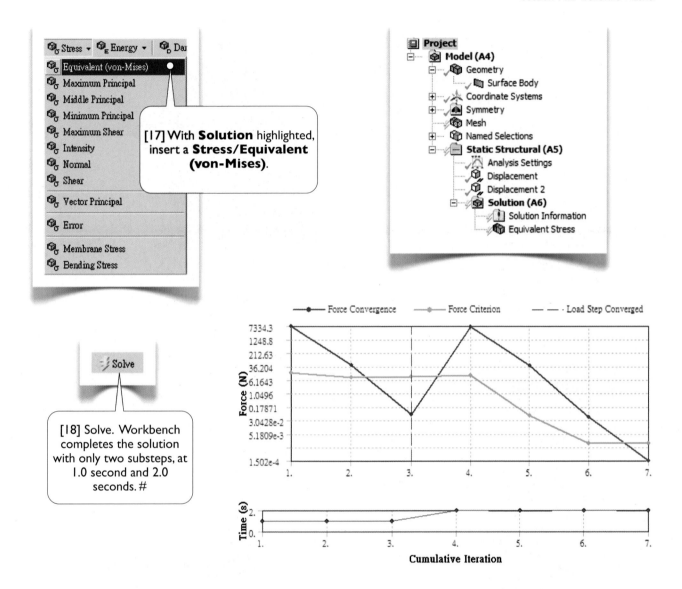

14.2-5 Results of the Linear Material Simulation

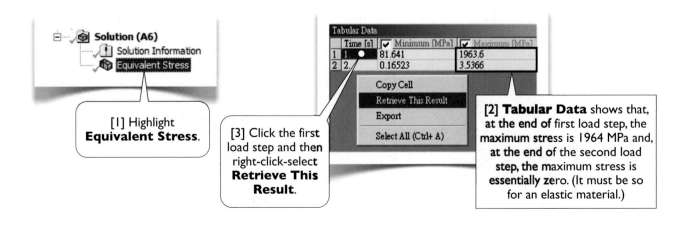

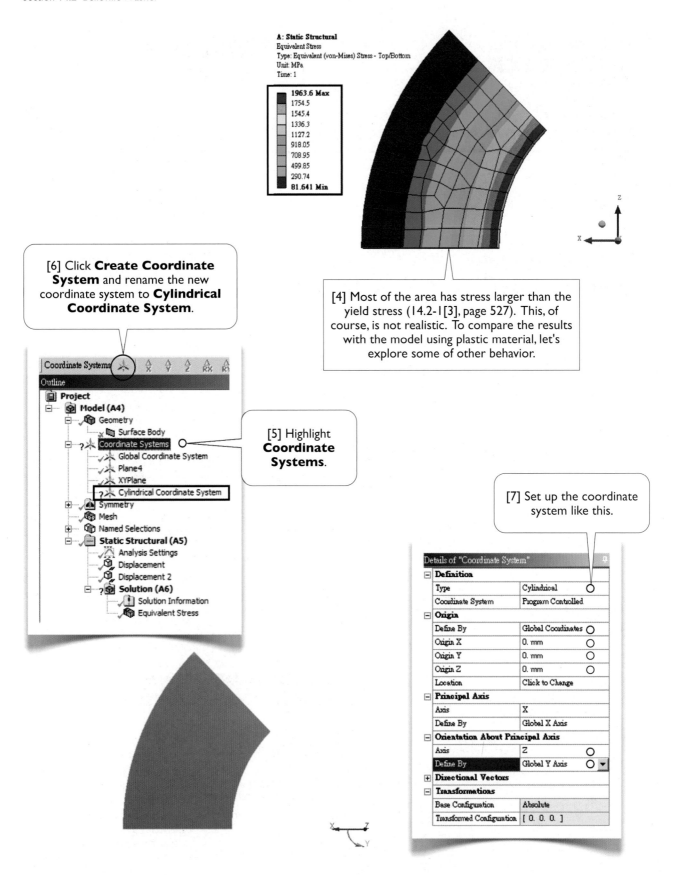

A: Static Structural
Equivalent Stress
Type: Equivalent (von-Mises) Stress - Top/Bottom
Unit: MPa
Time: 1

1963.6 Max
1754.5
1545.4
1336.3
1127.2
918.05
708.95
499.85
290.74
81.641 Min

[6] Click **Create Coordinate System** and rename the new coordinate system to **Cylindrical Coordinate System**.

[4] Most of the area has stress larger than the yield stress (14.2-1[3], page 527). This, of course, is not realistic. To compare the results with the model using plastic material, let's explore some of other behavior.

Coordinate Systems

Outline

Project
Model (A4)
  Geometry
    Surface Body
  Coordinate Systems
    Global Coordinate System
    Plane4
    XYPlane
    Cylindrical Coordinate System
  Symmetry
  Mesh
  Named Selections
  Static Structural (A5)
    Analysis Settings
    Displacement
    Displacement 2
    Solution (A6)
      Solution Information
      Equivalent Stress

[5] Highlight **Coordinate Systems**.

[7] Set up the coordinate system like this.

Details of "Coordinate System"	
Definition	
Type	Cylindrical
Coordinate System	Program Controlled
Origin	
Define By	Global Coordinates
Origin X	0. mm
Origin Y	0. mm
Origin Z	0. mm
Location	Click to Change
Principal Axis	
Axis	X
Define By	Global X Axis
Orientation About Principal Axis	
Axis	Z
Define By	Global Y Axis
Directional Vectors	
Transformations	
Base Configuration	Absolute
Transformed Configuration	[0. 0. 0.]

533

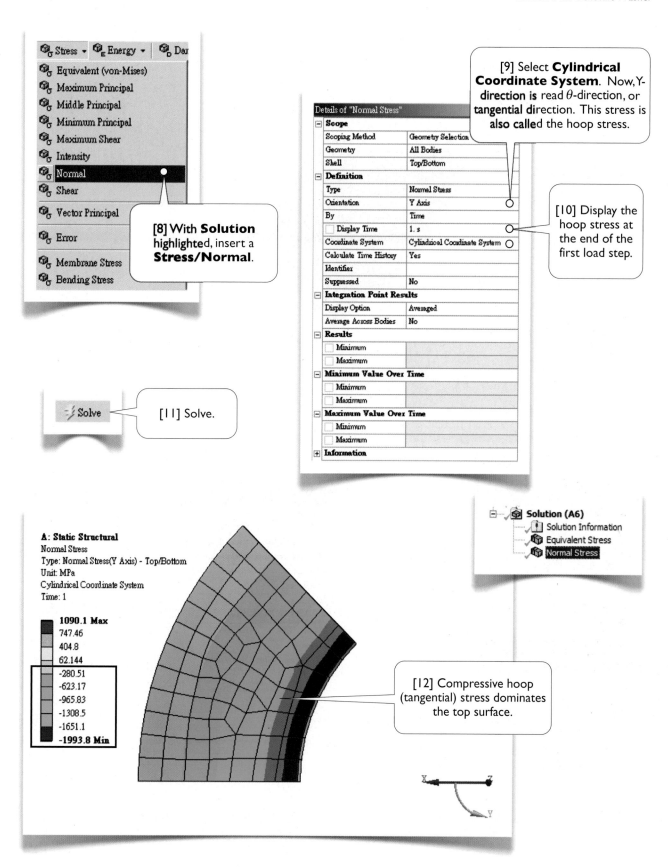

Stress ▾ Energy ▾ Dar

 Equivalent (von-Mises)
 Maximum Principal
 Middle Principal
 Minimum Principal
 Maximum Shear
 Intensity
 Normal
 Shear
 Vector Principal
 Error
 Membrane Stress
 Bending Stress

[8] With **Solution** highlighted, insert a **Stress/Normal**.

[9] Select **Cylindrical Coordinate System**. Now, Y-direction is read θ-direction, or tangential direction. This stress is also called the hoop stress.

Details of "Normal Stress"

Scope	
Scoping Method	Geometry Selection
Geometry	All Bodies
Shell	Top/Bottom
Definition	
Type	Normal Stress
Orientation	Y Axis
By	Time
Display Time	1. s
Coordinate System	Cylindrical Coordinate System
Calculate Time History	Yes
Identifier	
Suppressed	No
Integration Point Results	
Display Option	Averaged
Average Across Bodies	No
Results	
Minimum	
Maximum	
Minimum Value Over Time	
Minimum	
Maximum	
Maximum Value Over Time	
Minimum	
Maximum	
Information	

[10] Display the hoop stress at the end of the first load step.

Solve

[11] Solve.

Solution (A6)
 Solution Information
 Equivalent Stress
 Normal Stress

A: Static Structural
Normal Stress
Type: Normal Stress(Y Axis) - Top/Bottom
Unit: MPa
Cylindrical Coordinate System
Time: 1

1090.1 Max
747.46
404.8
62.144
-280.51
-623.17
-965.83
-1308.5
-1651.1
-1993.8 Min

[12] Compressive hoop (tangential) stress dominates the top surface.

X Z
Y

534

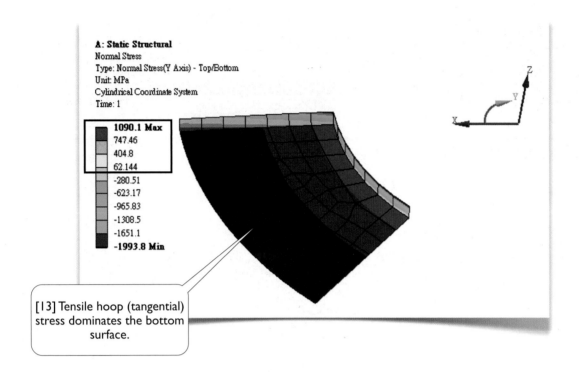

A: Static Structural
Normal Stress
Type: Normal Stress(Y Axis) - Top/Bottom
Unit: MPa
Cylindrical Coordinate System
Time: 1

| 1090.1 Max |
| 747.46 |
| 404.8 |
| 62.144 |
| -280.51 |
| -623.17 |
| -965.83 |
| -1308.5 |
| -1651.1 |
| -1993.8 Min |

[13] Tensile hoop (tangential) stress dominates the bottom surface.

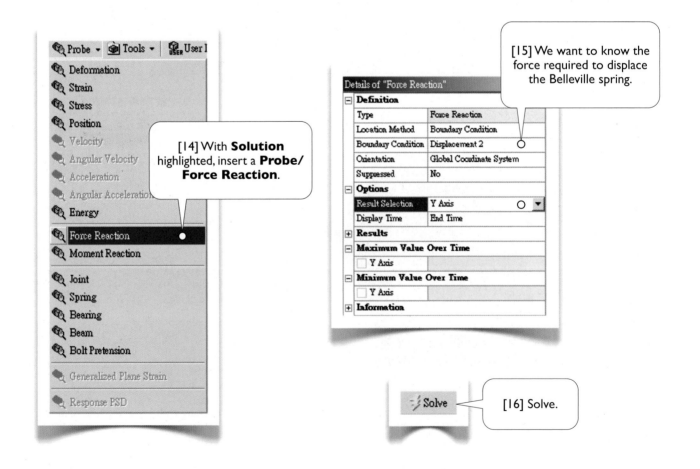

Probe ▾ Tools ▾ USER User

- Deformation
- Strain
- Stress
- Position
- Velocity
- Angular Velocity
- Acceleration
- Angular Acceleration
- Energy
- Force Reaction
- Moment Reaction
- Joint
- Spring
- Bearing
- Beam
- Bolt Pretension
- Generalized Plane Strain
- Response PSD

[14] With **Solution** highlighted, insert a **Probe/Force Reaction**.

[15] We want to know the force required to displace the Belleville spring.

Details of "Force Reaction"

Definition	
Type	Force Reaction
Location Method	Boundary Condition
Boundary Condition	Displacement 2
Orientation	Global Coordinate System
Suppressed	No
Options	
Result Selection	Y Axis
Display Time	End Time
Results	
Maximum Value Over Time	
☐ Y Axis	
Minimum Value Over Time	
☐ Y Axis	
Information	

Solve

[16] Solve.

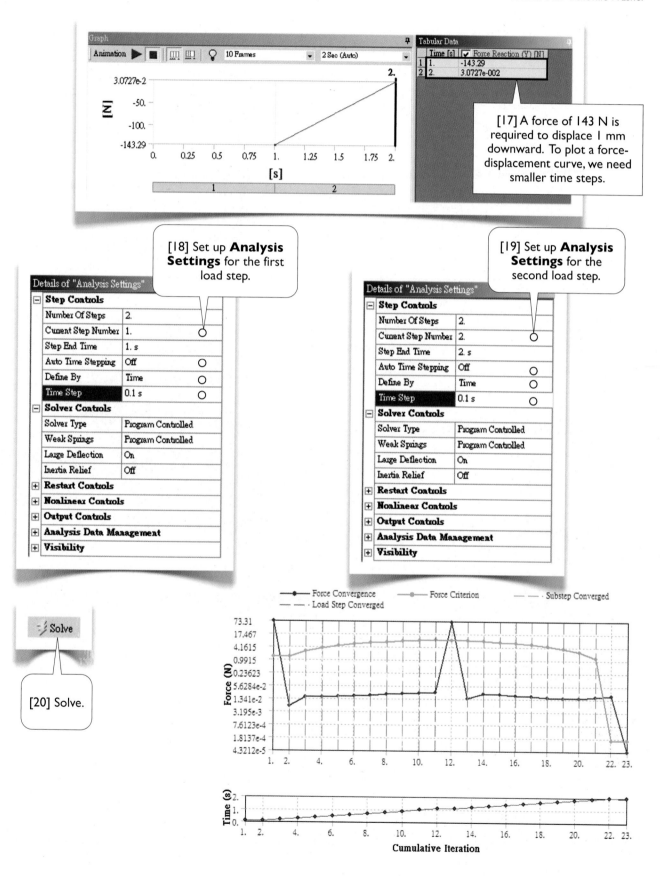

[17] A force of 143 N is required to displace 1 mm downward. To plot a force-displacement curve, we need smaller time steps.

[18] Set up **Analysis Settings** for the first load step.

Details of "Analysis Settings"

Step Controls	
Number Of Steps	2.
Current Step Number	1.
Step End Time	1. s
Auto Time Stepping	Off
Define By	Time
Time Step	0.1 s
Solver Controls	
Solver Type	Program Controlled
Weak Springs	Program Controlled
Large Deflection	On
Inertia Relief	Off
Restart Controls	
Nonlinear Controls	
Output Controls	
Analysis Data Management	
Visibility	

[19] Set up **Analysis Settings** for the second load step.

Details of "Analysis Settings"

Step Controls	
Number Of Steps	2.
Current Step Number	2.
Step End Time	2. s
Auto Time Stepping	Off
Define By	Time
Time Step	0.1 s
Solver Controls	
Solver Type	Program Controlled
Weak Springs	Program Controlled
Large Deflection	On
Inertia Relief	Off
Restart Controls	
Nonlinear Controls	
Output Controls	
Analysis Data Management	
Visibility	

Solve

[20] Solve.

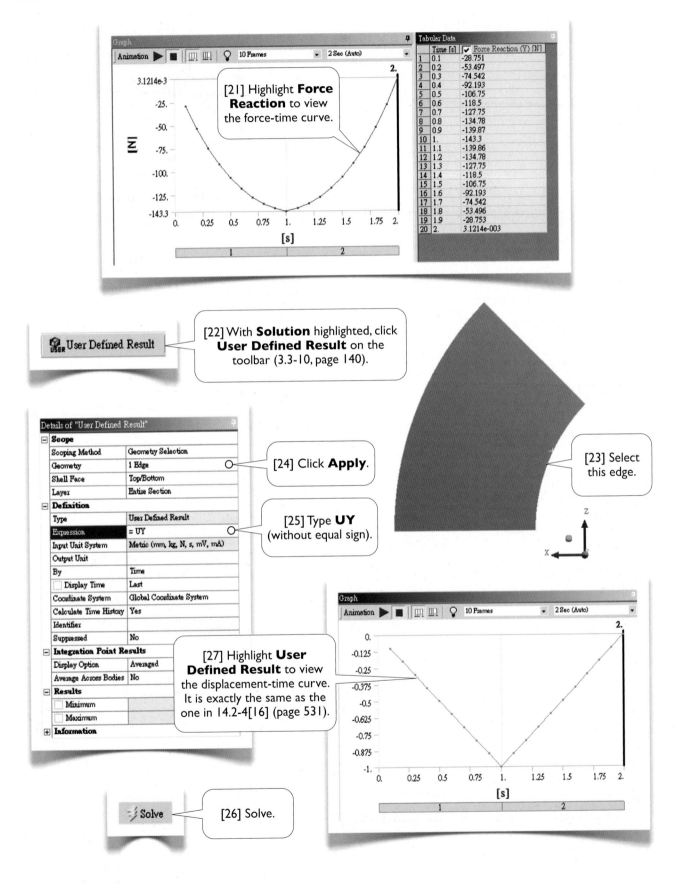

[21] Highlight **Force Reaction** to view the force-time curve.

	Time [s]	✓ Force Reaction (Y) [N]
1	0.1	-28.751
2	0.2	-53.497
3	0.3	-74.542
4	0.4	-92.193
5	0.5	-106.75
6	0.6	-118.5
7	0.7	-127.75
8	0.8	-134.78
9	0.9	-139.87
10	1.	-143.3
11	1.1	-139.86
12	1.2	-134.78
13	1.3	-127.75
14	1.4	-118.5
15	1.5	-106.75
16	1.6	-92.193
17	1.7	-74.542
18	1.8	-53.496
19	1.9	-28.753
20	2.	3.1214e-003

[22] With **Solution** highlighted, click **User Defined Result** on the toolbar (3.3-10, page 140).

[23] Select this edge.

[24] Click **Apply**.

[25] Type **UY** (without equal sign).

Details of "User Defined Result"

Scope	
Scoping Method	Geometry Selection
Geometry	1 Edge
Shell Face	Top/Bottom
Layer	Entire Section
Definition	
Type	User Defined Result
Expression	= UY
Input Unit System	Metric (mm, kg, N, s, mV, mA)
Output Unit	
By	Time
☐ Display Time	Last
Coordinate System	Global Coordinate System
Calculate Time History	Yes
Identifier	
Suppressed	No
Integration Point Results	
Display Option	Averaged
Average Across Bodies	No
Results	
☐ Minimum	
☐ Maximum	
Information	

[27] Highlight **User Defined Result** to view the displacement-time curve. It is exactly the same as the one in 14.2-4[16] (page 531).

[26] Solve.

Create a Force-Displacement Curve

In step [21], we have a force-time curve while, in step [27], we have a displacement-time curve. Now, we want to combine these information to create a force-displacement curve.

[28] Click **New Chart and Table** on the toolbar.

[29] Control-select **Force Reaction** and **User Defined Result** in the project tree and click **Apply**.

[30] A force-displacement curve.

Details of "Chart"		
Definition		
Outline Selection	2 Objects	O
Chart Controls		
X Axis	User Defined Result (Max)	O
Plot Style	Both	
Scale	Linear	
Gridlines	None	
Axis Labels		
X-Axis	Displacement	O
Y-Axis	Force	O
Report		
Content	Chart And Tabular Data	
Caption		
Input Quantities		
Time	Omit	O
Output Quantities		
[A] Force Reaction (Y)	Display	
[B] User Defined Result (Min)	Omit	O
User Defined Result (Max)	X Axis	

[31] This is a re-plot of the force-displacement curve in a more convenient way. The emphasis here is that the loading curve is the same as unloading curve; this is always true for elastic materials, but may not be true for plastic materials.
Close **Mechanical**. #

Close **Mechanical**.

14.2-6 Set Up for Simulation with Plastic Material Model

[1] Right-click here and select **Duplicate**.

[4] Double-click **Engineering Data**.

[5] Select **tonne-mm-s** as project units and select **Display Values in Project Units**.

[2] Rename it like this.

[3] Rename it like this.

[6] Double-click **Multilinear Isotropic Hardening** to include it to the **Structural Steel**.

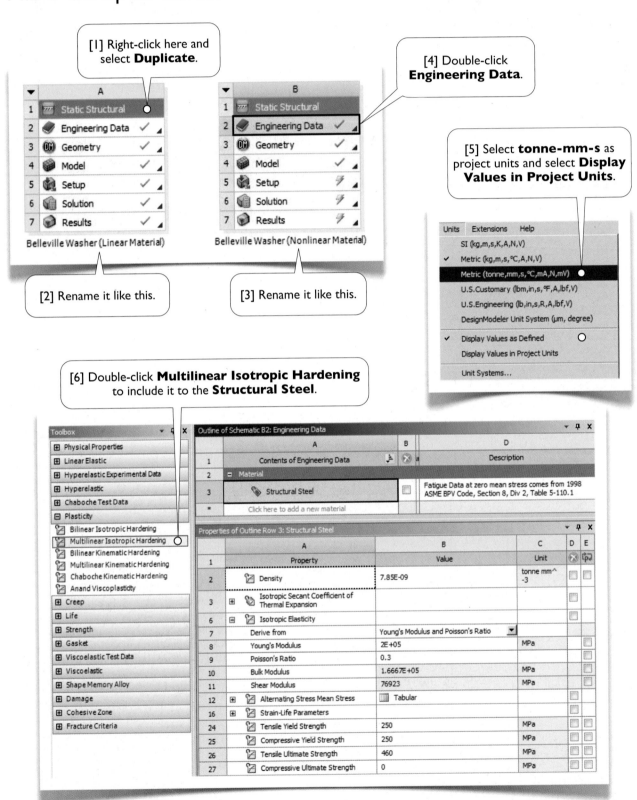

Belleville Washer (Linear Material)

Belleville Washer (Nonlinear Material)

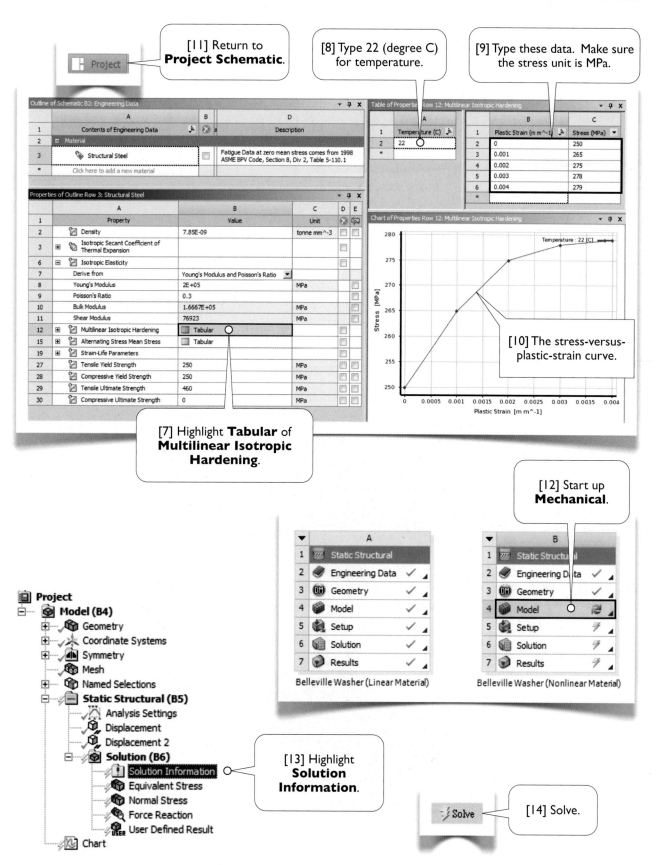

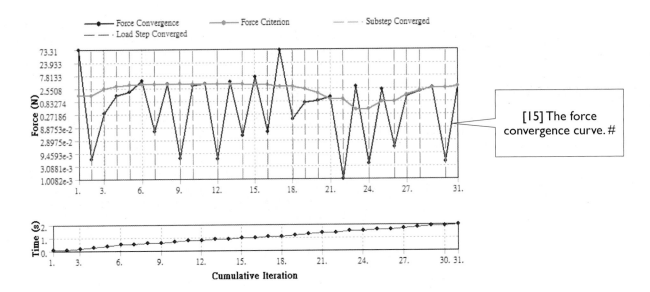

14.2-7 Results of the Nonlinear Material Simulation

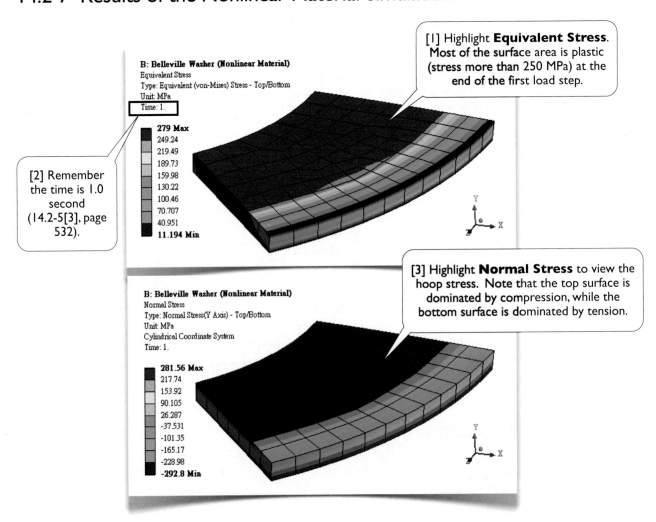

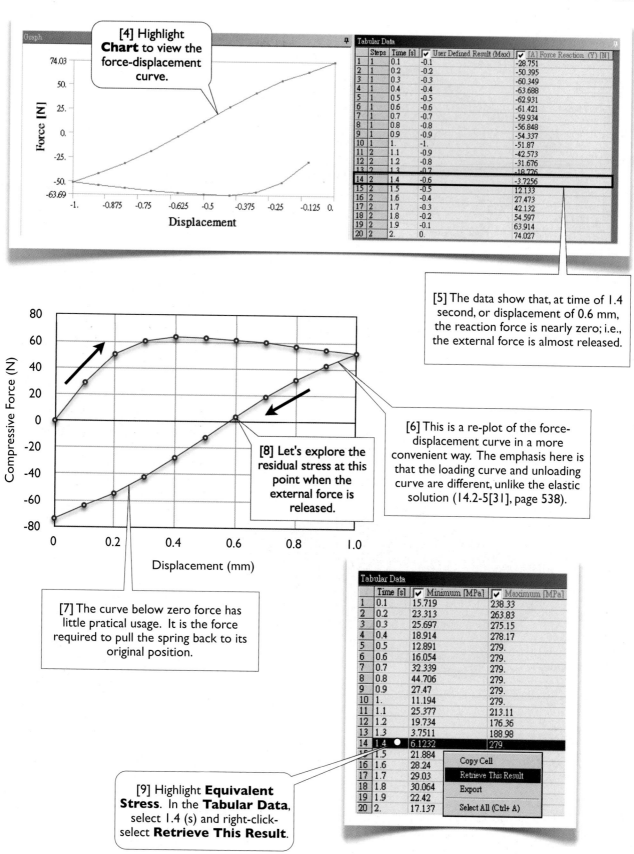

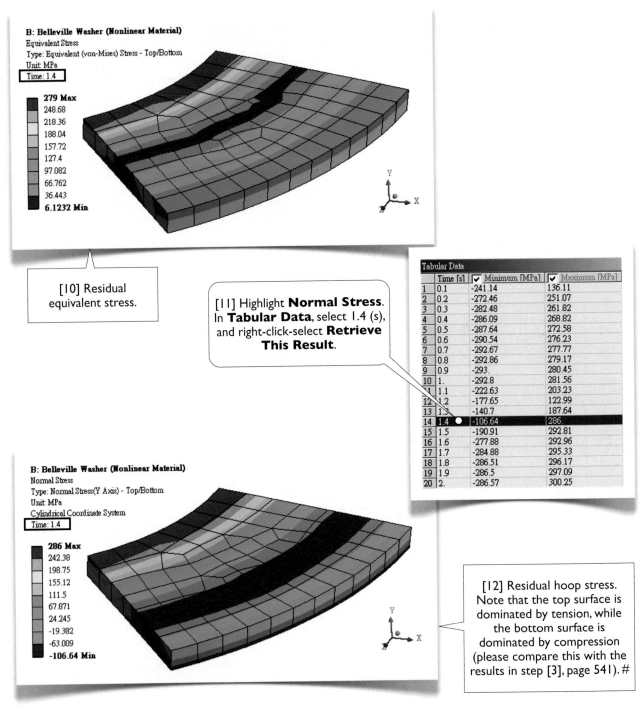

B: Belleville Washer (Nonlinear Material)
Equivalent Stress
Type: Equivalent (von-Mises) Stress - Top/Bottom
Unit: MPa
Time: 1.4

279 Max
248.68
218.36
188.04
157.72
127.4
97.082
66.762
36.443
6.1232 Min

[10] Residual equivalent stress.

[11] Highlight **Normal Stress**. In **Tabular Data**, select 1.4 (s), and right-click-select **Retrieve This Result**.

Tabular Data

	Time [s]	Minimum [MPa]	Maximum [MPa]
1	0.1	-241.14	136.11
2	0.2	-272.46	251.07
3	0.3	-282.48	261.82
4	0.4	-286.09	268.82
5	0.5	-287.64	272.58
6	0.6	-290.54	276.23
7	0.7	-292.67	277.77
8	0.8	-292.86	279.17
9	0.9	-293.	280.45
10	1.	-292.8	281.56
11	1.1	-222.63	203.23
12	1.2	-177.65	122.99
13	1.3	-140.7	187.64
14	1.4	-106.64	286.
15	1.5	-190.91	292.81
16	1.6	-277.88	292.96
17	1.7	-284.88	295.33
18	1.8	-286.51	296.17
19	1.9	-286.5	297.09
20	2.	-286.57	300.25

B: Belleville Washer (Nonlinear Material)
Normal Stress
Type: Normal Stress(Y Axis) - Top/Bottom
Unit: MPa
Cylindrical Coordinate System
Time: 1.4

286 Max
242.38
198.75
155.12
111.5
67.871
24.245
-19.382
-63.009
-106.64 Min

[12] Residual hoop stress. Note that the top surface is dominated by tension, while the bottom surface is dominated by compression (please compare this with the results in step [3], page 541). #

Wrap Up

Save the project and exit Workbench.

Reference

1. Wikipedia>Belleville Washer.

Section 14.3

Planar Seal

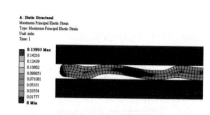

A: Static Structural
Maximum Principal Elastic Strain
Type: Maximum Principal Elastic Strain
Unit: in/in
Time: 1

0.15993 Max
0.14216
0.12439
0.10662
0.088851
0.071081
0.05331
0.03554
0.01777
0 Min

14.3-1 About the Planar Seal

The seal [1-4] is used in the door of a refrigerator. The seal is a long strip of rubber, and we will model it as a plane strain problem.   A series of material tests has been conducted, including a uniaxial tensile test, a biaxial tensile test, and a shear test [5-7].

The strain range of the original test data covers much more extent than the data shown here [5-7]. However, a preliminary study of the problem shows that the maximum strain does not exceed 0.3. Therefore, we decided to use the portion of data up to a strain of 0.3. The data fitting will be better if we use only the relevant data. A series of trials of data fitting shows that, for these material testing data, the two-parameter Mooney-Rivlin hyperelastic model fits the data better than other models. We decide to use two-parameter Mooney-Rivlin model.

The unit system used in this section is **in-lbm-lbf-s**.

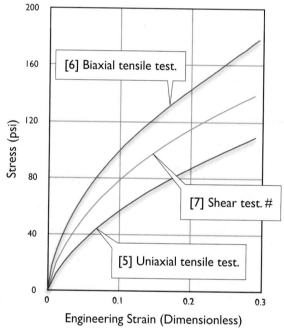

[6] Biaxial tensile test.

[7] Shear test. #

[5] Uniaxial tensile test.

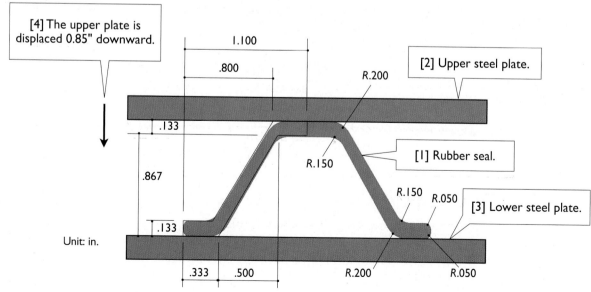

[4] The upper plate is displaced 0.85" downward.

[2] Upper steel plate.

[1] Rubber seal.

[3] Lower steel plate.

1.100

.800

R.200

.133

R.150

.867

R.150   R.050

.133

R.200   R.050

.333   .500

Unit: in.

544

14.3-2 Prepare Material Properties for the Rubber

Launch Workbench. Create a **Static Structural** analysis system. Save the project as **Seal**.

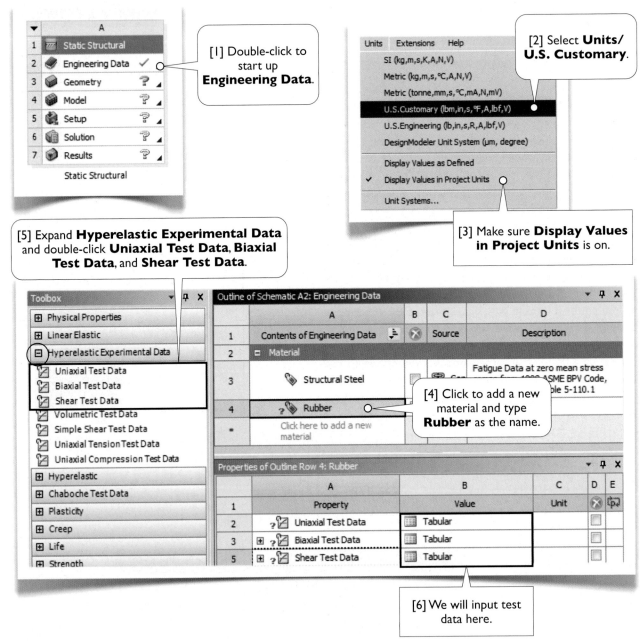

[1] Double-click to start up **Engineering Data**.

[2] Select **Units/U.S. Customary**.

[3] Make sure **Display Values in Project Units** is on.

[5] Expand **Hyperelastic Experimental Data** and double-click **Uniaxial Test Data**, **Biaxial Test Data**, and **Shear Test Data**.

[4] Click to add a new material and type **Rubber** as the name.

[6] We will input test data here.

Type or Copy Test Data?

In the next page, we will show you how to input the test data [7-15]. We purposely simplify the data so that you should be able to type the data manually in a few minutes. However, if you really hate to type these data manually, you may copy them from an Excel file, **Testdata.xls**, which can be download from the companion webpage or can be found in the companion DVD (see Preface, page 4).

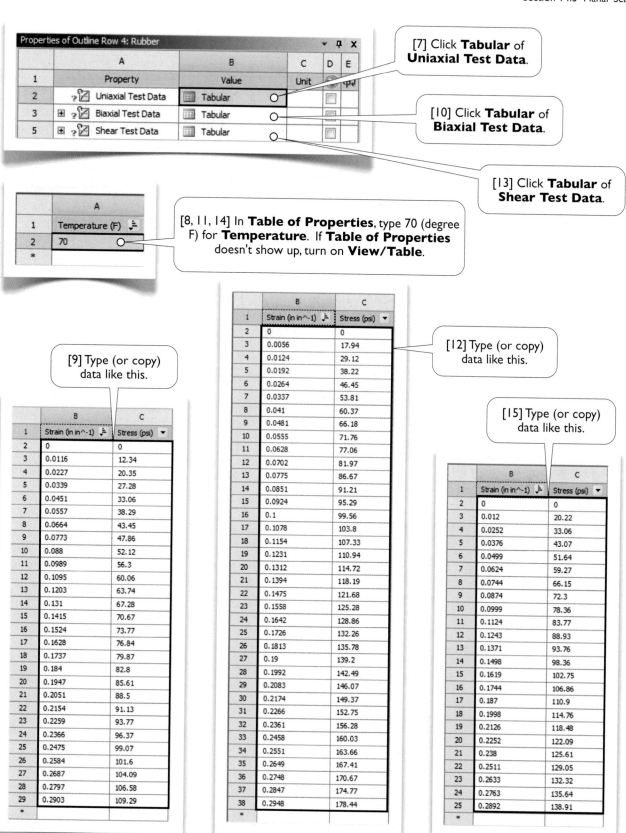

Properties of Outline Row 4: Rubber

	A	B	C	D	E
1	Property	Value	Unit		
2	? Uniaxial Test Data	Tabular ○			
3	⊞ ? Biaxial Test Data	Tabular ○			
5	⊞ ? Shear Test Data	Tabular ○			

[7] Click **Tabular** of **Uniaxial Test Data**.

[10] Click **Tabular** of **Biaxial Test Data**.

[13] Click **Tabular** of **Shear Test Data**.

	A
1	Temperature (F)
2	70 ○
*	

[8, 11, 14] In **Table of Properties**, type 70 (degree F) for **Temperature**. If **Table of Properties** doesn't show up, turn on **View/Table**.

[9] Type (or copy) data like this.

	B	C
1	Strain (in in^-1)	Stress (psi)
2	0	0
3	0.0116	12.34
4	0.0227	20.35
5	0.0339	27.28
6	0.0451	33.06
7	0.0557	38.29
8	0.0664	43.45
9	0.0773	47.86
10	0.088	52.12
11	0.0989	56.3
12	0.1095	60.06
13	0.1203	63.74
14	0.131	67.28
15	0.1415	70.67
16	0.1524	73.77
17	0.1628	76.84
18	0.1737	79.87
19	0.184	82.8
20	0.1947	85.61
21	0.2051	88.5
22	0.2154	91.13
23	0.2259	93.77
24	0.2366	96.37
25	0.2475	99.07
26	0.2584	101.6
27	0.2687	104.09
28	0.2797	106.58
29	0.2903	109.29
*		

[12] Type (or copy) data like this.

	B	C
1	Strain (in in^-1)	Stress (psi)
2	0	0
3	0.0056	17.94
4	0.0124	29.12
5	0.0192	38.22
6	0.0264	46.45
7	0.0337	53.81
8	0.041	60.37
9	0.0481	66.18
10	0.0555	71.76
11	0.0628	77.06
12	0.0702	81.97
13	0.0775	86.67
14	0.0851	91.21
15	0.0924	95.29
16	0.1	99.56
17	0.1078	103.8
18	0.1154	107.33
19	0.1231	110.94
20	0.1312	114.72
21	0.1394	118.19
22	0.1475	121.68
23	0.1558	125.28
24	0.1642	128.86
25	0.1726	132.26
26	0.1813	135.78
27	0.19	139.2
28	0.1992	142.49
29	0.2083	146.07
30	0.2174	149.37
31	0.2266	152.75
32	0.2361	156.28
33	0.2458	160.03
34	0.2551	163.66
35	0.2649	167.41
36	0.2748	170.67
37	0.2847	174.77
38	0.2948	178.44
*		

[15] Type (or copy) data like this.

	B	C
1	Strain (in in^-1)	Stress (psi)
2	0	0
3	0.012	20.22
4	0.0252	33.06
5	0.0376	43.07
6	0.0499	51.64
7	0.0624	59.27
8	0.0744	66.15
9	0.0874	72.3
10	0.0999	78.36
11	0.1124	83.77
12	0.1243	88.93
13	0.1371	93.76
14	0.1498	98.36
15	0.1619	102.75
16	0.1744	106.86
17	0.187	110.9
18	0.1998	114.76
19	0.2126	118.48
20	0.2252	122.09
21	0.238	125.61
22	0.2511	129.05
23	0.2633	132.32
24	0.2763	135.64
25	0.2892	138.91
*		

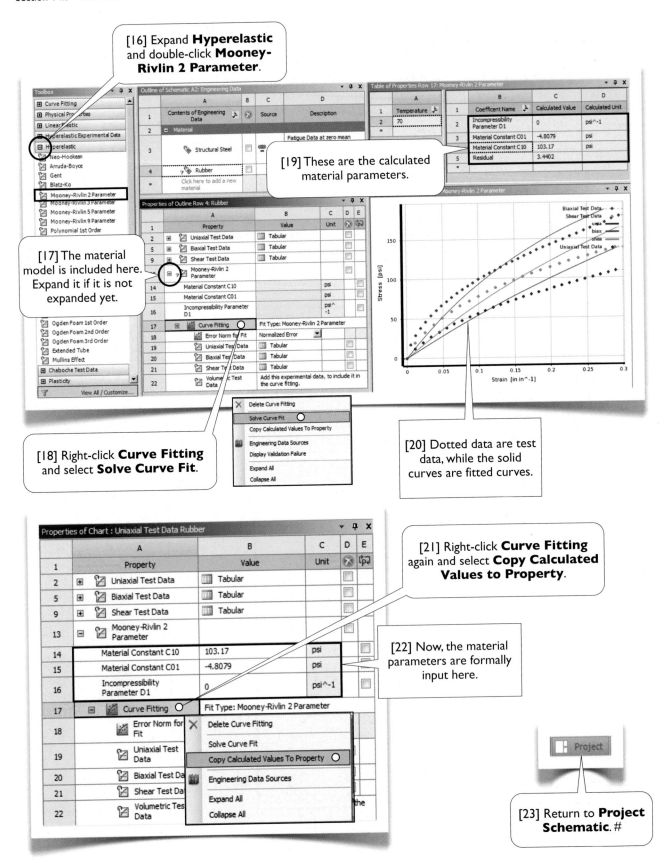

[16] Expand **Hyperelastic** and double-click **Mooney-Rivlin 2 Parameter**.

[17] The material model is included here. Expand it if it is not expanded yet.

[18] Right-click **Curve Fitting** and select **Solve Curve Fit**.

[19] These are the calculated material parameters.

[20] Dotted data are test data, while the solid curves are fitted curves.

[21] Right-click **Curve Fitting** again and select **Copy Calculated Values to Property**.

[22] Now, the material parameters are formally input here.

[23] Return to **Project Schematic**. #

14.3-3 Create 2D Geometry

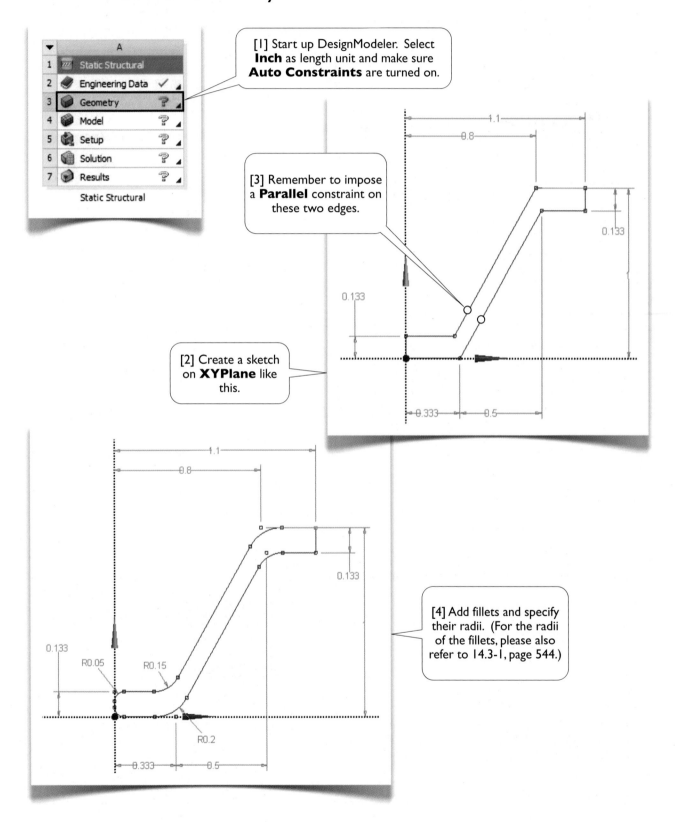

[1] Start up DesignModeler. Select **Inch** as length unit and make sure **Auto Constraints** are turned on.

[3] Remember to impose a **Parallel** constraint on these two edges.

[2] Create a sketch on **XYPlane** like this.

[4] Add fillets and specify their radii. (For the radii of the fillets, please also refer to 14.3-1, page 544.)

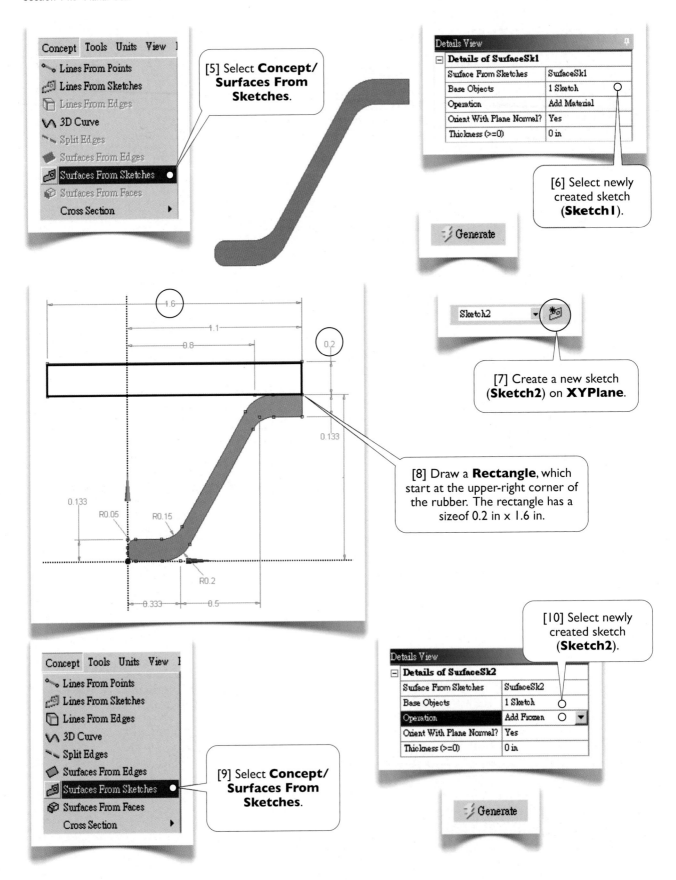

Concept Tools Units View
- Lines From Points
- Lines From Sketches
- Lines From Edges
- 3D Curve
- Split Edges
- Surfaces From Edges
- Surfaces From Sketches ●
- Surfaces From Faces
- Cross Section

[5] Select **Concept/ Surfaces From Sketches**.

Details View

□ **Details of SurfaceSk1**	
Surface From Sketches	SurfaceSk1
Base Objects	1 Sketch
Operation	Add Material
Orient With Plane Normal?	Yes
Thickness (>=0)	0 in

[6] Select newly created sketch (**Sketch1**).

↯ Generate

1.6

1.1

0.8

0.2

0.133

0.133

R0.05   R0.15

R0.2

0.333   0.5

Sketch2

[7] Create a new sketch (**Sketch2**) on **XYPlane**.

[8] Draw a **Rectangle**, which start at the upper-right corner of the rubber. The rectangle has a sizeof 0.2 in x 1.6 in.

[10] Select newly created sketch (**Sketch2**).

Details View

□ **Details of SurfaceSk2**	
Surface From Sketches	SurfaceSk2
Base Objects	1 Sketch
Operation	Add Frozen
Orient With Plane Normal?	Yes
Thickness (>=0)	0 in

↯ Generate

Concept Tools Units View
- Lines From Points
- Lines From Sketches
- Lines From Edges
- 3D Curve
- Split Edges
- Surfaces From Edges
- Surfaces From Sketches ●
- Surfaces From Faces
- Cross Section

[9] Select **Concept/ Surfaces From Sketches**.

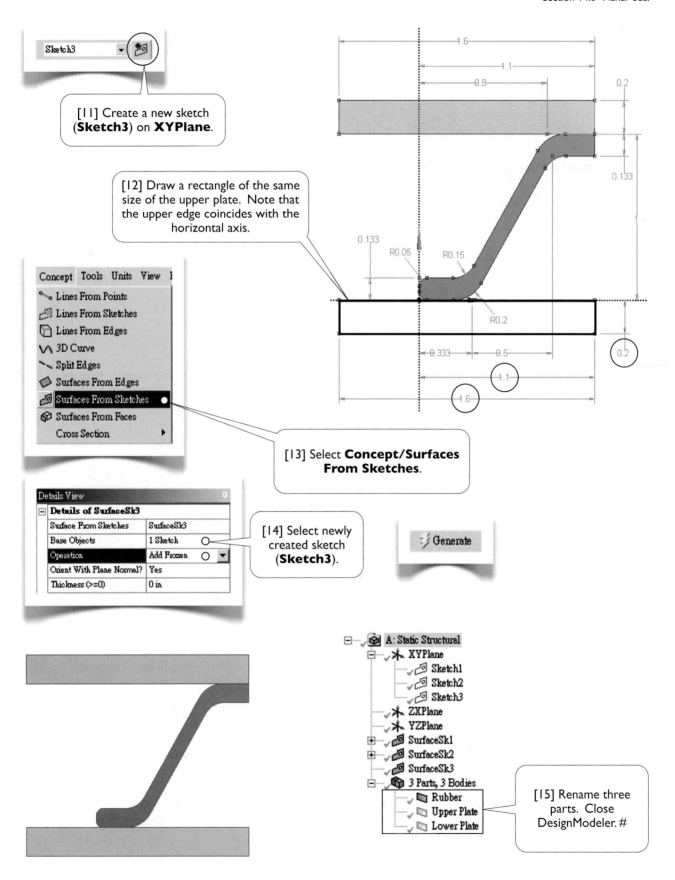

[11] Create a new sketch (**Sketch3**) on **XYPlane**.

[12] Draw a rectangle of the same size of the upper plate. Note that the upper edge coincides with the horizontal axis.

Concept Tools Units View I

- Lines From Points
- Lines From Sketches
- Lines From Edges
- 3D Curve
- Split Edges
- Surfaces From Edges
- Surfaces From Sketches ●
- Surfaces From Faces
- Cross Section ▶

[13] Select **Concept/Surfaces From Sketches**.

Details View

Details of SurfaceSk3	
Surface From Sketches	SurfaceSk3
Base Objects	1 Sketch
Operation	Add Frozen
Orient With Plane Normal?	Yes
Thickness (>=0)	0 in

[14] Select newly created sketch (**Sketch3**).

Generate

☐ A: Static Structural
 ☐ XYPlane
  Sketch1
  Sketch2
  Sketch3
 ZXPlane
 YZPlane
 ☐ SurfaceSk1
 ☐ SurfaceSk2
 SurfaceSk3
 ☐ 3 Parts, 3 Bodies
  Rubber
  Upper Plate
  Lower Plate

[15] Rename three parts. Close DesignModeler. #

550

14.3-4 Set Up for Simulation

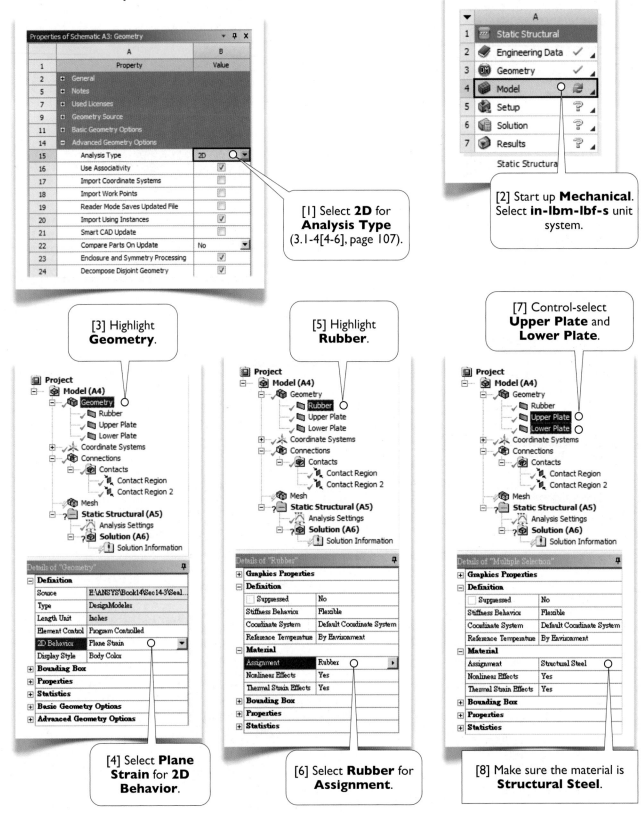

[1] Select **2D** for **Analysis Type** (3.1-4[4-6], page 107).

[2] Start up **Mechanical**. Select **in-lbm-lbf-s** unit system.

[3] Highlight **Geometry**.

[4] Select **Plane Strain** for **2D Behavior**.

[5] Highlight **Rubber**.

[6] Select **Rubber** for **Assignment**.

[7] Control-select **Upper Plate** and **Lower Plate**.

[8] Make sure the material is **Structural Steel**.

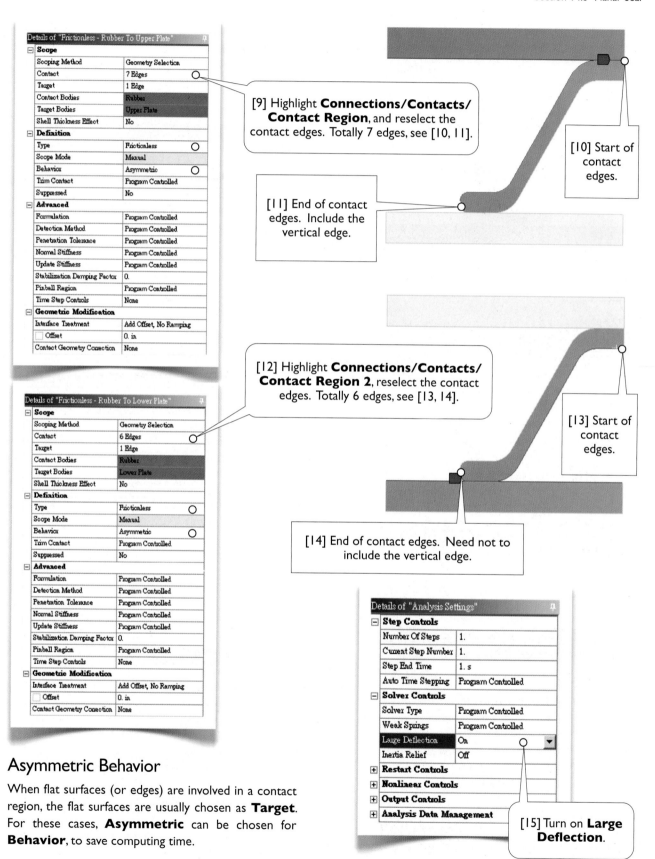

Details of "Frictionless - Rubber To Upper Plate"

Scope	
Scoping Method	Geometry Selection
Contact	7 Edges
Target	1 Edge
Contact Bodies	Rubber
Target Bodies	Upper Plate
Shell Thickness Effect	No
Definition	
Type	Frictionless
Scope Mode	Manual
Behavior	Asymmetric
Trim Contact	Program Controlled
Suppressed	No
Advanced	
Formulation	Program Controlled
Detection Method	Program Controlled
Penetration Tolerance	Program Controlled
Normal Stiffness	Program Controlled
Update Stiffness	Program Controlled
Stabilization Damping Factor	0.
Pinball Region	Program Controlled
Time Step Controls	None
Geometric Modification	
Interface Treatment	Add Offset, No Ramping
Offset	0. in
Contact Geometry Correction	None

[9] Highlight **Connections/Contacts/ Contact Region**, and reselect the contact edges. Totally 7 edges, see [10, 11].

[10] Start of contact edges.

[11] End of contact edges. Include the vertical edge.

[12] Highlight **Connections/Contacts/ Contact Region 2**, reselect the contact edges. Totally 6 edges, see [13, 14].

[13] Start of contact edges.

Details of "Frictionless - Rubber To Lower Plate"

Scope	
Scoping Method	Geometry Selection
Contact	6 Edges
Target	1 Edge
Contact Bodies	Rubber
Target Bodies	Lower Plate
Shell Thickness Effect	No
Definition	
Type	Frictionless
Scope Mode	Manual
Behavior	Asymmetric
Trim Contact	Program Controlled
Suppressed	No
Advanced	
Formulation	Program Controlled
Detection Method	Program Controlled
Penetration Tolerance	Program Controlled
Normal Stiffness	Program Controlled
Update Stiffness	Program Controlled
Stabilization Damping Factor	0.
Pinball Region	Program Controlled
Time Step Controls	None
Geometric Modification	
Interface Treatment	Add Offset, No Ramping
Offset	0. in
Contact Geometry Correction	None

[14] End of contact edges. Need not to include the vertical edge.

Details of "Analysis Settings"

Step Controls	
Number Of Steps	1.
Current Step Number	1.
Step End Time	1. s
Auto Time Stepping	Program Controlled
Solver Controls	
Solver Type	Program Controlled
Weak Springs	Program Controlled
Large Deflection	On
Inertia Relief	Off
Restart Controls	
Nonlinear Controls	
Output Controls	
Analysis Data Management	

[15] Turn on **Large Deflection**.

Asymmetric Behavior

When flat surfaces (or edges) are involved in a contact region, the flat surfaces are usually chosen as **Target**. For these cases, **Asymmetric** can be chosen for **Behavior**, to save computing time.

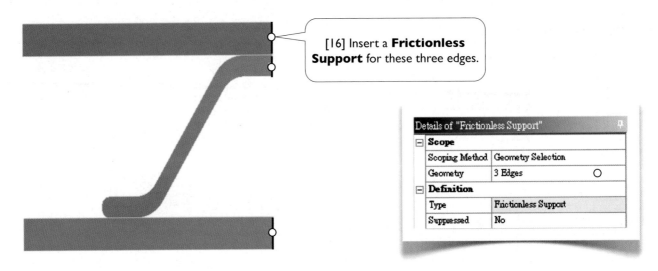

[16] Insert a **Frictionless Support** for these three edges.

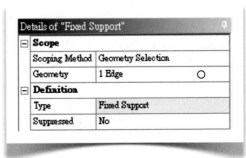

Details of "Frictionless Support"		
☐ **Scope**		
Scoping Method	Geometry Selection	
Geometry	3 Edges	○
☐ **Definition**		
Type	Frictionless Support	
Suppressed	No	

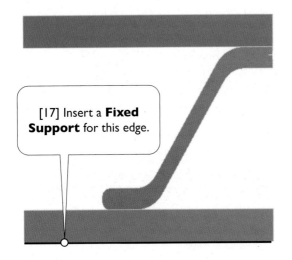

[17] Insert a **Fixed Support** for this edge.

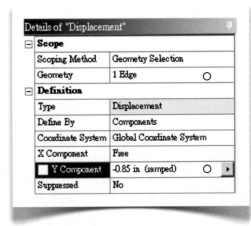

Details of "Fixed Support"		
☐ **Scope**		
Scoping Method	Geometry Selection	
Geometry	1 Edge	○
☐ **Definition**		
Type	Fixed Support	
Suppressed	No	

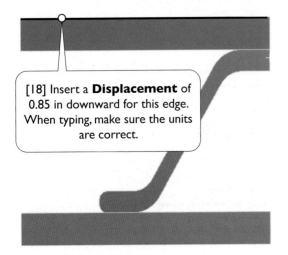

[18] Insert a **Displacement** of 0.85 in downward for this edge. When typing, make sure the units are correct.

Details of "Displacement"			
☐ **Scope**			
Scoping Method	Geometry Selection		
Geometry	1 Edge	○	
☐ **Definition**			
Type	Displacement		
Define By	Components		
Coordinate System	Global Coordinate System		
X Component	Free		
☐ Y Component	-0.85 in (ramped)	○	▶
Suppressed	No		

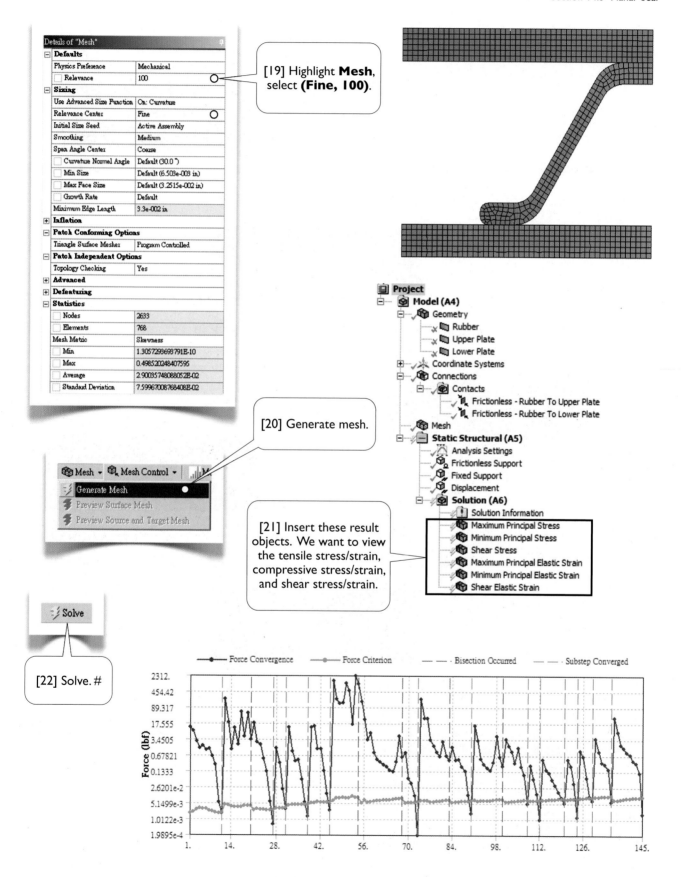

14.3-5 View the Results

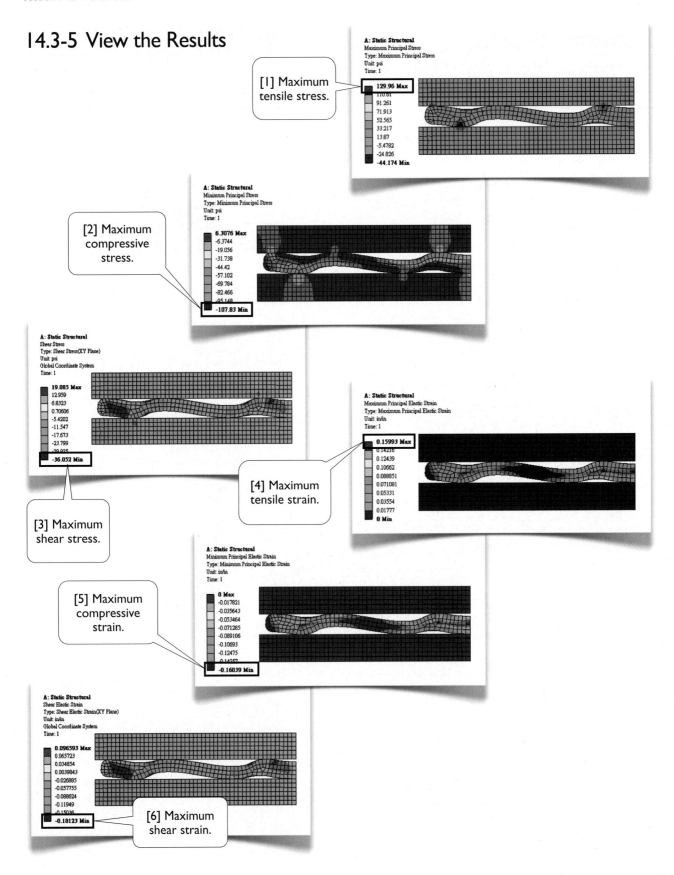

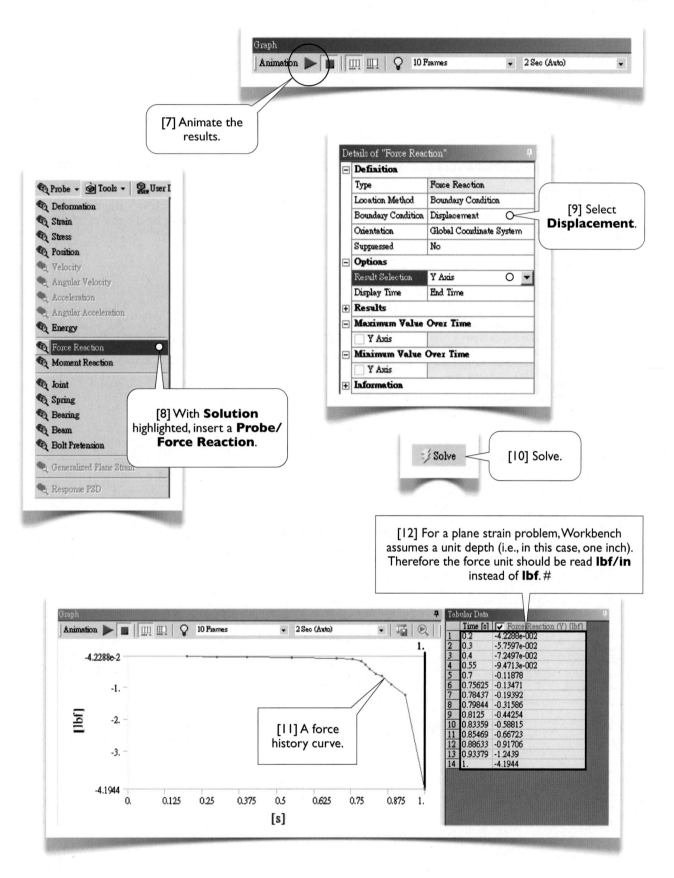

[7] Animate the results.

[9] Select **Displacement**.

[8] With **Solution** highlighted, insert a **Probe/Force Reaction**.

[10] Solve.

[12] For a plane strain problem, Workbench assumes a unit depth (i.e., in this case, one inch). Therefore the force unit should be read **lbf/in** instead of **lbf**. #

[11] A force history curve.

Details of "Force Reaction"

Definition	
Type	Force Reaction
Location Method	Boundary Condition
Boundary Condition	Displacement
Orientation	Global Coordinate System
Suppressed	No
Options	
Result Selection	Y Axis
Display Time	End Time
Results	
Maximum Value Over Time	
☐ Y Axis	
Minimum Value Over Time	
☐ Y Axis	
Information	

Probe ▾ Tools ▾ User I
- Deformation
- Strain
- Stress
- Position
- Velocity
- Angular Velocity
- Acceleration
- Angular Acceleration
- Energy
- Force Reaction
- Moment Reaction
- Joint
- Spring
- Bearing
- Beam
- Bolt Pretension
- Generalized Plane Strain
- Response PSD

Tabular Data

	Time [s]	☑ Force Reaction (Y) [lbf]
1	0.2	-4.2288e-002
2	0.3	-5.7597e-002
3	0.4	-7.2497e-002
4	0.55	-9.4713e-002
5	0.7	-0.11878
6	0.75625	-0.13471
7	0.78437	-0.19392
8	0.79844	-0.31586
9	0.8125	-0.44254
10	0.83359	-0.58815
11	0.85469	-0.66723
12	0.88633	-0.91706
13	0.93379	-1.2439
14	1.	-4.1944

14.3-6 Create a Force-versus-Displacement Chart

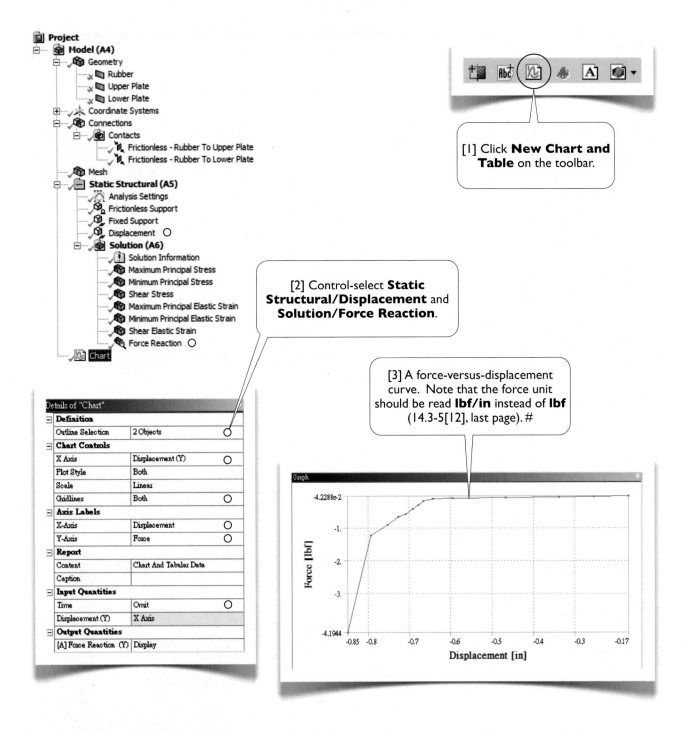

[1] Click **New Chart and Table** on the toolbar.

[2] Control-select **Static Structural/Displacement** and **Solution/Force Reaction**.

[3] A force-versus-displacement curve. Note that the force unit should be read **lbf/in** instead of **lbf** (14.3-5[12], last page). #

Wrap Up

Save the project and exit Workbench.

Section 14.4

Review

14.4-1 Keywords

Choose a letter for each keyword from the list of descriptions

1. (   ) Elastic Material

2. (   ) Hardening Rules

3. (   ) Hyperelastic Materials

4. (   ) Kinematic Hardening

5. (   ) Initial Yield Point

6. (   ) Initial Yield Surface

7. (   ) Isometric Hardening

8. (   ) Linear Materials

9. (   ) Nonlinear Materials

10. (   ) Plastic Materials

11. (   ) Rate-Dependent Materials

12. (   ) Subsequent Yield Surface

13. (   ) Time-Dependent Materials

14. (   ) Von Mises Yield Surface

15. (   ) Yield Surface

Answers:

1. (B)   2. (M)   3. (G)   4. (N)   5. (H)   6. (K)   7. (O)   8. (A)   9. (E)   10. (F)
11. (D)   12. (L)   13. (C)   14. (J)   15. (I)

List of Descriptions

(A) A material that has a linear stress-strain relation and can be described by the Hooke's law. In this book, a linear material also implies elasticity, time-independent, and rate-independent.

(B) Materials of which the strain returns to its original state when the stress is removed.

(C) Also called viscous materials. If you apply a constant stress on the materials, the occurrence of strain is time-dependent. The strain typically increases with time and stabilizes to a constant value after a period of time. This behavior is called creeping. On the other hand, if you apply a constant strain on the materials, the occurrence of stress is also time-dependent. The stress typically decreases with time and stabilizes to a constant value after a period of time. This behavior is called stress-relaxation.

(D) Materials of which the stress-strain relation is such that the magnitude of stress depends on not only the magnitude of strain but also the rate of the strain, and vice versa.

(E) In this book, all materials that are not described by the Hooke's law are categorized as nonlinear materials.

(F) Materials that exhibit plasticity behavior, which is characterized by that the strain does not return to its original state after the stress is removed. Rather, residual strain, or plastic strain, remains.

(G) Materials that remain elastic under a very large strain.

(H) In plasticity, yield point may change. Stress at the elastic limit is called the initial yield point.

(I) In a uniaxial test, yielding occurs at certain stress values; they are called yield points. In 3D cases, yielding occurs at certain stress states; these stress states form a "surface" (or hyper-surface) in the multiaxial stress space. The surface is called a yield surface. During stressing, the yield surface may change in size as well as location.

(J) Different yield criterion results in different yield surface. If von Mises yield criterion is used, the yield surface is a cylindrical surface in the σ_1-σ_2-σ_3 space.

(K) In plasticity, yield surface may change in size as well as location during stressing. The original yield surface before any yielding occurs (i.e., the behavior is still elastic) is called the initial yield surface.

(L) In plasticity, during stressing, yield surface may change in size as well as location. All possible yield surfaces except the initial yield surface are called the subsequent yield surfaces.

(M) Rules that describe how yield surface changes its size and location.

(N) Kinematic hardening assumes that the yield surface changes only the location but not the size.

(O) Isometric hardening assumes that the yield surface changes only the size but not the location.

14.4-2 Additional Workbench Exercises

Model the Belleville Washer as a 2D Problem

In Section 14.2, we modeled the Belleville washer as a surface body and meshed it with shell elements. It is possible to model it as a 2D problem and mesh it with 2D axisymmetric solid elements. Do it and draw the pros and cons for both modeling methods.

Pneumatic Finger

Replace the material of the pneumatic finger, simulated in Section 9.1, with the rubber used in the Section 14.3. Redo the simulation for the pneumatic finger.

Chapter 15

Explicit Dynamics

Many transient dynamic simulations require extremely small integration time steps, for examples, high-speed impact, drop test, or highly nonlinear problems. In such cases, the time steps may be as small as a few nanoseconds, and the use of **Transient Structural** analysis system becomes impractical, since the run time would be too enormous. The integration method used in **Transient Structural** analysis system is an *implicit method*.

Explicit Dynamics analysis system also deals with transient structural dynamics; however, it uses an *explicit integration method*. The *explicit method* is very efficient for each time step. It thus allows a large number of time steps to calculate within an acceptable time. A characteristics of explicit methods is that the integration time steps must be very small (e.g., microseconds or nanoseconds) in order to achieve stable solutions. If the dynamic behavior must be observed within a long duration (say, several seconds), then many millions of time steps are required to complete the simulation. In these situations, high-performance computing facilities are usually used to facilitate the computations.

Purpose of This Chapter

This chapter provides basics of explicit dynamics so that the students have enough background knowledge to perform some simple explicit dynamics simulations. A high-speed impact simulation and a drop test simulation are used to demonstrate the applications of **Explicit Dynamics**.

About Each Section

Section 15.1 provides the basics of explicit dynamics. Section 15.2 presents a high-speed impact example and Section 15.3 presents a drop test simulation. Both examples are simple yet instructional. We use default settings as much as possible, so that the students can appreciate the built-in capabilities of **Explicit Dynamics**.

Section 15.1

Basics of Explicit Dynamics

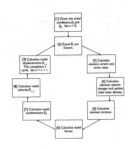

15.1-1 Implicit Integration Methods

As mentioned, transient dynamic simulations involve solving Eq. 12.1-4(1),

$$[M]\{\ddot{D}\}+[C]\{\dot{D}\}+[K]\{D\}=\{F\} \qquad \text{Copy of 12.1-4(1), page 418}$$

Consider a typical time step at t_n. Let $D_n, \dot{D}_n$, and $\ddot{D}_n$ be the displacement, velocity, and acceleration at t_n, and $D_{n+1}, \dot{D}_{n+1}$, and $\ddot{D}_{n+1}$ at t_{n+1}. Also, let $\Delta t = t_{n+1} - t_n$. We temporarily assume that the acceleration is linear over the time step (i.e., $\dddot{D}_n = \dddot{D}_{n+1} = 0$), then, by Taylor series expansions at t_n,

$$\dot{D}_{n+1} = \dot{D}_n + \Delta t \ddot{D}_n + \frac{\Delta t^2}{2}\dddot{D}_n \qquad (1)$$

$$D_{n+1} = D_n + \Delta t \dot{D}_n + \frac{\Delta t^2}{2}\ddot{D}_n + \frac{\Delta t^3}{6}\dddot{D}_n \qquad (2)$$

The quantity $\dddot{D}_n$ can be approximated by

$$\dddot{D}_n = \frac{\ddot{D}_{n+1} - \ddot{D}_n}{\Delta t} \qquad (3)$$

Substitution of Eq. (3) into Eqs. (1) and (2) respectively yields

$$\dot{D}_{n+1} = \dot{D}_n + \frac{\Delta t}{2}\left(\ddot{D}_{n+1} + \ddot{D}_n\right) \qquad (4)$$

$$D_{n+1} = D_n + \Delta t \dot{D}_n + \Delta t^2 \left(\frac{1}{6}\ddot{D}_{n+1} + \frac{1}{3}\ddot{D}_n\right) \qquad (5)$$

Eqs. (4) and (5) can be regarded as a special case of Newmark methods,

$$\dot{D}_{n+1} = \dot{D}_n + \Delta t\left[\gamma\ddot{D}_{n+1} + (1-\gamma)\ddot{D}_n\right] \qquad (6)$$

$$D_{n+1} = D_n + \Delta t \dot{D}_n + \frac{1}{2}\Delta t^2\left[2\beta\ddot{D}_{n+1} + (1-2\beta)\ddot{D}_n\right] \qquad (7)$$

If you substitute $\gamma = 1/2$ and $\beta = 1/6$ into Eqs. (6) and (7) respectively, you will come up with Eqs. (4) and (5).

Eqs. (6) and (7) are used in **Transient Structural** analysis system. The parameters γ and β are chosen to control characteristics of the algorithm such as accuracy, numerical stability, etc. It is called an *implicit method* because the calculation of $\dot{D}_{n+1}$ and D_{n+1} requires knowledge of $\ddot{D}_{n+1}$. That is, the response at the current time step depends on not only the historical information but also the current information; therefore, solving Eqs. (6) and (7) involves iterative process.

Calculation of the response at time t_{n+1} is depicted in [1-6]. In the beginning [1], the displacement D_n, velocity $\dot{D}_n$, and acceleration $\ddot{D}_n$ of the last step are already known (For $n = 0$, we may assume $\ddot{D}_0 = 0$). Since $\ddot{D}_{n+1}$ is needed in Eqs. (6) and (7), we use $\ddot{D}_n$ as an initial guess of $\ddot{D}_{n+1}$. Knowing D_n, $\dot{D}_n$, and $\ddot{D}_{n+1}$, the quantities $\dot{D}_{n+1}$ and D_{n+1} can be calculated according to Eqs. (6) and (7) [2]. The next step [3] is to substitute $\ddot{D}_{n+1}$, $\dot{D}_{n+1}$, and D_{n+1} into Eq. 12.1-4(1). If Eq. 12.1-4(1) is satisfied, then the calculation of the response at time t_{n+1} is complete [5], otherwise, $\ddot{D}_{n+1}$ is updated and another iteration is initiated [6]. Update of $\ddot{D}_{n+1}$ [6] is similar to the Newton-Raphson method described in 13.1-4, page 464.

With implicit methods, integration time step is typically about milliseconds; a typical simulation time is about 0.1 to 10 seconds, which requires hundreds to ten-thousands of integration time steps.

Implicit methods can be used for most of transient structural simulations. However, for highly nonlinear problems, it often fails due to convergence issues; for high-speed impact problems, the integration time is so small that the computing time becomes intolerable. In such cases, explicit methods are more applicable.

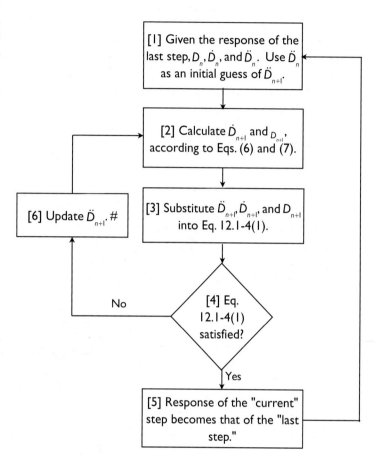

15.1-2 Explicit Integration Methods

The explicit method used in **Explicit Dynamics** analysis system is based on half-step central differences

$$\ddot{D}_n = \frac{\dot{D}_{n+\frac{1}{2}} - \dot{D}_{n-\frac{1}{2}}}{\Delta t}, \text{ or } \dot{D}_{n+\frac{1}{2}} = \dot{D}_{n-\frac{1}{2}} + \ddot{D}_n \Delta t \tag{1}$$

$$\dot{D}_{n+\frac{1}{2}} = \frac{D_{n+1} - D_n}{\Delta t}, \text{ or } D_{n+1} = D_n + \dot{D}_{n+\frac{1}{2}} \Delta t \tag{2}$$

Eqs. (1) and (2) are called *explicit methods* because the calculation of $\dot{D}_{n+\frac{1}{2}}$ and D_{n+1} requires knowledge of historical information only. That is, the response at the current time can be calculated explicitly; no iterations within a time step is needed. Therefore, it is very efficient to complete a time step, also called a *cycle*. One of the distinct characteristics of the explicit method is that its integration time step needs to be very small to achieve a stable solution.

The procedure used in **Explicit Dynamics** analysis system is illustrated in [1-9]. In the beginning of a cycle [1, 2], the displacement D_n and velocity $\dot{D}_n$ of the last cycle are already known. With these information, we can calculate the strain and strain rate for each element [3], using the relations such as Eqs. 1.3-2(2) (page 32) and 1.2-7(1) (page 27). The volume change for each element is then calculated, according to the equations of state, and the mass density is updated [4]. The volumetric information is needed for the calculation of stresses. With these information, the element stresses can be calculated [5] according to a constitutive model, relation between stresses and strains/strain rates, such as Eq. 1.2-8(1), page 27. The stresses are integrated over the elements, and the external loads are added to form the nodal forces F_n [6]. The nodal accelerations are then calculated [7] according to

$$\ddot{D}_n = \frac{F_n}{m} + \frac{b}{\rho} \tag{3}$$

where b is the body force (see Eq. 1.2-6(2), page 26), m is the nodal mass, and ρ is the mass density. The nodal velocities at $t_{n+\frac{1}{2}}$ are calculated [8] according to Eq. (1) and the nodal displacements at t_{n+1} are calculated [9] according to Eq. (2).

With explicit methods, a typical integration time step is about nanoseconds to microseconds; a typical simulation time is about 1 millisecond to 1 second, which will need many thousands or millions of cycles.

Explicit methods are useful for high-speed impact problems and highly nonlinear problems. For low-speed problems, where the durations are usually long, using explicit methods becomes impractical due to an enormous computing time, since it requires very small integration time steps.

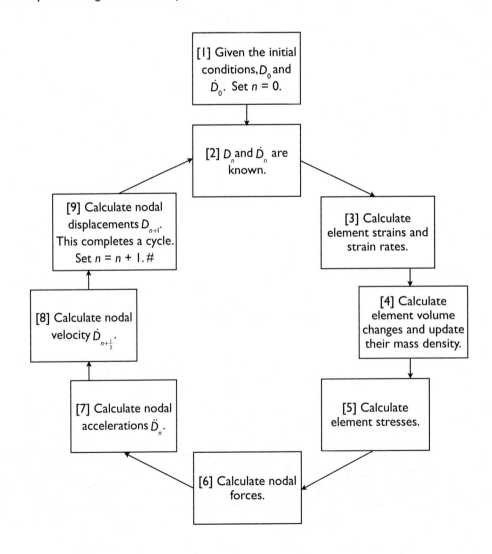

15.1-3 Solution Accuracy

In **Transient Structural**, in which an implicit method is used, convergence criteria are used to control the solution accuracy, similar to the Newton-Raphson method described in 13.1-4, page 464. Equilibrium iterations imply that force balance must be satisfied. In **Explicit Dynamics**, since no equilibrium iterations are involved, solution accuracy is not controlled with convergence criteria. Instead, it uses the *principle of conservation of energy* to monitor the solution accuracy. It calculates overall energy at each cycle. If the energy error, defined in the next paragraph, reaches a threshold, the solution is regarded as unstable and stops. The default threshold is 10% of a reference energy [1]. Energy statistics can be viewed by selecting **Energy Conservation** in **Solution Output** [2, 3].

At any time, *Current Energy* of the system can be calculated, including its kinetic energy and strain energy. The *principle of work and energy,* a form of the principle of conservation of energy, states

$$(\text{Reference Energy}) + (\text{Work Done})_{\text{Reference}\rightarrow\text{Current}} = (\text{Current Energy}) \tag{1}$$

Where the *Reference Energy* is the total energy of a reference time, default to the initial time, The *Energy Error* is defined by

$$\text{Energy Error} = \frac{\left|(\text{Current Energy})-(\text{Reference Energy}) - (\text{Work Done})_{\text{Reference}\rightarrow\text{Current}}\right|}{\max\left(\left|\text{Current Energy}\right|, \left|\text{Reference Energy}\right|, \left|\text{Kinetic Energy}\right|\right)} \tag{2}$$

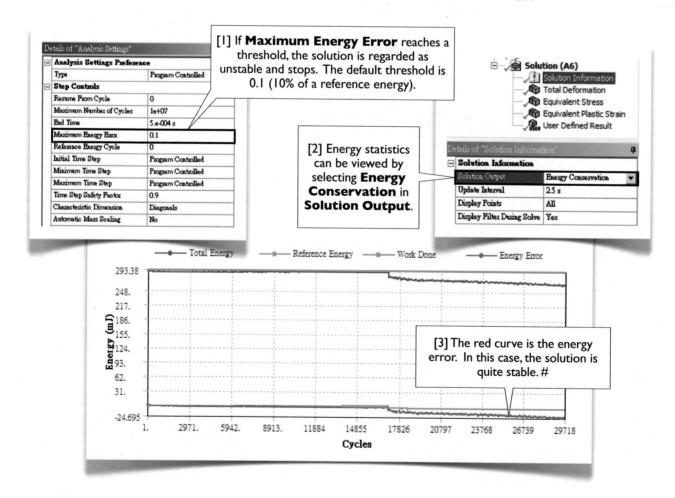

[1] If **Maximum Energy Error** reaches a threshold, the solution is regarded as unstable and stops. The default threshold is 0.1 (10% of a reference energy).

[2] Energy statistics can be viewed by selecting **Energy Conservation** in **Solution Output**.

[3] The red curve is the energy error. In this case, the solution is quite stable. #

564

15.1-4 Integration Time Steps

With explicit methods, the integration time step needs to be small enough to ensure stability and accuracy of the solution. How small should the time step be? The German mathematicians, Courant, Friedrichs, and Lewy[Refs 1, 2], suggested that, in a single time step Δt, a wave should not travel further than the smallest element size; i.e.,

$$\Delta t \leq \frac{h}{c} \tag{1}$$

where h is the smallest element size, c is the wave speed in the element. Eq. (1) is called the *CFL condition*. In **Explicit Dynamics**, a safety factor f is used to further ensure the solution stability [1, 2]; i.e.,

$$\Delta t \leq f \frac{h}{c} \tag{2}$$

Because of the CFL condition, when generating meshes for **Explicit Dynamics**, make sure that one or two very small elements do not control the time step. In general, a uniform mesh size is desirable in **Explicit Dynamics** simulations.

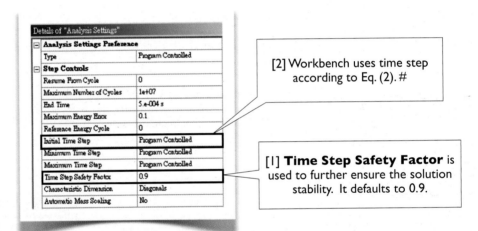

[2] Workbench uses time step according to Eq. (2). #

[1] **Time Step Safety Factor** is used to further ensure the solution stability. It defaults to 0.9.

15.1-5 Automatic Mass Scaling

The wave speed in a material is $c = \sqrt{E/\rho}$, where E is the Young's modulus and ρ is the mass density of the material. Further, $\rho = m/V$, where m is the mass and V is the volume of an element. Substitution of these into Eq. 15.1-4(2) yields

$$\Delta t \leq fh\sqrt{\frac{m}{VE}} \tag{1}$$

The idea of mass scaling is to artificially increase the mass of small elements, so that the stability time step can be increased. Mass scaling is applied only to those elements which have a calculated stability time step less than a specified value, default to 1e-20 s [1], which is to ensure that no mass scaling takes place. If a mesh contains very few small elements, this idea can be useful. Note that mass scaling changes the inertial properties of the model. Be careful to ensure that the model remains valid for the physical problem.

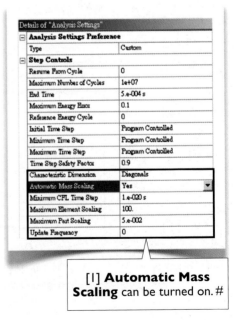

[1] **Automatic Mass Scaling** can be turned on. #

15.1-6 Static Damping

Explicit Dynamics is designed for solving transient dynamic problems. To solve a static problem, we may perform a transient dynamic analysis and find the steady-state solution. **Static Damping** option [1] is to facilitate the finding of the steady-state solution. The idea is to introduce a damping force, to critically damp the lowest mode of oscillation.

Value of **Static Damping** for critical damping of the lowest mode of vibration is

$$\frac{2f\,\Delta t}{1 + 2\pi f\,\Delta t} \tag{1}$$

where f is the lowest frequency of the system.

Using the critical damping may minimize simulation run time. If **Static Damping** is larger than the critical damping, the solution would take unnecessarily longer time (than critical damping) to reach a steady state. On the other hand, if **Static Damping** is smaller than the critical damping, the solution would oscillate unnecessarily many times to reach a steady state.

For some highly-nonlinear static problems that fail with **Static Structural** analysis system, you may want to try this idea.

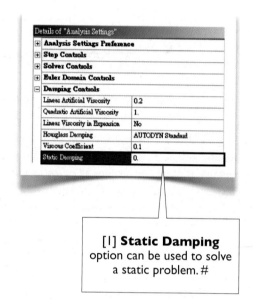

[1] **Static Damping** option can be used to solve a static problem. #

References

1. Wikipedia>Courant-Friedrichs-Lewy condition. (An English translation of the original paper can be downloaded from the webpage.)
2. Cook, R.D., Milkus, D. S., Plesha, M. E., and Witt, R. J., *Concepts and Applications of Finite Element Analysis, Fourth Edition*, John Wiley & Sons, Inc., 2002.

Section 15.2

High-Speed Impact

15.2-1 About the High-Speed Impact Simulation

Imagine that, during an explosion, an aluminum pipe blasts away under the explosive pressure, hitting a steel solid column, deforming, and finally torn to fragments due to excessive strain [1-6]. In this section, we will simulate this scenario. We will use the default settings as much as possible to demonstrate that a complicated simulation like this can be done in **Explicit Dynamic** analysis system with just a few input data.

Both the aluminum pipe and the steel solid column have a diameter of 50 mm and a length of 200 mm. The steel column is modeled as a rigid body and fixed in the space. The aluminum pipe has a thickness of 1 mm and, right before hitting the pipe, has a speed of 300 m/s, about the speed of sound in the air. The aluminum is modeled as a bilinear isotropic plasticity material (Section 14.1) using the material parameters stored in **Engineering Data** with a modification that the tangent modulus is set to zero; i.e., the aluminum is modeled as a perfectly elastic-plastic material. It is assumed that the aluminum will be torn apart when the plastic strain is larger than 75%.

The **Millimeter** will be used to create the geometry and the **SI** unit systems will be used in the simulation.

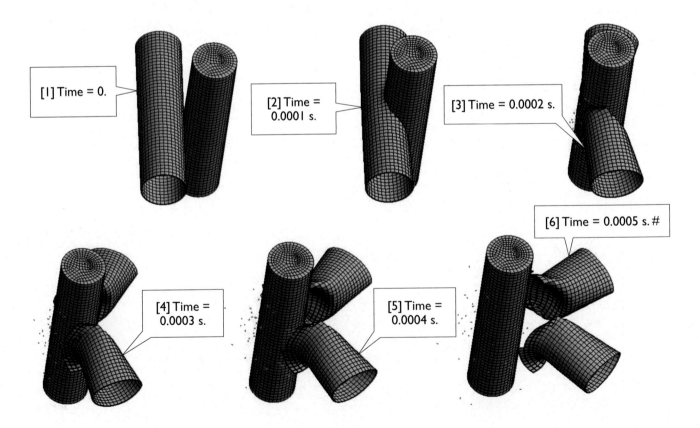

[1] Time = 0.

[2] Time = 0.0001 s.

[3] Time = 0.0002 s.

[4] Time = 0.0003 s.

[5] Time = 0.0004 s.

[6] Time = 0.0005 s. #

15.2-2 Start Up

Launch Workbench. Create an **Explicit Dynamics** analysis system by double-clicking it in **Toolbox**. Save the project as **Impact**.

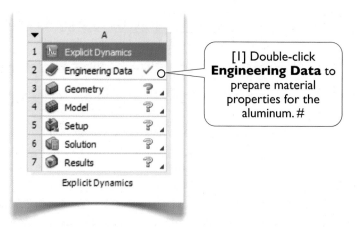

[1] Double-click **Engineering Data** to prepare material properties for the aluminum. #

15.2-3 Prepare Material Properties for Aluminum

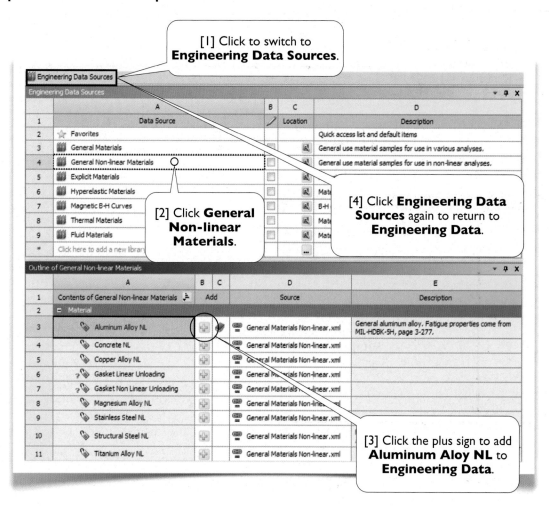

[1] Click to switch to **Engineering Data Sources**.

[2] Click **General Non-linear Materials**.

[4] Click **Engineering Data Sources** again to return to **Engineering Data**.

[3] Click the plus sign to add **Aluminum Aloy NL** to **Engineering Data**.

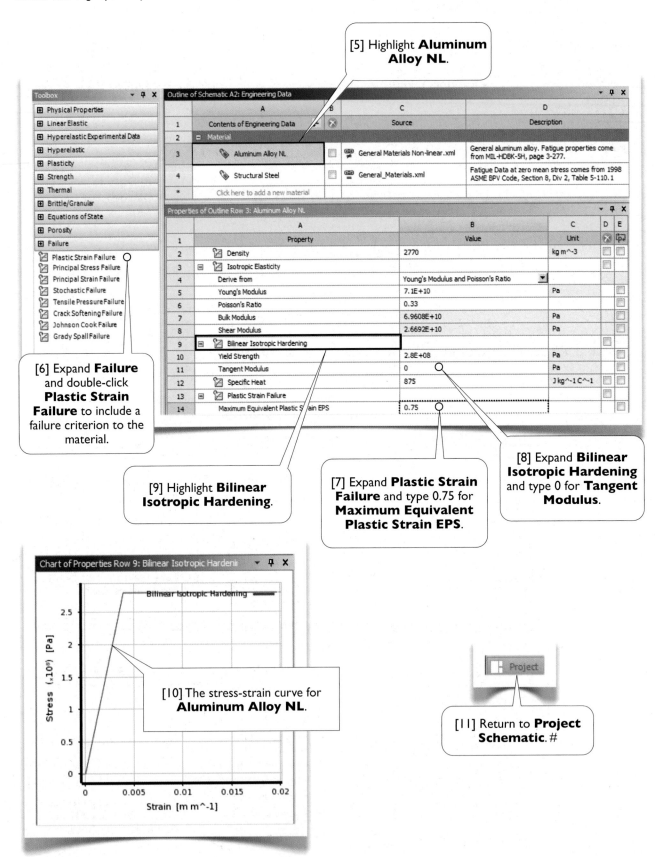

[5] Highlight **Aluminum Alloy NL**.

[6] Expand **Failure** and double-click **Plastic Strain Failure** to include a failure criterion to the material.

[9] Highlight **Bilinear Isotropic Hardening**.

[7] Expand **Plastic Strain Failure** and type 0.75 for **Maximum Equivalent Plastic Strain EPS**.

[8] Expand **Bilinear Isotropic Hardening** and type 0 for **Tangent Modulus**.

[10] The stress-strain curve for **Aluminum Alloy NL**.

[11] Return to **Project Schematic**. #

15.2-4 Create Geometry

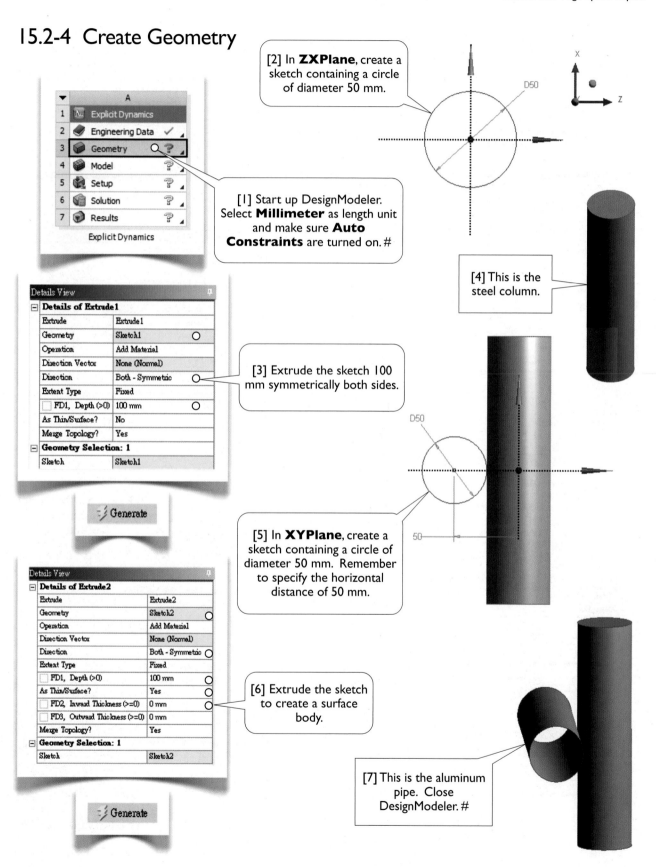

A			
1	🔷 Explicit Dynamics		
2	📄 Engineering Data	✓	
3	📦 Geometry	○ ?	
4	📦 Model	?	
5	📦 Setup	?	
6	📦 Solution	?	
7	📦 Results	?	

Explicit Dynamics

[2] In **ZXPlane**, create a sketch containing a circle of diameter 50 mm.

D50

[1] Start up DesignModeler. Select **Millimeter** as length unit and make sure **Auto Constraints** are turned on. #

Details View
Details of Extrude1

Extrude	Extrude1	
Geometry	Sketch1	○
Operation	Add Material	
Direction Vector	None (Normal)	
Direction	Both - Symmetric	○
Extent Type	Fixed	
FD1, Depth (>0)	100 mm	○
As Thin/Surface?	No	
Merge Topology?	Yes	
Geometry Selection: 1		
Sketch	Sketch1	

[4] This is the steel column.

[3] Extrude the sketch 100 mm symmetrically both sides.

Generate

D50

50

Details View
Details of Extrude2

Extrude	Extrude2	
Geometry	Sketch2	○
Operation	Add Material	
Direction Vector	None (Normal)	
Direction	Both - Symmetric	○
Extent Type	Fixed	
FD1, Depth (>0)	100 mm	○
As Thin/Surface?	Yes	○
FD2, Inward Thickness (>=0)	0 mm	○
FD3, Outward Thickness (>=0)	0 mm	
Merge Topology?	Yes	
Geometry Selection: 1		
Sketch	Sketch2	

[5] In **XYPlane**, create a sketch containing a circle of diameter 50 mm. Remember to specify the horizontal distance of 50 mm.

[6] Extrude the sketch to create a surface body.

[7] This is the aluminum pipe. Close DesignModeler. #

Generate

15.2-5 Set Up for Simulation

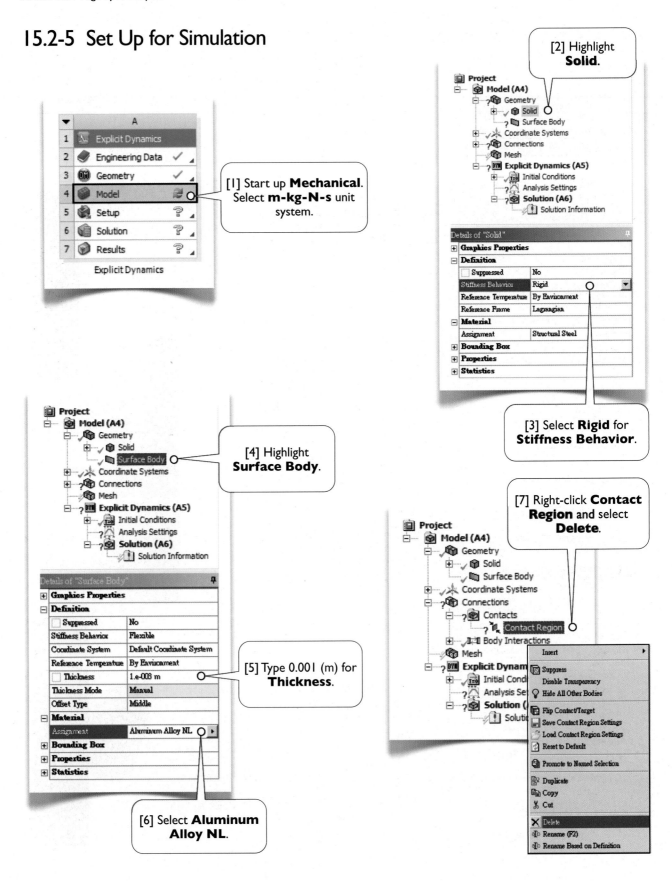

[1] Start up **Mechanical**. Select **m-kg-N-s** unit system.

[2] Highlight **Solid**.

[3] Select **Rigid** for **Stiffness Behavior**.

[4] Highlight **Surface Body**.

[5] Type 0.001 (m) for **Thickness**.

[6] Select **Aluminum Alloy NL**.

[7] Right-click **Contact Region** and select **Delete**.

571

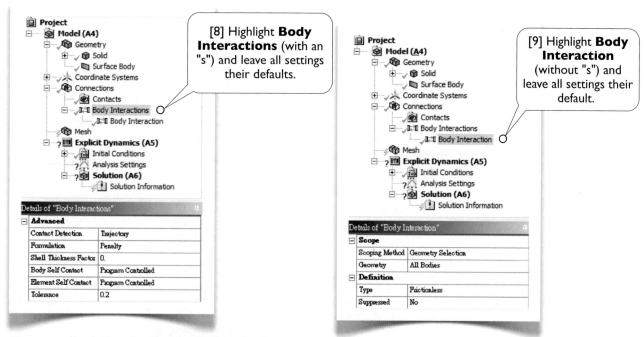

[8] Highlight **Body Interactions** (with an "s") and leave all settings their defaults.

[9] Highlight **Body Interaction** (without "s") and leave all settings their default.

Contact and **Body Interactions**?

Body Interactions is to specify contacts between bodies while **Contacts** is to specify contacts between surfaces. You can choose either way to specify the contact relations. **Body Interactions** is simpler, but **Contacts** may be more computationally efficient. By default, **Frictionless** body interactions are establishe among all bodies.

A feature of **Body Interactions** is that two bodies can be specified as both **Bonded** and **Frictionless** (or **Frictional**). In that case, two bodies are bonded initially. After the bond breaks during the simulation, the frictionless (or frictional) contact will take place.

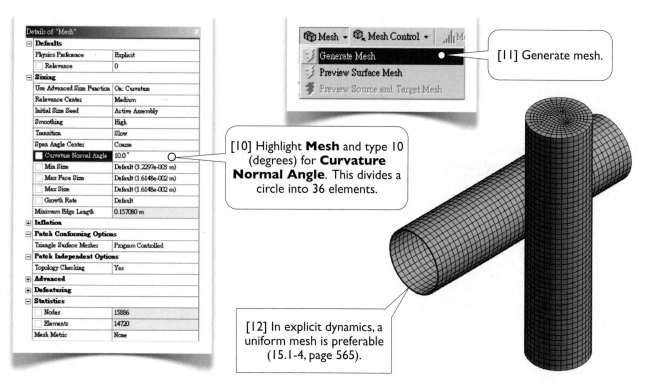

[10] Highlight **Mesh** and type 10 (degrees) for **Curvature Normal Angle**. This divides a circle into 36 elements.

[11] Generate mesh.

[12] In explicit dynamics, a uniform mesh is preferable (15.1-4, page 565).

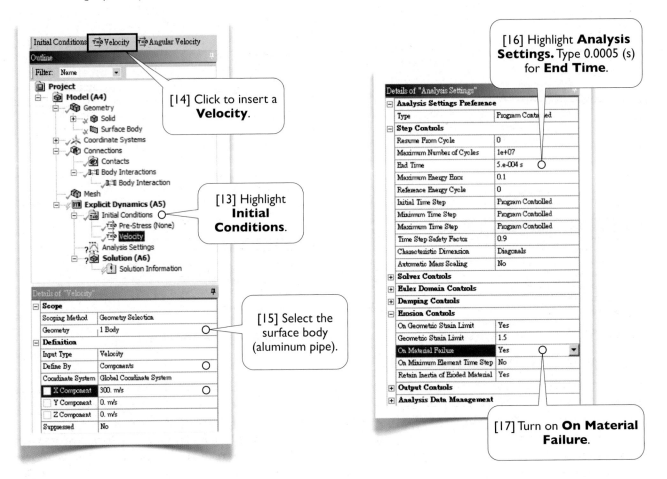

[14] Click to insert a **Velocity**.

[13] Highlight **Initial Conditions**.

[15] Select the surface body (aluminum pipe).

[16] Highlight **Analysis Settings.** Type 0.0005 (s) for **End Time**.

[17] Turn on **On Material Failure**.

Erosion Controls[Ref 1]

Erosion Controls in **Analysis Settings** determines the conditions that an element will be removed. The default condition is that an element is removed when its *geometric strain*, or *effective strain*[Ref 1], exceeds a limit of 150%. This value is large enough to assure that no elements are removed by default.

In this case, we add another failure condition: an element is removed when its plastic strain exceeds 75% ([17] and 15.2-3[6, 7], page 569).

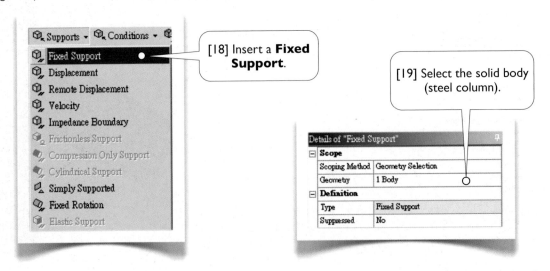

[18] Insert a **Fixed Support**.

[19] Select the solid body (steel column).

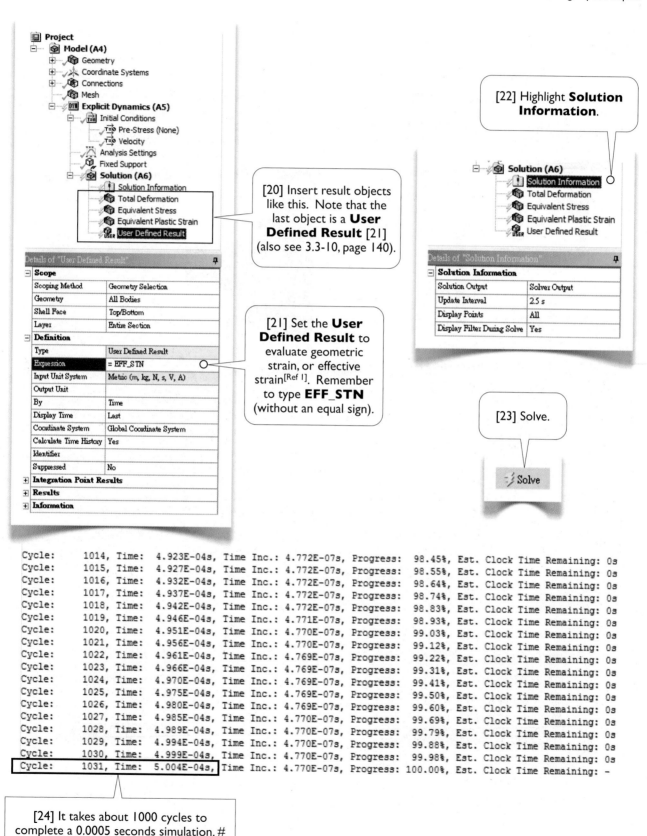

Project
Model (A4)
Geometry
Coordinate Systems
Connections
Mesh
Explicit Dynamics (A5)
Initial Conditions
Pre-Stress (None)
Velocity
Analysis Settings
Fixed Support
Solution (A6)
Solution Information
Total Deformation
Equivalent Stress
Equivalent Plastic Strain
User Defined Result

[20] Insert result objects like this. Note that the last object is a **User Defined Result** [21] (also see 3.3-10, page 140).

[22] Highlight **Solution Information**.

Solution (A6)
Solution Information
Total Deformation
Equivalent Stress
Equivalent Plastic Strain
User Defined Result

Details of "User Defined Result"

Scope	
Scoping Method	Geometry Selection
Geometry	All Bodies
Shell Face	Top/Bottom
Layer	Entire Section
Definition	
Type	User Defined Result
Expression	= EFF_STN
Input Unit System	Metric (m, kg, N, s, V, A)
Output Unit	
By	Time
Display Time	Last
Coordinate System	Global Coordinate System
Calculate Time History	Yes
Identifier	
Suppressed	No
Integration Point Results	
Results	
Information	

Details of "Solution Information"

Solution Information	
Solution Output	Solver Output
Update Interval	2.5 s
Display Points	All
Display Filter During Solve	Yes

[21] Set the **User Defined Result** to evaluate geometric strain, or effective strain[Ref 1]. Remember to type **EFF_STN** (without an equal sign).

[23] Solve.

Solve

```
Cycle:     1014, Time:  4.923E-04s, Time Inc.: 4.772E-07s, Progress:  98.45%, Est. Clock Time Remaining: 0s
Cycle:     1015, Time:  4.927E-04s, Time Inc.: 4.772E-07s, Progress:  98.55%, Est. Clock Time Remaining: 0s
Cycle:     1016, Time:  4.932E-04s, Time Inc.: 4.772E-07s, Progress:  98.64%, Est. Clock Time Remaining: 0s
Cycle:     1017, Time:  4.937E-04s, Time Inc.: 4.772E-07s, Progress:  98.74%, Est. Clock Time Remaining: 0s
Cycle:     1018, Time:  4.942E-04s, Time Inc.: 4.772E-07s, Progress:  98.83%, Est. Clock Time Remaining: 0s
Cycle:     1019, Time:  4.946E-04s, Time Inc.: 4.771E-07s, Progress:  98.93%, Est. Clock Time Remaining: 0s
Cycle:     1020, Time:  4.951E-04s, Time Inc.: 4.770E-07s, Progress:  99.03%, Est. Clock Time Remaining: 0s
Cycle:     1021, Time:  4.956E-04s, Time Inc.: 4.770E-07s, Progress:  99.12%, Est. Clock Time Remaining: 0s
Cycle:     1022, Time:  4.961E-04s, Time Inc.: 4.769E-07s, Progress:  99.22%, Est. Clock Time Remaining: 0s
Cycle:     1023, Time:  4.966E-04s, Time Inc.: 4.769E-07s, Progress:  99.31%, Est. Clock Time Remaining: 0s
Cycle:     1024, Time:  4.970E-04s, Time Inc.: 4.769E-07s, Progress:  99.41%, Est. Clock Time Remaining: 0s
Cycle:     1025, Time:  4.975E-04s, Time Inc.: 4.769E-07s, Progress:  99.50%, Est. Clock Time Remaining: 0s
Cycle:     1026, Time:  4.980E-04s, Time Inc.: 4.769E-07s, Progress:  99.60%, Est. Clock Time Remaining: 0s
Cycle:     1027, Time:  4.985E-04s, Time Inc.: 4.770E-07s, Progress:  99.69%, Est. Clock Time Remaining: 0s
Cycle:     1028, Time:  4.989E-04s, Time Inc.: 4.770E-07s, Progress:  99.79%, Est. Clock Time Remaining: 0s
Cycle:     1029, Time:  4.994E-04s, Time Inc.: 4.770E-07s, Progress:  99.88%, Est. Clock Time Remaining: 0s
Cycle:     1030, Time:  4.999E-04s, Time Inc.: 4.770E-07s, Progress:  99.98%, Est. Clock Time Remaining: 0s
Cycle:     1031, Time:  5.004E-04s, Time Inc.: 4.770E-07s, Progress: 100.00%, Est. Clock Time Remaining: -
```

[24] It takes about 1000 cycles to complete a 0.0005 seconds simulation. #

15.2-6 Animate the Deformation

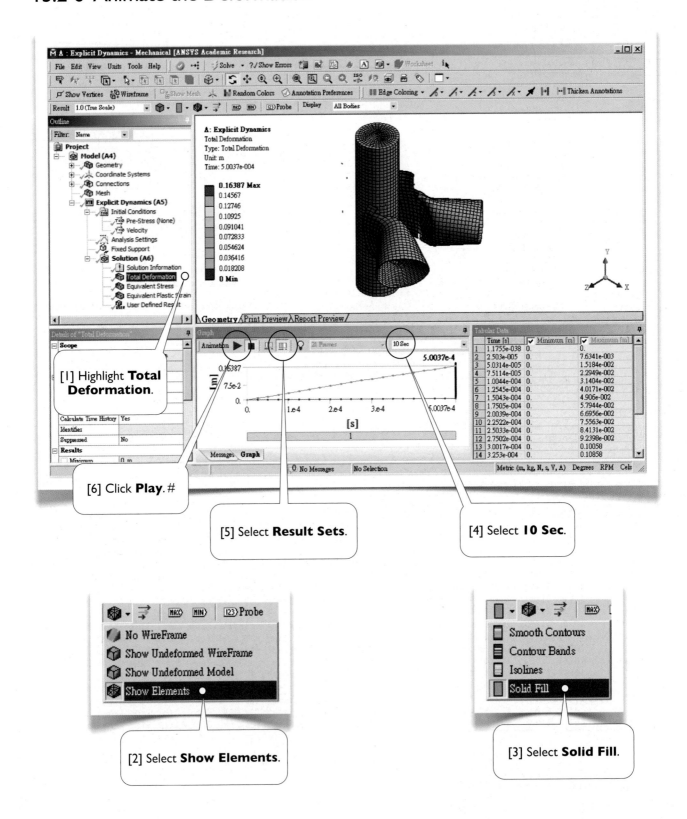

[1] Highlight **Total Deformation**.

[6] Click **Play**. #

[5] Select **Result Sets**.

[4] Select **10 Sec**.

[2] Select **Show Elements**.

[3] Select **Solid Fill**.

15.2-7 View Numerical Results

[1] Select **Smooth Contours**. #

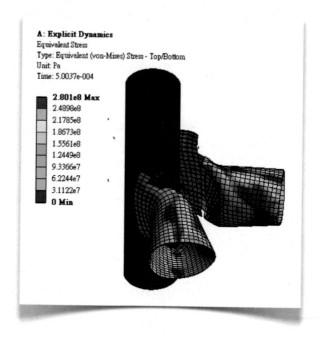

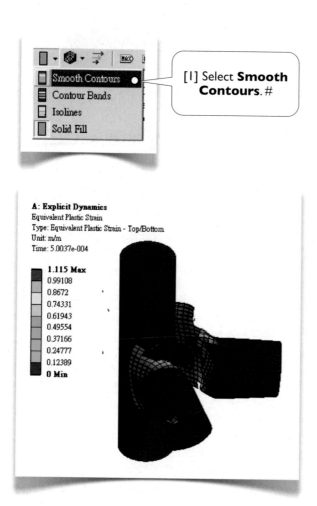

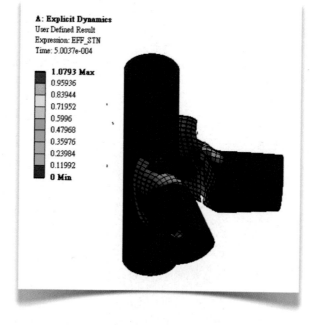

Wrap Up

Save the project and exit Workbench.

Reference

1. ANSYS Documentation//Mechanical Applications//Mechanical User's Guide//Appendix G. Explicit Dynamics Theory Guide// Analysis Settings//Erosion Controls

Section 15.3

Drop Test

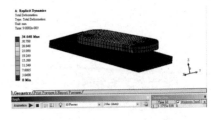

15.3-1 About the Drop Test Simulation

Drop test simulation is a special case of impact simulation, in which one of the impacting objects is a stationary floor, typically made of concrete, steel, or stone. In this section, we will simulate a scenario that a mobile phone falls off from your pocket and drops on a concrete floor. This kind of simulations typically takes hours of computing time. We learned from Section 15.2 that a typical integration time step in **Explicit Dynamics** is 10^{-7} to 10^{-8} seconds. It would take about 100,000 to 1,000,000 cycles to complete a 0.01 seconds of drop test. In this section, we will simplify the model to shorten the run time. A more realistic model will be suggested and left for the students as an exercise (15.4-2, page 590).

The phone body is a shell of thickness 0.5 mm and made of an aluminum alloy [1]. The concrete floor is modeled as an 160 mm x 80 mm x 10 mm block [2]. When the phone hits the floor, its velocity is 5 m/s, which is equivalent to a free fall from a height of 1.25 m. We will assume that the phone body forms an angle of 20° with the horizon when it hits the floor.

We will use **mm-kg-N-s** unit system in the simulation.

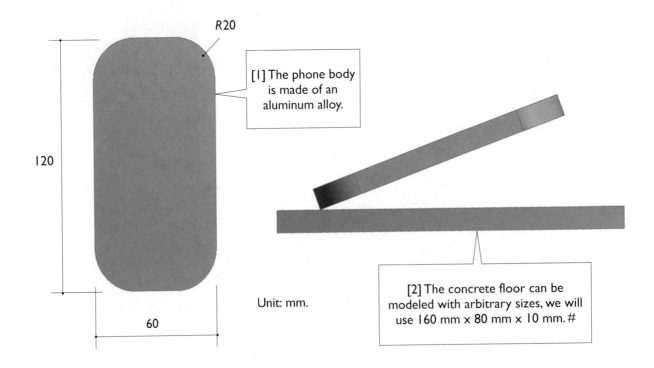

R20

[1] The phone body is made of an aluminum alloy.

120

60

Unit: mm.

[2] The concrete floor can be modeled with arbitrary sizes, we will use 160 mm x 80 mm x 10 mm. #

15.3-2 Start Up

Launch Workbench. Create an **Explicit Dynamics** analysis system by double-clicking it in **Toolbox**. Save the project as **Drop**.

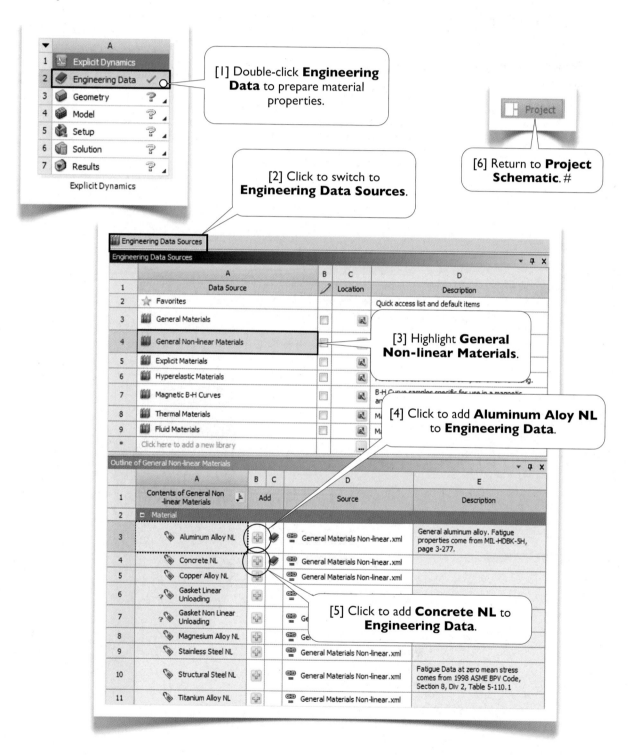

[1] Double-click **Engineering Data** to prepare material properties.

[2] Click to switch to **Engineering Data Sources**.

[6] Return to **Project Schematic**. #

[3] Highlight **General Non-linear Materials**.

[4] Click to add **Aluminum Aloy NL** to **Engineering Data**.

[5] Click to add **Concrete NL** to **Engineering Data**.

15.3-3 Create Geometry

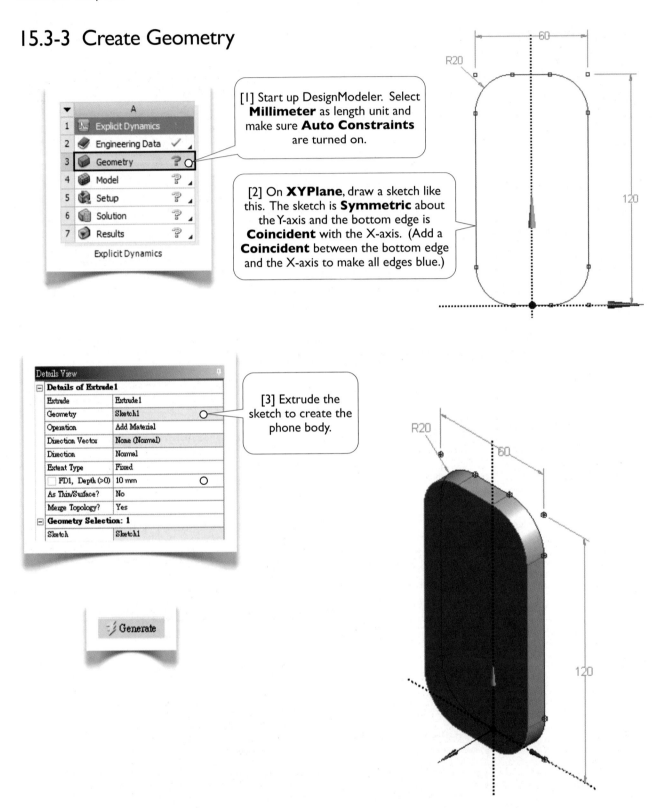

[1] Start up DesignModeler. Select **Millimeter** as length unit and make sure **Auto Constraints** are turned on.

[2] On **XYPlane**, draw a sketch like this. The sketch is **Symmetric** about the Y-axis and the bottom edge is **Coincident** with the X-axis. (Add a **Coincident** between the bottom edge and the X-axis to make all edges blue.)

[3] Extrude the sketch to create the phone body.

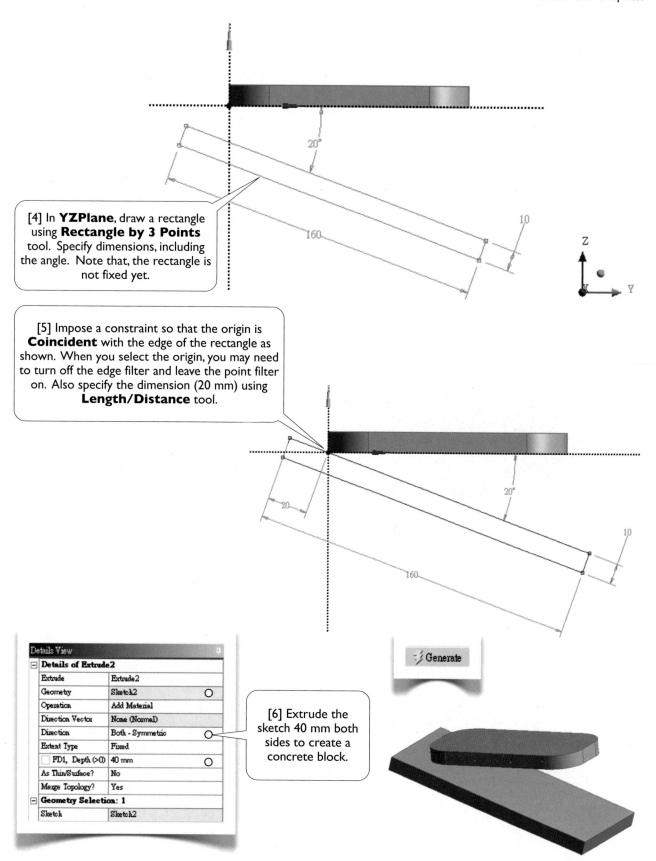

[4] In **YZPlane**, draw a rectangle using **Rectangle by 3 Points** tool. Specify dimensions, including the angle. Note that, the rectangle is not fixed yet.

[5] Impose a constraint so that the origin is **Coincident** with the edge of the rectangle as shown. When you select the origin, you may need to turn off the edge filter and leave the point filter on. Also specify the dimension (20 mm) using **Length/Distance** tool.

Details View

Details of Extrude2		
Extrude	Extrude2	
Geometry	Sketch2	O
Operation	Add Material	
Direction Vector	None (Normal)	
Direction	Both - Symmetric	O
Extent Type	Fixed	
☐ FD1, Depth (>0)	40 mm	O
As Thin/Surface?	No	
Merge Topology?	Yes	
Geometry Selection: 1		
Sketch	Sketch2	

Generate

[6] Extrude the sketch 40 mm both sides to create a concrete block.

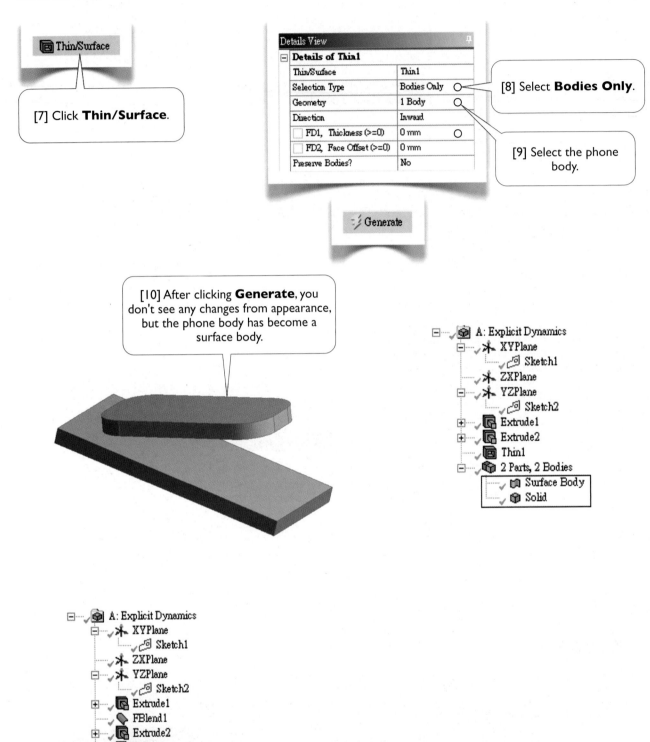

Thin/Surface

[7] Click **Thin/Surface**.

Details View

Details of Thin1	
Thin/Surface	Thin1
Selection Type	Bodies Only
Geometry	1 Body
Direction	Inward
FD1, Thickness (>=0)	0 mm
FD2, Face Offset (>=0)	0 mm
Preserve Bodies?	No

[8] Select **Bodies Only**.

[9] Select the phone body.

Generate

[10] After clicking **Generate**, you don't see any changes from appearance, but the phone body has become a surface body.

A: Explicit Dynamics
    XYPlane
        Sketch1
    ZXPlane
    YZPlane
        Sketch2
    Extrude1
    Extrude2
    Thin1
    2 Parts, 2 Bodies
        Surface Body
        Solid

A: Explicit Dynamics
    XYPlane
        Sketch1
    ZXPlane
    YZPlane
        Sketch2
    Extrude1
    FBlend1
    Extrude2
    Thin1
    2 Parts, 2 Bodies
        Phone
        Floor

[11] Rename the bodies. Close DesignModeler. #

15.3-4 Set Up for Simulation

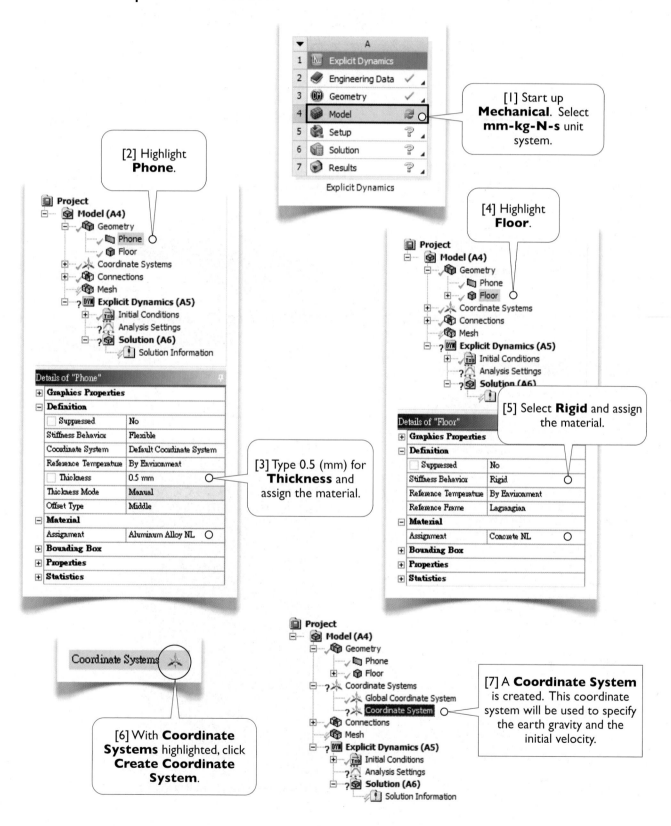

[1] Start up **Mechanical**. Select **mm-kg-N-s** unit system.

[2] Highlight **Phone**.

[4] Highlight **Floor**.

[3] Type 0.5 (mm) for **Thickness** and assign the material.

[5] Select **Rigid** and assign the material.

[6] With **Coordinate Systems** highlighted, click **Create Coordinate System**.

[7] A **Coordinate System** is created. This coordinate system will be used to specify the earth gravity and the initial velocity.

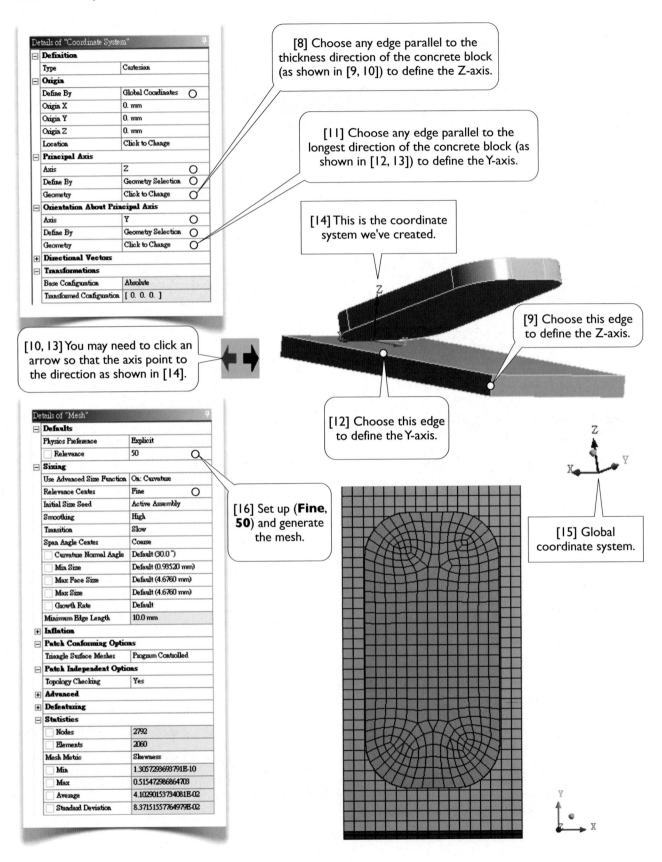

[8] Choose any edge parallel to the thickness direction of the concrete block (as shown in [9, 10]) to define the Z-axis.

[11] Choose any edge parallel to the longest direction of the concrete block (as shown in [12, 13]) to define the Y-axis.

[14] This is the coordinate system we've created.

[9] Choose this edge to define the Z-axis.

[10, 13] You may need to click an arrow so that the axis point to the direction as shown in [14].

[12] Choose this edge to define the Y-axis.

[16] Set up (**Fine**, **50**) and generate the mesh.

[15] Global coordinate system.

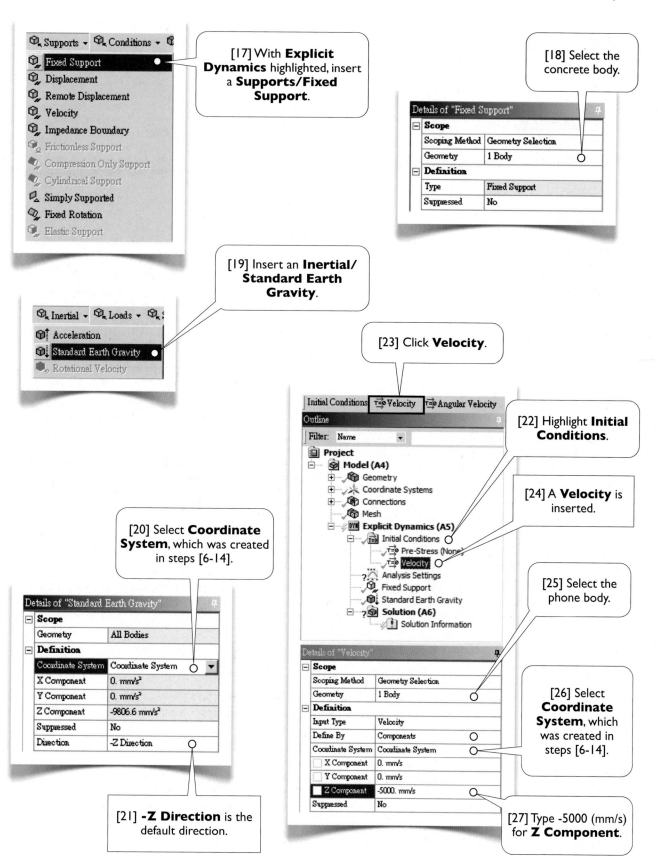

[17] With **Explicit Dynamics** highlighted, insert a **Supports/Fixed Support**.

[18] Select the concrete body.

Details of "Fixed Support"

Scope	
Scoping Method	Geometry Selection
Geometry	1 Body
Definition	
Type	Fixed Support
Suppressed	No

[19] Insert an **Inertial/ Standard Earth Gravity**.

[23] Click **Velocity**.

[22] Highlight **Initial Conditions**.

Initial Conditions   Velocity   Angular Velocity

Outline

Filter: Name

- Project
  - Model (A4)
    - Geometry
    - Coordinate Systems
    - Connections
    - Mesh
    - Explicit Dynamics (A5)
      - Initial Conditions
        - Pre-Stress (None)
        - Velocity
      - Analysis Settings
      - Fixed Support
      - Standard Earth Gravity
    - Solution (A6)
      - Solution Information

[24] A **Velocity** is inserted.

[25] Select the phone body.

[20] Select **Coordinate System**, which was created in steps [6-14].

Details of "Standard Earth Gravity"

Scope	
Geometry	All Bodies
Definition	
Coordinate System	Coordinate System
X Component	0. mm/s²
Y Component	0. mm/s²
Z Component	-9806.6 mm/s²
Suppressed	No
Direction	-Z Direction

[21] **-Z Direction** is the default direction.

Details of "Velocity"

Scope	
Scoping Method	Geometry Selection
Geometry	1 Body
Definition	
Input Type	Velocity
Define By	Components
Coordinate System	Coordinate System
X Component	0. mm/s
Y Component	0. mm/s
Z Component	-5000. mm/s
Suppressed	No

[26] Select **Coordinate System**, which was created in steps [6-14].

[27] Type -5000 (mm/s) for **Z Component**.

584

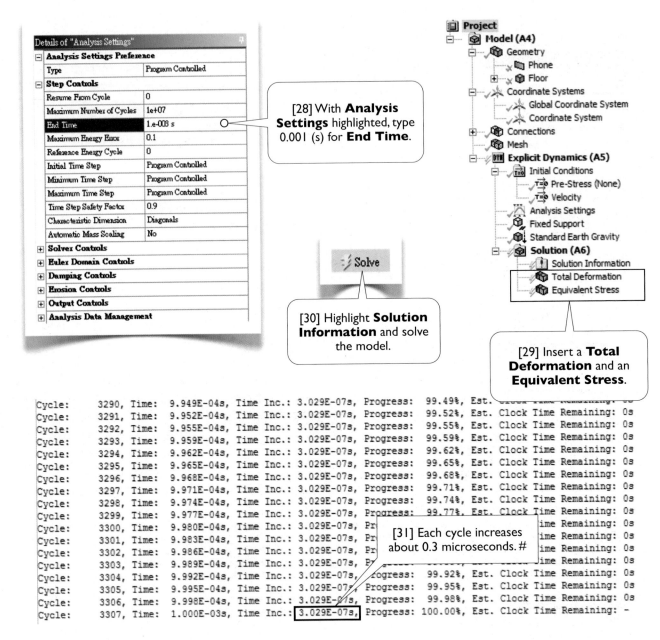

Details of "Analysis Settings"

Analysis Settings Preference	
Type	Program Controlled
Step Controls	
Resume From Cycle	0
Maximum Number of Cycles	1e+07
End Time	1.e-008 s
Maximum Energy Error	0.1
Reference Energy Cycle	0
Initial Time Step	Program Controlled
Minimum Time Step	Program Controlled
Maximum Time Step	Program Controlled
Time Step Safety Factor	0.9
Characteristic Dimension	Diagonals
Automatic Mass Scaling	No
Solver Controls	
Euler Domain Controls	
Damping Controls	
Erosion Controls	
Output Controls	
Analysis Data Management	

[28] With **Analysis Settings** highlighted, type 0.001 (s) for **End Time**.

Project
Model (A4)
  Geometry
    Phone
    Floor
  Coordinate Systems
    Global Coordinate System
    Coordinate System
  Connections
  Mesh
  Explicit Dynamics (A5)
    Initial Conditions
      Pre-Stress (None)
      Velocity
    Analysis Settings
    Fixed Support
    Standard Earth Gravity
    Solution (A6)
      Solution Information
      Total Deformation
      Equivalent Stress

Solve

[30] Highlight **Solution Information** and solve the model.

[29] Insert a **Total Deformation** and an **Equivalent Stress**.

```
Cycle:   3290, Time:  9.949E-04s, Time Inc.: 3.029E-07s, Progress:  99.49%, Est. Clock Time Remaining: 0s
Cycle:   3291, Time:  9.952E-04s, Time Inc.: 3.029E-07s, Progress:  99.52%, Est. Clock Time Remaining: 0s
Cycle:   3292, Time:  9.955E-04s, Time Inc.: 3.029E-07s, Progress:  99.55%, Est. Clock Time Remaining: 0s
Cycle:   3293, Time:  9.959E-04s, Time Inc.: 3.029E-07s, Progress:  99.59%, Est. Clock Time Remaining: 0s
Cycle:   3294, Time:  9.962E-04s, Time Inc.: 3.029E-07s, Progress:  99.62%, Est. Clock Time Remaining: 0s
Cycle:   3295, Time:  9.965E-04s, Time Inc.: 3.029E-07s, Progress:  99.65%, Est. Clock Time Remaining: 0s
Cycle:   3296, Time:  9.968E-04s, Time Inc.: 3.029E-07s, Progress:  99.68%, Est. Clock Time Remaining: 0s
Cycle:   3297, Time:  9.971E-04s, Time Inc.: 3.029E-07s, Progress:  99.71%, Est. Clock Time Remaining: 0s
Cycle:   3298, Time:  9.974E-04s, Time Inc.: 3.029E-07s, Progress:  99.74%, Est. Clock Time Remaining: 0s
Cycle:   3299, Time:  9.977E-04s, Time Inc.: 3.029E-07s, Progress:  99.77%, Est. Clock Time Remaining: 0s
Cycle:   3300, Time:  9.980E-04s, Time Inc.: 3.029E-07s, Pr                                    ime Remaining: 0s
Cycle:   3301, Time:  9.983E-04s, Time Inc.: 3.029E-07s, Pr                                    ime Remaining: 0s
Cycle:   3302, Time:  9.986E-04s, Time Inc.: 3.029E-07s, Pr                                    ime Remaining: 0s
Cycle:   3303, Time:  9.989E-04s, Time Inc.: 3.029E-07s, P                                     ime Remaining: 0s
Cycle:   3304, Time:  9.992E-04s, Time Inc.: 3.029E-07s,    rogress:  99.92%, Est. Clock Time Remaining: 0s
Cycle:   3305, Time:  9.995E-04s, Time Inc.: 3.029E-07s, Progress:  99.95%, Est. Clock Time Remaining: 0s
Cycle:   3306, Time:  9.998E-04s, Time Inc.: 3.029E-07s, Progress:  99.98%, Est. Clock Time Remaining: 0s
Cycle:   3307, Time:  1.000E-03s, Time Inc.: 3.029E-07s, Progress: 100.00%, Est. Clock Time Remaining: -
```

[31] Each cycle increases about 0.3 microseconds. #

Integration Time Steps

In this case, the integration time step is about 0.3 microseconds. This number, proportional to the size of the smallest element, controls the overall run time (15.1-4, page 565). In **Explicit Dynamics**, a mesh of uniform element size is the most efficient mesh.

Perform Simulation Incrementally

A drop test simulation typically takes hours of computing time. If you make a mistake, it will waste much of time. As a good practice, always perform the simulation incrementally. In this case, we perform a 0.001 seconds of simulation first to see if anything goes wrong. If everything is okay, then we can continue the simulation starting from the last cycle. This feature is called the **Restart** of the simulation.

15.3-5 View the Results

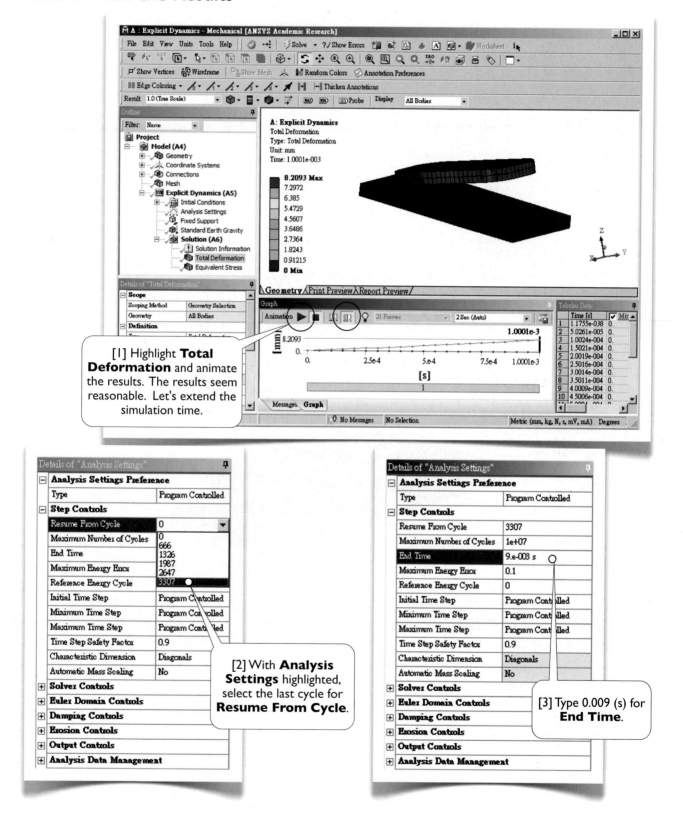

[1] Highlight **Total Deformation** and animate the results. The results seem reasonable. Let's extend the simulation time.

[2] With **Analysis Settings** highlighted, select the last cycle for **Resume From Cycle**.

[3] Type 0.009 (s) for **End Time**.

[4] Highlight **Solution Information** and solve again.

```
Cycle:    29704, Time:  8.995E-03s, Time Inc.: 3.029E-07s, Progress:   99.94%, Est. Clock Time Remaining: 0s
Cycle:    29705, Time:  8.995E-03s, Time Inc.: 3.029E-07s, Progress:   99.95%, Est. Clock Time Remaining: 0s
Cycle:    29706, Time:  8.996E-03s, Time Inc.: 3.029E-07s, Progress:   99.95%, Est. Clock Time Remaining: 0s
Cycle:    29707, Time:  8.996E-03s, Time Inc.: 3.029E-07s, Progress:   99.96%, Est. Clock Time Remaining: 0s
Cycle:    29708, Time:  8.996E-03s, Time Inc.: 3.029E-07s, Progress:   99.96%, Est. Clock Time Remaining: 0s
Cycle:    29709, Time:  8.997E-03s, Time Inc.: 3.029E-07s, Progress:   99.96%, Est. Clock Time Remaining: 0s
Cycle:    29710, Time:  8.997E-03s, Time Inc.: 3.029E-07s, Progress:   99.97%, Est. Clock Time Remaining: 0s
Cycle:    29711, Time:  8.997E-03s, Time Inc.: 3.029E-07s, Progress:   99.97%, Est. Clock Time Remaining: 0s
Cycle:    29712, Time:  8.997E-03s, Time Inc.: 3.029E-07s, Progress:   99.97%, Est. Clock Time Remaining: 0s
Cycle:    29713, Time:  8.998E-03s, Time Inc.: 3.029E-07s, Progress:   99.98%, Est. Clock Time Remaining: 0s
Cycle:    29714, Time:  8.998E-03s, Time Inc.: 3.029E-07s, Progress:   99.98%, Est. Clock Time Remaining: 0s
Cycle:    29715, Time:  8.998E-03s, Time Inc.: 3.029E-07s, Progress:   99.98%, Est. Clock Time Remaining: 0s
Cycle:    29716, Time:  8.999E-03s, Time Inc.: 3.029E-07s, Progress:   99.99%, Est. Clock Time Remaining: 0s
Cycle:    29717, Time:  8.999E-03s, Time Inc.: 3.029E-07s, Progress:   99.99%, Est. Clock Time Remaining: 0s
Cycle:    29718, Time:  8.999E-03s, Time Inc.: 3.029E-07s, Progress:   99.99%, Est. Clock Time Remaining: 0s
Cycle:    29719, Time:  9.000E-03s, Time Inc.: 3.029E-07s, Progress:  100.00%, Est. Clock Time Remaining: 0s
Cycle:    29720, Time:  9.000E-03s, Time Inc.: 3.029E-07s, Progress:  100.00%, Est. Clock Time Remaining: 0s
Cycle:    29721, Time:  9.000E-03s, Time Inc.: 3.029E-07s, Progress:  100.00%, Est. Clock Time Remaining: -
```

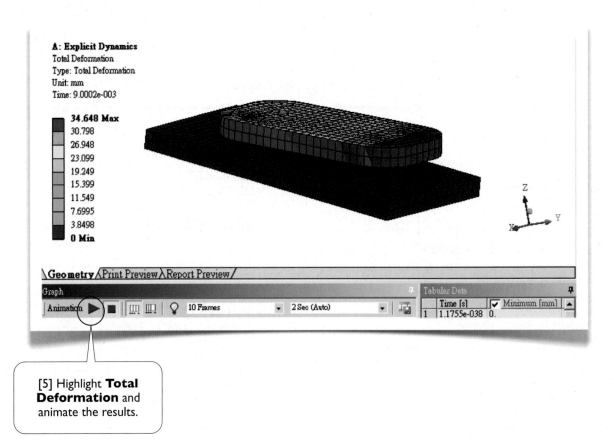

[5] Highlight **Total Deformation** and animate the results.

587

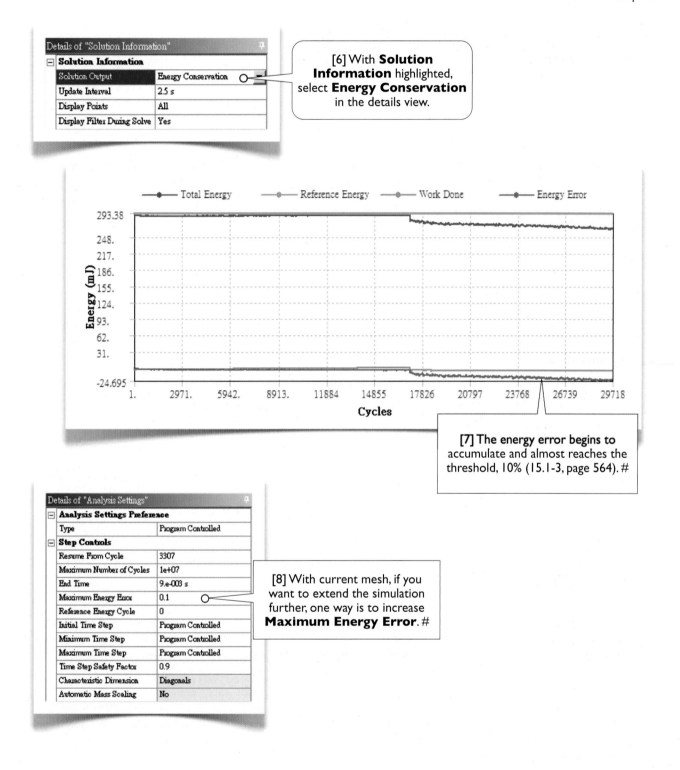

Details of "Solution Information"

Solution Information	
Solution Output	Energy Conservation
Update Interval	2.5 s
Display Points	All
Display Filter During Solve	Yes

[6] With **Solution Information** highlighted, select **Energy Conservation** in the details view.

[7] The energy error begins to accumulate and almost reaches the threshold, 10% (15.1-3, page 564). #

Details of "Analysis Settings"

Analysis Settings Preference	
Type	Program Controlled
Step Controls	
Resume From Cycle	3307
Maximum Number of Cycles	1e+07
End Time	9.e-003 s
Maximum Energy Error	0.1
Reference Energy Cycle	0
Initial Time Step	Program Controlled
Minimum Time Step	Program Controlled
Maximum Time Step	Program Controlled
Time Step Safety Factor	0.9
Characteristic Dimension	Diagonals
Automatic Mass Scaling	No

[8] With current mesh, if you want to extend the simulation further, one way is to increase **Maximum Energy Error**. #

Wrap Up

Save the project and exit Workbench.

Section 15.4

Review

15.4-1 Keywords

Choose a letter for each keyword from the list of descriptions

1. (   ) Automatic Mass Scaling
2. (   ) CFL Condition
3. (   ) Energy Error
4. (   ) Explicit Method
5. (   ) Implicit Method
6. (   ) Principle of Work and Energy
7. (   ) Static Damping

Answers:

1. (F)   2. (E)   3. (D)   4. (B)   5. (A)   6. (C)   7. (G)

List of Descriptions

(A)   A time integration method used in **Transient Structural** analysis system. It is so named because the method calculates the response at current time using implicit information. It thus requires iterations for a time step, implying an expensive runtime for a time step. It, however, allows relatively large time steps. Overall, it is suitable for most of transient simulations except high-speed or highly-nonlinear simulations.

(B)   A time integration method used in **Explicit Dynamics** analysis system. It is so named because the method calculates the response at current time using explicit information. It thus doesn't require iterations for a time step, implying an efficient runtime for a time step. It, however, requires very small time steps. Overall, it is suitable for high-speed or highly-nonlinear simulations.

(C)   The principle of work and energy states that the energy at a reference time plus the work done from the reference time to a specific time is equal to the energy at that specific time.

(D) **Explicit Dynamics** uses this value to monitor the solution stability. If the energy error reaches a threshold, the solution is regarded as unstable and the computation stops. According to the principle of work and energy, the energy at a reference time plus the work done from the reference time to the current time is equal to the energy at the current time. If not equal, the difference is called an energy error. The energy error is further transformed into a dimensionless value by dividing using the maximum energy.

(E) In a single time step, a wave should not travel further than the smallest element size. This condition is used by **Explicit Dynamics** to determine the integration time step.

(F) The idea of mass scaling is to artificially increase the mass of some small elements, so that the stability time step is increased, to reduce the overall runtime.

(G) **Explicit Dynamics** is primarily for solving transient dynamic problems. However, a steady-state solution may also be obtained by introducing a damping force to critically damp the lowest mode of oscillation.

15.4-2 Additional Workbench Exercises

Performing a More Realistic Drop Test

As mentioned in 15.3-1 (page 577), to reduce the runtime, the model is over-simplified. In reality, the housing of a mobile phone may include a battery compartment. Include a slide-in battery in the model and perform a drop test to see if the battery will fall off during the impact with the floor.

Index

Terms used in the ANSYS Workbench user interface remain their capitalized forms.